Chiral Ligands

New Directions in Organic and Biological Chemistry

Series Editor

Philip Page

Emeritus Professor, School of Chemistry, University of East Anglia

Biocontrol of Plant Diseases by Bacillus subtilis
Makoto Shoda

Advances in Microwave Chemistry
Bimal Krishna Banik and Debasish Bandyopadhyay

Carbocation Chemistry
Applications in Organic Synthesis
Jie Jack Li

Modern NMR Techniques for Synthetic Chemistry
Julie Fisher

Concerted Organic and Bio-organic Mechanisms
Andrew Williams

Capillary Electrophoresis: Theory and Practice
Patrick Camilleri

Chiral Sulfur Reagents
M. Mikolajczyk, J. Drabowicz, and P. Kielbasiński

Chemical Approaches to the Synthesis of Peptides and Proteins
Paul Lloyd-Williams, Fernando Albericio, and Ernest Giralt

Organozinc Reagents in Organic Synthesis
Ender Erdik

Chirality and the Biological Activity of Drugs
Roger J. Crossley

C-Glycoside Synthesis
Maarten Postema

The Anomeric Effect
Eusebio Juaristi and Gabriel Cuevas

Mannich Bases Chemistry and Uses
Maurilio Tramontini and Luigi Angiolini

Dianion Chemistry in Organic Synthesis
Charles M. Thompson

Chiral Ligands
Evolution of Ligand Libraries for Asymmetric Catalysis
Montserrat Diéguez

For more information about this series, please visit: https://www.crcpress.com/New-Directions-in-Organic--Biological-Chemistry/book-series/CRCNDOBCHE

Chiral Ligands

Evolution of Ligand Libraries
for Asymmetric Catalysis

Edited by
Montserrat Diéguez

CRC Press
Taylor & Francis Group
Boca Raton London New York

CRC Press is an imprint of the
Taylor & Francis Group, an **informa** business

First edition published 2021
by CRC Press
6000 Broken Sound Parkway NW, Suite 300, Boca Raton, FL 33487-2742

and by CRC Press
2 Park Square, Milton Park, Abingdon, Oxon, OX14 4RN

© 2021 Taylor & Francis Group, LLC

CRC Press is an imprint of Taylor & Francis Group, LLC

ISBN: 978-0-367-42848-8 (hbk)
ISBN: 978-0-367-76170-7 (pbk)
ISBN: 978-0-367-85573-4 (ebk)

Typeset in Times
by Deanta Global Publishing Services, Chennai, India

Contents

Preface

Enantiopure compounds play a key role in many technologically important and biologically relevant applications. In this respect, industry is always searching for more selective, straightforward, less costly, and environmentally friendly synthetic procedures. Asymmetric catalysis is crucial for achieving these goals, and research into improved catalysts is a very active field. The performance of enantioselective catalysts depends, to a large extent, on the selection of the appropriate chiral ligands. Among the thousands of chiral ligands which have been developed, a few stand out for their versatility. The most efficient ligands, known as "privileged chiral ligands", derive from a few core structures. Research into the development of improved catalysts for asymmetric catalysis commonly starts by evaluating these privileged structures, followed by their structural optimization, using information from the structure/catalytic performance relationship for these ligands. Despite significant advances, the optimization of chiral ligands still lacks predictive tools that could guide the discovery of improved catalysts and is mostly carried out on a trial-and-error basis. The excellent book "Privileged Chiral Ligands and Catalysts", by Qi-Lu Zhou (2011), describes 11 of these privileged ligands/catalysts, focusing on their structure and on the relationship between structure and catalytic success. This approach is not common, since most previous books/reviews focused only on the reaction performance. Most of the 11 ligands are bidentate homodonor ligands/catalysts, mostly with C_2 symmetry, with only one heterodonor bidentate phosphine–nitrogen P,N ligand, phosphine-oxazoline (PHOX), being included. Since 2011, many new heterodonor bidentate P,N and P,X ligands (where X = P', S, O) have been reported, which achieve remarkable results. Also, new tridentate-based, carbene-based, and monophosphorus ligands have been identified, the study of which was not so extensive previously, but which has improved greatly in the past decade. Modifications of the ligands, to achieve reactions in alternative, more sustainable media and for catalyst recovery, were also not considered prior to 2011. Finally, the modern design of new efficient catalysts combines structural information and computational studies, features often neglected in previous books. Placing the knowledge within a firm structure is crucial in the search for new ligands/catalysts for asymmetric processes.

This book organizes and discusses those new ligands/catalysts libraries which have appeared since 2011, as well as the relationship between structure and performance. As mentioned, organizing the available knowledge is crucial in the search for new ligands/catalysts for asymmetric processes. To that end, experts in each field have been invited to collaborate in this work and their contributions are collected here in this book.

The book is divided into eight chapters. Chapters 1 to 4 are devoted to heterodonor ligand libraries, grouped by type of donor atom (chiral bidentate heterodonor P–oxazoline ligands; chiral bidentate heterodonor P–other N ligands; chiral bidentate heterodonor P–S/O ligands; and chiral bidentate heterodonor P–P' ligands. Chapter 5 focuses on tridentate-based ligands, whereas Chapters 6 and 7 discuss the evolution of and prospects for carbene- and monophosphorus-based ligands. Chapter 8 presents an overview of how the proper choice of solvent is key to the success of the asymmetric reaction. Besides classical solvents also unusual solvents like fluorous solvents, ionic liquid, supercritical fluids, and water will be evaluated always in relation to the chiral ligands and catalysts used. Benefits of unusual solvent and problems will be critically analyzed. For each family of ligands, an overview of the historical context at the time of the discovery of the ligand library is presented initially, followed, in some cases, by describing the synthetic route and then its application in various catalytic asymmetric reactions. For each particular reaction, an overview of the state-of-the-art is first provided, followed by discussions of the key mechanistic aspects and ligand parameters needed to achieve high catalytic performance.

We believe that this book will become the principal reference source for chemistry researchers (in the disciplines of organic, catalytic, organometallic, and theoretical chemistry) involved in this field, as well as for industrial researchers involved in the preparation of high- value compounds (e.g., pharmaceutical drugs, agrochemical products, specialty materials ...). The book also has an academic style, that would make it a perfect entry point for new researchers in the field, as well as a text-book resource for use in advanced teaching courses.

I would like to thank Prof. Philip Page for the invitation to prepare this book and who helped me to initiate this book project, together with Hilary LaFoe, and Jessica Poile, whose help during the preparation of the book, editing and production process was invaluable. We thank the Spanish Ministry of Economy and Competitiveness (CTQ2016-74878-P and PID2019-104904GB-I00), European Regional Development Fund (AEI/FEDER, UE), the Catalan Government (2017SGR1472), and the ICREA Foundation (ICREA Academia award to Montserrat Diéguez). Finally, I would also like to thank all the authors who contributed to this project.

Montserrat Diéguez

Editor

Montserrat Diéguez studied chemistry at the Rovira i Virgili University (URV) in Tarragona (Spain), where she earned her Ph.D. in 1997, working in the group of Prof. C. Claver. After completing a postdoctoral fellowship where she worked with Prof. R.H. Crabtree at Yale University, in New Haven (USA), she returned to Tarragona in 1999 and accepted a lectureship position at the URV, becoming part of the permanent staff in 2002. In 2011 she was promoted to full Professor in Inorganic Chemistry at the URV.

She has been involved in more than 60 research projects in the fields of organometallic chemistry, steroselective synthesis, and asymmetric catalysis. She is the author/co-author of more than 150 articles in SCI-indexed journals and book chapters, and of several contributions to conferences. She received distinction from the Generalitat de Catalunya for the promotion of University Research in 2004 and from the URV in 2008. She has also been awarded the ICREA Academia Prize from the Catalan Institution for Research and Advanced Studies 2009–14 and 2015–20, for research excellence, facilitating research priority dedication. Her main research interests are focused on the sustainable design, synthesis, and screening of highly active and selective chiral catalysts for reactions of interest to the biological, pharmaceutical, and organic nanotechnological industries. Her areas of interest include organometallic chemistry, stereoselective synthesis, and asymmetric catalysis, using combinatorial and biotechnological approaches.

List of Contributors

Uchchhal Bandyopadhyay
CNRS, LCC (Laboratoire de Chimie de
Coordination)
Université de Toulouse
Toulouse, France

Marina Bellido
Institut Institute e for Research in
Biomedicine (IRB Barcelona)
Barcelona Institute of Science and
Technology
Barcelona, Spain

Maria Biosca
Departament de Química Física i
Inorgànica
Universitat Rovira i Virgili
Tarragona, Spain

Armin Börner
Leibniz-Institut für Katalyse
Universität Rostock
Rostock, Germany

Vincent César
CNRS, LCC (Laboratoire de Chimie
de Coordination)
Université de Toulouse
Toulouse, France

Montserrat Diéguez
Departament de Química Física i
Inorgànica
Universitat Rovira i Virgili
Tarragona, Spain

Jordi Faiges
Departament de Química Física i
Inorgànica
Universitat Rovira i Virgili
Tarragona, Spain

Christophe Fliedel
CNRS, LCC (Laboratoire de Chimie de
Coordination)
Université de Toulouse
Toulouse, France

Agnès Labande
CNRS, LCC (Laboratoire de Chimie de
Coordination)
Université de Toulouse
Toulouse, France

Eric Manoury
CNRS, LCC (Laboratoire de Chimie de
Coordination)
Université de Toulouse
Toulouse, France

Jèssica Margalef
Departament de Química Física i
Inorgànica
Universitat Rovira i Virgili,
Tarragona, Spain

Oscar Pàmies
Departament de Química Física i
Inorgànica
Universitat Rovira i Virgili
Tarragona, Spain

Miquel A. Pericàs
Institute of Chemical Research of
Catalonia (ICIQ)
Tarragona, Spain

and

Barcelona Institute of Science and
Technology
Barcelona, Spain

Antonio Pizzano
Instituto de Investigaciones Químicas
CSIC-Universidad de Sevilla
Seville, Spain

Rinaldo Poli
CNRS, LCC (Laboratoire de Chimie de
    Coordination)
Université de Toulouse
Toulouse, France

Antoni Riera
Institute for Research in Biomedicine
    (IRB Barcelona)
Barcelona Institute of Science and
    Technology
Barcelona, Spain

Pep Rojo
Institute for Research in Biomedicine
    (IRB Barcelona)
Barcelona Institute of Science and
    Technology
Barcelona, Spain

Basker Sundararaju
Department of Chemistry
Indian Institute of Technology
Kanpur, India.

Wenjun Tang
Shanghai Institute of Organic
    Chemistry
Chinese Academy of Science
Shanghai, China

Xavier Verdaguer
Institute for Research in Biomedicine
    (IRB Barcelona)
Barcelona Institute of Science and
    Technology
Barcelona, Spain

Feng Wan
Shanghai Institute of Organic
    Chemistry
Chinese Academy of Science
Shanghai, China

He Yang
Shanghai Institute of Organic
    Chemistry
Chinese Academy of Science
Shanghai, China

Abbreviations and Acronyms

A³ coupling	Aldehyde-alkyne-amine-coupling
Ac	Acetyl
Alk	Alkyl
Anth	Anthracenyl
ao-QMs	aza-*o*-quinone methides
ax	axial
BAr$_F$	tetrakis[3,5-bis(trifluoromethyl)phenyl]borate
BIBOP	3,3''-Di-*tert*-butyl-4,4''-disubstituted-2,2'',3,3''-tetrahydro-2,2''-bibenzo[*d*][1,3]oxaphosphole
BI-DIME	3-(*tert*-Butyl)-4-(2,6-dimethoxyphenyl)-2,3-dihydrobenzo[*d*][1,3]oxaphosphole
BINAP	2,2'-Bis(diphenylphosphino)-1,1'-binaphthyl; Binepine = Phenylbinaphthophosphepine
BIPI	Boehringer-Ingelheim Phosphinoimidazolines
Bn	Benzyl
Boc	*tert*-Butyloxycarbonyl
BoQPhos	Bo Qu Phosphine
Bpin	Boronic acid pinacol ester
Bz	Benzoyl
Cbz	Benzyloxycarbonyl
cod	1,5-cyclooctadiene
CuTC	Copper (I) thiophene-2-carboxylate
Cy	Cyclohexyl
dba	Dibenzylideneacetone
Dec	Decyl
DFT	Density-functional theory
DIOP	2,3-O-isopropylidene-2,3-dihydroxy-1,4-bis(diphenylphosphino)butane
DKR	Dynamic kinetic resolution
DMF	*N,N*-Dimethylformamide
dr	diastereomeric ratio
dtbbpy	4,4'-di-*tert*-butyl-2,2'-bipyridine
EDG	Electron donating group
ee	enantiomeric excess
EWG	Electron withdrawing group
Fc	Ferrocene
IAN	Isoquinoline-amino naphthalene
MAP	2'-(Diphenylphosphino)-*N,N*-disubstituted[1,1'-binaphthalen]-2-amine

MOP	2-(Diphenylphosphino)-2′-methoxy-1,1′-binaphthyl
MPFA	α-[2-Dimethylphosphinoferrocenyl]ethyldimethylamine
MS	Molecular sieves
MTBE	Methyl *tert*-butyl ether
NMR	Nuclear Magnetic Resonance
NOBIN	2-Amino-2′-hydroxy-1,1′-binaphthyl
Phen	Phenanthryl; PHIM = Phosphino-imidazoline
PHOX	Phosphinooxazoline
PINAP	4-[2-(Diphenylphosphino)-1-naphthalenyl]-*N*-[1-phenylethyl]-1-phthalazinamine
Piv	Pivaloyl
PMP	*p*-Methoxy-phenyl
POP	3-(*tert*-Butyl)-2-(di-*tert*-butylphosphino)-4-substituted-2,3-dihydrobenzo[*d*][1,3]oxaphosphole
PPFA	α-[-2-diphenylphosphinoferrocenyl]ethyldimethylamine
ppy	2-phenylpyridine
***p*-QMs**	*p*-Quinone methides
Py	Pyridine; QUINAP = 1-(2-Diphenylphosphino-1-naphthyl)isoquinoline
S/C	Substrate/catalyst ratio
SAMP	(*S*)-1-Amino-2-(methoxymethyl)-pyrrolidine
SPINOL	1,1′-Spirobiindane-7,7′-diol
SpiroAP	Spiro aminophosphines
SpiroBAP	Spiro benzylamino-phosphine
Tf	Trifluoromethanesulfonate (triflate)
THF	Tetrahydrofuran
TMS	Trimethylsilyl
Tol	Tolyl
Ts	4-Toluenesulfonyl (tosyl).
CAMP	Scheme 1a **L1** cyclohexyl(2-methoxyphenyl)(methyl)phosphane

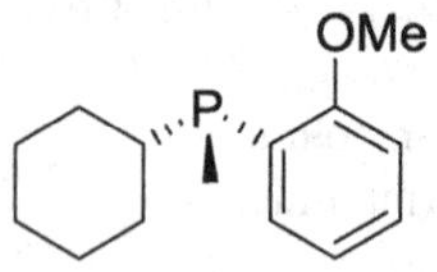

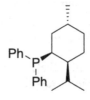

NMDPP	neomenthyldiphenylphosphine

BINOL   1,1'-bi-2,2'-naphthol

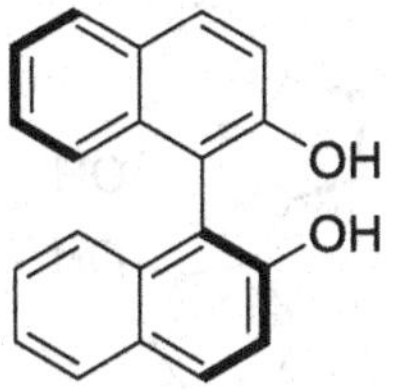

TADDOL   4,5-bis(diphenylhydroxymethyl)-2,2-dimethyl-1,3-dioxolane

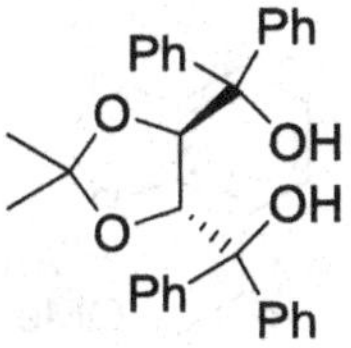

THQphos   Scheme 1b **L22**

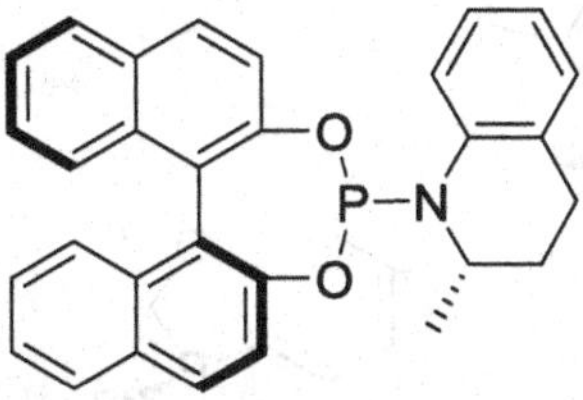

BHQphos   Scheme 1b **L23**

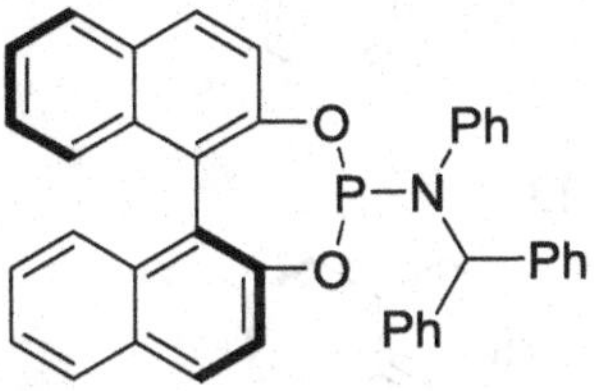

SIPHOS-PE Scheme 1b **L25**

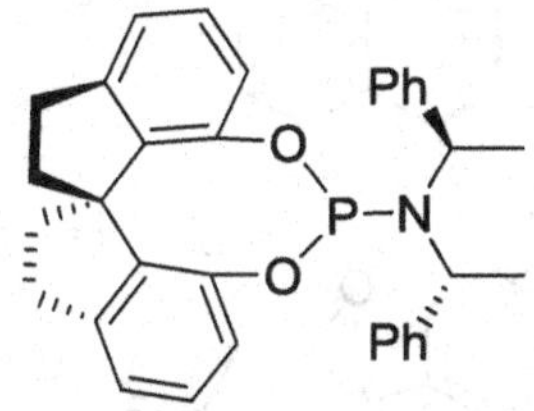

DFT   density functional theory
TMM   trimethylenemethane

SPINOL   1,1′-spirobiindane-7,7′-diol

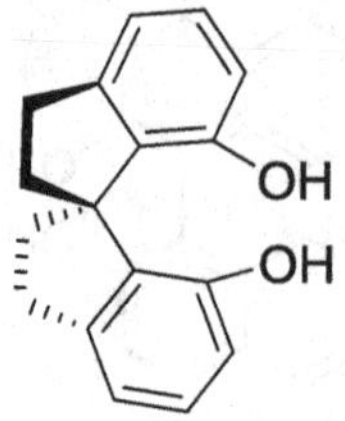

BaryPhos   Scheme 1a **L8**

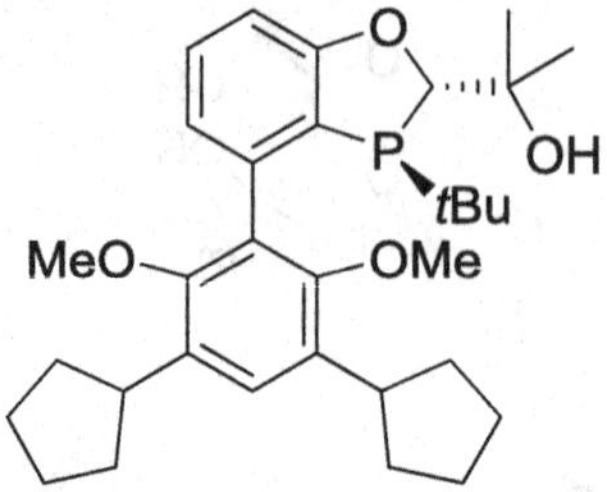

BI-DIME   Scheme 1a **L5**

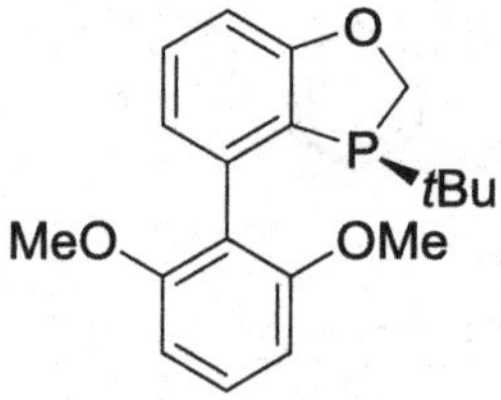

MOP   Scheme 1b **L13**

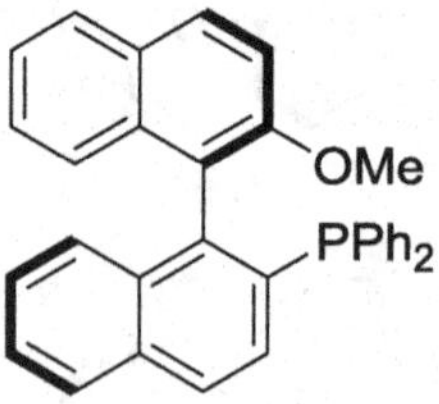

JoshPhos   Scheme 29 **L48**

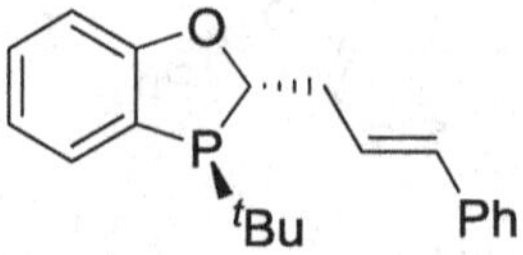

AQ   8-aminoquinoline

1 Chiral Bidentate Heterodonor P-Oxazoline Ligands

Maria Biosca, Jordi Faiges, Montserrat Diéguez, and Oscar Pàmies

CONTENTS

1.1 INTRODUCTION

The performance of enantioselective metal catalysts depends mainly on the selection of the most appropriate chiral ligand.[1] Among the thousands of chiral ligands which have been developed, a few stand out for their broad applicability. The most efficient, called "privileged chiral ligands", derive from a few core structures.[2] Surprisingly, most of them possess C_2 symmetry (Figure 1.1). The reason for initially choosing bidentate ligands with C_2-symmetry was to reduce the number of catalyst/substrate arrangements and transition states, facilitating mechanistic studies, and the elucidation of the relationship between structure and catalytic performance. So, for a long time, the research focused on developing C_2-symmetric ligands. However, the intermediate transition metal ligand complex that arises during a catalytic cycle may not be symmetric and, in these cases, the desymmetrization of the ligand, by tuning each donor atom to accommodate a specific purpose in the catalytic cycle, has been shown to achieve better enantiocontrol for some reactions. One of the most effective methods of desymmetrizing a ligand is by using different donor atoms. In the last few decades, heterodonor ligands, containing dual, strongly and weakly donor heteroatom pairs, have emerged as an increasingly useful ligand class, since the different electronic and steric properties of these heteroatoms are powerful stereocontrol elements.[3] The two functionalities also facilitate catalyst optimization, because both functionalities can be independently modified for improved performance. Among them, P–N ligands have been the most-widely-used P–oxazolines, as

1

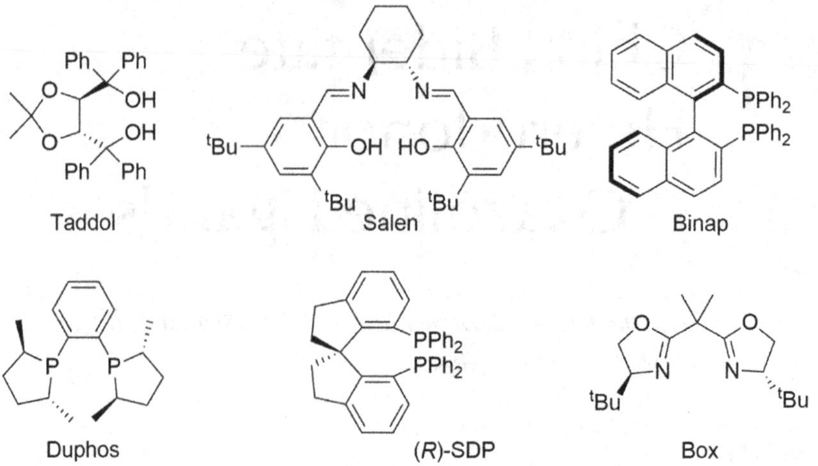

FIGURE 1.1   Selected C_2-privileged ligand structures.

a result of being the most- studied combination, due to their ready accessibility and modular construction. [3a-e] The vast majority of P-oxazoline ligands are derived from readily available chiral amino alcohols, in short and efficient synthetic sequences. [3a-e]

1.2   APPLICATION OF BIDENTATE HETERODONOR P-OXAZOLINE LIGANDS IN ASYMMETRIC CATALYTIC TRANSFORMATIONS

The origin of P-oxazoline ligands can be traced back to 1993, with three independent publications from Helmchem, Pfaltz, and Williams, with the introduction of a new versatile class of ligands, the phosphine-oxazoline PHOX ligands **1** (Figure 1.2).[4] PHOX ligands have been successfully applied to many metal-catalyzed asymmetric reactions, such as hydrogenation, inter- and intramolecular Heck reactions, allylic substitutions and decarboxylative allylation reactions, conjugate additions to enones, Diels-Alder and aza-Diels-Alder reactions, among others.[2] Due to this extensive success, PHOX is the only heterodonor ligand included in the family of "*privileged ligands*".[2] Despite having been described more than 25 years ago, they continue to be used in new asymmetric transformations, underlining their status as a privileged chiral ligand class.[5]

Inspired by the PHOX ligands, several variations on P-oxazoline ligands have been made by changing either the ligand scaffold or the properties of the phosphine group, or by substituing the phosphine moiety by other P-donor groups.[3a-e] These specific modifications of the simple PHOX ligands achieved improvements in enantioselectivity in some cases. However, only a few of them have been successfully applied to several, mechanistically unrelated, asymmetric reactions, with a broad substrate/reagent range. A broad range of reactions, and a wide substrate/reagent scope are desirable to minimize the time necessary to achieve ligand discovery and preparation. Figure 1.3 presents a selection of the most representative

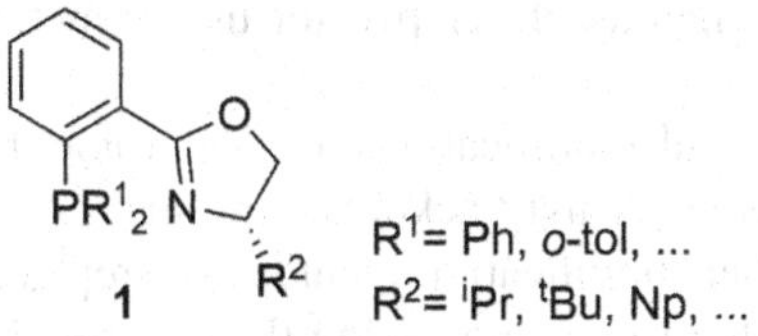

R^1= Ph, o-tol, ...
R^2= ^{i}Pr, ^{t}Bu, Np, ...

FIGURE 1.2 Phosphine-oxazoline PHOX ligands **1**.

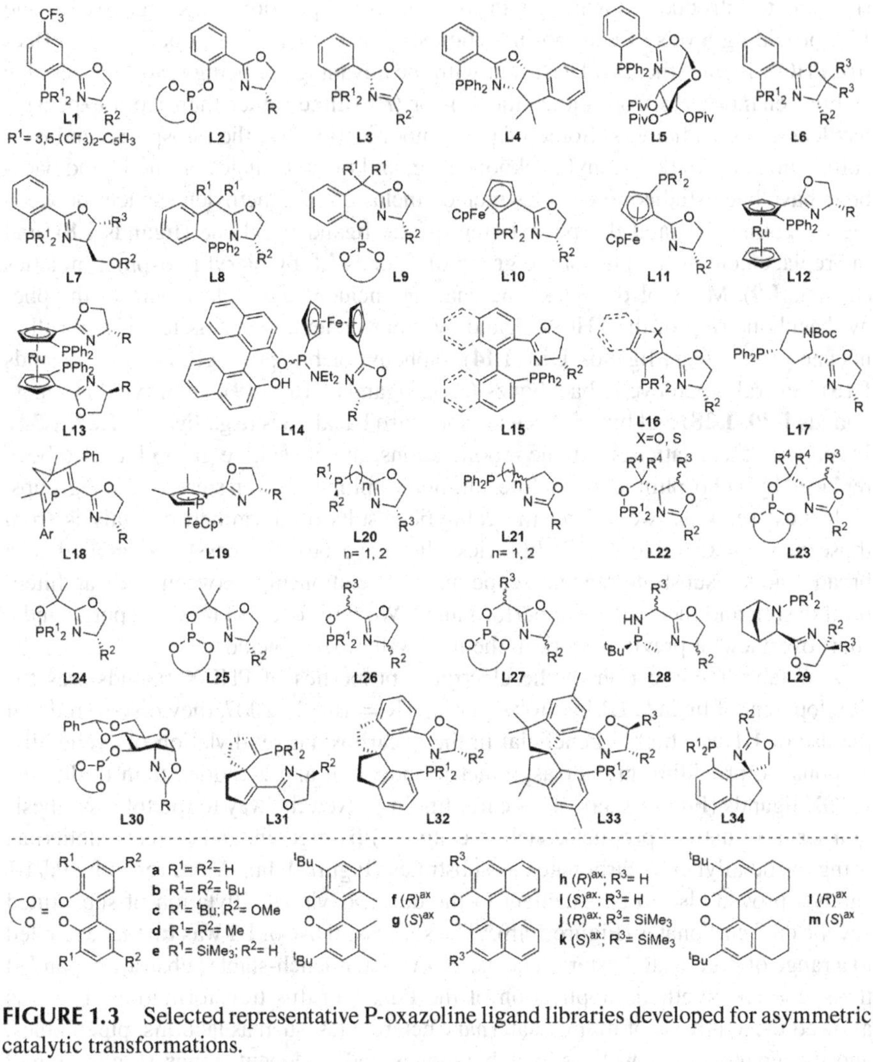

FIGURE 1.3 Selected representative P-oxazoline ligand libraries developed for asymmetric catalytic transformations.

families of P-oxazoline ligands, developed for use in metal-catalyzed asymmetric transformations.

Some of these successful modifications are electronic, by introducing electron-withdrawing groups in the phenyl backbone ring and/or in the phosphine moiety (e.g., ligands **L1**). Another modification includes the replacement of the phosphine group by biaryl phosphite groups, modifying the electronic and steric properties of the ligand (ligands **L2**). Other modifications are on the oxazoline group, by attaching the phenyl backbone ring to the stereogenic center next to the oxazoline (ligands **L3**), or introducing other oxazoline substituents, such as ferrocene, tricyclic, and sugar oxazoline groups (e.g., ligands **L4**–**L5**). Another modification of the oxazoline ring was to introduce substituents in the 5 and/or 5' positions (e.g., ligands **L6** and **L7**), providing levels of enantio-induction similar to that of the usually most effective PHOX ligand, the 'BuPHOX, but with the advantage of being readily accessible as both enantiomers from either the (*S*)- or (*R*)-valine rather than from expensive *tert*-leucinol enantiomers. Some of these modifications on the phosphine and oxazoline moieties, in the phenyl backbone ring, and many changes on the ligand backbone have been studied. Examples include inclusion of a methylene spacer between the oxazoline ring and the phenyl ring on the ligand backbone (ligands **L8**), and the replacement of the phosphine group of ligands **L8** by biaryl phosphite moieties (ligands **L9**). Many of the backbone changes includes the replacement of the phenyl backbone ring of the PHOX ligand by other moieties, such as ferro- and ruthenocene groups (e.g., ligands **L10**–**L14**), biphenyl or binapthyl groups (e.g., ligands **L15**), several heterocyclic backbones (e. g., ligands **L16**–**L19**), an alkyl chain (e.g., ligands **L20**–**L28**) and byciclic, sugar, and spiro backbones (e.g., ligands **L29**–**L34**). In many of these latter backbone modifications, the phosphine group has also been replaced by a phosphinite, phosphite, aminophosphine, or even stereogenic P groups.

In next sections, we collect the catalytic results on asymmetric catalysis from those new P-oxazoline ligand libraries, that have been successfully applied in a broad reaction/substrate/reagent scope and the relationship between their architectural design and their catalytic performance. We will focus on recent reports, and a short overview of previous work in the field will also be included.

A notable modification of the electronic properties of PHOX ligands was the development of ligands **L1** by Stoltz's group ($R^2 =$ 'Bu). In 2007, they discovered that the use of **L1** was highly beneficial in the decarboxylative allylation of cyclic allyl carbonates, providing greater asymmetric induction and activities than traditional PHOX ligands (Figure 1.4a).[6] This early finding paved the way to the total synthesis of a series of natural products, such as elatol, (+)-hamigeran, and (−)-cephalotaxine, using cyclic allyl enol carbonates as substrates (Figure 1.4a).[7] More recently, Pd/**L1** catalyst proved also to be excellent in the decarboxylative allylation of substituted acyclic enol carbonates (Figure 1.4b).[8] The successful use of **L1** was further extended to a range of β-keto allyl esters. The use of this later bench-stable substrate expanded the scope and synthetic application of the Pd/**L1** in this transformation. This has allowed the synthesis of many quaternary heterocyles, such as lactams, piperazines, and diazepanones, as well as cyclobutanones and cyclopentanones (Figure 1.4c).[9] This methodology has been successfully applied to the synthesis of many natural

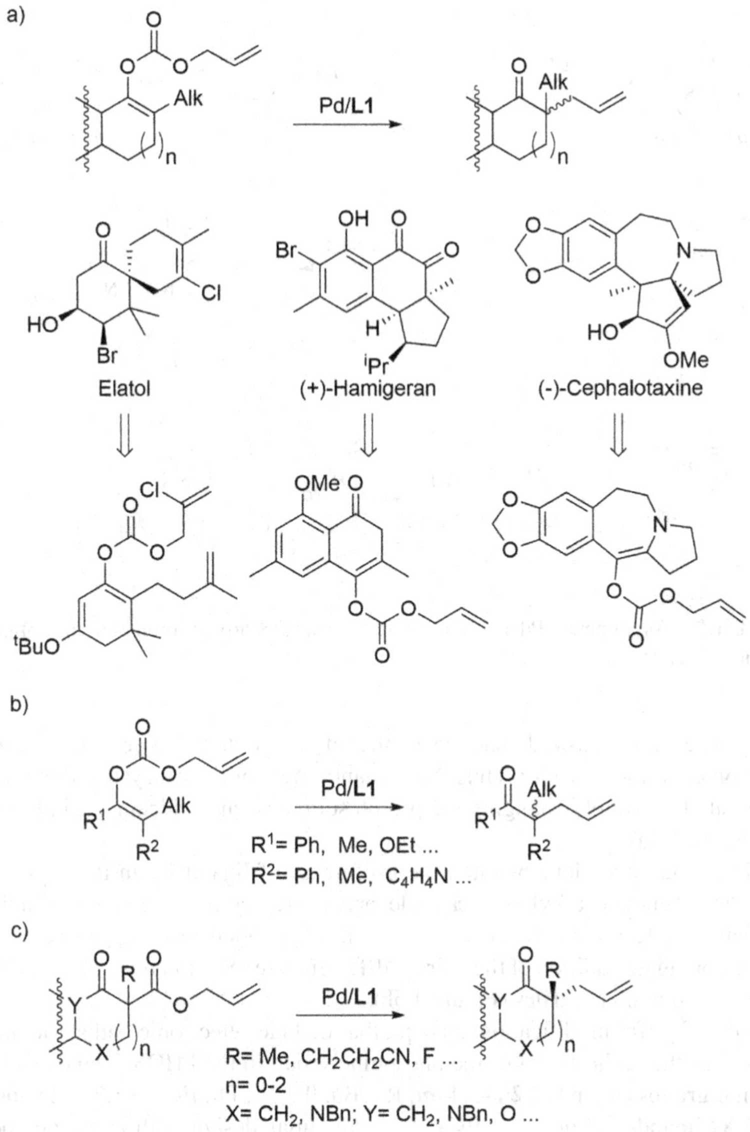

FIGURE 1.4 Representative applications of Pd/**L1** catalytic system to the asymmetric decarboxylative allylation reactions.

products, such as (+)-sibirinine, (−)-goniomitine, (+)-limaspermidine, and nigelladine A, among others.[10]

Guiry's group also demonstrated the benefits of using **L1** in the construction of sterically hindered tertiary α-aryl ketones, such as isoflavones and α-aryl-1-indanones, *via* Pd-catalyzed decarboxylative protonation of α-aryl-β-keto allyl

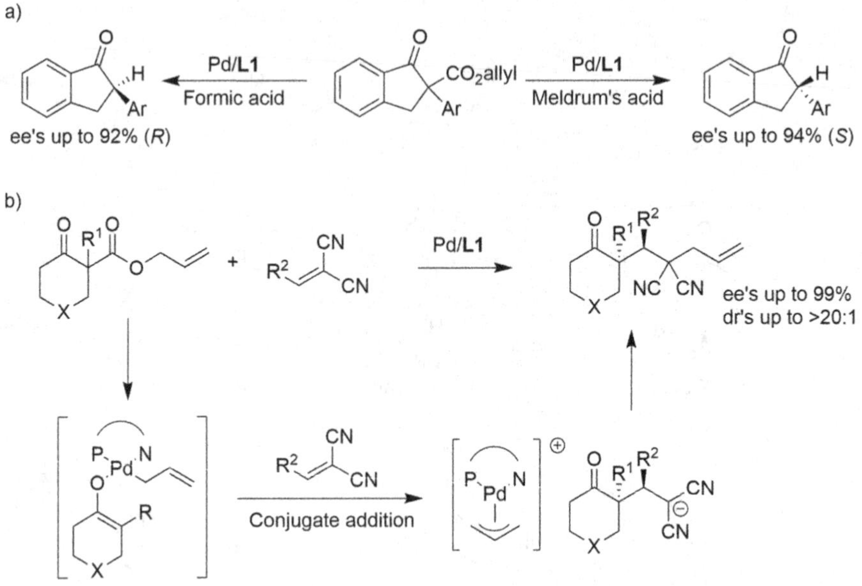

a)

ee's up to 92% (R)   ee's up to 94% (S)

b)

FIGURE 1.5 Asymmetric Pd/L1-catalyzed a) decarboxylative protonation and b) enolate alkylation cascade reactions.

esters.[11] They also showed that the nature of the proton source had a dramatic impact on enantioselectivity. Thus, both enantiomers of the α-aryl-1-indanones can be generated by simply changing the proton source from Meldrum's acid to formic acid (Figure 1.5a).

Stolz's group also demonstrated the usefulness of ligand **L1** in the asymmetric Pd-catalyzed enolate alkylation cascade procedure for the formation of adjacent quaternary and tertiary stereocenters.[12] This multiple bond-forming procedure takes place by conjugate addition of the chiral Pd/**L1**-enolate, generated *in situ* from β-keto allyl esters, to malononitriles (Figure 1.5b).

Another fruitful modification concept, that includes electronic and steric modifications, was the replacement of the phosphine group in the PHOX ligands by biaryl phosphite groups (ligands **L2a–c, l–m**, R= 'Bu, 'Pr, Et, Ph; Figure 1.3).[13] In contrast to PHOX, ligands **L2** do not only show, a modular design with numerous potential phosphite groups available, but also they are air-stable solids, which are made in the same number of synthetic steps as PHOX. Phosphite-containing ligands are particularly useful for asymmetric catalysis. They show greater resistance to oxidation than do phosphines, they are easily synthesized from readily available chiral alcohols, and their modular constructions are easy.[14] The success with these new air stable phosphite-oxazoline ligands **L2** (Figure 1.3) has been attributed to the flexibility conferred by the biaryl phosphite moiety that is able to adequately control the size of the chiral pocket to meet the demands of the reaction.[13b] In addition, its π-acceptor capacity has been shown to have a positive effect on activity in many of the asymmetric catalytic reactions studied.[13,15,16] Another advantage of the new phosphite-oxazolines **L2** over the PHOX ligands is that the best ligands are derived from affordable (*S*)-phenyl glycinol or (*S*)-valinol (R= Ph or ^{i}Pr), rather than from expensive (*S*)-*tert*-leucinol found in PHOX ligands.

Although ligands **L2** were initially developed to overcome the limited substrate range of PHOX for the Pd-catalyzed allylic substitutions,[13,17] they were successfully applied to other metal-catalyzed asymmetric reactions.[15,16] In Pd-catalyzed allylic substitution reactions, ligands that tolerate a wide range of substrates and nucleophiles are certainly rare.[1c, 18] Pd/**L2** catalysts proved to be very effective catalysts for the allylic substitution of both hindered and unhindered linear and cyclic substrates,[13] outperforming Pd-PHOX catalysts, which achieve outstanding enantioselectivities with *rac*-(*E*)-1,3-diarylallyl substrates, moderate to high enantioselectivities with 1,3-dialkylallyl substrates, but provide essentially racemic products with cyclic substrates[2a,3a]. Excellent activities turnover frequencies (TOFs> 2,400 mol substrate × (mol Pd × h)${}^{-1}$, regio- (up to 99%) and enantioselectivities (enantiomeric excess (ee) values up to 99%) were therefore achieved for mono-, di-, and trisubstituted substrates (Figure 1.6).[13] The highest enantioselectivities in the allylic substitution of the benchmark substrate *rac*-1,3-diphenylallyl acetate (**S1**), were achieved using the simple tropoisomeric ligand **L2b,** independent of the oxazoline substituent (R= Ph, Et, ^{i}Pr, or ^{t}Bu). Pd/**L2b** was also very tolerant of variation of the nucleophile sources (Figure 1.6). Thus, excellent enantioselectivities were achieved with allyl-, butenyl-, pentenyl-, and propargyl-substituted malonates (up to >99% ee; Figure 1.6), the products of which are key intermediates in the synthesis of more complex chiral products.[19] Ee values up to 99% could also be obtained in the allylic fluorobis(phenylsulfonyl)methylation of **S1** and up to 97% ee using 4-(trifluoromethyl)benzyl alcohol as O-nucleophile (Figure 1.6). Even more remarkable are the almost perfect enantioselectivities (ee values up to 99%) obtained in the etherification of **S1** with silanols. Pd/**L2b** was also successfully applied to other symmetrical linear substrates with steric and electronic requirements (**S2–S7**) different from those of **S1** (Figure 1.6), using highly interesting nucleophiles such as those α-substituted with methyl-, allyl-, and butenyl- groups. A wide range of C-nucleophiles, including the less studied α-substituted malonates and acetylacetone, can also react efficiently with more demanding cyclic substrates to provide the corresponding compounds at high yields and enantioselectivities (up to >99 % ee; Figure 1.6). Remarkably, excellent enantioselectivities (ee values between 96% and >99%) were obtained, even with **S9**, which usually generates products with much lower enantioselectivities than **S8**. In contrast to previous substrates, for cyclic substrates the highest enantioselectivities were obtained with ligands **L2b** and **L2m**. The high performance of Pd/**L2b** and Pd/**L2m** also extended to the allylic substitution of challenging unsymmetrical monosubstituted allylic substrates **S11–S12** (Figure 1.6). Most Pd catalysts favor the formation of the usually undesired achiral linear product.[17,20] The increases in regioselectivity toward the desired branched isomer in monosubstituted linear substrates can be explained by the large π-acceptor ability of the phosphite moiety, which decreases the electron density of the most substituted allylic terminal carbon atom *via* the *trans*-influence, favoring the nucleophilic attack of this carbon atom. Furthermore, excellent enantioselectivities were achieved with 1,3,3-trisubstituted allylic substrates **S13–S14** (Figure 1.6).

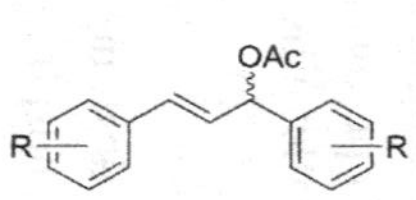

S1
S1 R= H
S2 R= 4-Me
S3 R= 4-Br
S4 R= 3-OMe
S5 R= 2-Me

H-Nu	% Yield	% ee
H-CH(CO$_2$Me)$_2$	94	>99 (S)
H-CH(CO$_2$Et)$_2$	93	>99 (S)
H-CH(CO$_2$Bn)$_2$	95	>99 (S)
H-CMe(CO$_2$Me)$_2$	91	99 (R)
H-Callyl(CO$_2$Me)$_2$	92	>99 (R)
H-Cbutenyl(CO$_2$Et)$_2$	89	>99 (R)
H-Cpentenyl(CO$_2$Et)$_2$	93	93 (R)
H-propargyl(CO$_2$Et)$_2$	91	>99 (R)
H-CH(COMe)$_2$	89	98 (S)
H-CF(SO$_2$Ph)$_2$	76	99 (R)
H-OCH$_2$(p-CF$_3$-C$_6$H$_4$)	93	97 (-)
H-OSiMe$_2$Ph	79	98 (R)
H-OSiPh$_3$	91	99 (R)

S2

H-Nu	% Yield	% ee
H-CH(CO$_2$Me)$_2$	93	99 (S)
H-Callyl(CO$_2$Me)$_2$	91	99 (R)
H-Cbutenyl(CO$_2$Et)$_2$	92	94 (R)

S3

H-Nu	% Yield	% ee
H-CH(CO$_2$Me)$_2$	89	99 (S)

S4

H-Nu	% Yield	% ee
H-CH(CO$_2$Me)$_2$	91	99 (S)

S5

H-Nu	% Yield	% ee
H-CH(CO$_2$Me)$_2$	90	99 (S)

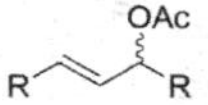

S6 R= Me   **S7** R= iPr

S6

H-Nu	% Yield	% ee
H-CH(CO$_2$Me)$_2$	89	93 (S)
H-CMe(CO$_2$Me)$_2$	86	80 (S)
H-Callyl(CO$_2$Me)$_2$	89	90 (S)
H-Cbutenyl(CO$_2$Et)$_2$	90	87 (S)

S7

H-Nu	% Yield	% ee
H-CH(CO$_2$Me)$_2$	92	>95 (S)

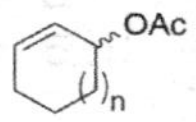

S8 n= 1
S9 n= 0
S10 n= 2

S8

H-Nu	% Yield	% ee
H-CH(CO$_2$Me)$_2$	92	99 (S)
H-CH(CO$_2$Et)$_2$	93	>99 (S)
H-CH(CO$_2$Me)$_2$	90	97 (S)
H-CMe(CO$_2$Me)$_2$	89	99 (+)
H-Callyl(CO$_2$Me)$_2$	91	>99 (-)
H-propargyl(CO$_2$Et)$_2$	88	92 (S)
H-CH(COMe)$_2$	93	99 (-)

S9

H-Nu	% Yield	% ee
H-CH(CO$_2$Me)$_2$	86	>95 (-)
H-propargyl(CO$_2$Et)$_2$	87	96 (S)

S10

H-Nu	% Yield	% ee
H-CH(CO$_2$Me)$_2$	92	>99 (S)
H-propargyl(CO$_2$Et)$_2$	65	>99 (R)

S11
91% yield
68% branched
86% (S) ee

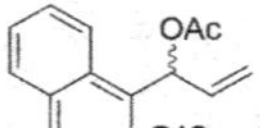

S12
92% yield
>99% branched
92% (S) ee

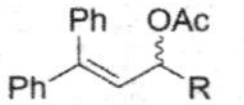

R	% Yield	% ee
S13	Me	84	>99 (R)
S14	Ph	95	>99 (R)

FIGURE 1.6 Summary of the catalytic results in the Pd-catalyzed allylic substitution with Pd/**L2b** catalyst for substrates **S1–S7** and **S11–S14**, and with Pd/**L2m** for **S8–S10**. Reactions were carried out using 0.5 mol% [Pd(η^3-C$_3$H$_5$)Cl]$_2$, 1.1 mol% ligand, CH$_2$Cl$_2$ as solvent, bis(trimethylsilyl) acetamide (BSA)/KOAc as base (except for O-nucleophiles, that were carried out using Cs$_2$CO$_3$ as the base), room temperature (except for substrate **S6** that was incubated at 0°C). For substrates **S11–S14**, reactions were carried out using 1 mol% [Pd(η^3-C$_3$H$_5$)Cl]$_2$ and 2.2 mol% ligand.

The wide substrate scope of the Pd/**L2b** catalyst was rationalized by NMR studies and density-functional theory (DFT) calculations of its Pd-η^2-olefin and Pd-η^3-allyl complexes.[13b] It was found that the biaryl phosphite group in ligand **L2b** adopts an (*S*)-configuration in the complexes, mimicking the product olefin complexes of hindered as well as unhindered substrates. Although the olefins coordinated with the same face to Pd in complexes with the corresponding rigid ligands **L2l** and **L2m**, products with opposite absolute configurations were obtained, due to the different energies of the transition states of nucleophilic attack at the Pd-η^3-allyl intermediates. These findings confirm that the broad substrate scope of Pd/biaryl phosphite-oxazoline systems results from their capacity to adjust the size of the binding pocket to the substrate type, a feature that also explains their excellent performance in other asymmetric reactions, such as asymmetric hydrogenation of unfunctionalized olefins[15] and intermolecular Heck reactions, with results comparable to that of the Pd/PHOX catalyst, using ligand **L2b** (R= Ph)[16a]. In the Ir-catalyzed hydroboration of challenging 1,1'-disubstituted olefins, the Pd/**L2b** (R= ^{i}Pr) can efficiently hydroborate a broader range of olefins, with ee values up to 94% and perfect regioselectivity,[16b] compared with previous phosphine-oxazoline PHOX ligands.[21] Compared with the hydroboration of monosubstituted and internal 1,2-disubstituted olefins, the control of regio- and enantioselectivity becomes a much greater challenge with 1,1'-disubstituted alkenes.[22] This is because the chiral transition-metal catalyst has difficulty in controlling not only the specific boration at the desired terminal β-position rather than at the more substituted α-position but also the face-selectivity coordination (due to the presence of the two relatively similar substituents at the geminal position). Particularly, Pd/**L2b** successfully hydroborates a wide range of α-*tert*-butylstyrenes, with aryl substituents that have different electronic and steric properties, thus complementing the results of Cu/N-heterocyclic carbene catalysts, the only other system reported to date to have been used to attempt these reactions.[23]

Another effective modification of PHOX ligands was to introduce a methylene spacer between the oxazoline and the phenyl ring of the ligand backbone, forming with the metal a higher seven-membered chelate ring (phosphine-oxazoline ligands **L8**, R^1= Me, H; R^2= Me, ^{i}Pr, ^{t}Bu, Figure 1.3). Thus, Hou and Wu and coworkers used these phosphine-oxazoline ligands **L8** in the Pd-catalyzed intermolecular Heck reaction (Figure 1.7).[24] Compared with the intramolecular asymmetric Heck reaction, the intermolecular version is less developed and its synthetic utility remains limited.[25] This is due to regioselectivity issues caused by the possible displacement of the C=C double bond, which leads to mixtures of products. Pfaltz showed, for the first time, that the phosphine-oxazoline ^{t}BuPHOX ligand minimized the double

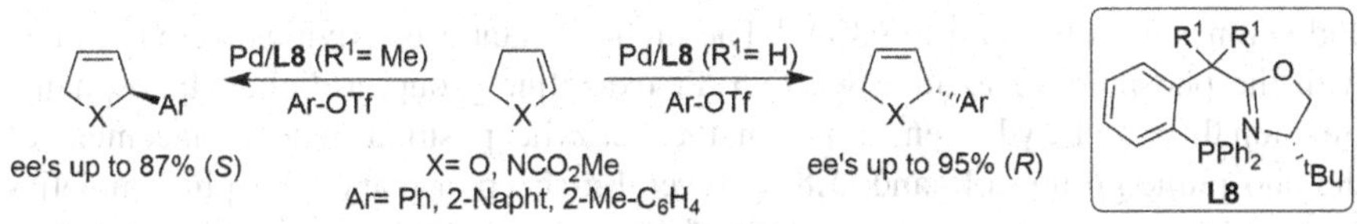

ee's up to 87% (*S*)   X= O, NCO$_2$Me   ee's up to 95% (*R*)

Ar= Ph, 2-Napht, 2-Me-C$_6$H$_4$

Pd/**L8** (R^1= Me)   Ar-OTf   Pd/**L8** (R^1= H)   Ar-OTf   **L8**

FIGURE 1.7 Application of ligands **L8** in the Pd-catalyzed intermolecular Heck reaction.

bond isomerization, achieving high regio- and enantioselectivity to the desired product, although still at low activity with long reaction times (3–7 days to achieve full conversion).[26] Ligands **L8** (R^1= Me, H, and R^2= ^{t}Bu) were screened in the reaction between 2,3-dihydrofuran and a range of aryl triflates with high regio- and enantioselectivities with results comparable to those of PHOX ligands (up to 95% ee, Figure 1.7).[24] Interestingly, ligands that did not contain any substitution in the benzylic position of the ligand produced the *R*-enantiomer, whereas ligands that contained *gem*-methyl substituents produced the *S*-enantiomer (Figure 1.7).

More recently our research group decided to study whether the biaryl phosphite moiety is still as effective when other type of ligand backbones are present. We decided to replace the phosphine group in ligands **L8** with several π-acceptor biaryl phosphite moieties (ligands **L9b, g, j–k, m**; R^1= Me, H; R^2= ^{i}Pr, ^{t}Bu, Ph; Figure 1.3).[27] This change was found to be highly advantageous in terms of activity, but also allowed extension of the range of substrates and triflate sources beyond that which could be coupled (ee values up to 98% and regioselectivities up to 99%). The highest results were obtained with the ligand that contained biaryl phosphite groups **b, g**, and **m**, and an ^{i}Pr substituent in the oxazoline moiety, instead of the expensive ^{t}Bu substituent found in the related phosphine-oxazoline **L8** and PHOX ligands, and a hydrogen in the R^1 positions.[27] Kinetic studies in a practical regime indicated that migratory insertion of the alkene is rate limiting, and that the more favorable formation of alkene adducts makes the phosphite-based catalysts more active than the phosphine-based ones.[27]

Interesting, the same family of phosphine/phosphite-oxazoline ligands **L8–L9** also provided excellent results in the hydrogenation of unfunctionalized olefins or with poorly coordinative groups. This was relevant because the reduction of these types of alkenes is underdeveloped, compared with the reduction of functionalized olefins.[28] The reason is that most catalysts are too specific for a certain geometry and substitution pattern of the olefin. For example, the most successful cases have been reported for trisubstituted *E*-unfunctionalized alkenes and, to a lesser extent, for *Z*-trisubstituted and 1,1'-disubstituted alkenes. The asymmetric hydrogenation of tetrasubstituted unfunctionalized substrates is still underdeveloped. A breakthrough in their hydrogenation came in 1997 when Pfaltz *et al.* used phosphine-oxazoline PHOX ligands **1** to design [Ir(PHOX)(cod)]BAr$_F$ (cod = 1,5-cyclooctadiene and BAr$_F$= 3,5-$(F_3C)_2$-$C_6H_3)_4$B) catalysts that could, for the first time, produce active, enantioselective, and stable catalysts for the reduction of this type of olefin.[29] Despite this success, its scope was limited, to mainly some *E*-trisubstituted olefins.[30] Phosphine-oxazoline ligands **L8** allows extension of the type of substrates which can be successfully hydrogenated. High enantioselectivities were achieved with *E*-trisubstituted aryl/alkyl alkenes, allylic alcohols, and α,β-unsaturated esters and ketones (ee values up to 98%).[31] The highest enantioselectivities were obtained with the phosphine-oxazoline with an ^{i}Pr oxazoline group, a diphenylphosphanyl functionality, and hydrogen atoms in the benzylic position. The replacement of the phosphine group in ligands **L8** by several π-acceptor biaryl phosphite moieties extended the range of substrates successfully hydrogenated even further, including more challenging 1,1'-disubstiuted olefins, thus overcoming another of the limitation of the PHOX ligands. The highest enantioselectivities were obtained with the ligands than contain Ph or ^{i}Pr oxazoline substituents and hydrogen atoms in the benzylic position.[15] The phosphite group depended on the substrate to be hydrogenated (Figure 1.8). No loss of enantioselectivity was observed when using 1,2-propylene carbonate (PC) as an environmentally friendly solvent. A summary of the successfully asymmetric hydrogenation of 55 olefins with the Ir/**L9** catalyst is shown in Figure 1.8. Excellent enantioselectivities were obtained in the reduction of trisubstituted olefins, even in the more challenging triarylsubstituted substrates. They also worked well for olefins with a variety of relevant neighboring polar groups such as α,β-unsaturated ketones, amide, lactones, lactams, alkenyl boronic esters, and enol phosphinates (ee values up to >99%). In addition, for each type of neighboring group, the enantioselectivities were quite independent of the electronic and steric nature of the substituents decorating such motifs. The effective hydrogenation of such a range of olefins is of great importance since their reduced products are key structural chiral units found in many high-value chemicals. For example, α- and β-chiral ketones, and carboxylic acid derivatives are ubiquitous in natural products, fragrances, agrochemicals, and drugs, while chiral borane compounds are useful building blocks in organic synthesis, because the C–B bond can be readily converted to C–O, C–N, and C–C bonds, with retention of the chirality. Unlike trisubstituted olefins, 1,1'-disubstituted substrates had not been successfully hydrogenated until

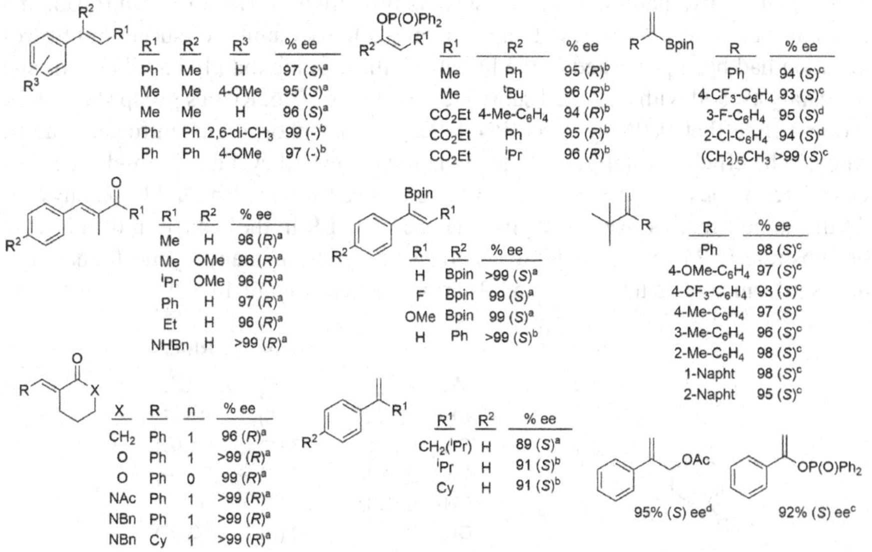

R^1	R^2	R^3	% ee
Ph	Me	H	97 (S)a
Me	Me	4-OMe	95 (S)a
Me	Me	H	96 (S)a
Ph	Ph	2,6-di-CH$_3$	99 (-)b
Ph	Ph	4-OMe	97 (-)b

R^1	R^2	% ee
Me	Ph	95 (R)b
Me	^{t}Bu	96 (R)b
CO$_2$Et	4-Me-C$_6$H$_4$	94 (R)b
CO$_2$Et	Ph	95 (R)b
CO$_2$Et	^{i}Pr	96 (R)b

R	% ee
Ph	94 (S)c
4-CF$_3$-C$_6$H$_4$	93 (S)c
3-F-C$_6$H$_4$	95 (S)d
2-Cl-C$_6$H$_4$	94 (S)d
(CH$_2$)$_5$CH$_3$	>99 (S)c

R^1	R^2	% ee
Me	H	96 (R)a
Me	OMe	96 (R)a
^{i}Pr	OMe	96 (R)a
Ph	H	97 (R)a
Et	H	96 (R)a
NHBn	H	>99 (R)a

R^1	R^2	% ee
H	Bpin	>99 (S)a
F	Bpin	99 (S)a
OMe	Bpin	99 (S)a
H	Ph	>99 (S)b

R	% ee
Ph	98 (S)c
4-OMe-C$_6$H$_4$	97 (S)c
4-CF$_3$-C$_6$H$_4$	93 (S)c
4-Me-C$_6$H$_4$	97 (S)c
3-Me-C$_6$H$_4$	96 (S)c
2-Me-C$_6$H$_4$	98 (S)c
1-Napht	98 (S)c
2-Napht	95 (S)c

X	R	n	% ee
CH$_2$	Ph	1	96 (R)a
O	Ph	1	>99 (R)a
O	Ph	0	99 (R)a
NAc	Ph	1	>99 (R)a
NBn	Ph	1	>99 (R)a
NBn	Cy	1	>99 (R)a

R^1	R^2	% ee
CH$_2$(^{i}Pr)	H	89 (S)a
^{i}Pr	H	91 (S)b
Cy	H	91 (S)b

95% (S) ee^d 92% (S) ee^c

FIGURE 1.8 Summary of the catalytic results for the hydrogenation of unfunctionalized trisubstituted olefins with [Ir(**L9**)(cod)]BAr$_F$ catalyst precursors. Reaction conditions: 2 mol% of catalyst, CH$_2$Cl$_2$ as solvent, 50 bar of H$_2$, 4 h and room temperature incubation. ^{a} Reactions using Ir/**L9k** (R^2= ^{i}Pr) catalytic system. ^{b} Reactions using Ir/**L9j** (R^2= ^{i}Pr) catalytic system. ^{c} Reactions using Ir/**L9b** (R^2= Ph) catalytic system. ^{d} Reactions using Ir/**L9j** (R^2= Ph) catalytic system.

recently[32] because the catalyst must control not only the face-selectivity coordination (only two substituents compared with the three of the trisubstituted olefins), but also control the isomerization of the olefins to form the more stable E-trisubstituted substrates, which are hydrogenated to form the opposite. It was very gratifying to obtain ee values up to 98% in a broad range of *tert*-butyl-aryl 1,1'-disubstituted alkenes (Figure 1.8) with different electronic and steric properties in the aryl substituent. The reduction of α-alkyl-styrenes with less bulky alkyl substituents proceeded with somewhat lower enantioselectivities (ee values from 83% to 91%), as with other successful Ir-catalysts reported.[28,32] Nevertheless, we found that enantioselectivity could be maximized by choosing the ligand parameters. Deuterium-labeling experiments confirm the existence of a competing isomerization pathway, which is responsible for the decrease in enantioselectivity. A positive correlation was observed between the extension of the D-incorporation in the allylic position and the loss of enantiocontrol. Note that the excellent catalytic performance obtained in the reduction of trisubstituted alkenyl boronic esters and enol phosphinates could also be extended to their 1,1'-disubstituted analogs (Figure 1.8). Similarly, the reduction of allylic acetate also proceeded with high enantioselectivity.

In addition, we found that the phosphite-oxazoline ligand **L9** was also successfully used in the asymmetric hydrogenation of another challenging type of substrate, the cyclic β-enamides.[33] Their hydrogenated products (such as 2-aminotetralines and 3-aminochromanes) are key structural units in many therapeutic agents and biologically active natural products (such as rotigotine, robalzotan, terutroban, and alnespirone).[34] Nevertheless, at that time, only a few examples of successful hydrogenation had been published in the literature, these being mainly based on Rh- and Ru-catalysts, and with a limited substrate scope.[35] In 2016, Riera's group showed for the first time that Ir-PN systems (PN = heterodonor phosphorus–nitrogen ligands) can also be efficient catalysts for the reduction of several cyclic β-enamides derived from 2-tetralones, with results that surpassed those from the Rh- and Ru-catalysts.[36] At the same time, we were studying the use of Ir/**L9** in the reduction of this kind of substrate.[33] We obtained high enantioselectivities in many cyclic β-enamides derived from both 2-tetralones and 3-chromanones (Figure 1.9). It is important to

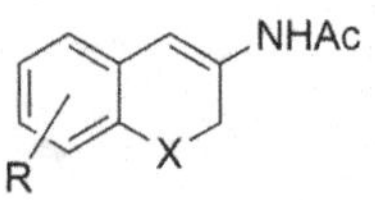

X	R	Ir-cat % ee	Rh-cat % ee
CH$_2$	H	98 (S)	94 (R)
CH$_2$	6-Br	97 (S)	96 (R)
CH$_2$	5-OMe	97 (S)	93 (R)
CH$_2$	6-OMe	98 (S)	92 (R)
CH$_2$	7-OMe	99 (S)	89 (R)
CH$_2$	8-OMe	99 (S)	88 (R)
O	H	98 (R)	91 (S)

FIGURE 1.9 Summary of the catalytic results for the hydrogenation of cyclic β-enamides with [Ir(**L9j**)(cod)]BAr$_F$ and [Rh(**L9j**)(cod)]BF$_4$ catalyst precursors. Reaction conditions: 1 mol% of catalyst, DCM as solvent, 50 bar of H$_2$ for Ir and 10 bar of H$_2$ for Rh, 20 h for Ir and 50 h for Rh, and room temperature for Ir and 5°C for Rh.

note the high enantioselectivity obtained in the reduction of N-(5-methoxy-3,4-dihydronaphthalen-2-yl)acetamide, the hydrogenated product of which is a key intermediate for the synthesis of rotigotine (Figure 1.9).[33] The best enantioselectivities were achieved with ligands containing an R-binaphthyl phosphite moiety, such as **L9j** (R[1]= H; R[2]= [i]Pr). We also found that, by switching from Ir to Rh, both enantiomers of the hydrogenated product could be obtained at high enantioselectivity. Again, the use of propylene carbonate had no effect on the enantioselectivities.

Many of other notable modifications of PHOX ligands involve the replacement of the phenyl group of the ligand backbone by other groups. A group of relevant modified PHOX ligands are those involving two carbon atoms between the two donor functionalities. We can highlight the development of phosphinite-oxazoline ligands **L22** and **L24** (Figure 1.3) by the Pfaltz group, where the *ortho*-phenylene tether of the PHOX has been replaced by a branched alkyl chain, with the aim to reach a wider substrate scope in the reduction of unfunctionalized olefins.

The phosphinite–oxazolines **L22** (Figure 1.3; R[1]= Ph, *o*-Tol, Cy; R[2]= [t]Bu, Ph, ferrocenyl, 2-Naph; R[3]= H, Me, 3,5-Me$_2$-C$_6$H$_3$, and R[4]= Me, [i]Pr, [t]Bu, Bn), which are derived from serine and threonine, constitute one of the most privileged ligands for the Ir-catalyzed hydrogenation of unfunctionalized olefins.[22] In contrast to the PHOX ligands, the phosphorus unit is attached to the stereogenic center next to the oxazoline nitrogen atom. Starting from different carboxylic acid derivatives, chlorophosphines, and Grignard reagents, a highly diverse library of ligands was easily prepared. This ligand class allowed E- and Z-2-aryl-2-butenes to be hydrogenated with very high enantioselectivities for the first time (Figure 1.10).[37] By variation of the substituents at the oxazoline ring and ligand backbone, it was possible to systematically optimize the enantioselectivity for each substrate. The greatest enantioselectivities were achieved with ligands containing a methyl substituent at R[3], a benzyl substituent at R[4], and a phenyl at R[1]. However, the appropriate substituent at the oxazoline and the configuration of the carbon at R[3] depend on the substrate to be hydrogenated. For E-trisubstituted olefins, ee values are highest with ligands containing a Ph, a 3,5-Me$_2$-Ph, and a S-configuration, whereas, for Z-olefins, the highest enantioselectivities were achieved using ligands with Ph and an R-configuration. Further fine-tuning allowed hydrogenation of a limited range of the more challenging terminal olefins and 1,1'-disubstituted enamines for the first time (up to 99% ee;

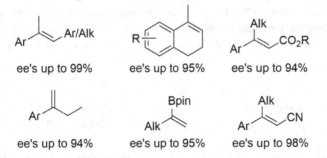

FIGURE 1.10   Selected catalytic hydrogenation results with [Ir(**L22**)(cod)]BAr$_F$ catalyst.

Figure 1.10), with the ligand that contains a methyl substituent at R^3, a benzyl substituent at R^4, but a cyclohexenyl at R^1.[37c,f] More recently, Ir catalysts containing ligands **L22** have also been successfully used in the reduction of α,β-unsaturated nitriles (ee values up to 98%; Figure 1.10).[37e] In addition, these catalysts work efficiently in propylene carbonate as an environmentally friendly solvent, and this allowed the Ir catalysts to be reused, while maintaining the excellent enantioselectivities.[38]

Based on ligands **L22**, Pfaltz's group developed ligands **L24** (Figure 1.3; R^1= Ph, o-Tol, and R^2= ^{i}Pr, ^{t}Bu), where the alkyl chain is bonded in the C-2, instead of the C-4, of the oxazoline moiety, which shifts the chirality from the alkyl chain to the oxazoline moiety, as in PHOX ligands.[39] The scope of these ligands in carrying out asymmetric hydrogenation of unfunctionalized olefins is narrower than with the privileged phosphinite-oxazoline ligands **L22**, although they are complementary. Ligands **L24** achieved high enantioselectivities for the reduction of allylic alcohols and alkenes with heteroaromatic substituents. In the hydrogenation of the cyclic substrate 6-methoxy-1-methyl-3,4-dihydronaphthalene, the observed ee of 95% was one of the highest values recorded for this substrate. Harmata and Hong have used this catalyst in the total synthesis of the antibiotic pseudopteroxazole. The catalyst was able to hydrogenate only the internal double bond, but not the exocyclic C=C bond, with near-perfect regioselectivity in 90% yield, along with only traces of over-hydrogenated product.[40]

Later, our group also prepared the corresponding phosphite-based analogs of ligands **L22**. The introduction of biaryl phosphite groups produced Ir/**L23** catalysts that allowed extension of the number of substrates that could be successfully reduced with Ir/**L22** and Ir/**L24** catalysts.[41] Selecting the ligand parameters, high enantioselectivities were achieved for the reduction of a broad range of trisubstituted unfunctionalized olefins or those with poorly coordinative groups.[41] Whereas the Ir/**L23i** catalyst provided the highest enantioselectivities for linear and cyclic olefins, and α,β-unsaturated esters, the highest ee values for the more demanding Z-isomers, and allylic alcohol and acetate were obtained with the Ir/**23b** catalyst. This high catalytic performance was also extended for the first time to include the hydrogenation of a broader range of the more challenging 1,1'-disubstituted olefins (29 compounds, Figure 1.11), becoming one of the best catalysts for the reduction of this type of substrate and surpassing the performance of previous successful Ir/**L9** and Ir/**L22** catalysts.[32] High enantioselectivities were achieved in a broad range of aryl-alkyl (ee values up to >99%), even with substrates bearing decreasingly sterically alkyl substituents, and heteroaromatic-alkyl (up to >99% ee) olefins. These catalyst precursors also tolerate the presence of neighboring polar groups very well. High

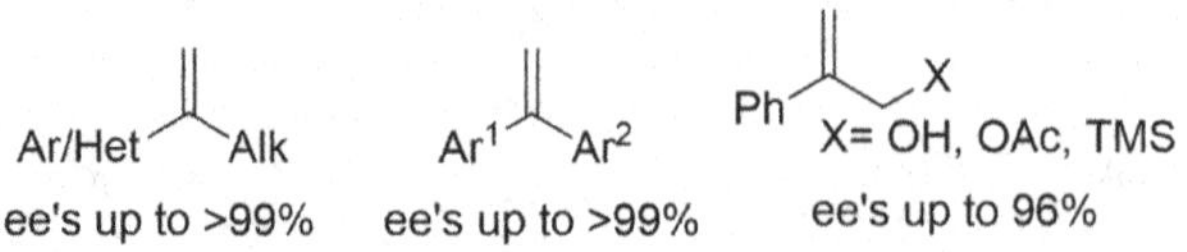

Ar/Het⎯Alk   Ar^1⎯Ar^2 Ph⎯X   X= OH, OAc, TMS

ee's up to >99%   ee's up to >99%   ee's up to 96%

FIGURE 1.11 Selected results achieved with [Ir(**L23**)(cod)]BAr$_F$ catalysts in the hydrogenation of 1,1'-disubstituted substrates.

enantioselectivities were achieved in the reduction of allylic alcohols and an allylic silane. Interestingly, the reaction showed no loss of enantioselectivity when dichloromethane was replaced by propylene carbonate. In addition, the use of propylene carbonate allowed the catalysts to be recycled up to five times by a simple two-phase extraction, while maintaining the excellent enantioselectivities.

Interestingly, the same family of ligands **L23** were also used successfully in the Pd-catalyzed allylic substitution of a broad range of mono, di-, and trisubstituted linear hindered and unhindered linear substrates (ee values up to 99%), using a range of C- and N-nucleophiles.[42] In addition, both enantiomers of substitution products can be obtained with high enantioselectivities by simply changing either the absolute configuration of the alkyl backbone chain or the absolute configuration of the biaryl phosphite moiety. The results are superior to those of PHOX ligands and are comparable to the best reported with the previous family of phosphite-oxazoline ligands **L2,** except for the allylic substitution of cyclic substrate (ee values up to 83%). To address this limitation, it was necessary to replace the oxazoline group with a thiazoline moiety.[43] The introduction of a thiazoline moiety creates a smaller chiral pocket, more suited to unhindered cyclic substrates, while maintaining the flexibility conferred by the biaryl phosphite group. With this simple modification, the enantioselectivities for unhindered cyclic substrates improved considerably (94% ee). So, the two families of ligands (phosphite-oxazoline/thiazoline) are complementary.

Finally, ligands **L23** were also successfully applied in the Pd-catalyzed Heck reaction. Comparable high regio- and enantioselectivity could be achieved for a wide range of substrates and triflate sources (Figure 1.12).[16a] The best results were obtained with ligands **L23b** ($R^2 = p$-CH_3-Ph or ^tBu, $R^3 =$ H, $R^4 =$ CH_3). Under microwave-irradiation conditions, reaction times were considerably shorter (from the 24 h reported for phosphine-oxazoline PHOX to 10 min with both ligands **L23b**), while regio- and enantioselectivities were still excellent (ee values up to 99%). Therefore, it was found that, by selecting the ligand parameters, the results achieved with the

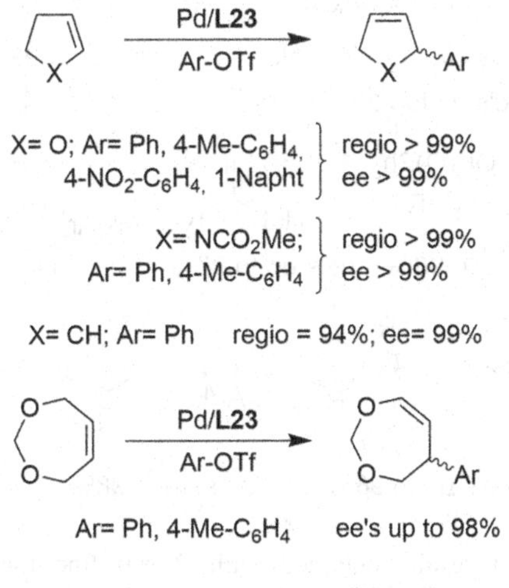

X= O; Ar= Ph, 4-Me-C_6H_4,   regio > 99%
4-NO_2-C_6H_4, 1-Napht   ee > 99%

X= NCO_2Me;   regio > 99%
Ar= Ph, 4-Me-C_6H_4 ee > 99%

X= CH; Ar= Ph   regio = 94%; ee= 99%

Ar= Ph, 4-Me-C_6H_4 ee's up to 98%

FIGURE 1.12 Application of **L23** in the Pd-catalyzed intermolecular Heck reaction.

previous Pd/**L8–L9** catalysts and the PHOX ligands could be considerably improved. As a result, a larger number of substrates could be successfully coupled, using a larger number of triflate sources (Figure 1.12). These results compare favorably with the best ones published in the literature.[25]

More recently, new families of P-oxazolines (ligands **L25–L27**), related to the Pfaltz ones, and still with two carbon atoms between the P- and N-donor functionalities, have been reported. In contrast to previous ligands, **L22–L23**, the chirality is again found in the oxazoline substituent (R^2 group). These new air-stable phosphite/phosphinite-oxazoline ligands **L25–L27** were successfully used in the asymmetric Ir-catalyzed hydrogenation and Pd-catalyzed allylic substitution reactions. With respect to the hydrogenation, ligands **L26** and **L27b, f–g, j–k** (R^1= Ph, o-Tol, Cy; R^2= Ph, ^{i}Pr, ^{t}Bu; R^3= Ph, Me, ^{i}Pr; Figure 1.3) represented the first Ir/P-oxazoline catalytic family capable of successfully hydrogenating di-, tri-, and tetrasubstituted minimally functionalized olefins (ee values up to 99%).[44] From a common skeleton, the correct choice of either a phosphite group or a phophinite group gives ligands that are suitable for di-, tri-, and tetrasubstituted olefins (62 examples; Figure 1.13). The asymmetric hydrogenation of tetrasubstituted olefins is of particular interest because the products may contain two new stereogenic centers. As already mentioned, among the most challenging substrates to date are unfunctionalized tetrasubstituted olefins.[28f,45] High enantioselectivities has been reported in few publications and with a limited substrate range, except for a recent report on the application of Ir/**L21** catalyst[46]

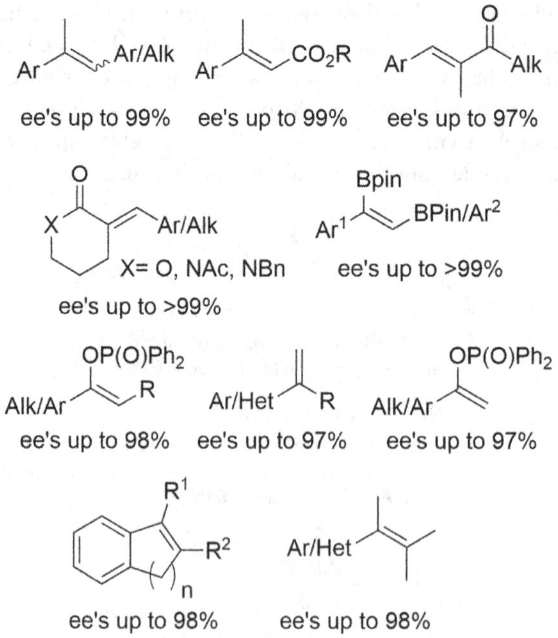

FIGURE 1.13 Selected results obtained with P-oxazoline ligands **L26–L27** in the Ir-catalyzed hydrogenation of di-, tri-, and tetrasubstituted olefins.

(Figure 1.3). Improving the previous results reported,[47] Ir/**L26** catalysts are able to efficiently reduce a range of indenes and the challenging 1,2-dihydro-napthalene (up to 98% ee), as well as a range of the most elusive acyclic tetrasubstituted olefins, with enantioselectivities up to 98% ee under mild reaction conditions.[44] For the hydrogenation of these tetrasubstituted olefins, the highest enantioselectivities were obtained with the phosphinite-oxazoline ligands **L26**. It should be noted that the more rigid the tetrasubstituted olefin is, the less bulky are the phosphinite moieties required to achieve the maximum enantioselectivity. Thus, for the more rigid cyclic indene derivatives, the best catalytic performance is achieved by the phosphinite-based ligand with a Ph phosphinite substituent, whereas for the less rigid cyclic substrate, the phosphinite ligand with a bulkier *o*-tolyl group is needed. Finally, the even less rigid acyclic substrates require the ligand with the bulkiest cyclohexyl phosphinite group. Interestingly, maintaining the same skeleton of the ligand by simply changing the phosphinite functionality by using the correct phosphite group (ligands **L27**), we could also efficiently reduce many unfunctionalized tri- and disubstituted olefins. The catalysts not only exhibited an unprecedented high tolerance to the geometry and steric constraints of the olefin, but they could also tolerate different functional groups very well. Thus, a broad range of olefins, containing both minimally coordinative groups (e.g., α,β-unsaturated carboxylic esters, enones, lactames, vinyl boronates, and enol phosphinates) and coordinative groups (the challenging β-enamides) could be hydrogenated, with high levels of enantioselectivity.[44]

Compared with the Pd/**L2**, the catalyst Pd/**L27k** ($R^2=R^3=$ Ph) achieved higher activities (TOF up to 8,000 h^{-1}) and even greater enantiocontrol in the Pd-allylic substitution of a wider range of mono- and symmetrically disubstituted substrates (ee values up to >99%, 74 examples in total; Figure 1.14).[48] High enantioselectivities were achieved in a wide range of symmetrically disubstituted linear allylic

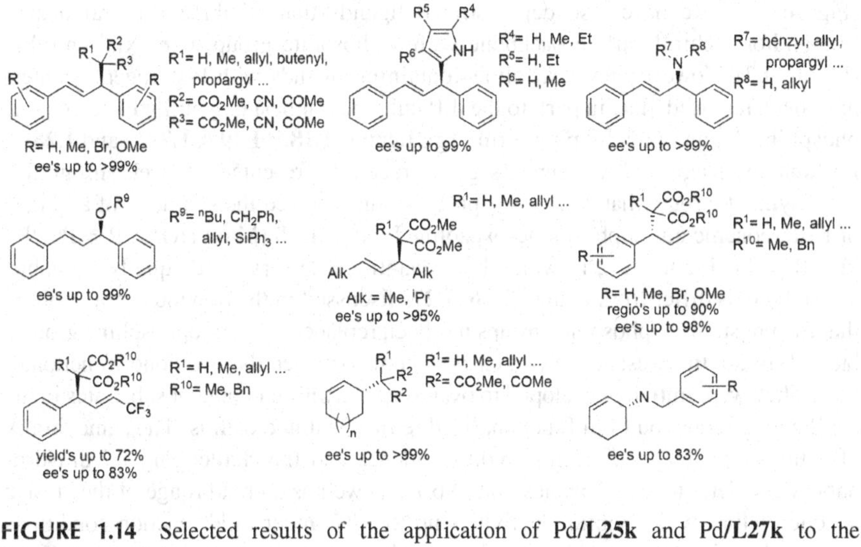

FIGURE 1.14 Selected results of the application of Pd/**L25k** and Pd/**L27k** to the Pd-catalyzed allylic substitution.

acetates, containing alkyl or aryl substituents, with many C-nucleophiles, including α-substituted malonates, malononitrile, diketones, 2-cyanoacetates, and pyrroles. The Pd/**L27k** catalyst also showed excellent enantioselectivities with various primary and secondary amines, containing either alkyl or aryl groups, and using benzylic, allylic, and alkylic alcohols, as well as silanols. With ligand **L25k**, a modification of ligand **L27k** with different substituents at the alkyl backbone chain, ee' values could be improved up to >99% in the allylic alkylation of cyclic substrates. Moreover, the Pd/**L27k** catalyst is one of the few catalytic systems that can deracemize unsymmetrically disubstituted substrates, such as 1,1,1-trifluoro-4-phenylbut-3-en-2-yl acetate *via* dynamic kinetic asymmetric transformation with a range of malonates (yields up to 72% and ee values up to 80%). This family of catalyst precursors was also used in the Pd-catalyzed allylic alkylation of 1-arylallyl acetates with malonates (regioselectivities up to 90% and up to 98% ee). However, the regioselectivities in favor of the branched product diminished when using α-substituted malonates (e.g., regioselectivities dropped from 83%, using dimethyl malonate, to 60%, using dimethyl 2-methylmalonate).

Mechanistic studies by NMR spectroscopy and DFT calculations showed an early transition state (TS), in which the stereochemistry of the reaction is governed by the relative populations of the *exo* and *endo* Pd-η3-allyl complexes and the electrophilicity of the allylic terminal carbon atoms. These studies also showed that the ratio of the Pd-allyl intermediates, which provide the two enantiomeric substitution products, is influenced by the ligand design. Thus, while the ratio of *endo* and *exo* isomers of cyclic substrates is mostly controlled by the configuration of the phosphite moiety and the substituent in the alkyl backbone chain, the oxazoline substituent also plays a key role in linear substrates.[48]

Most heterodonor P-oxazoline ligands developed have the chirality in the stereogenic carbon centers located on the oxazoline ring and/or in the carbon backbone (Figure 1.3). We have also demonstrated ligands that combine a chiral oxazoline and/or a chiral carbon backbone with a phosphite group with axial chirality (Figure 1.3). However, few oxazoline-containing ligands with P-sterogenic centers have been reported, due, in part, to the difficulty of synthesizing bulky P-stereogenic phosphines in an optically pure form (e.g., ligands **L18**,[49] **L19**[50], **L20**[51], and **L28**[52]). Fortunately, Riera and Verdaguer's group recently presented a novel, straightforward synthetic route that solved this problem and allowed the synthesis of a library of P-stereogenic aminophosphine-oxazoline ligands **L28** (MaxPHOX ; R² = Ph, ⁱPr, ⁱBu; R³ = ⁱPr; Figure 1.3), in which both enantiomeric series are equally available. These ligands differ from ligands **L26–L27**, discussed in the previous paragraph, in that the phosphinite/phosphite groups has been replaced by aminophosphine groups, but still retain the presence of two carbon atoms between the two donor functionalities. They were initially developed to overcome the limited range of substrates in the challenging reduction of unfunctionalized tetrasubstituted olefins. The same family of catalysts is able to efficiently reduce indenes and the challenging 1,2-dihydronapthalene derivatives (ee values up to 96%), as well as a broad range of the elusive acyclic olefins, with enantioselectivities up to 99% under mild reaction conditions (Figure 1.15).[52] Moreover, the excellent catalytic performance is maintained for a

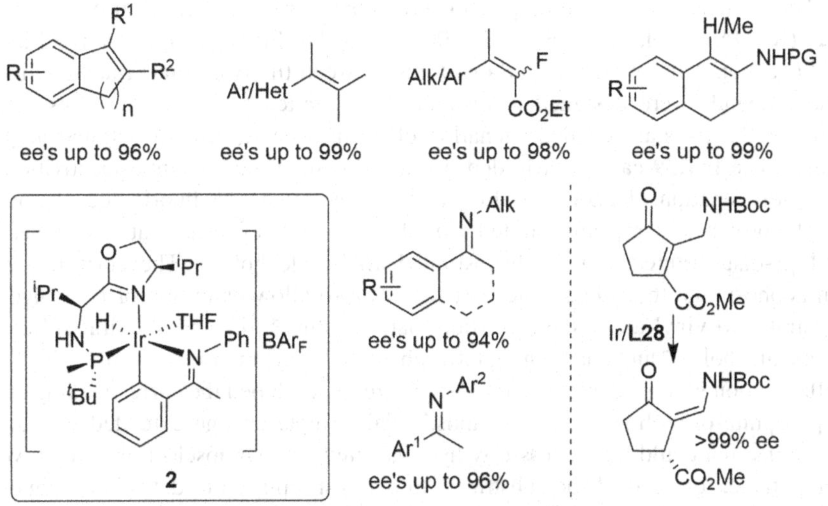

FIGURE 1.15 Representative uses of **L28** in several catalytic asymmetric transformations.

range of aryl and alkyl vinyl fluorides (dr values > 99% and ee values up to 98%), where two vicinal stereogenic centers are created.[52] Catalysts Ir/**L28** (R^2= (R)-^{t}Bu or (R)-^{i}Pr, and R^3= (S)-^{i}Pr), which have the oxazoline substituent and the bulky group at the P-center *cis* to each other, also achieved excellent results (>99% ee) in the reduction of cyclic β-enamides (10 examples; Figure 1.15).[36] Lowering the hydrogen pressure (3 bar) resulted in an increase in enantioselectivity. Moreover, the process can be carried out in more environmentally friendly solvents, such as methanol and ethyl acetate, with no loss of selectivity. The Ir/**L28** (R^2= (S)-^{i}Pr and R^3= (R)-^{i}Pr) catalyst was also demonstrated to be as active and selective as the best Ir−P,N system reported to date in the reduction of challenging N-aryl imines (Figure 1.15).[53] The reaction was run using a balloon of H$_2$ at −20 °C, achieving up to 96% ee. The authors found that, although the configuration at the P*-center had a small influence on the selectivity of the catalyst, it had a great impact on the activity of the catalyst. Interestingly, they were able to isolate the true catalyst, complex **2**, with a tetrahydrofuran (THF) molecule attached to iridium. Complex **2** was an effective catalyst for the direct hydrogenation of N-methyl ketimines (Figure 1.15).[54] Evaluation of the substrate range showed that both N-methyl imines and N-alkyl imines were tolerated and reduced with excellent enantioselectivity (up to 94%), under mild reaction conditions (1 mol % of **2**, 3 bar of H$_2$ at 0 °C). Asymmetric hydrogenation of this class of substrates had not yet been achieved with useful levels of enantioselectivity, probably due to the higher basicity of N-methyl amines, when compared with N-aryl amines, which may lead to catalyst deactivation.[55] Finally, the Ir/**L28** (R^2= (R)-Ph and R^3= (R)-^{i}Pr) catalyst was successfully used in the asymmetric isomerization of N-allyl amides to enamides (Figure 1.15). This reaction allowed the development of the shortest synthesis route reported to date for the enantioselective preparation of the antibiotic R-sarkomycin methyl ester.[56]

The Andersson group developed an aminophosphine-oxazoline family of ligands **L29** (R¹= Ph, *o*-Tol, Cy; R²= H, ᵗBu, Ph; R³= H, Ph; Figure 1.3) with a rigid bicyclic backbone, and with two carbon atoms between the two donor functionalities. These ligands were designed to overcome the limited range of substrates available for the Ir-catalyzed hydrogenation of unfunctionalized olefins found at that time.[57] The Ir/**L29** catalyst afforded, for the first time, high enantioselectivities in the hydrogenation of enol phosphinates,[57b,c] vinyl silanes,[57d] fluorinated olefins,[57e] vinyl boronates,[57f] α,β-unsaturated acryclic esters,[57g] α,β-unsaturated lactones,[57g] and γ,γ-disubstituted and β,γ-disubstituted allylic alcohols[57h]. The corresponding aminophosphine-thiazole counterpart ligands also allow extension of the range of substrates to vinyl allyl silanes,[57d] fluorinated olefins,[57e] *E*- and *Z*-chiral sulfones, allylic alcohols,[57h] and challenging tetrasubstituted vinyl fluorides (Figure 1.16).[58] In collaboration with Andersson, our research group developed the corresponding aminophosphite-oxazoline/thiazole ligands.[59] This simple change extended the range of olefins that could be successfully hydrogenated, at enantioselectivities that were comparable, for most of the substrates, to the best reported to date. These ligands have been able to successfully reduce a range of unfunctionalized *E*- and *Z*-tri- and disubstituted substrates, vinyl silanes, enol phosphinates, tri- and disubstituted alkenylboronic esters, and α,β-unsaturated enones, with enantioselectivities up to 99% and with high conversions.

Another family of P-oxazoline ligands, used with success in several unrelated asymmetric transformations, and also with two carbon atoms between the two donor functionalities, are pyranoside ligands **L30b–e,h–k** (R= Me, ᶦPr, ᵗBu, Ph, Bn; Figure 1.3).[60] As with the previous ligands **L22–L23**, the phosphorus unit is attached to the stereogenic center next to the oxazoline nitrogen atom but differs by being in the presence of a more rigid sugar backbone. Ligands **L30** were efficiently synthesized from D-glucosamine, an inexpensive natural feedstock. A systematic optimization of the ligand parameters led to its efficient application in the hydrogenation of unfunctionalized olefins, allylic substitution, and intermolecular Heck reactions.[60] Phosphite-oxazoline ligands **L30** represented the first successful application of phosphite-containing ligands in the challenging Ir-hydrogenation of unfunctionalized olefins.[60a] It was found that, to achieve high enantioselectivities, the presence

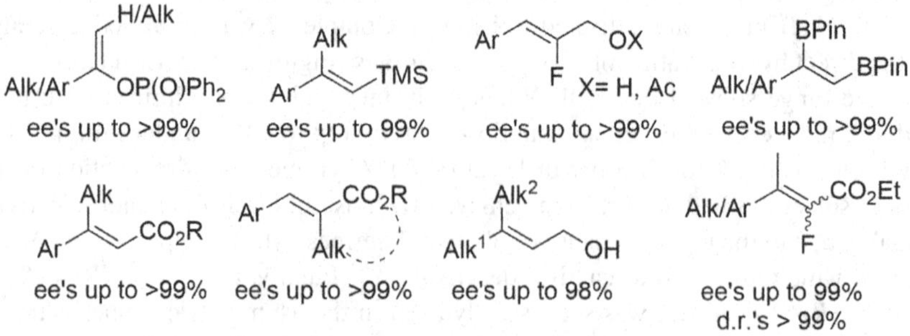

FIGURE 1.16 Representative hydrogenation results with Ir/**L29** catalysts and its corresponding Ir/aminophosphine-thiazole analogs.

of bulky substituents in the biaryl phosphite group and less sterically demanding substituents in the oxazoline moiety was required.[60a,b] Thus, it was possible to identify two general ligands, **L30b** and **L30e** (R= Ph), that provide high enantioselectivities. For comparative purposes, the phosphinite-oxazoline analogs were also tested, generating lower enantioselectivities. With the two best ligands, high enantioselectivities and activities (ee values between 91% and >99%) in many trisubstituted olefins (25 examples; Figure 1.17), even in the reduction of the more challenging Z-isomers, and triarylsubstituted substrates, which provides an easy entry point to diarylmethine chiral centers, that are present in several important drugs and natural products. Moreover, high enantioselectivities could also be achieved in the hydrogenation of many trisubstituted substrates with poorly coordinative groups, such as α,β-unsaturated esters and ketones, vinyl silanes, allylic alcohols, and acetates. The excellent enantioselectivities obtained in the hydrogenation of vinyl boronates (ee values ranging from 92% to >99%) should also be noted. Furthermore, high enantioselectivities were obtained in the reduction of a broad range of unfunctionalized 1,1'-disubstituted olefins (19 examples; Figure 1.17), including heteroaromatic terminal olefins (up to 99% ee).

With the aim of understanding the catalytic performance of these catalysts, a detailed computational study was carried out in collaboration with Prof. P.-O. Norrby.[60b] DFT studies confirmed that the preferred reaction path is an Ir^{III}/Ir^{V} cycle, with migratory insertion of a hydride as the selectivity-determining step. The calculated structure also allowed the formulation of a quadrant model, which explains the effect of the ligand parameters on enantioselectivities (Figure 1.18). In this model, the phenyl group of the oxazoline substituent occupies the upper left quadrant, and one of the aryls of the biaryl phosphite moiety partly blocks the lower right quadrant, while the other two quadrants are open (Figure 1.18a). This calculated structure shows a chiral pocket that fit with E-olefins (Figure 1.18b). This quadrant model also explains the change of ligand (from **L30e** to **L30b**) to obtain a high

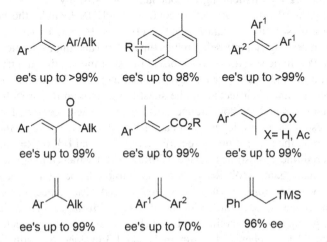

FIGURE 1.17 Selected hydrogenation results with [Ir(**L30b,e**)(cod)]BAr$_F$ catalysts.

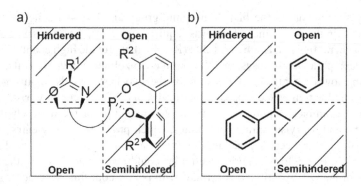

FIGURE 1.18 a), b) Model diagram describing the enantioselective substrate–ligand interactions.

enantioselectivity in Z-olefins. Ligand **L30b** has bulky substituents in the *para* position, which increase the dihedral angle of the biaryl group and result in lower occupancy of the lower right quadrant than with ligand **L30e**. Therefore, the substituent in the biphenyl group can tune the occupancy of the lower right quadrant, so that Z-alkenes can also be successfully hydrogenated. The same explanation accounts for the triaryl- and disubstituted substrates. In conclusion, the DFT studies confirm that the flexibility of the biaryl phosphite group is a crucial parameter in the achievement of high enantioselectivities for substrates with different geometries and steric requirements.

Ligands **L30e** (R= Me) and **L30b** (R= Ph) were also successfully used in the intermolecular Heck reaction.[60e,f] They provided high activities (full conversions in hours using thermal conditions, or in minutes using microwave conditions) and high regio- and enantioselectivities (up to 99% in both cases) for a range of substrates (including the challenging cyclopentene and 4,7-dihydro-1,3-dioxepin) and triflate sources. Advantageous and in contrast to PHOX ligands, the presence of bulky oxazoline substituents had a negative effect on both activities and selectivities. Furthermore, high enantioselectivities (up to 99%) and high activities have been achieved in Pd-allylic substitution of a broad range of mono-, di-, and trisubstituted hindered and unhindered linear and cyclic substrates.[60c,d] The results indicate that catalytic performance is affected by the substituents in both the oxazoline and the phosphite moieties and the cooperative effect between stereocenters. However, the effect of these parameters depends on each substrate class. Thus, for the hindered linear substrate, the best results were achieved with ligand **L30b** (R= Ph), whereas, for unhindered linear substrates, ligand **L30k** (R= Me) achieved the greatest enantioselectivity, and ligand **L30j** (R= ^{i}Pr) was found to be the best for cyclic substrates. High enantio- and regioselectivities (up to 96% ee and 85%, respectively), combined with high activities, were obtained for mono- and trisubstituted substrates. The corresponding phosphinite-oxazoline analogs achieved lower enantioselectivities. The study of the Pd-1,3-diphenyl, 1,3-dimethyl, and 1,3-cyclohexenyl allyl intermediates by NMR spectroscopy indicates that, for enantioselectivities to be high, the substituents in the biaryl phosphite moiety and the electronic and steric properties of the oxazoline substituents need to be correctly combined in order to predominantly produce the isomer that reacts faster with the nucleophile and to avoid the formation of species with ligands coordinated in a monodentate fashion.[60d] This study also indicates that the nucleophilic attack takes place predominantly at the allylic terminal carbon atom located *trans* to the phosphite moiety.

Since the pioneering work of Chan and coworkers, the spiro backbone has been recognized as a privileged structure for chiral ligand families and catalysts.[61, 62] Four main kinds of spiro phosphine-oxazoline ligands have been reported (Figure 1.3), that include SpinPHOX[63] (**L31**) by Ding, SIPHOX[64] (**L32**) by Zhou, HMSI-PHOX[65] (**L33**) by Lin, and SMIPHOX[66] (**L34**) by Teng. They are all modular and have achieved excellent catalytic performances in metal-catalyzed transformations. The SMIPHOX shows some distinct features, compared with the other three, such as an spiro indane-based P,N ligand, with non-C_2-symmetric skeleton and greater rigidity, and only one chiral center, avoiding the complex stereochemistry.

Among these ligands, we can highlight the work of Zhou and coworkers with the spiro phosphine-oxazoline (SIPHOX) Ir-catalysts (Figure 1.19), for the successful hydrogenation of imines and the first Ir-catalysts that could efficiently hydrogenate unsaturated carboxylic acids in the presence of a base[67], overcoming the limitations of most studied Rh- and Ru-catalysts, the scope of which is limited to acrylic and cinnamic acids.[68] The addition of a base results in the formation of a carboxylate anion, which acts as a strong coordinating group. Under mild reaction conditions, excellent yields (90–97%) and enantioselectivities (96–>99% ee) could be achieved for a broad range of α-aryloxy and α-alkyloxy substituted α,β-unsaturated acids, with TONs up to 10,000 (Figure 1.19). It was found that the most effective catalysts contained a ligand with a bulky P-aryl group (R^1= 3,5-^{t}Bu$_2$Ph), which was the best choice for most of the substrates studied later. The hydrogenation was used in the synthesis of α-benzyloxy-carboxylic acid, a key intermediate in the preparation of rhinovirus protease inhibitor, rupintrivir.[69] In contrast to previous Ir-catalysts,[70] the rigidity and bulkiness of the spiro scaffold on SIPHOX ligands seemed to prevent the Ir-catalysts from trimerization under hydrogenation conditions. Interestingly,

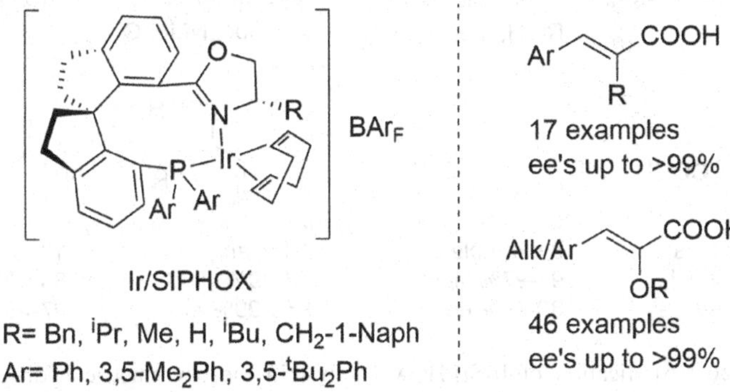

Ir/SIPHOX

R= Bn, ^{i}Pr, Me, H, ^{i}Bu, CH$_2$-1-Naph
Ar= Ph, 3,5-Me$_2$Ph, 3,5-^{t}Bu$_2$Ph

Ar $\diagup$ COOH
R
17 examples
ee's up to >99%

Alk/Ar $\diagup$ COOH
OR
46 examples
ee's up to >99%

FIGURE 1.19 Ir/SIPHOX catalysts and their use in asymmetric hydrogenation.

whereas Ir-SIPHOX catalysts were not effective in the reduction of α,β-unsaturated esters,[71] the Ir-PHOX analog achieved excellent enantioselectivities. Both catalyst types have therefore complementary ranges of substrates.

With the use of Ir-SIPHOX (spiro phosphine-oxazoline) catalysts, Zhou and co-workers have significantly increased the number of unsaturated carboxylic acid derivatives that can be successfully hydrogenated,[71,72] and for which Rh- and Ru-catalysts were not effective. They further expanded its application to the reduction of other trisubstituted α,β-unsaturated carboxylic acids, such as α-aryl- and α-oxymethyl-substituted cinnamic acids (Figure 1.20). Their hydrogenations were used as a key step for the total synthesis of two natural products, namely (S)-equol and (S)-(+)-homoisoflavone.[73] Similarly, a range of N- and O-heterocycles of different ring sizes could be also hydrogenated, with enantioselectivities ranging from 89 to 99% ee (Figure 1.20). The methodology allowed the direct preparation of (R)-nipecotic acid and (R)-tiagabine, with excellent yields and enantioselectivities.[74] A range of tetrasubstituted acrylic acids, with α-aryl, α-alkyl, α-aryloxy, or α-alkyloxy substituents, were also hydrogenated at high enantioselectivities (90-99% ee) (Figure 1.20). Some of the reduced products were used for the preparation of chiral drugs, such as mibefradil and fenvalerate.[75] The excellent results of SIPHOX ligands were also achieved in the reduction of several β,γ-unsaturated acids, which gives access to molecules with a chiral center at the γ-position. 4-Alkyl-4-aryl-3-butenoic acids could be hydrogenated up to 97% ee, with a ligand containing an α-naphthylmethyl group on the oxazoline ring and a 3,5-Me₂Ph as a phosphine substituent (Figure 1.20). This hydrogenation was also used in a key step, for the total syntheses of the natural products (R)-aristelegone-A, (R)-curcumene, and (R)-xanthorrhizol.[76] The use of a carboxy-directing group was also extended to the hydrogenation of terminal 1,1-dialkyl, 1,1-diaryl, and 1-aryl-1-alkyl γ,δ-unsaturated acids (Figure 1.20; ee values up to 99%).[77] It was shown that the directing carboxy group on these substrates can be

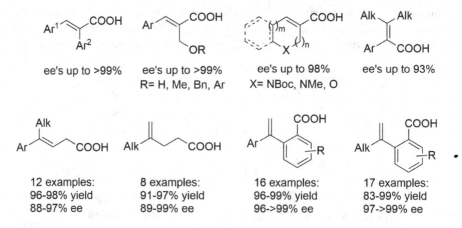

FIGURE 1.20 Application of Ir/SIPHOX catalyst in the asymmetric hydrogenation of α-alkyl, α-aryloxy-, and α-alkyloxy-substituted α,β-unsaturated carboxylic acids, β,γ-unsaturated acids, and γ,δ-unsaturated acids and acrylic acids.

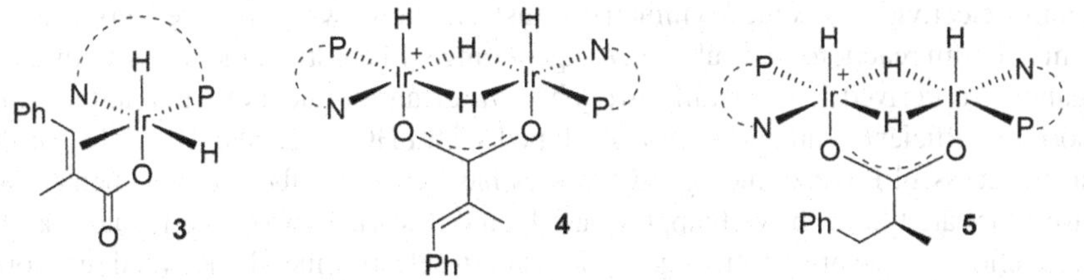

FIGURE 1.21 Structures of isolated intermediates **3–5** in the model reaction of sodium (*E*)-2-methyl-3-phenyl acrylate with the Ir-**L32** catalyst precursor (Ar= 3,5-^{t}Bu$_2$Ph, R= H).

subsequently removed or easily transformed to other useful functional groups, if needed.[77a] A range of α-alkyl-α-aryl terminal olefins were reduced at excellent enantioselectivities (from 98 to >99% ee; Figure 1.20), yielding valuable compounds with a chiral benzylmethyl center. Their asymmetric reduction was also successfully used as a key step in the synthesis of (*S*)-curcudiol and (*S*)-curcumene.[77b] Finally, the authors further confirmed the role of the carboxylate moiety as a directing group, by showing that, when no free carboxylic acid was present or no basic conditions were used, the reaction did not proceed.[77a] Moreover, it was found that, for substrates having an extra C=C double bond in the alkyl side chain, the presence of the carboxy-directing group makes the reaction chemoselective towards the α-alkyl-α-aryl double bond (Figure 1.20), even when the additional double bond was placed in the terminal position.[77b]

The authors also performed a mechanistic study, including DFT studies, that agrees with an Ir(III)/Ir(V) catalytic cycle. The high stability of the chiral spiro-iridium catalysts allowed the trapping of the active intermediates.[78] To mimic the basic conditions used in the hydrogenation reactions, the authors used sodium (*E*)-2-methyl-3-phenyl acrylate as a model substrate. The isolation of the monohydride intermediate **3** (Figure 1.21) was key to understanding the mechanism. Dimeric species **4** and **5** (Figure 1.21), which are off-cycle species, were also isolated and characterized by X-ray diffraction. The isolation of intermediates **3–5** confirms the coordination of the carboxy group to Ir, when the reaction is performed under basic conditions. It should be noted that, in contrast to Ir-hydrogenation of unfunctionalized olefins, the Ir-dihydride olefin complex **3** can undergo migratory insertion in the absence of H$_2$. This mechanistic divergence could indicate that the mechanism of Ir-catalyzed hydrogenation of alkenes may vary, depending on the type of substrate and/or catalyst.

1.3 CONCLUSIONS

The success with the phosphine-oxazoline ligands PHOX has inspired the development of many successful P-oxazoline families of ligands by modifying the ligand backbone, by modifying the electronic/steric properties of the phosphine group, or by replacing the phosphine moiety with a phosphinite or phosphite group. These specific modifications of the simple PHOX ligands achieved improvement of the enantioselectivities in some asymmetric transformations, with a broader versatility, in terms of both reactions and substrate/reagent range. In addition, the majority of these ligands are derived from readily available chiral amino alcohols, maintaining the short and efficient synthetic routes developed with PHOX ligands. The utility of the most successful P-oxazoline ligand families has been described in this chapter. We have illustrated how, through appropriate ligand design, P-oxazoline ligands can be an excellent source of versatile ligands for enantioselective metal-catalyzed reactions, with superior catalytic performance in many reactions than has been achieved by the best C_2-symmetric N,N- and P,P-ligands reported so far. Their excellent results, together with their ease of synthesis, are expected to lead to a new generation of P-oxazoline ligands and a further expansion of the scope of metal-mediated processes catalyzed by them. This will drive the growth of asymmetric catalysis as a key tool for the sustainable preparation of enantiopure compounds in the coming years.

1.4 ACKNOWLEDGEMENTS

We gratefully acknowledge financial support from the Spanish Ministry of Economy and Competitiveness (CTQ2016-74878-P and PID2019-104904GB-I00), European Regional Development Fund (AEI/FEDER, UE), the Catalan Government (2017SGR1472) and the ICREA Foundation (ICREA Academia award to M.D).

REFERENCES

1. a) H.-U. Balser, H.-J. Federsel, *Asymmetric Catalysis in Industrial Scale: Challenges, Approaches and Solutions*, 2nd Ed, Wiley, Weinheim, 2010; b) I. Ojima, *Catalytic Asymmetric Synthesis*, 3rd Ed, John Wiley & Sons, Inc., Hoboken, 2010; c) J. M. Brown, *Comprehensive Asymmetric Catalysis*, ed. E. N. Jacobsen, A. Pfaltz and H. Yamamoto, Springer-Verlag, Berlin, 1999; d) R. Noyori, *Asymmetric Catalysis in Organic Synthesis*, Wiley, New York, 1994; e) B. Cornils and W. A. Herrmann, *Applied Homogeneous Catalysis with Organometallic Compounds*, 2nd Ed, Wiley-VCH, Weinheim, 2002; f) A. Börner, *Phosphorus Ligands in Asymmetric Catalysis. Synthesis and Applications*, Wiley-VCH, Weinheim, 2008.
2. a) A. Pfaltz, W. J. Drury III, *PNAS* 2004, *101*, 5723–5726; b) T. P. Yoon, E. N. Jacobsen, *Science* **2003**, *299*, 1691–1693; c) W. Sommer, D. Weibel, Asymmetric Catalysis, Privileged Ligands and Complexes, *Sigma Aldrich's Chemfiles* **2008**, *2*, 1–91; d) Q. Zhou, *Privileged Chiral Ligands and Catalysts*, John Wiley& Sons Inc., New York, 2011.
3. a) G. Helmchen, A. Pfaltz, *Acc. Chem. Res.* **2000**, *33*, 336–345; b) H. A. McManus, P. J. Guiry, *Chem. Rev.* **2004**, *101*, 4151–4202; c) G. C. Hargaden, P. J. Guiry, *Chem. Rev.* **2009**, *109*, 2505–2550; d) M. P. Carroll, P. J. Guiry, *Chem. Soc. Rev.* **2014**, *43*, 819–833; e) M. Biosca, O. Pàmies, M. Diéguez, *Catal. Sci. Technol.* **2020**, *10*, 613–624; f) J. C. Carretero, J. Adrio, M. Rodríguez Rivero, Chiral Ferrocene in Asymmetric Catalysis, ed. L.-X. Dai and X.-L. Hou, *Sulfur– and Selenium–Containing Ferrocenyl Ligands in Chiral Ferrocenes in Asymmetric Catalysis*, Wiley-VCH, Weiheim, 2010, 257–282; g) J. Margalef, O. Pàmies, M. A. Pericàs, M. Diéguez, *Chem. Commun.* **2020**, *56*, 10795–10808.
4. a) J. Sprinz, G. Helmchen, *Tetrahedron Lett.* **1993**, *34*, 1769–1772; b) P. von Matt, A. Pfaltz, *Angew. Chem. Int. Ed.* **1993**, *32*, 566–568; c) G. J. Dawson, C. G. Frost, J. M. J. Williams, S. J. Coote, *Tetrahedron Lett.* **1993**, *34*, 3149–3150.

5. For recent succesful applications, see: a) É. Bélanger, C. Houzé, N. Guimond, K. Cantin, J. F. Paquin, *Chem. Commun.* **2008**, 3251–3253; b) D. Audisio, G. Gopakumar, L.-X. Xie, L. G. Alves, C. Wirtz, A. M. Martins, W. Thiel, C. Farès, N. Maulide, *Angew. Chem. Int. Ed.* **2013**, *52*, 6313–6316; c) C. M. Reeves, D. C. Behenna, B. M. Stoltz, *Org. Lett.* **2014**, *16*, 2314–2317; d) F. Nahra, Y. Macé, A. Boreux, F. Billard, O. Riant, *Chem. Eur. J.* **2014**, *20*, 10970–10981; e) K. Balaraman, C. Wolf, *Angew. Chem. Int. Ed.* **2017**, *56*, 1390–1395; f) N. J. Adamson, K. C. E. Wilbur, S. J. Malcolmson, *J. Am. Chem. Soc.* **2018**, *140*, 2761–2764; g) T. Song, X. Zhao, J. Hu, W. Dan, *Eur. J. Org. Chem.* **2018**, *2008*, 1141–1144; h) S. Park, N. J. Adamson, S. J. Malcolmson, *Chem. Sci.* **2019**, *10*, 5176–5182; i) M. Faltracco, S. Cotogno, C. M. L. Vande Velde, E. Ruijter, *J. Org. Chem.* **2019**, *84*, 12058–12070; j) C. C. Lynch, K. Balaraman, C. Wolf, *Org. Lett.* **2020**, *22*, 3180–3184.

6. a) K. Tani, D. C. Behenna, R. M. McFadden, B. M. Stoltz, *Org. Lett.* **2007**, *9*, 2529–2531; b) D. C. Behenna, J. T. Mohr, N. H. Sherden, S. C. Marinescu, A. M. Harned, K. Tani, M. Seto, S. Ma, Z. Novák, M. R. Krout, R. McFadden, J. L. Roizen, J. A. Enquist, D. E. White, S. R. Levine, K. V. Petrova, A. Iwashita, S. C. Virgil, B. M. Stoltz, *Chem. Eur. J.* **2011**, *17*, 14199–14223; c) J. Liu, S. Mishra, A. Aponick, *J. Am. Chem. Soc.* **2018**, *140*, 16152–16158.

7. See for instance: a) D. E. White, I. C. Stewart, R. H. Grubbs, B. M. Stoltz, *J. Am. Chem. Soc.* **2008**, *130*, 810–811; b) H. Mukherjee, N. T. McDougal, S. C. Virgil, B. M. Stoltz, *Org. Lett.* **2011**, *13*, 825–827; c) Z.-W. Zhang, C.-C. Wang, H. Xue, Y. Dong, J.-H. Yang, S. Liu, W.-Q. Liu, W.-D. Z. Li, *Org. Lett.* **2018**, *20*, 1050–1053.

8. See for example: a) E. J. Alexy, H. Zhang, B. M. Stoltz, *J. Am. Chem. Soc.* **2018**, *140*, 10109–10112; b) E. J. Alexy, T. J. Fulton, H. Zhang, B. M. Stoltz, *Chem. Sci.* **2019**, *10*, 5996–6000; c) R. Lavernhe, E. J. Alexy, H. Zhang, B. M. Stoltz, *Org. Lett.* **2020**, *22*, 4272–4275.

9. a) C. M. Reeves, C. Eidamshaus, J. Kim, B. M. Stoltz, *Angew. Chem. Int. Ed.* **2013**, *52*, 6718–6721; b) Y. Lu, E. L. Goldstein, B. M. Stoltz, *Org. Lett.* **2018**, *20*, 5657–5660; c) R. A. Craig, S. A. Loskot, J. T. Mohr, D. C. Behenna, A. M. Harned, B. M. Stoltz, *Org. Lett.* **2015**, *17*, 5160–5163; d) D. C. Behenna, Y. Liu, T. Yurino, J. Kim, D. E. White, S. C. Virgil, B. M. Stoltz, *Nat. Chem.* **2012**, *4*, 130–133; e) K. M. Korch, C. Eidamshaus, D. C. Behenna, S. Nam, D. Horne, B. M. Stoltz, *Angew. Chem. Int. Ed.* **2015**, *54*, 179–183; f) Y. Numajiri, G. Jiménez-Osés, B. Wang, K. N. Houk, B. M. Stoltz, *Org. Lett.* **2015**, *17*, 1082–1085; g) Z. P. Sercel, A. W. Sun, B. M. Stoltz, *Org. Lett.* **2019**, *21*, 9158–9161; h) M. M. Yamano, R. R. Knapp, A. Ngamnithiporn, M. Ramirez, K. N. Houk, B. M. Stoltz, N. K. Garg, *Angew. Chem. Int. Ed.* **2019**, *58*, 5653–5657; i) A. W. Sun, S. N. Hess, B. M. Stoltz, *Chem. Sci.* **2019**, *10*, 788–792.

10. See for instance: a) B. P. Pritchett, J. Kikuchi, Y. Numajiri, B. M. Stoltz, *Angew. Chem. Int. Ed.* **2016**, *55*, 13529–13532; b) K. E. Kim, B. M. Stoltz, *Org. Lett.* **2016**, *18*, 5720–5723; c) B. P. Pritchett, E. J. Donckele, B. P. Pritchett, *Angew. Chem. Int. Ed.* **2017**, *56*, 12624–12627; d) S. A. Loskot, D. K. Romney, F. H. Arnold, B. M. Stoltz, *J. Am. Chem. Soc.* **2017**, *139*, 10196–10199; e) T. J. fulton, A. Y. Chen, M. D. Bartberger, B. M. Stoltz, *Chem. Sci.* **2020**, *11*, 10802–10806; f) M. Mizutani, S. Yasuda, C. Mukai, *Chem. Comm.* **2014**, *50*, 5782–5785; g) Z. Xu, X. Bao, Q. Wang, J. Zhu, *Angew. Chem. Int. Ed.* **2015**, *54*, 14937–14940; h) Y. Numajiri, B. P. Pritchett, K. Chiyoda, B. M. Stoltz, *J. Am. Chem. Soc.* **2015**, *137*, 1040–1043.

11. a) M. P. Carroll, H. Müller-Bunz, P. J. Guiry, *Chem. Commun.* **2012**, *48*, 11142–11144; b) R. Doran, M. P. Carroll, R. Akula, B. F. Hogan, M. Martins, S. Fanning, P. J. Guiry, *Chem. Eur. J.* **2014**, *20*, 15354–15359; c) C. Kingston, P. J. Guiry, *J. Org. Chem.* **2017**, *82*, 3806–3819.

12. J. Streuff, D. E. White, S. C. Virgil, B. M. Stoltz, *Nat. Chem.* **2010**, 2, 192–196.

13. a) O. Pàmies, M. Diéguez, C. Claver, *J. Am. Chem. Soc.* **2005**, *127*, 3646–3647; b) R. Bellini, M. Magre, M. Biosca, P.-O. Norrby, O. Pàmies, M. Diéguez, C. Moberg, *ACS Catalysis* **2016**, *6*, 1701–1712.
14. For reviews on phosphite-containing ligands, see: a) P. W. N. M. van Leeuwen, P. C. J. Kamer, C. Claver, O. Pàmies, M. Diéguez, *Chem. Rev.* **2011**, *111*, 2077–2118; b) M. Diéguez, O. Pàmies, *Isr. J. Chem.* **2012**, *52*, 572–581; c) O. Pàmies, M. Magre, M. Diéguez, *Chem. Rec.* **2016**, *16*, 1578–1590; d) O. Pàmies, M. Diéguez, *Chem. Rec.* **2016**, *16*, 2460–2481.
15. M. Biosca, M. Magre, M. Coll, O. Pàmies, M. Diéguez, *Adv. Synth. Catal.* **2017**, *359*, 2801–2814.
16. a) J. Mazuela, O. Pàmies, M. Diéguez, *Chem. Eur. J.* **2010**, *16*, 3434–3434; b) M. Magre, M. Biosca, O. Pàmies, M. Diéguez, *ChemCatChem* **2015**, *7*, 114–120.
17. The benefits of a biaryl phosphite group in P-oxazoline ligands for Pd-allylic alkylation was first described by Pfaltz's group, with a specific type of substrate, the monosubstituted ones. Similar phosphite-oxazoline ligands to **L2** were specifically designed to improve the regioselectivity of the Pd-allylic substitution of monosubstituted substrates. By selecting the the oxazoline's substituent and the configuration in the biaryl phosphite group, excellent regioselectivity to the desired branched isomer (up to 95%), with a high enantioselectivity (up to 94% ee) could be achieved in Pd-catalyzed alkylation of some monosubstituted substrates. See: R. Prétôt, A. Pfaltz, *Angew. Chem. Int. Ed.* **1998**, *37*, 323–325.
18. See for example the following reviews: a) J. Tsuji, *Palladium Reagents and Catalysis: Innovations in Organic Synthesis*, Wiley, New York, 1995; b) M. Johannsen, K. A. Jorgensen, *Chem. Rev.* **1998**, *98*, 1689–1708; c) M. Diéguez, O. Pàmies, *Acc. Chem. Res.* **2010**, *43*, 312–322; d) E. Martin, M. Diéguez, *C. R. Chim.* **2007**, *10*, 188–205; e) Z. Lu, S. Ma, *Angew. Chem. Int. Ed.* **2008**, *47*, 258–297.
19. a) J. Mazuela, O. Pàmies, M. Diéguez, *Chem. Eur. J.* **2013**, *19*, 2416–2432; b) M. Coll, O. Pàmies, M. Diéguez, *Org. Lett.* **2014**, *16*, 1892–1895; c) J. Mazuela, O. Pàmies, M. Diéguez, *ChemCatChem* **2013**, *5*, 1504–1516; d) M. Biosca, J. Margalef, X. Caldentey, M. Besora, C. Rodríguez-Escrich, J. Saltó, X. C. Cambeiro, F. Maseras, O. Pàmies, M. Diéguez, M. A. Pericàs, *ACS Catal.* **2018**, *8*, 3587–3601.
20. For other successful applications of Pd catalysts, see: a) R. Hilgraf, A. Pfaltz, *Synlett* **1999**, *1999*, 1814–1816; b) S.-L. You, X.-Z. Zhu, Y.-M. Luo, X.-L. Hou, L.-X. Dai, *J. Am. Chem. Soc.* **2001**, *123*, 7471–7472; c) R. Hilgraf, A. Pfaltz, *Adv. Synth. Catal.* **2005**, *347*, 61–77; d) J.-P. Chen, C.-H. Ding, W. Liu, X.-L. Hou, L.-X. Dai, *J. Am. Chem. Soc.* **2010**, *132*, 15493–15495.
21. C. Mazet, D. Gérard, *Chem. Commun.* **2011**, *47*, 298–300.
22. For reviews, see: a) K. Burgess, M. J. Ohlmeyer, *Chem. Rev.* **1991**, *91*, 1179–1191; b) T. Hayashi, *Comprehensive Asymmetric Catalysis*, ed. E. N. Jacobsen, A. Pfaltz and H. Yamamoto, Springer, Berlin, 1999, Vol. 1, pp. 351–366; c) C. M. Crudden, D. Edwards, *Eur. J. Org. Chem.* **2003**, 4695–4712; d) A.-M. Carroll, T. P. O'Sullivan, P. J. Guiry, *Adv. Synth. Catal.* **2005**, *347*, 609–631; e) C. M. Crudden, B. W. Glasspoole, C. J. Lata, *Chem. Commun.* **2009**, 6704–6716; f) S. P. Thomas, V. K. Aggarwal, *Angew. Chem. Int. Ed.* **2009**, *48*, 1896–1898.
23. R. Corberán, N. W. Mszar, A. H. Hoveyda, *Angew. Chem. Int. Ed.* **2011**, *50*, 7079–7082.
24. W.-Q. Wu, Q. Peng, D-X. Dong, X.-L. Hou, Y-D. Wu, *J. Am. Chem. Soc.* **2008**, *130*, 9717–9725.
25. a) M. Oestreich, *Angew. Chem. Int. Ed.* **2014**, *53*, 2282–2285; b) M. Diéguez, O. Pàmies, *Carbohydrates-Tools for Stereoselective Synthesis*, ed. M. M. K. Boysen, Wiley-VCH Verlag, Weinheim, Germany, 2013, Chapter 11, 245–251; c) D. Mc Cartney, P. J. Guiry, *Chem. Soc. Rev.* **2011**, *40*, 5122–5150; d) V. Coeffard, P. J. Guiry, *Curr. Org. Chem.*

2010, *14*, 212–229; e) M. Oestreich, *Mizoroki-Heck Reaction*, Wiley-VCH, Weinheim, 2009; f) L. T. Tietze, H. Ila, H. P. Bell, *Chem Rev.* **2004**, *104*, 3453–3516; g) L. X. Dai, T. Tu, S. L. You, W. P. Deng, X. L. Hou, *Acc. Chem. Res.* **2003**, *36*, 659–667; h) C. Bolm, J. P. Hildebrand, K. Muñiz, N. Hermanns, *Angew. Chem. Int. Ed.* **2001**, *44*, 3284–3308; i) M. Shibasaki, E. M. Vogl, *Comprehensive Asymmetric Catalysis*, ed. E. N. Jacobsen, A. Pfaltz, H. Yamamoto, Springer, Heidelberg, 1999; j) O. Loiseleur, M. Hayashi, M. Keenan, N. Schemees, A. Pfaltz, *J. Organomet. Chem.* **1999**, *576*, 16–22; k) M. Beller, T. H. Riermeier, G. Stark, *Transition Metals for Organic Synthesis*, ed. M. Beller, C. Bolm, Wiley-VCH, Weinheim, 1998; l) F. Diederich, P. J. Stang, *Metal-Catalyzed Cross-Coupling Reactions*, Wiley-VCH, Weinheim, 1998; m) J. Magano, J. Dunetz, *Chem. Rev.* **2011**, *111*, 2177–2250; n) F. Ozawa, A. Kubo, T. Hayashi, *J. Am. Chem. Soc.* **1991**, *113*, 1417–1419.

26. a) O. Loiseleur, P. Meier, A. Pfaltz, *Angew. Chem. Int. Ed. Engl.* **1996**, *35*, 200–202; b) O. Loiseleur, M. Hayashi, N. Schmees, A. Pfaltz, *Synthesis* **1997**, 1338–1345.

27. Z. Mazloomi, M. Magre, E. Del Valle, M. A. Pericàs, O. Pàmies, P. W. N. M. Leeuwen, M. Diéguez,*Adv. Synth. Catal.* **2018**, *360*, 1650–1664.

28. For reviews see: a) X. Cui, K. Burgess, *Chem. Rev.* **2005**, *105*, 3272–3296; b) S. Roseblade, A. Pfaltz, *Acc. Chem. Res.* **2007**, *40*, 1402–1411; c) D. H. Woodmansee, A. Pfaltz, *Chem. Commun.* **2011**, *47*, 7912–7916; d) Y. Zhu, K. Burgess, *Acc. Chem. Res.* **2012**, *45*, 1623–1636; e) J. J. Verendel, O. Pàmies, M. Diéguez, P. G. Andersson, *Chem. Rev.* **2014**, *114*, 2130–2169; f) C. Margarita, P. G. Andersson, *J. Am. Chem. Soc.* **2017**, *139*, 1346–1356.

29. P. Schnider, G. Koch, R. Prétôt, G. Wang, F. M. Bohnen, C. Krüger, A. Pfaltz, *Chem. Eur. J.* **1997**, *3*, 887–892.

30. A. Lightfoot, P. Schnider, A. Pfaltz, *Angew. Chem. Int. Ed.* **1998**, *37*, 2897–2899.

31. a) W. J. Lu, Y. W. Chen, X. L. Hou, *Adv. Synth. Catal.* **2010**, *352*, 103–107; b) W. J. Lu, Y. W. Chen, X. L. Hou, *Angew. Chem. Int. Ed.* **2008**, *47*, 10133–10136.

32. O. Pàmies, P. G. Andersson, M. Diéguez, *Chem. Eur. J.* **2010**, *16*, 14232–14240.

33. M. Biosca, M. Magre, O. Pàmies, M. Diéguez, *ACS Catal.* **2016**, *6*, 5186–5190.

34. a) D. Q. Pharm, A. Nogid, *Clin. Ther.* **2008**, *30*, 813–824 (Rotigotine); b) J. I. Osende, D. Shimbo, V. Fuster, M. Dubar, J. J. Badimon, *J. Thromb. Haemostasis* **2004**, *2*, 492–497 (Terutroban); c) S. B. Ross, S.-O. Thorberg, E. Jerning, N. Mohell, C. Stenfors, C. Wallsten, I. G. MilChert, G. A. Ojteg, *CNS Drug Rev.* **1999**, *5*, 213–232 (Robalzotan); d) B. Astier, L. Lambás-Señas, F. Soulière, P. Schmitt, N. Urbain, N. Rentero, L. Bert, L. Denoroy, B. Renaud, M. Lesourd, C. Muñoz, G. Chouvet, *Eur. J. Pharmacol.* **2003**, *459*, 17–26 (Alnespirone).

35. See, for example: a) J. L. Renaud, P. Dupau, A.-E. Hay, M. Guingouain, P. H. Dixneouf, C. Bruneau, *Adv. Synth. Catal.* **2003**, *345*, 230–238; b) R. Hoen, M. van den Berg, H. Bernsmann, A. J. Minnaard, J. G. de Vries, B. L. Feringa, *Org. Lett.* **2004**, *6*, 1433–1436; c) X.-B. Jiang, L. Lefort, P. E. Goudriaan, A. H. M. de Vries, P. W. N. M. van Leeuwen, J. N. H. Reek, *Angew. Chem. Int. Ed.* **2006**, *45*, 1223–1227; d) A. J. Sandee, A. M. van der Burg, J. N. H. Reek, *Chem. Commun.* **2007**, 864–866; e) M. Revós, C. Ferrer, T. León, S. Doran, P. Etayo, A. Vidal-Ferran, A. Riera, X. Verdaguer, *Angew. Chem. Int. Ed.* **2010**, *49*, 9452–9455; f) Z. Wu, T. Ayad, V. Ratovelomanana-Vidal, *Org. Lett.* **2011**, *13*, 3782–3785; g) L. Pignataro, M. Boghi, M. Civera, S. Carboni, U. Piarulli, C. Gennari, *Chem. Eur. J.* **2012**, *18*, 1383–1400; h) D. J. Frank, A. Franzke, A. Pfaltz, *Chem. Eur. J.* **2013**, *19*, 2405–2415; i) M. J. Bravo, R. M. Ceder, G. Muller, M. Rocamora, *Organometallics* **2013**, *32*, 2632–2462; j) I. Arribas, M. Rubio, P. Kleman, A. Pizzano, *J. Org. Chem.* **2013**, *78*, 3997–4005; k) G. Liu, X. Liu, Z. Cai, G. Jiao, G. Xu, W. Tang, *Angew. Chem. Int. Ed.* **2013**, *52*, 4235–4238.

36. E. Salomó, S. Orgué, A. Riera, X. Verdaguer, *Angew. Chem. Int. Ed.* **2016**, *55*, 7988–7992.

37. a) J. Blankenstein, A. Pfaltz, *Angew. Chem. Int. Ed.* **2001**, *40*, 4445–4447; b) F. Menges, A. Pfaltz, *Adv. Synth. Catal.* **2002**, *344*, 40–44; c) S. McIntyre, E. Hörmann, F. Menges, S. P. Smidt, A. Pfaltz, *Adv. Synth. Catal.* **2005**, *347*, 282–288; d) A. Baeza, A. Pfaltz, *Chem. Eur. J.* **2009**, *15*, 2266–2269; e) M. A. Müller, A. Pfaltz, *Angew. Chem. Int. Ed.* **2014**, *53*, 8668–8671; f) F. Maurer, V. Huch, A. Ullrich, U. Kazmaier, *J. Org. Chem.* **2012**, *77*, 5139–5143.

38. S. Verevkin, A. Preetz, A. Börner, *Angew. Chem. Int. Ed.* **2007**, *46*, 5971–5974; b) S. P. Verevkin, V. N. Emelyanenko, J. Bayardon, B. Schäffner, W. Baumann, A. Börner, *Ind. Eng. Chem. Res.* **2011**, *51*, 126–132.

39. a) S. P. Smidt, F. Menges, A. Pfaltz, *Org. Lett.* **2004**, *6*, 2023–2026; b) A. Ganič, A. Pfaltz. *Chem. Eur. J.* **2012**, *18*, 6724–6728.

40. M. Harmata, X. Hong, *Org. Lett.* **2005**, *7*, 3581–3583.

41. a) J. Mazuela, J. J. Verendel, M. Coll, B. Schäffner, A. Börner, P. G. Andersson, O. Pàmies, M. Diéguez, *J. Am. Chem. Soc.* **2009**, *131*, 12344–12353; b) M. Diéguez, J. Mazuela, O. Pàmies, J. J. Verendel, P. G. Andersson, *Chem. Commun.* **2008**, 3888–3890.

42. M. Diéguez, O. Pàmies, *Chem. Eur. J.* **2008**, *14*, 3653–3669.

43. J. Mazuela, O. Pàmies, M. Diéguez, *ChemCatChem* **2013**, *5*, 1504–1516.

44. M. Biosca, M. Magre, O. Pàmies, M. Diéguez, *ACS Catal.* **2018**, *8*, 10316–10320.

45. S. Kraft, K, Ryan, R. B. Kargbo, *J. Am. Chem. Soc.* **2017**, *139*, 11630–11641.

46. R. Bigler, K. A. Mack, J. Shen, P. Tosatti, C. Han, S. Bachmann, H. Zhang, M. Scalone, A. Pfaltz, S. E. Denmark, S. Hildbrand, F. Gosselin, *Angew. Chem. Int. Ed.* **2020**, *59*, 2844–2849.

47. a) M. G. Schrems, E. Neumann, A. Pfaltz, *Angew. Chem. Int. Ed.* **2007**, *46*, 8274–8276; b) C. A. Busacca, B. Qu, N. Grět, K. R. Fandrick, A. K. Saha, M. Marsini, D. Reeves, N. Haddad, M. Eriksson, J. P. Wu, N. Grinberg, H. Lee, Z. Li, B. Lu, D. Chen, Y. Hong, S. Ma, C. H. Senanayake, *Adv. Synth. Catal.* **2013**, *355*, 1455–1463; c) S. Ponra, W. Rabten, J. Yang, H. Wu, S. Kerdphon, P. G. Andersson, *J. Am. Chem. Soc.* **2018**, *140*, 13878–13883.

48. For successful Ir/P-N catalyst, see: M. Biosca, J. Saltó, M. Magre, P.-O. Norrby, O. Pàmies, M. Diéguez, *ACS Catal.* **2018**, *9*, 6033–6036.

49. S. R. Gilbertson, D. G. Genov, A. L. Rheingold, *Org. Lett.* **2000**, *2*, 2885–2888.

50. a) R. Shintani, M.-C. M. Lo, G. C. Fu, *Org. Lett.* **2000**, *2*, 3695–3697; b) R. Shintani, G. C. Fu, *Org. Lett.* **2002**, *4*, 3699–3702; c) A. Suárez, W. Downey, G. C. Fu, *J. Am. Chem. Soc.* **2005**, *127*,11244–11245.

51. H. Danjo, M. Higuchi, M. Yada, T. Imamoto, *Tetrahedron Lett.* **2004**, *45*, 603–606.

52. M. Biosca, E. Salomó, P. de la Cruz-Sánchez, A. Riera, X. Verdaguer, O. Pàmies, M. Diéguez, *Org. Lett.* **2019**, *21*, 807–811.

53. E. Salomó, P. Rojo, P. Hernández-Lladó, A. Riera, X. Verdaguer, *J. Org. Chem.* **2018**, *83*, 4618–4627.

54. E. Salomó, A. Gallen, G. Sciortino, G. Ujaque, A. Grabulosa, A. Lledós, A. Riera, X. Verdaguer, *J. Am. Chem. Soc.* **2018**, *140*, 16967–16970.

55. The most successful approaches for the asymmetric reduction of *N*-methyl imines are hydrosilylation and Brönsted acid-catalyzed reduction, using Hantzsch ester in the presence of Boc_2O. In these approaches, the final imine is protected *in situ*, thus avoiding the basicity issue. See: a) X. Verdaguer, U. E. W. Lange, M. T. Reding, S. L. Buchwald, *J. Am. Chem. Soc.* **1996**, *118*, 6784–6785; b) V. N. Wakchaure, P. S. J. Kaib, M. Leutzsch, B. List, *Angew. Chem. Int. Ed.* **2015**, *54*, 1852–11856.

56. A. Cabré, H. Khaizourane, M. Garçon, X. Verdaguer, A. Riera, *Org. Lett.* **2018**, *20*, 3953–3957.

57. a) A. Trifonova, J. S. Diesen, P. G. Andersson, *Chem. Eur. J.* **2006**, *12*, 2318–2328; b) P. Cheruku, J. Diesen, P. G. Andersson, *J. Am. Chem. Soc.* **2008**, *130*, 5595–5599; c) P. Cheruku, S. Gohil, P. G. Andersson, *Org. Lett.* **2007**, *9*, 1659–1661; d) K. Källström, I. J. Munslow, C. Hedberg, P. G. Andersson, *Adv. Synth. Catal.* **2006**, *348*, 2575–2578; e) M. Engman, J. S. Diesen, A. Paptchikhine, P. G. Andersson, *J. Am. Chem. Soc.* **2007**, *129*, 4536–4537; f) A. Paptchikhine, P. Cheruku, M. Engman, P. G. Andersson, *Chem. Commun.* **2009**, 5996–5998; g) J. J. Verendel, J. Q. Li, X. Quan, B. Peters, T. Zhou, O. R. Gautun, T. Govender, P. G. Andersson, *Chem. Eur. J.* **2012**, *18*, 6507–6513; h) J. Q. Li, J. Liu, S. Krajangsri, N. Chumnanvej, T. Singh, P. G. Andersson, *ACS Catal.* **2016**, *6*, 8342–8349; i) J. Zheng, J. Jongcharoenkamol, B. B. C. Peters, J. Guhl, S. Ponra, M. S. G. Ahlquist, P. G. Andersson, *Nat. Catal.* **2019**, *2*, 1093–1100; j) S. K. Chakka, B. K. Peters, P. G. Andersson, G. E. M. Maguire, H. G. Kruger, T. Govender, *Tetrahedron: Asymmetry* **2010**, *21*, 2295–2231.
58. S. Ponra, W. Rabten, J. Yang, H. Wu, S. Kerdphon, P. G. Andersson, *J. Am. Chem. Soc.* **2018**, *140*, 13878–13883.
59. M. Biosca, A. Paptchikhine, O. Pàmies, P. G. Andersson, M. Diéguez, *Chem. Eur. J.* **2015**, *21*, 3455–3464.
60. a) M. Diéguez, J. Mazuela, O. Pàmies, J. J. Verendel, P. G. Andersson, *J. Am. Chem. Soc.* **2008**, *130*, 7208–7209; b) J. Mazuela, P.-O. Norrby, P. G. Andersson, O. Pàmies, M. Diéguez, *J. Am. Chem. Soc.* **2011**, *133*, 13634–13645; c) Y. Mata, M. Diéguez, O. Pàmies, C. Claver, *Adv. Synth. Catal.* **2005**, *347*, 1943–1947; d) Y. Mata, O. Pàmies, M. Diéguez, *Adv. Synth. Catal.* **2009**, *351*, 3217–3234; e) Y. Mata, M. Diéguez, O. Pàmies, C. Claver, *Org. Lett.* **2005**, *7*, 5597–5599; f) Y. Mata, O. Pàmies, M. Diéguez, *Chem. Eur. J.* **2007**, *13*, 3296–3304.
61. a) J.-H. Xie, Q.-L. Zhou, *Acc. Chem. Res.* **2008**, *41*, 581–593; b) G. B. Bajracharya, M. A. Arai, P. S. Koranne, T. Suzuki, S. Takizawa, H. Sasai, *Bull. Chem. Soc. Jpn.* **2009**, *82*, 285–302; c) K. Ding, Z. Han, Z. Wang, *Chem. Asian J.* **2009**, *4*, 32–41; d) Y. Liu, W. Li, J. Zhang, *Natl. Sci. Rev.* **2017**, *4*, 326–358.
62. A. S. C. Chan, W. Hu, C.-C. Pai, C.-P. Lau, Y. Jiang, A. Mi, M. Yan, J. Sun, R. Lou, J. Deng, *J. Am. Chem. Soc.* **1997**, *119*, 9570–9571.
63. a) Z. B. Han, Z. Wang, X. M. Zhang, K. L. Ding, *Angew. Chem. Int. Ed.* **2009**, *48*, 5345–5349; b) Y. Zhang, Z. B. Han, F. Y. Li, K. L. Ding, A. Zhang, *Chem. Commun.* **2010**, *46*, 156–158; c) J. Shang, Z. B. Han, Y. Li, Z. Wang, K. L. Ding, *Chem. Commun.* **2012**, *48*, 5172–5174; d) X. M. Wang, Z. B. Han, Z. Wang, K. L. Ding, *Angew. Chem. Int. Ed.* **2012**, *51*, 936–940.
64. a) S. F. Zhu, J. B. Xie, Y. Z. Zhang, S. Li, Q. L. Zhou, *J. Am. Chem. Soc.* **2006**, *128*, 12886–12891; b) S. Li, S. F. Zhu, C. M. Zhang, S. Song, Q. L. Zhou, *J. Am. Chem. Soc.* **2008**, *130*, 8584–8585; c) S. Li, S. F. Zhu, J. H. Xie, S. Song, C. M. Zhang, Q. L. Zhou, *J. Am. Chem. Soc.* **2010**, *132*, 1172–1179; d) S. Song, S. F. Zhu, S. Yang, S. Li, Q. L. Zhou, *Angew. Chem. Int. Ed.* **2012**, *51*, 2708–2711; e) S. Song, S. F. Zhu, L. Y. Pu, Q. L. Zhou, *Angew. Chem. Int. Ed.* **2013**, *52*, 6072–6075; f) S. Song, S. F. Zhu, Y. B. Yu, Q. L. Zhou, *Angew. Chem. Int. Ed.* **2013**, *52*, 1556–1559.
65. W. Y. Sun, H. R. Gu, X. F. Lin, *J. Org. Chem.* **2018**, *83*, 4034–4043.
66. a) Z. X. Qiu, R. Sun, D. W. Teng, *Org. Biomol. Chem.* **2018**, *16*, 7717–7724; b) Y. F. Gao, Z. X. Qiu, R. Sun, D. W. Teng, *Tetrahedron Lett.* **2018**, *59*, 3938–3941.
67. S. Li, S. F. Zhu, C. M. Zhang, S. Song, Q. L. Zhou, *J. Am. Chem. Soc.* **2008**, *130*, 8584–8585.
68. See for example: a) J. P. Genêt, *Modern Reduction Methods*, ed. P. G. Andersson and I. J. Munslow, Wiley-VCH, Weinheim, 2008; b) W. Tang, X. Zhang, *Chem. Rev.* **2003**, *103*, 3029–3070; c) M. Kitamura, R. Noyori, *Ruthenium in Organic Synthesis*, ed. S.-I. Murahashi, Wiley-VCH, Weinheim, 2004; d) B. Weiner, W. Szymanski, D. B. Janssen, A. J. Minnaard, B. L. Feringa, *Chem. Soc. Rev.* **2010**, *39*, 1656–1691.

69. S. Li, S. F, Zhu, J. H. Xie, S. Song, C. M. Zhang, Q. L. Zhou, *J. Am. Chem. Soc.* **2010**, *132*, 1172–1179.
70. a) A. Scrivanti, S. Bovo, A. Ciappa, U. Matteoli, *Tetrahedron Lett.* **2006**, *47*, 9261–9265; b) J. Zhou, J. W. Ogle, Y. Fan, V. Banphavichit, Y. Zhu, K. Burgess, *Chem. Eur. J.* **2007**, *13*, 7162–7170; c) S. P. Smidt, A. Pfaltz, E. Martínez-Viviente, P. S. Pregosin, A. Albinati, *Organometallics* **2003**, *22* 1000–1009.
71. S. F. Zhu, Q. L. Zhou, *Acc. Chem. Res.* **2017**, *50*, 988–1001.
72. S. Yang, W. Che, H. L. Wu, S. F. Zhu, Q. L. Zhou, *Chem. Sci.* **2017**, *8*, 1977–1980.
73. Z. Y. Li, S. Song, S. F. Zhu, N. Guo, L. X. Wang, Q. L. Zhou, *Chin. J. Chem.* **2014**, *32*, 783–787.
74. S. Song, S. F. Zhu, L. Y. Pu, Q. L. Zhou, *Angew. Chem. Int. Ed.* **2013**, *52*, 6072–6075.
75. S. Song, S. F. Zhu, Y. Li, Q. L. Zhou, *Org. Lett.* **2013**, *15*, 3722–3725.
76. S. Song, S. F. Zhu, S. Yang, S. Li, Q. L. Zhou, Angew. Chem. Int. Ed. **2012**, *51*, 2708–2711.
77. (a) S. Song, S. F. Zhu, Y. B. Yu, Q. L. Zhou, *Angew. Chem. Int. Ed.* **2013**, *52*, 1556–1559; (b) S. Yang, S. F. Zhu, N. Guo, S. Song, Q. L. Zhou, *Org. Biomol. Chem.* **2014**, *12*, 2049–2052.
78. M. L. Li, S. Yang, X. C. Su, H. L. Wu, L. L. Yang, S. F. Zhu, Q. L. Zhou, *J. Am. Chem. Soc.* **2017**, *139*, 541–547.

2 Chiral Bidentate Heterodonor *P-N*-Other-Ligands

*Pep Rojo, Marina Bellido,
Xavier Verdaguer, and Antoni Riera*

CONTENTS

2.1 INTRODUCTION

The design of phosphorus-nitrogen ligands is a cornerstone of homogeneous asymmetric catalysis.[1] In this family of ligands, those with an oxazoline ring as N-donor constitute the most relevant subgroup, as described in Chapter 1. Nevertheless, a wide array of heterodonor bidentate ligands, containing nitrogenated motifs other than oxazolines (so-called "P-N-other-ligands"), have been reported to achieve remarkable results in asymmetric catalytic transformations. Several reviews and books have acknowledged the importance of these types of P,N-ligands.[2–7] The present chapter will cover the relationship between the architectural design of P-N-other-ligands and its performance in asymmetric catalysis. We will describe recent reports as well as provide a quick overview of previous work in the field. The ligands are organized according to the nature of their N-donor atom, rather than by reaction type or metal coordination, in order to facilitate monitoring of the relationship between structure and catalytic performance.

The origin of P,N-ligands can be traced back to the discovery of Crabtree's catalyst, [Ir(cod)(py)(PCy$_3$)]PF$_6$ (Figure 2.1), in 1977.[8] This reagent showed high catalytic activity in the hydrogenation of tri- and tetrasubstituted olefins bearing non-coordinative groups.[9] A few years before the discovery of Crabtree's catalyst, Kagan reported that bidentate diphosphines, such as (−)-DIOP,[10] were superior to two monodentate ligands; the backbone rigidity allowed better control of the metal–substrate binding modes, affording better discrimination between the re- and the si-face of the prochiral substrate. For many applications, diphosphines displaying C$_2$-symmetry were switched to C$_1$-symmetric P,N-ligands with two electronically different donor atoms (Figure 2.1).[11–14]

2.2 PROPERTIES OF P,N-LIGANDS

Nitrogen ligands have a hard donor character, compared with phosphorus ligands, which are electronically soft. Phosphines behave as σ-donor and π-acceptor ligands, as they donate their sigma (nonbonding) electrons to the metal, while accepting electron density from the metal to the empty 3d atomic orbitals. This effect, known as backbonding, results in a stronger metal–phosphorus bond compared with the metal–nitrogen bond, in which the nitrogen mainly contributes as a σ-donor (Figure 2.2).

The common adjustable parameters of P,N-ligands are: backbone, chiral centers, linkers, substituents at the phosphorus atom, ring incorporation, and hybridization at the nitrogen atom. In most cases, the chirality is located at the backbone or in the N-heterocycle rather than at the P atom. However, substituents at the phosphorus fine-tune the steric and electronic properties of the ligand. Additionally, a heteroatom directly bound to phosphorus will increase its π-acceptor properties (stronger backbonding effect). On the other hand, changing the hybridization of the nitrogen from an amino to an imino group (sp^2-hybridization) would also increase its π-acceptor properties, avoiding reactive N–H bonds (Figure 2.2). Finally, the incorporation of heterocyclic N-donors increases the rigidity of the chelate. These different, tunable features, associated with each donor fragment of the ligand, provide a unique reactivity to the different metal complexes of P,N-ligands.

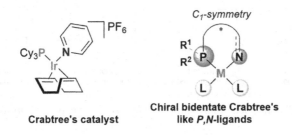

Crabtree's catalyst

Chiral bidentate Crabtree's like *P,N*-ligands

FIGURE 2.1   Crabtree's catalyst and chiral bidentate Crabtree-like coordination sphere.

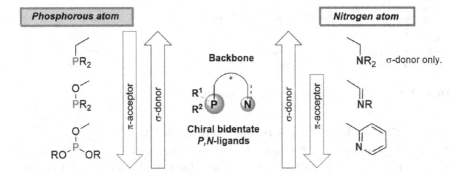

FIGURE 2.2   *P,N*-ligand structural parameters.

Regarding their catalytic performance, asymmetric allylic substitution and asymmetric hydrogenation are the most common applications of this type of ligands. Nevertheless, they have also shown excellent results in other asymmetric transformations, such as conjugate additions or aldehyde–alkyne–amine couplings (A^3 couplings). Future studies will pave the way to new types of ligands that will expand their catalytic applicability to other asymmetric transformations.

2.3   STRUCTURE-PERFORMANCE CHARACTERISTICS OF *P,N*-LIGANDS

In order to shed light on the structure-performance relationship of the most successful *P,N*-ligands, we have classified the ligands according to the nature of their *N*-donor atom: a) group 1: amino *N*-donors (N-sp^3); b) group 2: imino *N*-donors (N-sp^2); c) group 3: cyclic imino *N*-donors (N-sp^2) and d) group 4: pyridino *N*-donors (N-sp^2), as represented in Figure 2.3.

2.3.1   GROUP 1: *P,N*-AMINO LIGANDS

This group of ligands shares an aliphatic amino group with sp^3 hybridization and a chiral backbone. Some of them include cyclic amines to increase conformational rigidity. Despite their early discovery, successful applications of *P,N*-sp^3 bidentate ligands are still limited.

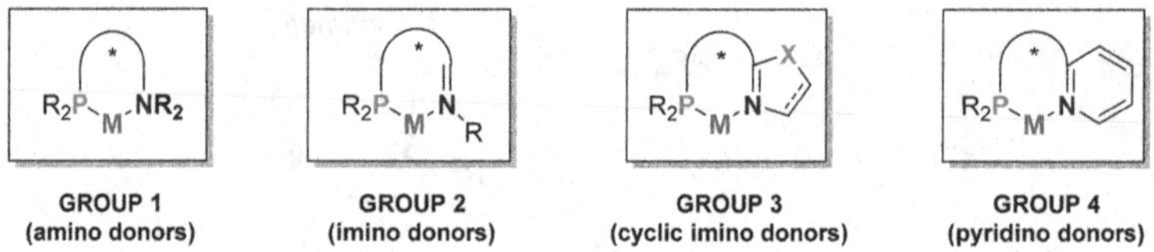

GROUP 1
(amino donors)

GROUP 2
(imino donors)

GROUP 3
(cyclic imino donors)

GROUP 4
(pyridino donors)

FIGURE 2.3   Classification of P,N-chiral ligands based on the type of nitrogen donor.

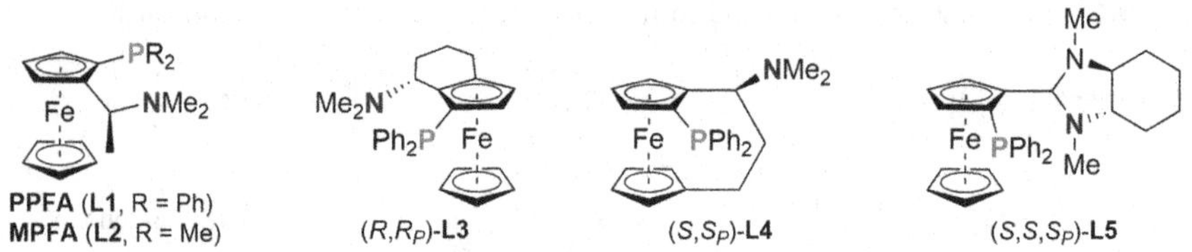

PPFA (L1, R = Ph)
MPFA (L2, R = Me) (R,R_P)-**L3** (S,S_P)-**L4** (S,S,S_P)-**L5**

FIGURE 2.4   Structures of phosphinoferrocenyl ligands.

2.3.1.1   Diphenylphosphinoferrocenylamine (PPFA) and Analogs

Many of the first reports of P,N-ligands can be attributed to the work of Hayashi and Kumada back in the late 1970s. They developed the first planar chiral P,N-ligands, PPFA and MPFA (**L1** and **L2**, respectively; Figure 2.4), both being used in the Rh-catalyzed hydrosilylation of ketones, with moderate yields and enantioselectivities.[15] PPFA was also used in the Ni-catalyzed asymmetric Grignard cross coupling between (1-phenylethyl)magnesium bromide and vinyl bromide.[16] More constrained PPFA-type ligands **L3** and **L4**, reported by Weissensteiner, showed greater performance in this reaction, affording enantiomeric excess (ee) values up to 79% and 63%, respectively.[17]

Later, Sestelo and Sarandeses reported the synthesis of 1,1'-binaphthyls by cross-coupling reactions of triorganoindium reagents, using PPFA as the chiral ligand, with ee values up to 86%.[18] Wang reported the application of both enantiomers of **L1** to the Cu-catalyzed addition of diethylzinc to imines.[19] The use of cyclic amines as nitrogen donors increases the rigidity of the chelate, which usually provides greater asymmetric induction. For instance, Jin's group synthesized a chiral imidazoline-derived PPFA **L5** that gave an outstanding performance in allylic substitution with dimethylmalonate (99.6% ee),[20] similar to the P,N-oxazoline analog.[21]

More recently, Guo also reported the use of PPFA in the tandem [3+2] cycloaddition, followed by a 1,4-addition, between aza-o-quinone methides (ao-QMs) **S1** and azomethine ylides, generating imidazolidine derivatives with ee values up to 99% (Scheme 2.1).[22]

2.3.1.2   β-Aminoalkylphosphines

The success of the Fc-derived ligands, as well as the studies that confirmed the contribution of the dimethylamino moiety to the high degree of asymmetric induction,[16] set the stage for the design and development of new phosphine ligands for the

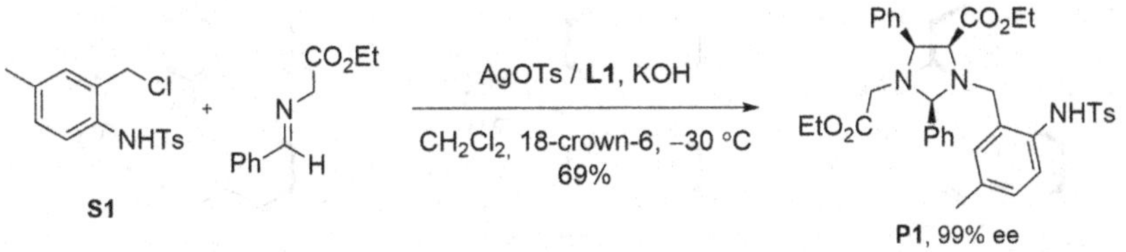

SCHEME 2.1 Tandem [3+2] cyclization/conjugate addition of ao-QMs and azomethine ylide.

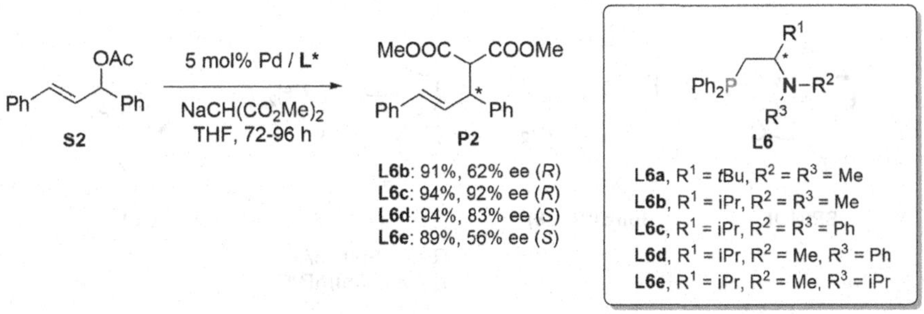

SCHEME 2.2 Allylic alkylation catalyzed by Pd/(*S*)-valine derivative ligand (**L6a-e**) complexes.

asymmetric Grignard cross coupling. Hayashi and Kumada created a new library of ligands, the β-aminoalkylphosphines **L6**, readily prepared from amino acids. These ligands promoted the Ni-catalyzed Grignard cross coupling of (1-phenylethyl)magnesium bromide and vinyl bromide, for which the bulkiest *tert*-Leuphos ligand **L6a** achieved the highest ee value (up to 94%).[23] It was proved that the hemilability that these ligands offered was crucial to improving the catalytic cycle, thereby enhancing both the catalytic activity and the asymmetric induction.

A wide array of analogs of these chiral β-aminoalkylphosphines has been developed over the years, due to their high stability, low toxicity, and ease of handling.[6] As described in Chapter 1, these nitrogen-phosphorus chelate chiral ligands are amongst the most successful in terms of palladium-catalyzed asymmetric allylic substitution.[24-26] Anderson *et al.* employed *P,N*-ligands **L6c-e** derived from valine in the palladium-catalyzed allylic substitution of **S2** with dimethyl malonate to obtain chiral **P2**, with enantioselectivities that ranged from 56% ee (*R*) to 92% ee (*S*) (Scheme 2.2).[27] The substitution at the nitrogen in **L6d** and **L6e** constitutes a key factor in the stereochemical outcome. The coordination to the metal induces a preferred orientation of the two substituents, producing a stereogenic nitrogen that leads to the opposite enantiomer to Valphos **L6b**.[28]

2.3.1.3 Diarylphosphino-1,1′-Binaphthyl Amines (MAP Family)

Kočovský and co-workers developed the MAP (2′-(diphenylphosphino)-*N,N*-disubstituted[1,1′-binaphthalen]-2-amine) family of ligands **L7**,[29] the first axially chiral

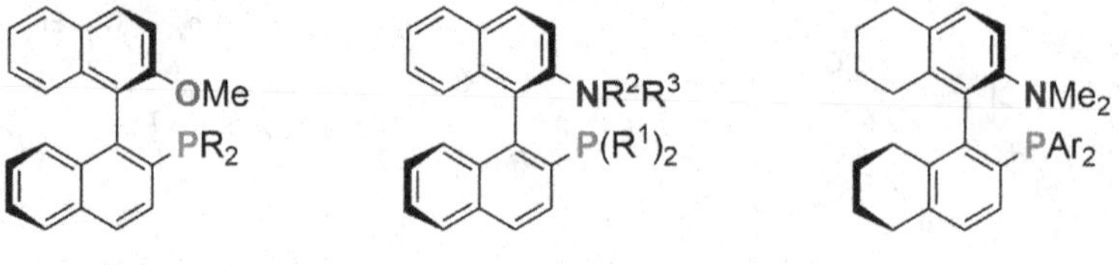

MOP

MAP (L7)

L7a, R^1 = Cy, R^2 = R^3 = Me

H$_8$-L7

FIGURE 2.5   MOP, MAP, and its octahydro analog ligands.

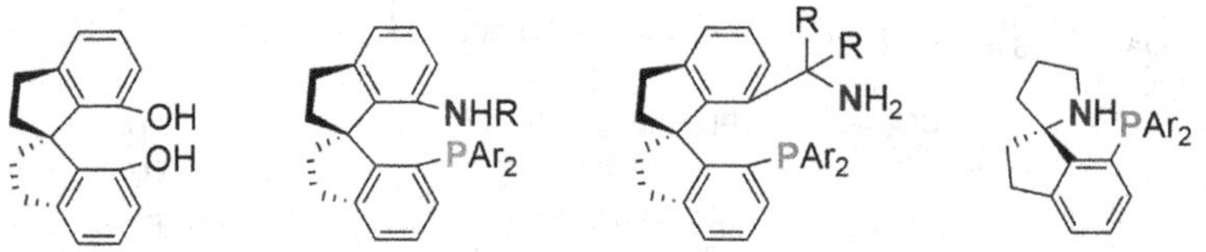

SPINOL

SpiroAP (L8)

L9

R = H, **SpiroBAP**

R = Alk, **SpiroBAP-R**

L10

FIGURE 2.6   Spiranic ligands and a synthetic precursor.

P,N-sp^3-ligands, which were nitrogen analogs of Hayashi's MOP (2-(diphenylpho sphino)-2′-methoxy-1,1′-binaphthyl) ligands (Figure 2.5).[30]

Although these P,N-ligands were designed to act as bidentate P,N-donor ligands, NMR studies of the PdCl$_2$/MAP complex showed a mixture of three species, where the major one was a cyclometalated complex.[31] The application of these complexes ranged from Pd-catalyzed asymmetric allylic alkylation to both Hartwig-Buchwald aminations and Suzuki cross-couplings.[32,33]

Buchwald subsequently reported the use of **L7a**, so-called (S)-KenPhos, the dicy-clohexylphosphine version of the MAP ligand, to prepare axially chiral biaryls at up to 92% ee *via* the asymmetric Suzuki-Miyaura coupling.[34] These results encouraged the broadening of the substrate range with high tolerance for different functional groups and the development of computational studies in order to explain the observed selectivity.[35] Furthermore, Ding prepared the octahydro analogs H$_8$-MAP (**H$_8$-L7**), which gave higher enantioselectivities than **L7** in allylic alkylation of **S2** (83% ee *vs.* 73% ee).[36] The enhancement of the reaction's enantioselectivity can be attributed to the larger bite angle of the biaryl backbone in the Pd/H$_8$-MAPs (**H$_8$-L7**) complexes.[37]

2.3.1.4   Spiranic Ligands

In 2010, Zhou and co-workers reported a wide family of chiral spiranic ligands SpiroAP, which stands for spiro aminophosphines, prepared from optically pure 1,1′-spirobiindane-7,7′-diol (SPINOL) (Figure 2.6).[38] These ligands were used in the Ir-catalyzed asymmetric hydrogenation of exocyclic α,β-unsaturated ketones to produce exocyclic allylic alcohols.

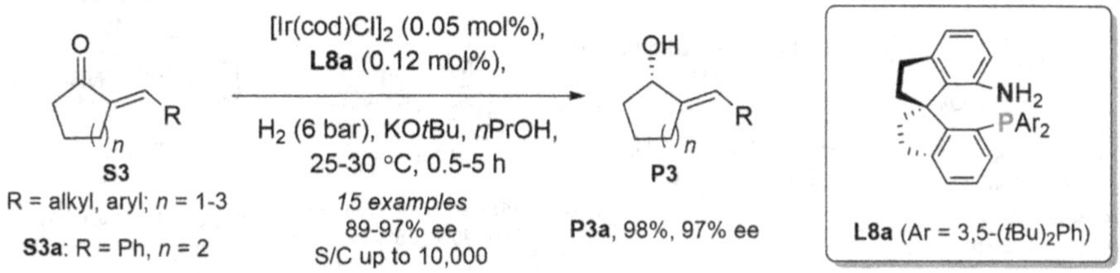

SCHEME 2.3 Ir-catalyzed asymmetric hydrogenation of exo-cyclic α,β-unsaturated ketones with SpiroAP ligand.

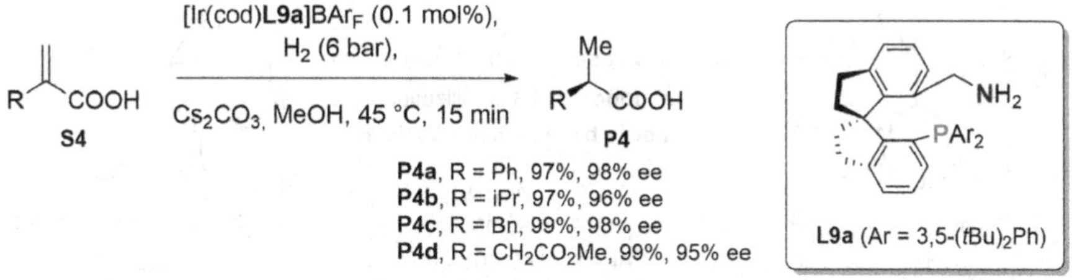

SCHEME 2.4 Asymmetric hydrogenation of acrylic acids with the SpiroBAP **L9a** ligand.

Ligand **L8a**, which has bulky 3,5-di-*tert*-butylphenyl groups at the P atom, gave the highest activity and enantioselectivity (98% yield, 97% ee) for **S3a** (R = Ph, *n* = 2) (Scheme 2.3). The scope of the reaction was then broadened to enones bearing different electron donating groups (EDG) or electron withdrawing groups (EWG) on the phenyl ring and various 7-, 6-, and 5-membered rings, with little effect on both the reactivity and enantioselectivity for 14 additional examples. Consequently, this robust methodology was applied to the synthesis of a key intermediate in the preparation of the anti-inflammatory, loxoprofen.[38]

Based on the remarkable results in the hydrogenation of ketones **S3**, Zhou and co-workers expanded the substrate range to the asymmetric hydrogenation of olefins. To this end, SpiroAP ligands were modified by the incorporation of a -CH$_2$- before the primary amino moiety, affording chiral spiro benzylamino-phosphine ligands **L9** (SpiroBAP; Figure 2.6). Among them, ligand **L9a** exhibited the highest enantioselectivity in the asymmetric hydrogenation of α-substituted acrylic acids **S4a–d** (Scheme 2.4).[39]

Surprisingly, SpiroAP **L8a**, bearing the anilino group, gave no conversion at all toward **P4**, indicating that the benzylamino moiety in **L9a** was necessary to catalyze the hydrogenation reaction. This observation encouraged the exploration of additional SpiroBAP-R ligands, such as the **L9b** ligand, bearing a dimethyl group at the benzylic position (Scheme 2.5).[40] Their corresponding iridium complexes catalyzed the enantioconvergent hydrogenation of diastereomeric mixtures of *E*- and *Z*-nitroalkenes **S5** to **P5** with high enantioselectivity (up to 95% ee) (Scheme 2.5). The enantioconvergent outcome was explained by an initial isomerization to the terminal alkene, which was detected during the hydrogenation of **S5**. The substrate

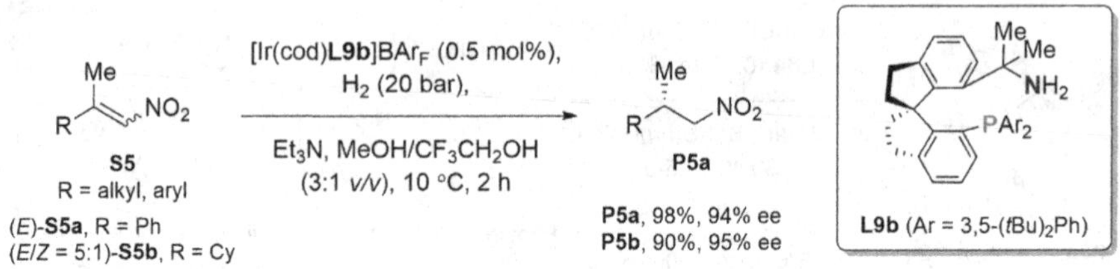

SCHEME 2.5 Hydrogenation of diastereomeric mixtures of nitroalkenes **S5**.

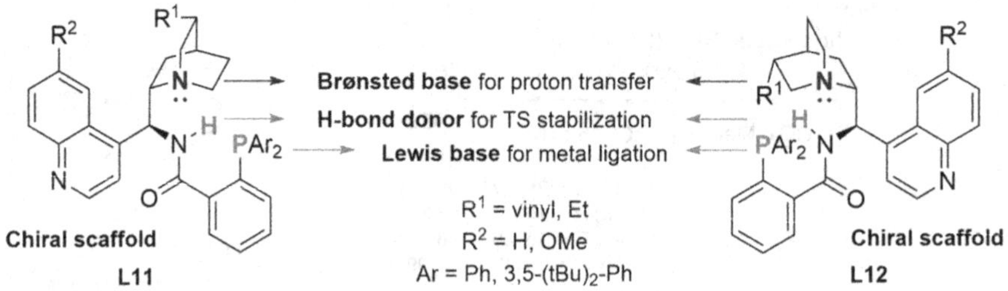

FIGURE 2.7 Active site features and function description of Dixon's ligands.

scope was expanded from β-aryl-β-methyl-nitroalkenes (91–98% ee values) to β-alkyl-β-methyl-nitroalkenes (77–95% ee values).

More recently, Jiao and co-workers reported on the air-stable **L10** ligands bearing a spiro[indane-1,2′-pyrrolidine] backbone (Figure 2.6).[41] As with many aminophosphine ligands, their application to the Pd-catalyzed asymmetric allylic substitution of **S2,** with malonate-, alcohol-, and amine-type nucleophiles, achieved 99% yield and up to 97% ee. The rigid backbone and structural simplicity make these compounds privileged ligands for enantioselective catalytic reactions.

2.3.1.5　Cinchona-Derived Ligands

In 2011, Dixon and co-workers developed a new family of cinchona-derived aminophosphine ligands **L11** and **L12**.[42] This family of ligands was readily prepared from commercially available *ortho*-diphenylphosphino benzoic acids and a selection of 9-amino(9-deoxy)*epi* cinchona alkaloids. The particular feature of these ligands is the close proximity of their active sites: a Brønsted base, a Lewis base and H-bond-donor groups are located around the chiral pocket of the cinchona backbone (Figure 2.7).

Ligands **L11–12** were first used in the cooperative Brønsted base/Lewis acid Ag(I)-catalyzed asymmetric aldol reaction of isocyanoacetate **S6** with different aldehydes (Scheme 2.6). A remarkable feature of these Ag-**L11/L12** complexes is that the catalytic species could be generated *in situ*, avoiding prior catalyst isolation. The substrate range of the reaction was explored with respect to aromatic and aliphatic aldehydes. Aromatic and branched aliphatic aldehydes achieved good asymmetric induction (61–98% ee).[42]

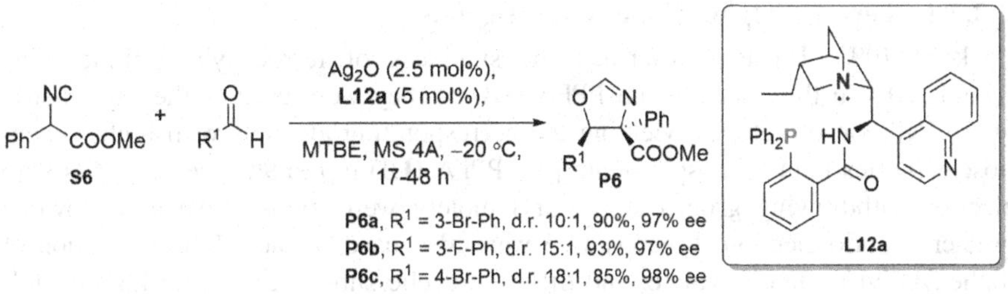

SCHEME 2.6 Dixon's silver-catalyzed isocyanoacetate aldol reaction.

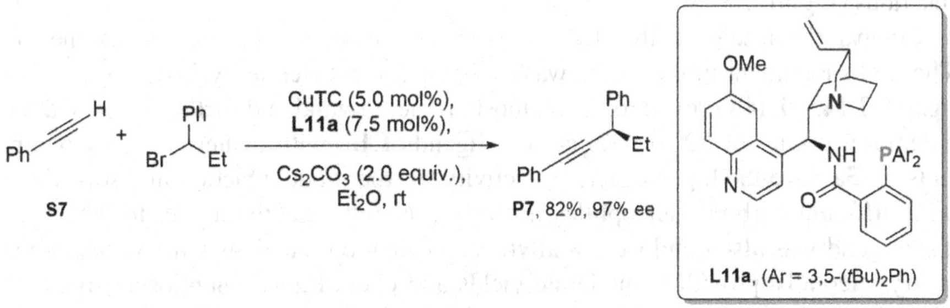

SCHEME 2.7 Asymmetric Sonogashira Csp³–Csp cross coupling, catalyzed by the cinchona *P,N*-ligand **L11a**.

Since its first discovery and application, the Dixon's catalyst system, i.e., cinchona-derived aminophosphine ligands **L11** and an Ag(I) salt, has been used almost exclusively in the isocyanide chemistry. The broad application range consists of several active methylene isocyanides with aldehydes,[43,44] aldimines,[45] ketones,[46,47] ketimines,[48–50] allenoates,[51] alkynyl ketones,[52] other carbon–carbon double bonds containing EWG,[53–57] or *p*-quinone methides (*p*-QMs).[58] Moreover, recent work by Liu and co-workers showed the utility of the quinine-derived ligand **L11a** in enantioconvergent Cu-catalyzed Sonogashira Csp³-Csp cross coupling – *via* radical intermediates – of **S7** and racemic alkyl bromide (Scheme 2.7).[59,60] The methodology was applied to a broad range of terminal alkynes and racemic alkyl halides.

2.3.2 Group 2: *P,N*-Imino Ligands

An imino group is less σ-donating than an amino group. However, the imino π* orbitals overlap with the d-orbitals of the transition metal, leading to a π-backbonding effect that strengthens the metal–nitrogen bond and stabilizes low-oxidation-state metals.[14] In Pd-catalyzed asymmetric allylic substitution reactions, the shift from an amino to an imino *N*-donor results in a stabilization of the zerovalent palladium complex and also in an increase in the electrophilicity of the η³-allyl intermediate species. For this reason, iminophosphine ligands were initially used in Pd-catalyzed asymmetric allylic substitution reactions, with remarkable results.

2.3.2.1 Ferrocenylphosphine-Imino Ligands

Back in 1995, Hayashi described the synthesis of ferrocenylphosphine-imine ligands **L13** in three steps from PPFA (**L1**).[61] They were used in the asymmetric hydrosilylation of ketones, yielding the corresponding alcohols with higher enantioselectivities than its N-sp³ counterpart PPFA (**L1**) (up to 90% ee). Ligands with electron-withdrawing groups at the aryl moiety were more active and provided greater enantioselection. Ligands **L13** were also used in the allylic alkylation of diphenylallyl acetate, pivalate, and different cyclic substrates, giving high enantioselectivities (up to 96% ee).[62,63] Since this work, many ferrocenylphosphine-imine ligands have been reported and used in Pd-catalyzed asymmetric allylic alkylation reactions (Figure 2.8).

Comparative results for the allylic alkylation of **S2** are summarized in Scheme 2.8. The aryl substituent in the imine was exchanged for different pyridyl rings, giving ligands **L14**.[64] **L14a** proved to be the most efficient ligand and further increased the enantioselectivity of **S2** to 99% ee. For ligands **L15**, with quaternary ammonium salts, **L15a** gave the highest stereoselectivity for the model reaction and was tested with different carbon nucleophiles, providing enantioselectivities up to 94% ee.[65] This ligand was also employed in allylic etherification reactions with a wide range of aryl alcohols, providing moderate yields and overall good enantioselectivities.[66] Ligands bearing a longer cationic chain achieved higher enantioselectivity in allylic amination. Phosphine-imidate **L16a** also showed very high enantioselectivities in the allylic substitution of **S2** and other substrates with different carbon nucleophiles.[67] It was proven that this unusually broad range in allylic alkylation was due to

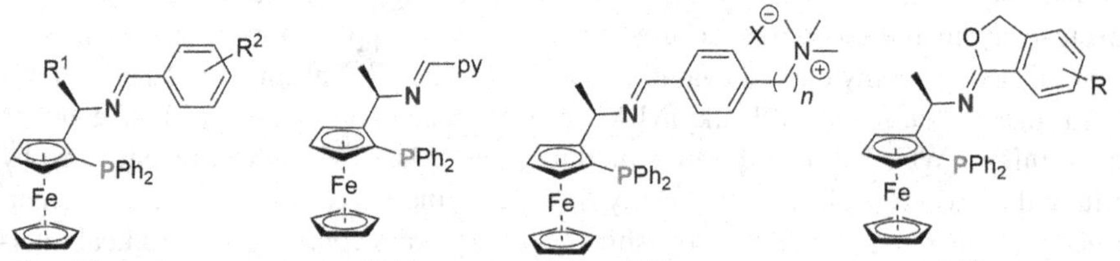

FIGURE 2.8 Ferrocenylphosphine-imino ligands.

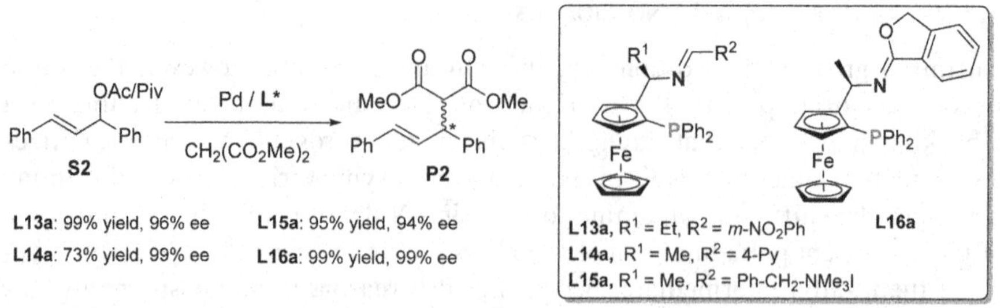

SCHEME 2.8 Allylic alkylation with phosphine-imine ligands.

the presence of the imido nitrogen. An extended family of these ligands was applied to the Ir-catalyzed asymmetric hydrogenation of di-, tri-, and tetrasubstituted olefins, with moderate results.[68]

2.3.2.2 α-Phosphino-β-Imine-Arene Ligands

In 1997, Mino and co-workers prepared hydrazone ligands **L17**, reacting with commercial 2-diphenylphosphinobenzaldehyde with chiral auxiliaries of the SAMP ((*S*)-1-amino-2-(methoxymethyl)-pyrrolidine) family.[69] In the following years, other ligands with α-phosphino-β-imine-substituted arene cores were prepared by the same method, such as **L18–L21** (Figure 2.9). These ligands have different functionalizations on the imino nitrogen atom and/or have planar chiral arenes, such as ferrocenes or phenyl-chromium tricarbonyls.

Ligands **L17** and **L18**, with little functionalization in the pyrrolidine side-arm (**L17a** and **L18a**, R = H), gave excellent enantioselectivities in the allylic alkylation of **S2** (Scheme 2.9).[69–71] When benzylamine was used as a nucleophile, **L18b** gave

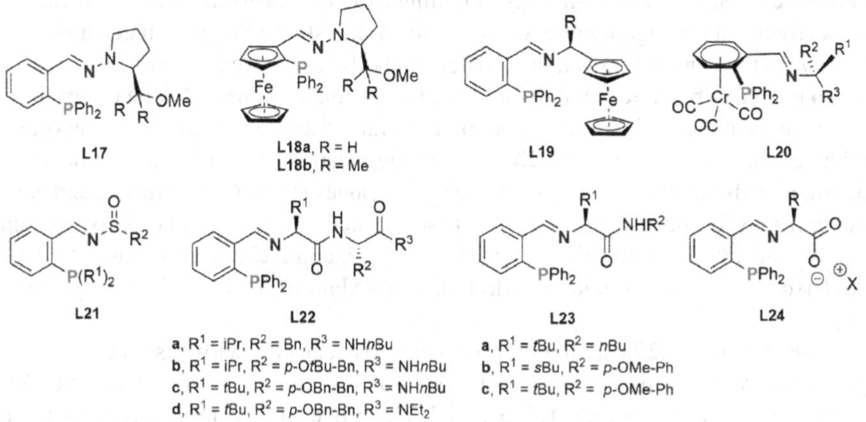
L17
L18a, R = H
L18b, R = Me
L19
L20
L21
L22
L23
L24
a, R1 = iPr, R2 = Bn, R3 = NHnBu
b, R1 = iPr, R2 = p-OtBu-Bn, R3 = NHnBu
c, R1 = tBu, R2 = p-OBn-Bn, R3 = NHnBu
d, R1 = tBu, R2 = p-OBn-Bn, R3 = NEt2
a, R1 = tBu, R2 = nBu
b, R1 = sBu, R2 = p-OMe-Ph
c, R1 = tBu, R2 = p-OMe-Ph

FIGURE 2.9 Structures of the main α-phosphino-β-imine-substituted arene-core ligands.

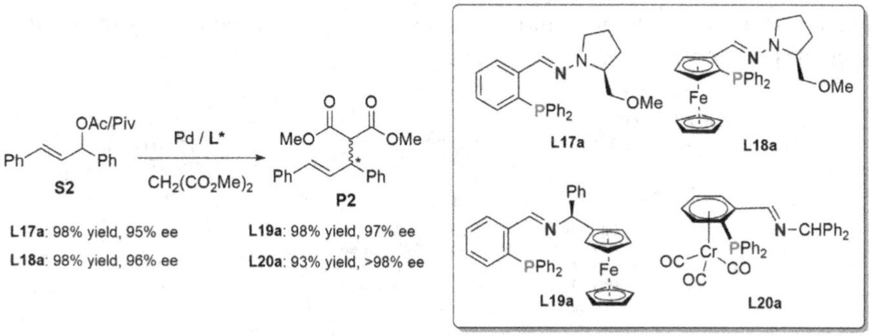
OAc/Piv
Pd / L*
CH2(CO2Me)2
S2
P2
L17a: 98% yield, 95% ee
L18a: 98% yield, 96% ee
L19a: 98% yield, 97% ee
L20a: 93% yield, >98% ee
L17a
L18a
L19a
L20a

SCHEME 2.9 Allylic alkylation with α-phosphino-β-imine-substituted arene-core ligands.

greater enantioselectivity (93% ee). Ligands **L19**, with a ferrocenyl substituent, led to an improvement in the enantioselectivity up to 97% ee.[72] Phenyl-chromium tricarbonyl ligands **L20** further increased the enantioselectivity to >98% ee.[73] Studies showed that, for these ligands, the asymmetric induction is mainly controlled by the planar chirality of the ferrocene motif and increases with the bulkiness of the *N*-substituents of the imine. Ligands **L21**, which include a chiral sulphur atom attached to the imino *N*-donor and bulky substituents, achieved high enantioselectivity in the allylic alkylation of **S2** (ee values up to 96%).[74] On the other hand, the use of an aryl substituent in the sulfinyl imine had a negative impact on the enantioselectivity. Most of these ligands were also used in the hydrogenation of trisubstituted olefins, achieving moderate enantioselectivities.[75,76]

Hoveyda prepared the peptide-based ligands **L22** and **L23** (Figure 2.9) by reacting readily available peptides to 2-(diphenylphosphino)benzaldehyde.[77] Their ease of production allowed the synthesis of many ligands with modifications in the peptidic core, which were then screened for the desired catalytic transformation to find the most suitable ligand for each case.[78] These compounds have been widely used in carbon–carbon and carbon–nitrogen bonding reactions, providing very high enantioselectivities for a large number of substrates. **L22a** showed very high ee values (up to 98%) in the conjugate addition of different alkylzincs to cyclic enones,[77,79] whereas **L22b** proved to be more suitable for disubstituted linear enones.[80,81] The use of these two ligands in the conjugate addition to unsaturated lactones has also been reported, with ee values up to >98%.[82] **L22c** showed high enantioselectivity in the conjugate addition of alkylzincs to cyclic nitroalkenes (ee values up to 96%),[83] while **L22d** gave better results for the disubstituted[84] and trisubstituted[85,86] linear counterparts, with up to 95% and 98% ee, respectively (Scheme 2.10). Ligand **L22d** was also used by J. C. Anderson in a tandem conjugate addition–nitro-Mannich reaction for the synthesis of β-nitroamines.[87]

Monopeptidic **L23a** achieved high enantioselectivity with respect to conjugate addition to cyclic enones[88] and unsaturated lactones (up to >98% and 96% ee, respectively).[82] Ligands **L23b** and **L23c** proved to be more efficient for a wide range of carbon–nitrogen bonding reactions, such as aza-Diels-Alder,[89,90] Mannich[91] and vinylogous Mannich (Scheme 2.11).[90,92–95] Inspired by these ligands, Mauduit and Crévisy reported phosphino-azomethinylate salts **L24**, which also gave excellent enantiomeric excess values in the conjugate addition to enones (up to 99% ee).[96]

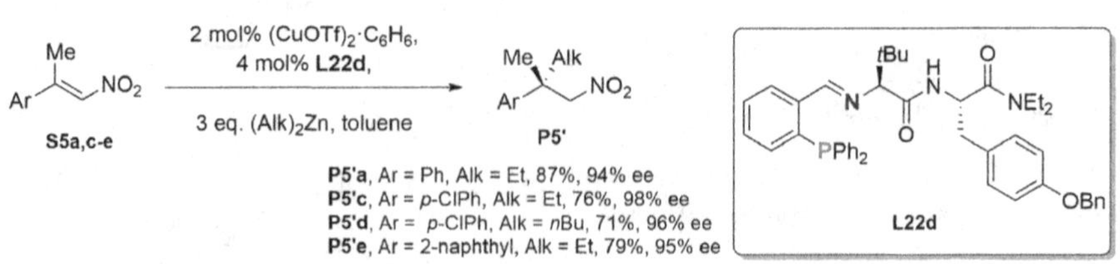
Me
Ar
NO_2
S5a,c-e
2 mol% $(CuOTf)_2 \cdot C_6H_6$,
4 mol% L22d,
3 eq. $(Alk)_2Zn$, toluene
Me Alk
Ar
NO_2
P5'
P5'a, Ar = Ph, Alk = Et, 87%, 94% ee
P5'c, Ar = p-ClPh, Alk = Et, 76%, 98% ee
P5'd, Ar = p-ClPh, Alk = nBu, 71%, 96% ee
P5'e, Ar = 2-naphthyl, Alk = Et, 79%, 95% ee
tBu
H
N
O
N
O
NEt_2
PPh_2
OBn
L22d

SCHEME 2.10 Cu-catalyzed conjugate addition to trisubstituted nitroalkenes with **L22d**.

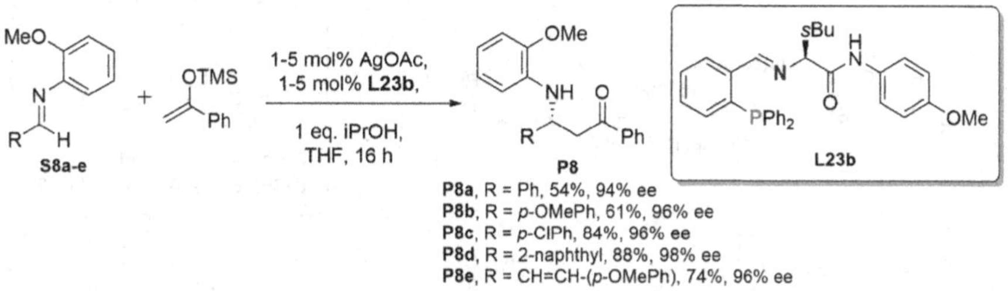

SCHEME 2.11 Ag-catalyzed Mannich reactions of an enol ether with different imines, using **L23b**.

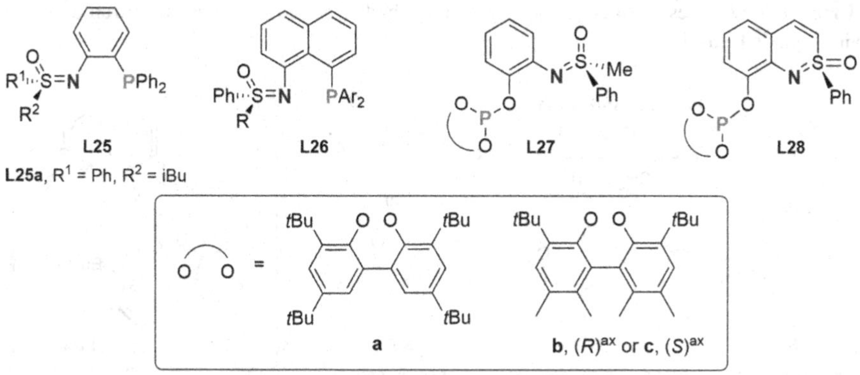

FIGURE 2.10 Phosphinosulfoximine ligands.

2.3.2.3 Phosphinosulfoximine Ligands

In 2005, Bolm synthesized a new class of sulphur-chiral phosphinosulfoximine ligands **L25**, bearing alkyl aryl substituents at the *S*-center (Figure 2.10).[97] These ligands were applied to the asymmetric hydrogenation of imines, using iodine as a promoter. Bulkier alkyl groups, such as isopropyl or cyclopentyl substituents at the sulphur atom, decreased both the activity and the enantioselectivity. However, ligand **L25a**, bearing a β-branched isobutyl substituent, achieved the best results. Electron-rich aryl substituents led to active catalysts but gave lower enantioselectivities than did phenyl substituents. Ligand **L25a** was used for the Ir-catalyzed asymmetric hydrogenation of a large number of *N*-aryl imines, leading to ee values greater than 90% for most substrates. Later, analogous bicyclic ligands **L26** were prepared from 1,8-diiodonaphtalene and tested in the Ir-catalyzed asymmetric hydrogenation of quinolines.[98]

Phosphinosulfoximines **L25a** and **L26** were also tested in the olefin hydrogenation of α,β-unsaturated trisubstituted ketones **S9** (Scheme 2.12).[99] Ligands **L26** proved to be more effective for this process than **L25a** or phosphinooxazoline (PHOX) ligands, leading to very high enantiomeric excesses for different enones.

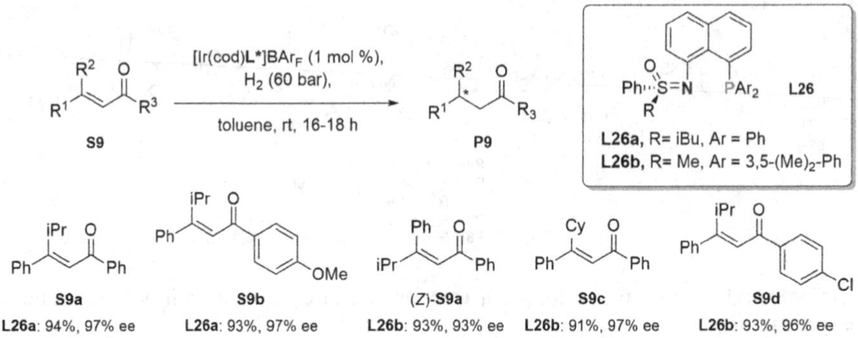

SCHEME 2.12 Best results of the asymmetric hydrogenations of α,β-unsaturated enones with ligands **L26a-b**.

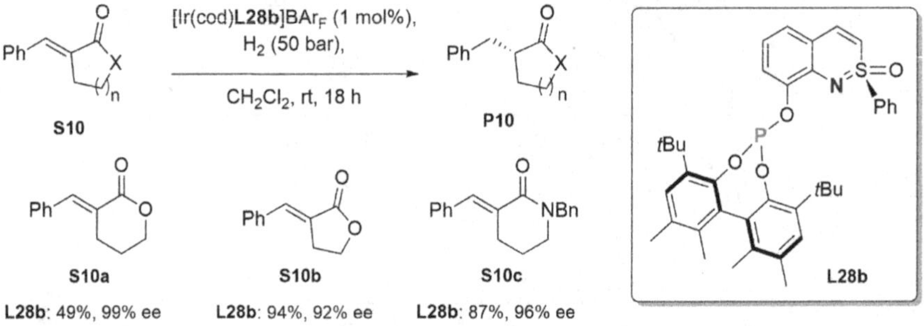

SCHEME 2.13 Highlighted results of the asymmetric hydrogenations with ligand **L28b**.

More recently, Diéguez and Pàmies synthesized analogous sulfoximines with biarylphosphite fragments (**L27**) as well as the more rigid benzothiazine derivatives **L28** (Figure 2.10).[100] The Ir-complexes of **L27** and **L28** increased the enantioselectivity obtained by **L26** in the hydrogenation of α,β-unsaturated enones, and expanded the range of reducible substrates to other olefins bearing poorly coordinative groups, such as α,β-unsaturated esters, lactones, lactams, and diphenyl alkenylboronic esters. Scheme 2.13 highlights the most important examples. For most substrates, the presence of a chiral *R*-biaryl phosphite moiety in the ligand (**L27b** and **L28b**) led to the highest enantioselectivities, suggesting a cooperative effect between the configurations of the sulfoximine and phosphite groups.

2.3.3 GROUP 3: *P,N*-CYCLIC IMINO LIGANDS

Cyclic iminophosphine ligands constitute one of the most widely employed *P,N*-ligand families in asymmetric catalysis. The rigidity of the heterocycle makes these ligands appealing in terms of catalytic performance and stereoselective control. As reviewed in Chapter 1, PHOX ligands made a breakthrough in the field due to their synthetic versatility and their broad catalytic applicability.[101] On account of the remarkable success of the phosphinooxazolines, many useful *P,N*-ligands with other five-membered nitrogen heterocycles, such as oxazoles, thiazoles, imidazoles, and imidazolines, have been developed.

2.3.3.1 Phosphinyl Oxazole- and Thiazole-Based Ligands

The Andersson group has developed a large library of ligands that gave outstanding results in several catalytic reactions, with a specific focus on asymmetric hydrogenations.[102] After undertaking a kinetic and computational study of the hydrogenation of (*E*)-1,2-diphenyl-1-propene with the Ir-PHOX complex,[103] Andersson and co-workers rationally designed a new class of bicyclic oxazole-based *P,N*-ligands **L29** (Figure 2.11).[104] These ligands meet specific requirements, such as maintaining the P and the N as donor atoms to achieve the *trans* effect, the generation of a six-membered chelate ring, containing a rigid backbone fused to the heterocycle, on complexation, and the formation of a chiral environment for asymmetric induction. From this perspective, they developed ligands **L29** with an oxazole moiety fused to a cyclohexane bearing a diarylphosphinite group. Iridium/**L29** complexes were successfully applied to the hydrogenation of disubstituted styrenes with outstanding enantioselectivities (93–99% ee).

These initial ligands evolved into a new set of thiazole-based ligands **L30** with the following modifications: (i) oxygen replacement by a sulphur atom in the heterocyclic moiety to increase ligand stability and to diversify the structure at a late stage, starting from one common chiral intermediate; (ii) replacement of the phosphinite by a phosphane unit to gain structural stability; and (iii) contraction or expansion of the fused ring to a five- or seven-membered saturated ring.[105] This research group thoroughly studied the effect of variation in this ligand in both the stereochemical outcome of the Ir-catalyzed hydrogenation of olefins and in the calculated bond distance between the iridium and the *ipso*-carbon in the 2-position on the heterocycle. Interestingly, calculations revealed that the replacement of the oxygen atom by a sulphur led to a more compact ligand with a shorter *ipso*-C–Ir distance. In addition,

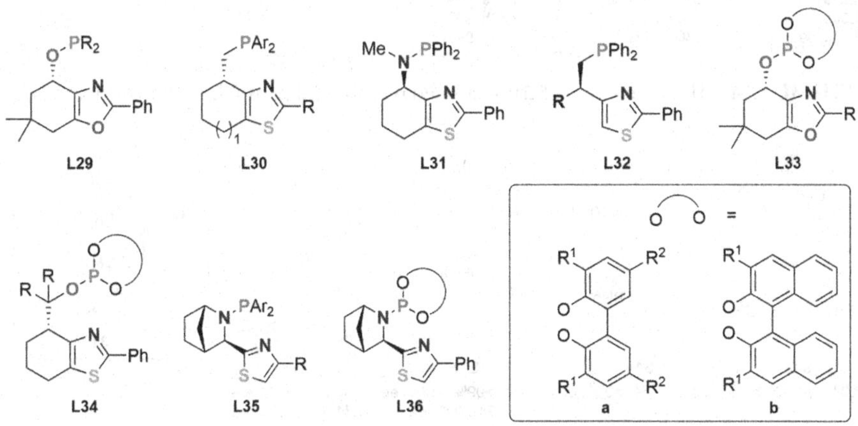

FIGURE 2.11 Overview of the oxazole- and thiazole-based ligands.

varying the ring size resulted in differences in the same bond distance. Experimental results confirmed that the thiazole-based ligands afforded greater asymmetric induction than did their oxazole analogs, and that the six-membered ring backbone was the best (Scheme 2.14).

A year later, the same research group reported a new class of thiazole-based ligands (**L31**, Figure 2.11), containing a phosphanamine unit. **L31** showed remarkable results in the Ir-catalyzed asymmetric hydrogenation of fluorinated olefins, such as **S13**, and 1,1-disubstituted vinylphosphonates, such as **S14** (Scheme 2.15).[106,107] Moreover, **L31** demonstrated enantioselectivities comparable to the ones obtained with **L30b** in the hydrogenation of trisubstituted olefins **S15** and β,β-diaryl methyl acrylate **S16**.[108]

To study the influence of the backbone in the catalytic activity, Andersson's group developed open-chain-backbone analogs **L32** (Figure 2.11).[109] However, their corresponding iridium catalysts, [Ir(cod)**L32**]BArF, were slightly less successful than their more rigid counterparts in the hydrogenation of trisubstituted olefins, allylic alcohols, and imines.

As in other ligand families, the replacement of the diarylphosphine by a biarylphosphite was performed by a collaboration between the Andersson and the Diéguez-Pàmies groups.[110] The oxazole- and thiazole-based family of ligands **L33–34** showed greater stability, versatility, and tunability than the previous ones

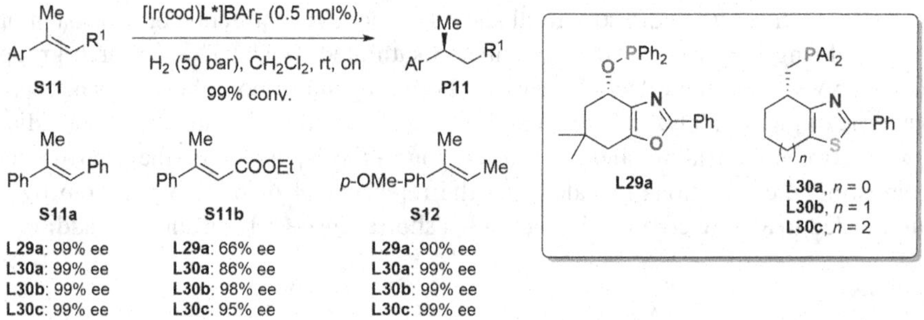

SCHEME 2.14 Hydrogenation of trisubstituted olefins with phosphinyl thiazole ligands.

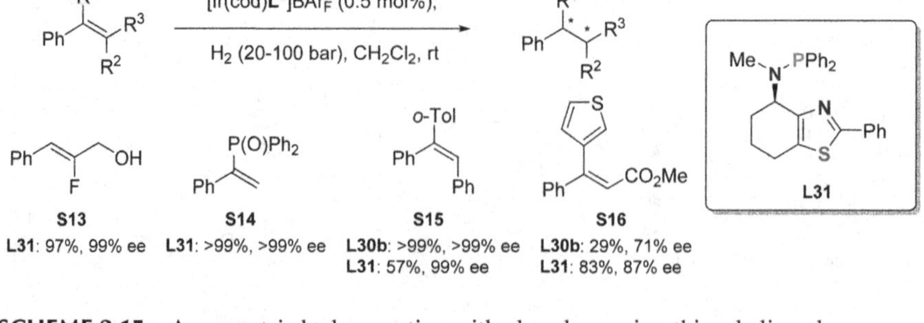

SCHEME 2.15 Asymmetric hydrogenation with phosphanamine-thiazole ligand.

(Figure 2.11). This modification expanded the substrate range of the enantioselective hydrogenation of both *E*- and *Z*-trisubstituted olefins and 1,1-disubstituted terminal alkenes (ee values up to 99%).[111] In general, ligand **L34a**, with bulky *tert*-butyl groups at the *ortho* and *para* positions of the biphenyl moiety, showed higher ee values than its phosphinite–oxazole and phosphine-thiazole analogs. Ligand **L34a** gave outstanding results and tolerated several functional groups, such as silanes (**S11d** and **S18c**), alcohols (**S17a**), and esters (**S11b** and **S17b**) (Scheme 2.16).

An important skeletal modification to these ligands, introduced by Andersson, produced ligands **L35** (Figure 2.11); the backbone was replaced by a bicyclic amine linking the phosphorus fragment.[112] Ligands **L35** gave air-stable iridium complexes that were assessed in the asymmetric hydrogenation of several olefins. The enantioselectivities obtained when using **L35a** in the hydrogenation of substrates **S11a**, **S11b**, **S11c**, and **S17** were almost as good as the ones obtained with **L30b**, but still better than its oxazoline analog. However, it showed low asymmetric induction for minimally functionalized 1,1-disubstituted terminal alkenes (**S18a** and **b**). The group used ligand **L35a** in the hydrogenation of vinyl boronates, obtaining excellent enantoslectivities (up to 98% ee).[113] A further application of these ligands was in the asymmetric hydrogenation of (*E*)-β,β-disubstituted α,β-unsaturated esters **S19** and a wide array of vinylic, allylic, and homoallylic sulfones **S20**.[114] Moreover, the Ir/**L35a** catalyst application range was broadened to the asymmetric hydrogenation of γ-substituted cinnamyl alcohols **S21** (ee values up to 99%)[115] and a wide range of α,β-unsaturated α-fluoro aryl and alkyl ketones **S22** (ee values up to >99%) (Scheme 2.17).[116]

The combination of the last two aforementioned modifications, i.e., the biarylphosphite moiety of **L34** and the nitrogenated bicyclic backbone of **L35**, led to the development of a novel family of ligands **L36**, with a π-acceptor biaryl phosphoroamidite moiety (Figure 2.11).[117] Ligand **L36a** (with (*S*)^{ax}-binaphthol) expanded the substrate range of the Ir-catalyzed hydrogenation of a broad array of minimally functionalized olefins, such as trisubstituted alkenylboronic esters **S23**, *Z*-trisubstituted alkenes **S24**, and α,β-unsaturated enones **S25**, with outstanding conversion and ee values (Scheme 2.18).

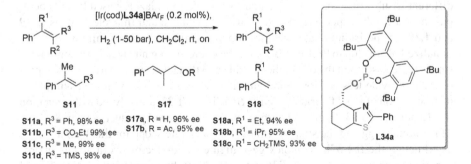

SCHEME 2.16 Catalytic hydrogenation with ligand **L34a**.

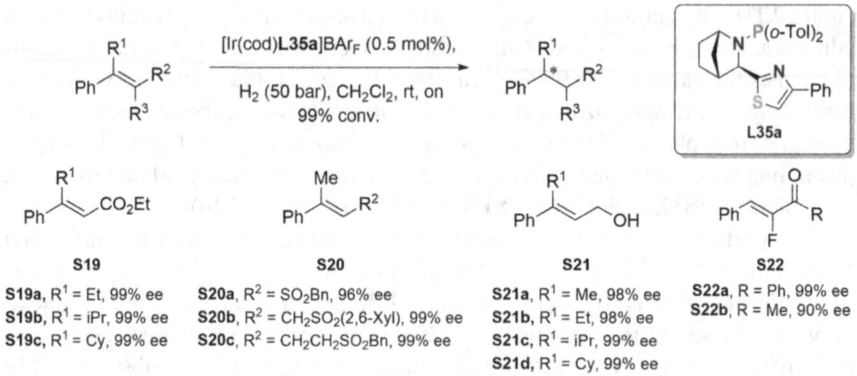

SCHEME 2.17 Substrate scope outline in the catalytic hydrogenation of trisubstituted olefins with ligand **L35a**.

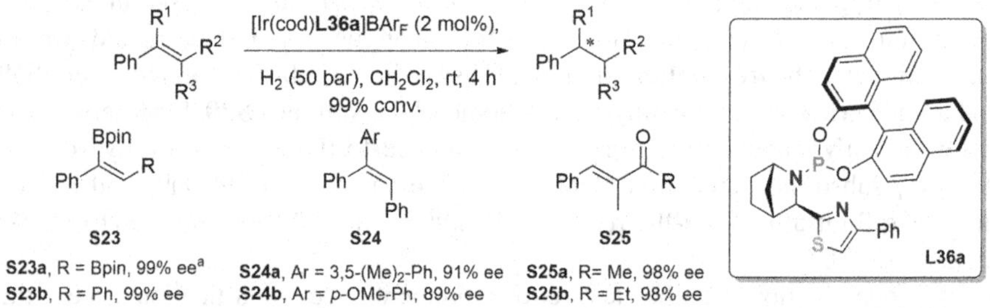

SCHEME 2.18 Ir-catalyzed hydrogenation with phosphoroamidite-thiazole ligands **L36a**. [a]Ligand **L36b** ((R)[ax]-binaphthol) was used instead of **L36a**.

2.3.3.2 Phosphinyl Imidazole-Based Ligands

The ease of access to imidazoles has facilitated the preparation of many imidazole-based *P,N*-ligands. Particularly successful examples were found by Andersson's group, who developed the phosphine-imidazole ligands **L37** (Scheme 2.19).[118] At first, the group evaluated the activity of **L37** in the asymmetric hydrogenation of ten different trisubstituted olefins as standard substrates. Imidazole-based ligands demonstrated a catalytic performance similar to its oxazole and thiazole analogs (**L29** and **L30**) in the iridium-catalyzed asymmetric hydrogenation of minimally functionalized olefins (up to 98% ee). Because the coordinating nitrogen in the imidazole ring is more basic than the one in the oxazole and thiazole counterparts,[106] it was proposed to provide less defluorination and more enantioselectivity in the asymmetric hydrogenation of vinyl fluorides, such as **S13**. The Ir/**L37ca**-catalyzed hydrogenation of **S13** achieved a 93/7 ratio of hydrogenated/defluorinated product, and an 86% ee. This finding triggered the study to enlarge the substrate scope for this ligand.

Although the Ir-**L37ba** complex gave high enantioselectivities in alkene hydrogenation of endocyclic enones (**P26**, up to 94% ee), hydrogenation of the carbonyl also took place, with long reaction times.[119] Experimental results revealed the influence of the aromatic substituent at the imidazole unit in the chirality transfer. As a

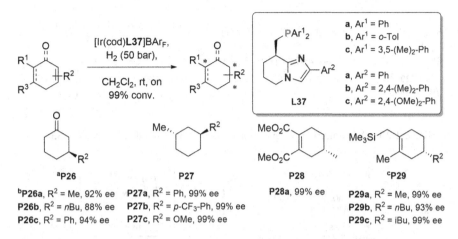

SCHEME 2.19 Summary of **L37ab** catalytic application. [a]Reactions were performed using ligand **L37ba**. [b]Reaction was performed in 2,2,2-trifluoroethanol under 100 bar of hydrogen pressure. [c]Reactions were performed in α,α,α-trifluorotoluene under 10 bar of hydrogen pressure and using ligand **L37ac**.

a)

	Oxazole	Thiazole	Imidazole
pK_aH	0.8	2.5	7.0

more basic heterocycle

b)

increasing acidity

FIGURE 2.12 a) pK_aH values of the unsubstituted heterocycles.[122] b) Increasing acidity of iridium hydrides with different *N*-donor heterocycles.

general feature, 2,4-aryl substitution, as occurred in ligands **L37ab** (2,4-diMePh) and **L37ac** (2,4-di(OMe)Ph), gave the highest enantioselectivities. This observation was also found in the hydrogenation of minimally functionalized alkenes and cyclic 1,4-dienes (**P27** and **P28** in Scheme 2.19), among many other examples, using ligand **L37ab**.[120] High hydrogen pressures and long reaction times (17 h) produced the corresponding fully saturated compounds with excellent enantioselectivities with **L37ba**, whereas 5 bar hydrogen pressure and short reaction times (0.5 h) only hydrogenated the less-hindered double bond, with no change in the enantioselectivity for **P27b**. More recently, the asymmetric hydrogenation of allylsilanes (**P29**) was reported, with excellent yields and ee values ranging from 92 to 99%, using **L37ac** (Scheme 2.19).[121]

Andersson's group also reported a comprehensive study of the hydrogenation of allylic alcohols, supported by density-functional theory (DFT) calculations, showing the influence of the Ir-*P,N*-complex acidity in the stereochemical outcome. Metal hydrides of the Ir-**L37ba** complex were found to be the least acidic of the series (oxazole-, thiazole-, and imidazole-based ligands; Figure 2.12), showing 98% ee in the hydrogenation of **S30** (Scheme 2.20).[115]

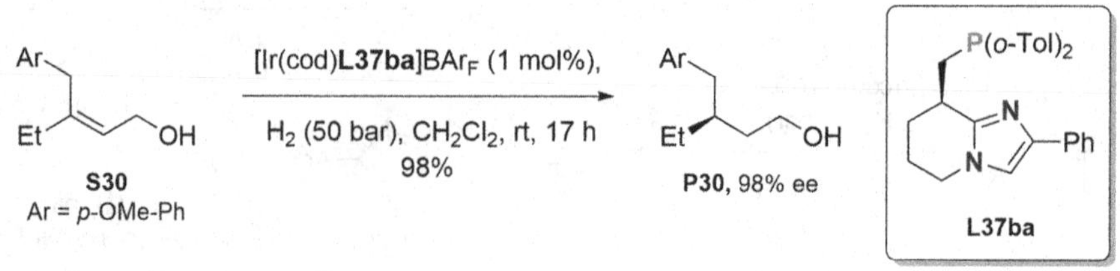

SCHEME 2.20   Asymmetric hydrogenation of dialkyl substituted allylic alcohol **S30**.

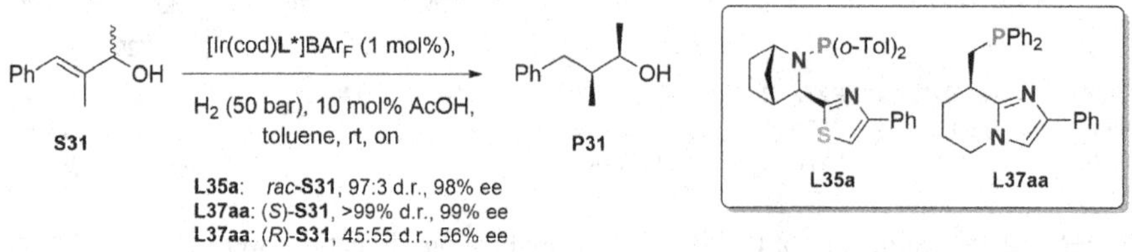

SCHEME 2.21   Asymmetric hydrogenation of chiral and racemic allylic alcohol **S31** *via* DKR.

Based on these results, the iridium-catalyzed asymmetric hydrogenation of allylic alcohols was studied for several ligands.[123] The imidazole-based ligand **L37aa** showed high conversion and enantioselectivity only in the asymmetric hydrogenation of (*S*)-**S31**, whereas the (*R*)-enantiomer showed a low diastereomeric ratio and low enantioselectivity (Scheme 2.21). This observation was explained by racemization at the C_2 position, due to the slight acidity of the hydrogenated imidazole-based Ir-*P,N* complexes (Figure 2.12).[115] As a result, they switched to more acidic complexes, such as its thiazole analog **L35a** to favour the allylic carbocation formation and thereby the dynamic kinetic resolution (DKR) of the **S31** racemic mixture. Ligand **L35a** led to the highest diastereomeric ratio (dr) (97:3) and enantioselectivity (98% ee) in the hydrogenation of racemate **S31** *via* DKR, when using 10 mol% of acetic acid (Scheme 2.21). Taking advantage of this finding, the scope of the reaction was expanded to include several secondary allylic alcohols, and the mechanism was revealed.

A recent report of Riera, Verdaguer, and co-workers described the synthesis of a new class of phosphino-imidazole ligands **L38**.[124] In this case, the chirality lies on both the P atom and on the C-backbone, as in its oxazoline analogs, the MaxPHOX family of ligands.[125] The hydrogenation of the cyclic enamide **S32** catalyzed by the iridium complex of **L38** showed high conversion and very high enantioselectivity (98% and 89% ee, respectively; Scheme 2.22). However, both catalytic activity and asymmetric induction were lower than its oxazoline analog.

2.3.3.3   Phosphinyl Imidazoline-Based Ligands

Phosphino-imidazolines are nitrogenated analogs of PHOX ligands. Compounds **L39**, so-called BIPI (Boehringer-Ingelheim Phosphinoimidazolines), were patented in 2001 by Busacca and co-workers (Figure 2.13).[126] These ligands were first used

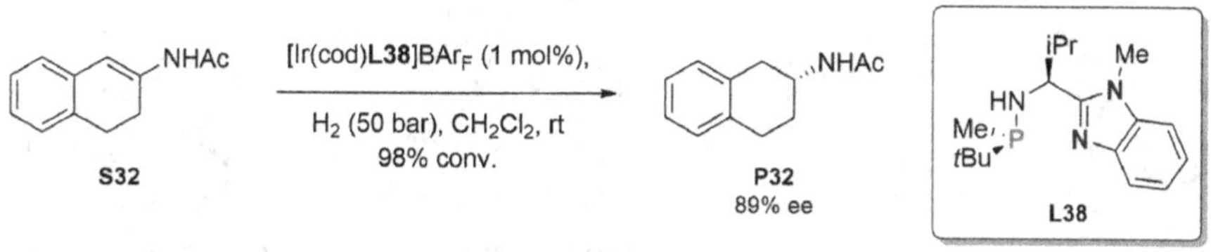

SCHEME 2.22   Example of Ir-catalyzed hydrogenation of cyclic enamides with ligand **L38**.

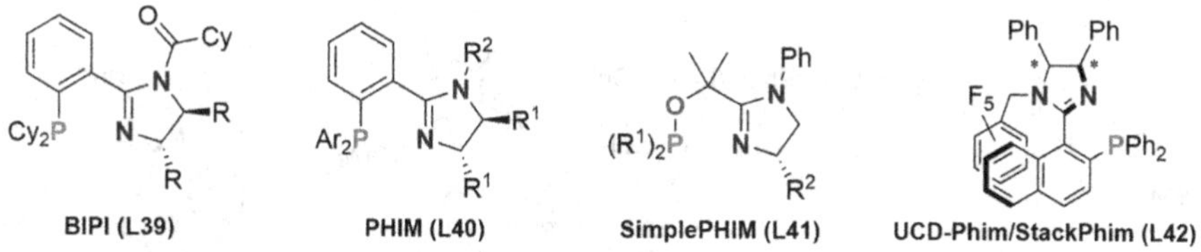

BIPI (L39)   **PHIM (L40)**   **SimplePHIM (L41)**   **UCD-Phim/StackPhim (L42)**

FIGURE 2.13   Phosphino-imidazoline ligands.

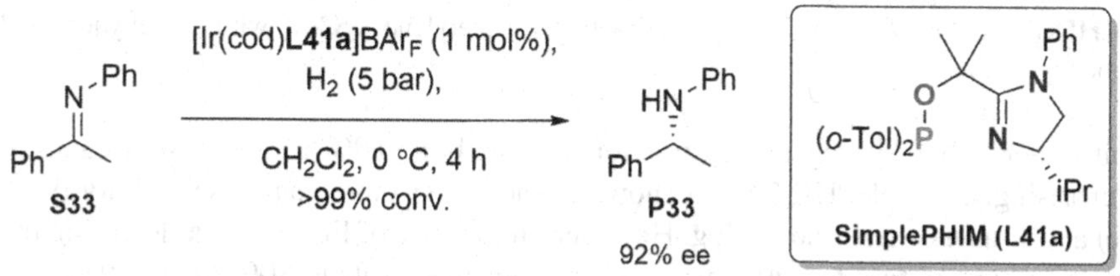

SCHEME 2.23   Asymmetric hydrogenation of imine **S33** with Pfaltz's ligand **L41a**.

in asymmetric Heck reactions.[127,128] A year later, Pfaltz and co-workers reported the phosphino-imidazoline (PHIM) ligands **L40** and their use in the Ir-catalyzed enantioselective hydrogenation of various unfunctionalized olefins (Figure 2.13).[129] The main feature that the PHIM ligands offer, compared with the PHOX ligands, is the additional substituent at the new sp^3 nitrogen, which thereby affords a bulkier ligand. Throughout the years, many uses of BIPI/PHIM ligands have been reported, such as the hydrogenation of imines,[130] unsaturated ureas,[131] urea esters, tert-butoxycarbonyl (Boc-) and benzyloxycarbonyl (Cbz-)-enecarbamates, enamides,[132] unfunctionalized olefins[133] and trialkoxy(vinyl)silanes.[134] We can also highlight the large-scale asymmetric synthesis of a cathepsin S inhibitor.[135] These excellent results paved the way for the syntheses of novel imidazoline-based ligands, such as the SimplePHIM **L41** and the StackPhim ligand **L42**.

This SimplePHIM ligand **L41**, developed and patented by Pfaltz in 2005, is a simplified version of the previous imidazoline analogs of PHOX.[136] These ligands showed excellent results in many catalytic processes. In 2010, a comprehensive catalyst screening for the asymmetric hydrogenation of imines was evaluated, as chiral amines are structural elements in numerous biologically active natural and non-natural products.[137] They first assessed the Ir-catalyzed hydrogenation of acetophenone N-phenylimine **S33** as the model substrate (Scheme 2.23). The reaction was carried

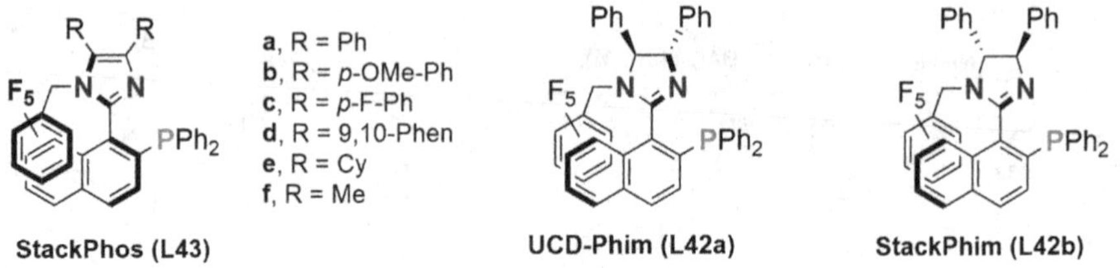

FIGURE 2.14 The StackPhos and the UCD-Phim/StackPhim ligands.

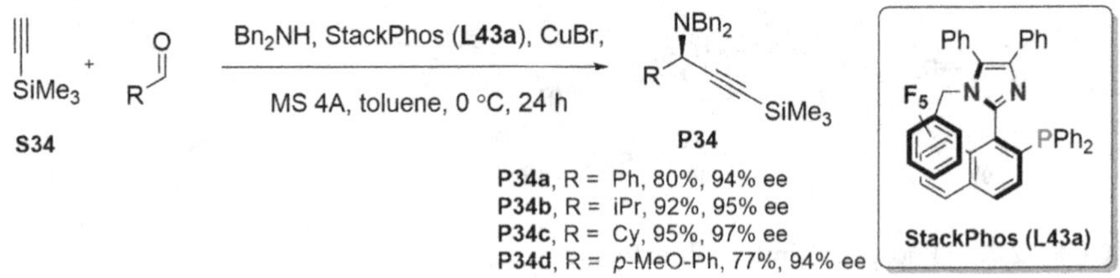

SCHEME 2.24 A^3 coupling of trimethylsilylacetylene (**S34**) with aldehydes and dibenzylamine.

out under 5 bar hydrogen pressure for 4 h, with 1 mol% iridium/**L41a** catalyst. Surprisingly, SimplePHIM **L41a** showed exactly the same enantioselectivity (92% ee) at 0°C as its oxazoline analog. However, in the case of ligand **L41a**, lowering the reaction temperature to −20°C reduced the conversion rate to 80% (with a 96% ee), whereas SimplePHOX gave complete conversion. Nevertheless, the SimplePHIM ligand showed excellent enantioselectivities in the asymmetric hydrogenation of terminal vinyl boronates (up to 95% ee) and vinylsilanes (up to 88% ee).[134,138]

The axially chiral phosphino-imidazole StackPhos ligand (**L43**) was first developed by Aponick and co-workers in 2013.[139] These ligands have the peculiarity of bearing an atropoisomeric backbone. The conformational stabilization was achieved by means of π-π stacking interactions between the electron-rich naphthalene ring and the electron-poor pentafluorophenyl group at the non-coordinating nitrogen of the imidazole ring (Figure 2.14).

StackPhos ligand was first applied to the enantioselective copper-catalyzed aldehyde-alkyne-amine coupling (A^3 coupling) with excellent yields and ee values up to 97% (Scheme 2.24). In addition to this first use, Aponick and co-workers have proven the usefulness of axially chiral ligands **L43** in several catalytic transformations.[140–144]

The development of the StackPhim ligands (**L42**), imidazoline analogs of StackPhos (**L43**), arose from the need to overcome the limitation of StackPhos preparation by racemic resolution.[143] From this perspective, Guiry and Aponick independently reported a new class of ligand, combining both axial and central chirality, namely UCD-PHIM (**L42a**) and StackPhim (**L42b**).[145,146] Common features of ligand **L42** are configurational stability, due to the π-stacking of the pentafluorobenzyl moiety (at the non-ligating N atom) with the electron-rich naphthalene core, and the additional central chirality, that allows resolution by fractional crystallization.[147]

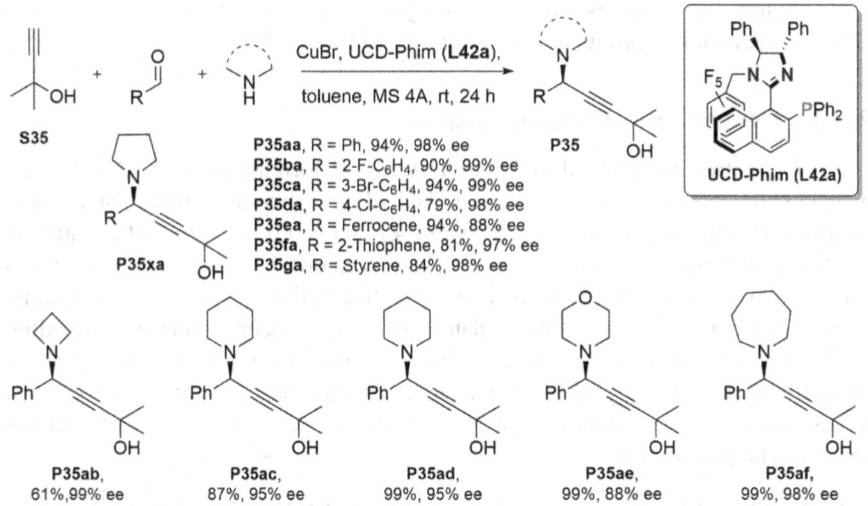

SCHEME 2.25 Expanded substrate range in A³ coupling, using UCD-Phim.

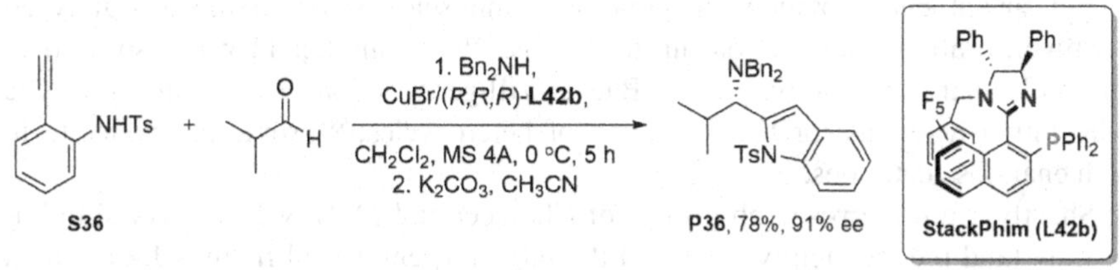

SCHEME 2.26 Application of StackPhim in A³ coupling/cyclization sequence.

These recent ligands, UCD-Phim/StackPhim, have shown outstanding activity in several catalytic reactions. Guiry and co-workers applied the UCD-Phim ligand in the enantioselective copper-catalyzed A³ coupling reaction.[145] Although low reactivity and only moderate enantioselectivity were observed initially when using benzaldehyde, a new strategy was published that overcame these problems.[148] The reaction proceeded under mild conditions, low catalytic loading, and at room temperature, producing propargylamines **P35** in excellent yields, up to 99%, and enantioselectivities of up to 99% ee. The reaction scope was thereby broadened to include several aromatic aldehydes and cyclic amines, which exhibited outstanding enantioselectivities, as shown in Scheme 2.25.

The StackPhim **L42b** (diastereomer of UCD-Phim **L42a**) was applied to the Cu-catalyzed enantioselective synthesis of the C₂-aminoalkyl five-membered heterocycle building blocks, with high enantioselectivity.[146] The alkynylation/cyclization sequence is convergent, highly modular, and generates the product **P36** with good yield and high enantioselectivity (Scheme 2.26). The substrate scope was expanded to include aliphatic branched, straight chain and aromatic aldehydes, as well as symmetrical and less common non-symmetrical secondary amines to generate the corresponding products at excellent ee values (84–94%).

2.3.4   GROUP 4: *P,N*-PYRIDINO LIGANDS

In 1977, Crabtree *et al.* introduced complexes of the formula $[Ir(py)(cod)(PR_3)]PF_6$, which exhibited high catalytic activity in the hydrogenation of olefins.[8,9] Since then, many efforts have focused on designing chiral *P,N*-ligands for asymmetric catalytic reactions. Among them, the ones that emulate the coordination sphere of Crabtree's catalyst most precisely are those that bear pyridine *N*-donor atoms. Over the years, a great number of *P,N*-pyridino chelate ligands have been described, providing excellent results for a large number of catalytic reactions, especially for asymmetric hydrogenation. *P,N*-Pyridino ligands, together with their oxazoline counterparts, represent the most important type of *P,N*-ligands in terms of number of compounds and catalytic performance.

2.3.4.1   Phosphino-Pyridine and Terpene-Based *P,N*-Pyridino Ligands

In 1999, Katsuki pioneered the work on *P,N*-pyridino ligands, describing ligands **L43** (Figure 2.15).[149] Ligand **L43a**, with a five-membered fused ring and an isopropyl substituent, provided the greatest enantioselectivities in the Pd-catalyzed asymmetric allylic substitution (up to 98% ee).[149,150] This ligand was also used in intramolecular allylic amination,[151] Baeyer-Villiger reactions[152] and tandem allylic substitution reactions for the formation of heterocycles,[153] providing moderate to high enantioselectivities.

Shortly after, the research groups of Chelucci and Malkov-Kočovský simultaneously (and independently) reported the monoterpene-based ligands **L44**, which are synthetically more accessible than **L43** (Figure 2.15).[154,155] Ligands **L44** could be prepared by Kröhnke's reaction between the terpene-derived α,β-unsaturated ketone and a pyridinium salt, followed by phosphination. The bulkier ligand *ent*-**L44a**, with an isopropyl substituent, increased the scope and enantioselectivity provided by **L43a** in the Baeyer-Villiger reaction of cyclobutanones.[156] Later, this

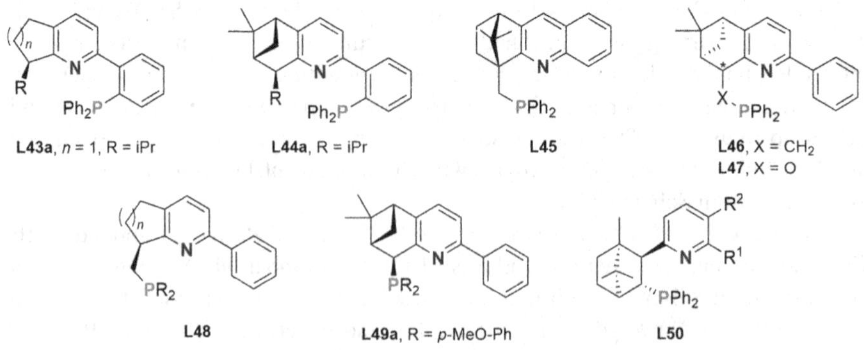

L43a, *n* = 1, R = iPr   **L44a**, R = iPr   **L45**   **L46**, X = CH₂
                                                                    L47, X = O

L48   **L49a**, R = *p*-MeO-Ph   **L50**

FIGURE 2.15   Phosphino-pyridine ligands.

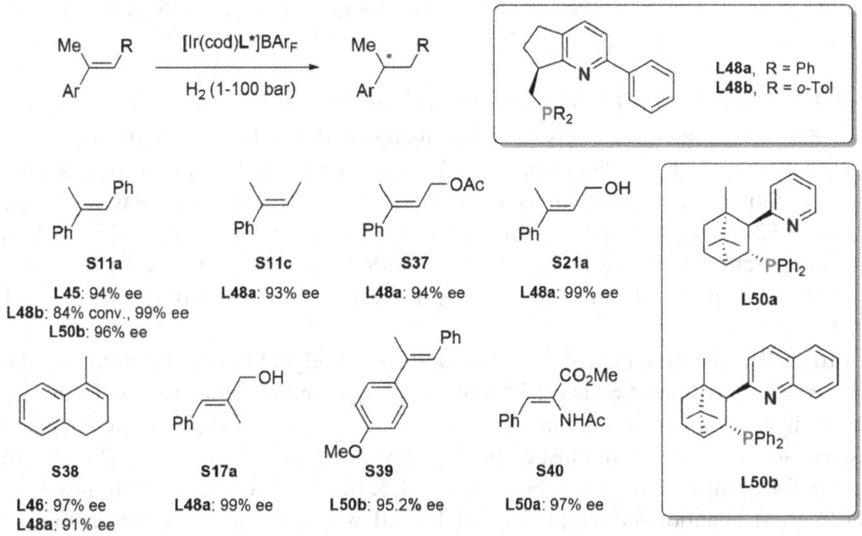

SCHEME 2.27 Iridium-catalyzed hydrogenation of trisubstituted olefins with ligands **L45–L50**. Unless otherwise specified, conversion was higher than 90%.

family was extended to include (+)-camphor-derived tetrahydroacridines, such as **L45,** and tested in the asymmetric hydrogenation of different substrates.[157] Ligand **L45** gave excellent results in the reduction of *trans*-α-methylstilbene **S11a** (Scheme 2.27).

Following the same synthetic methodology, Andersson reported bicyclic ligands **L46** and **L47**, with the phosphorus donor atom attached to the pinene chiral motif.[106] Ligand **L46** gave higher ee values than **L47** in the hydrogenation of different trisubstituted olefins, such as methylstilbenes, α,β-unsaturated esters and allylic alcohols (up to 97% ee), but the activities were poor, even at 100 bar of hydrogen pressure. Therefore, Andersson designed ligands **L48**, analogous to **L46** but without the pinene fragment.[158] Ligands **L48a** and **L48b** showed very high enantioselectivities in the hydrogenation of several trisubstituted olefins, performing better than **L46** in most cases (Scheme 2.27).

Li *et al.* described ligands **L49,** very similar to **L46,** but with the phosphine group directly attached to the pinene ring.[159] **L49a** (R = *p*-MeO-Ph) proved to be effective in the reduction of 1,1-disubstituted enol phosphonates (up to 90% ee). Knochel also contributed to the monoterpene-based ligand family with the synthesis of ligands **L50**, derived from D-(+)-camphor.[160,161] These ligands were used in the Ir-catalyzed hydrogenation of *trans*-α-methylstilbenes, with bicyclic **L50b** providing the highest enantioselectivities (up to 96% ee) (Scheme 2.27). However, the hydrogenation of trisubstituted alkenes functionalized with alcohol, acetate, and ester groups only attained moderate to high enantioselectivities. **L50** were also the first ligands that gave good results in the Ir-catalyzed hydrogenation of dehydroamino acids, studied extensively studied using Rh and Ru catalysts, due to the coordination ability of the amide group.[162] Ligand **L50a** gave 97% ee in the hydrogenation of **S40** under 1 bar of hydrogen pressure (Scheme 2.27).[160]

2.3.4.2 Pfaltz's Phosphine- and Phosphinite-Pyridines

In 2004, Pfaltz and co-workers designed ligands **L51** and **L52,** sterically similar to PHOX (Figure 2.16).[157] Phosphinites **L52** were superior to **L51** in the Ir-catalyzed hydrogenation of *trans*-α-methylstilbene **S11a** (97% ee *vs.* 88% ee), with the bulky ligand **L52a** being the best (Scheme 2.28). These ligands also provided very high enantioselectivities in the hydrogenation of other interesting olefins, such as (*E*)- and (*Z*)-**S12**, allylic alcohol **S17a**, conjugated ester **S11b**, and tetrasubstituted **S41** (Scheme 2.28).

To further increase the rigidity of the aliphatic bridge, Pfaltz introduced bicyclic phosphinite-pyridines **L53** and **L54**. Initially, they gave excellent enantioselectivities in the hydrogenation of dihydronaphthalenes, allylic alcohols, and α,β-conjugated esters (Scheme 2.29).[163] In most cases, 2-phenyl substituted ligands, with bulky *t*Bu and *o*-Tol groups on the phosphorus atom (**L53bb, L53bd,** and **L54bb**), provided the highest enantioselectivities. Ligand **L53bd** was used in the hydrogenation of the dihydronaphthalene core of antitumoral (−)-mutisianthol.[164] Ligands bearing bulkier substituents at the 2-pyridino position (**L53db, L53ea,** and **L53fa**) proved to be even more convenient for the hydrogenation of dihydronaphthalene substrates.[165] Ligand **L53bb** was used in the total synthesis of (+)-torrubiellone C and (−)-pyridovericin.[166,167]

Similar ligand (**L55a**) with a disubstituted pyridine scaffolds **L55** showed better performance in the hydrogenation of dialkyl- and alkylaryl-substituted α,β-unsaturated esters, with ee values up to 99% (Scheme 2.29).[165,168] For a few dialkyl-substituted esters, diazaphospholidine analogs **L56** (Figure 2.16) showed higher enantioselectivities. Ligand (1*S*,2*S*,2'*S*)-**L56a** was used in the total synthesis of antibiotic platensimycin.[169] The hydrogenation of different maleic and fumaric acid diesters with ligands **L55** was also attempted.[170] Ligand **L55b** (Scheme 2.29), bearing a more electron-withdrawing substituent at the 2-pyridino site, gave greater enantioselection than did the analogous phosphinite pyridines and PHOX, with ee values's up to 99%.

Phosphinite-pyridine ligands **L53** have also shown excellent enantioselectivity in the hydrogenation of purely alkyl-substituted olefins, such as **S47–S52** (Scheme 2.30).[171,172] The use of this efficient catalytic system in the hydrogenation

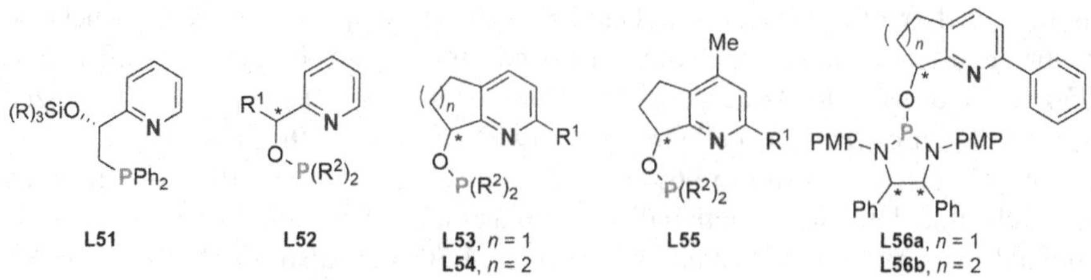

FIGURE 2.16   Pfaltz's phosphine- and phosphinite-pyridine ligands. PMP = *p*-MeO-Ph.

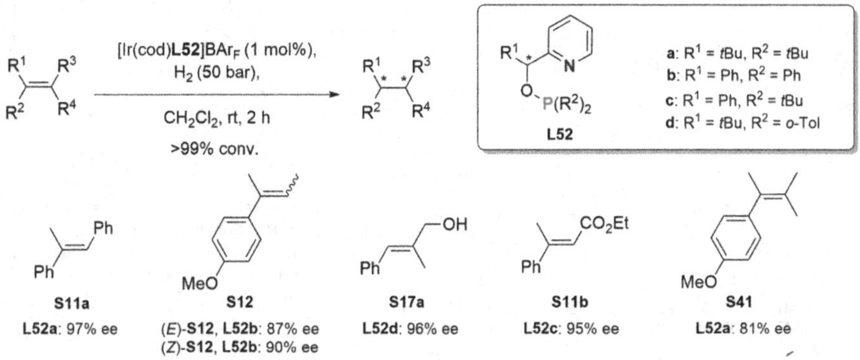

[Ir(cod)**L52**]BAr$_F$ (1 mol%),
H$_2$ (50 bar),

CH$_2$Cl$_2$, rt, 2 h

>99% conv.

a: R^1 = tBu, R^2 = tBu
b: R^1 = Ph, R^2 = Ph
c: R^1 = Ph, R^2 = tBu
d: R^1 = tBu, R^2 = o-Tol

L52

P(R^2)$_2$

S11a
L52a: 97% ee

S12
(E)-**S12**, **L52b**: 87% ee
(Z)-**S12**, **L52b**: 90% ee

S17a
L52d: 96% ee

S11b
L52c: 95% ee

S41
L52a: 81% ee

SCHEME 2.28 Highlighted results of the iridium-catalysed hydrogenation, using **L52**.

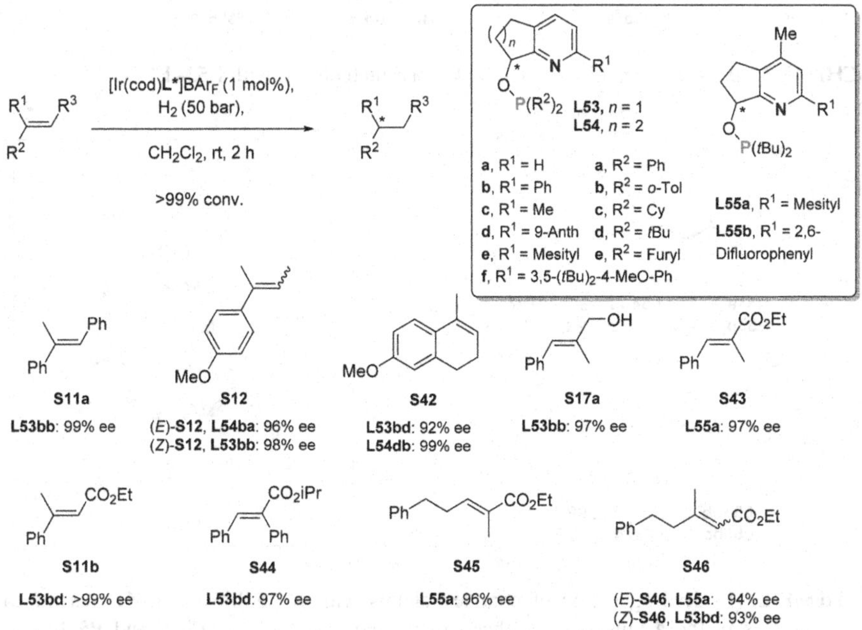

[Ir(cod)**L***]BAr$_F$ (1 mol%),
H$_2$ (50 bar),

CH$_2$Cl$_2$, rt, 2 h

>99% conv.

P(R^2)$_2$ **L53**, n = 1
L54, n = 2

a, R^1 = H   a, R^2 = Ph
b, R^1 = Ph   b, R^2 = o-Tol
c, R^1 = Me   c, R^2 = Cy
d, R^1 = 9-Anth   d, R^2 = tBu
e, R^1 = Mesityl e, R^2 = Furyl
f, R^1 = 3,5-(tBu)$_2$-4-MeO-Ph

P(tBu)$_2$

L55a, R^1 = Mesityl
L55b, R^1 = 2,6-
Difluorophenyl

S11a
L53bb: 99% ee

S12
(E)-**S12**, **L54ba**: 96% ee
(Z)-**S12**, **L53bb**: 98% ee

S42
L53bd: 92% ee
L54db: 99% ee

S17a
L53bb: 97% ee

S43
L55a: 97% ee

S11b
L53bd: >99% ee

S44
L53bd: 97% ee

S45
L55a: 96% ee

S46
(E)-**S46**, **L55a**: 94% ee
(Z)-**S46**, **L53bd**: 93% ee

SCHEME 2.29 Hydrogenation of different trisubstituted olefins with ligands **L53–L55**.

of these challenging alkenes has permitted the synthesis of different biologically relevant products. For example, α- and γ-tocopheryl acetates, precursors of α- and γ-tocopherol, which are main components of vitamin E, were synthesized by asymmetric hydrogenation of their triene precursors, employing ligand **L54bb** (Scheme 2.31).[172] More recently, Schmalz and co-workers have reported another route for the synthesis of α-tocopherol, involving the hydrogenation of a monoalkene spiro-cyclobutanol intermediate with the same ligand, that occurred with a dr ≥ 97:3.[173] Another related long-chain triene of natural importance, hexahydrofarnesol, was

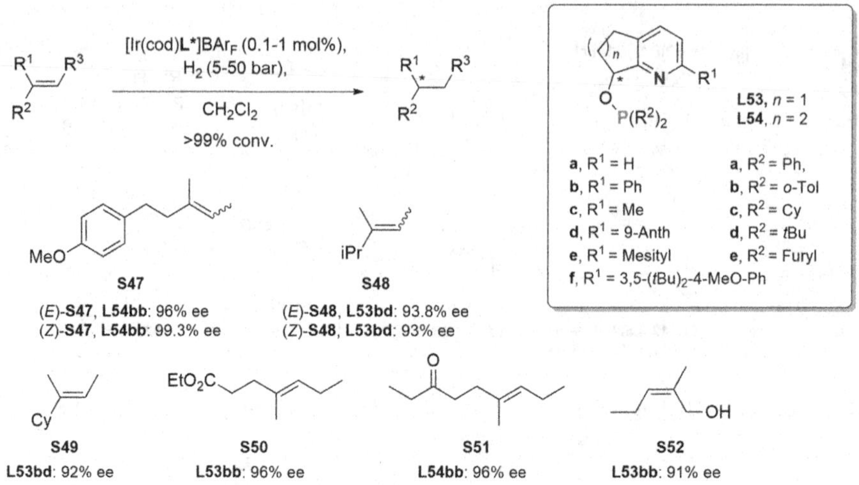

SCHEME 2.30   Hydrogenations of trialkyl-substituted olefins with **L53–L54**.

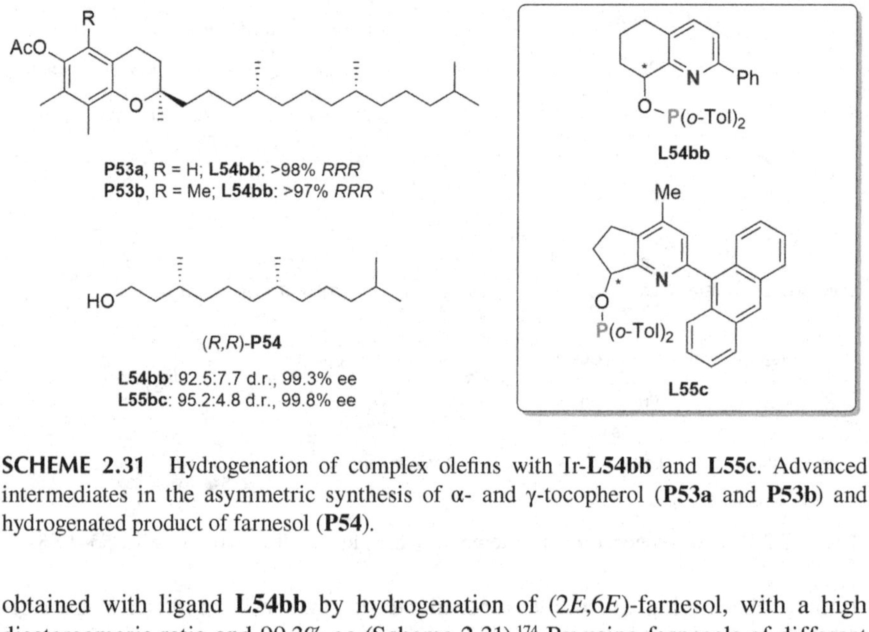

SCHEME 2.31   Hydrogenation of complex olefins with Ir-**L54bb** and **L55c**. Advanced intermediates in the asymmetric synthesis of α- and γ-tocopherol (**P53a** and **P53b**) and hydrogenated product of farnesol (**P54**).

obtained with ligand **L54bb** by hydrogenation of (2E,6E)-farnesol, with a high diastereomeric ratio and 99.3% ee (Scheme 2.31).[174] By using farnesols of different E/Z configurations at the double bonds, all four diastereomers of hexahydrofarnesol could be obtained at very high enantiopurities. Disubstituted pyridine ligand **L55c** increased the diastereoselectivity and enantioselectivity of this process up to 95.2:4.8 dr and 99.8% ee, respectively (Scheme 2.31).[165] Ligands of type **L53–L54** have also allowed the synthesis of other important natural products, such as macrocidin A and long-chain polydeoxypropionates.[175,176]

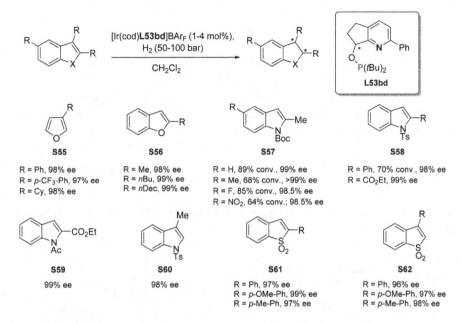

SCHEME 2.32 Selected results of the hydrogenation of furans, benzofurans, indoles, and benzo[*b*]thiophene 1,1-dioxides with **L53bd**. Unless otherwise specified, conversion is higher than 90%.

Pfaltz extended the substrate range of these ligands to include different mono-substituted and disubstituted furans and benzofurans,[177] as well as other hetero-cycles, such as 2- and 3-substituted indoles or benzo[*b*]thiophene 1,1-dioxides (Scheme 2.32).[178,179] Overall, **L53bd** was the best ligand with moderate to excellent conversions and ee values up to 99%. More recently, ligands of type **L55** demonstrated very high asymmetric induction in the hydrogenation of a dihydroquinoline core of agrochemical importance.[180]

Phosphinite-pyridines of this type were also used in the hydrogenation of alkenyl-boronic esters.[138] Whereas related phosphine-imidazole ligands were good for terminal boronic esters, **L53bd** was better for the trisubstituted substrates, providing very high enantioselectivities. More recently, this family of ligands **L53** has been applied to the hydrogenation of vinylsilanes.[134] Some ligands provided higher enantioselectivities than the related PHOX family, with ee values up to >99%.

2.3.4.3 Phosphite-Pyridine Ligands

Diéguez and Pàmies replaced the phosphinite moiety of ligand **L52** with different biaryl phosphite groups, generating the library of ligands **L57–L59** (Figure 2.17).[181] Their catalytic system proved to be capable of being fine-tuned to suit different reactions and substrates. Many of these ligands were screened in Pd-catalyzed allylic substitution reactions involving different allylic substrates and carbon-, oxygen-, and nitrogen-based nucleophiles, leading to high enantioselectivities. The allylic

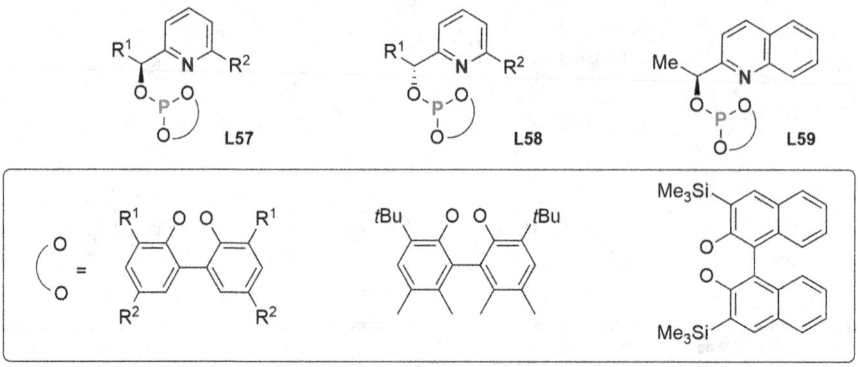

FIGURE 2.17 Phosphite-pyridine ligands **L57–L59,** developed by the Diéguez and Pàmies group.

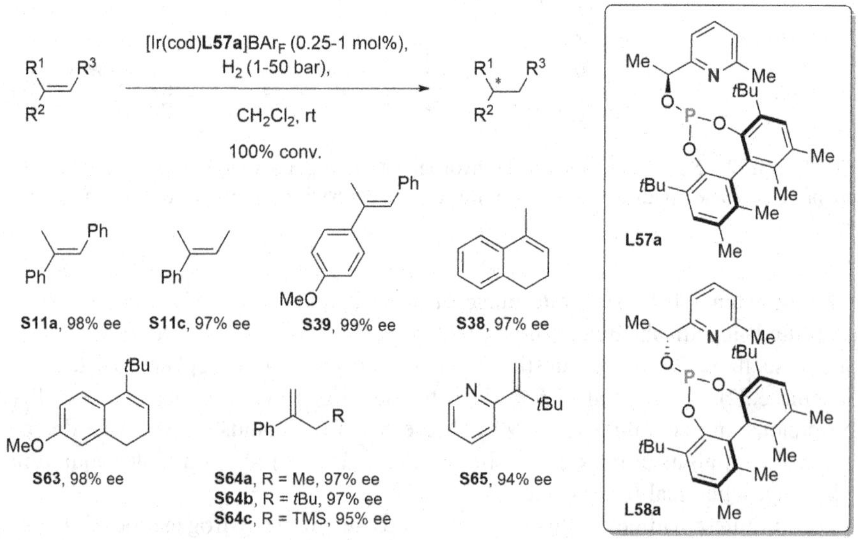

SCHEME 2.33 Highlighted results of **L57a** in the hydrogenation of 1,1-disubstituted and trisubstituted olefins.

substitution of diphenylallyl acetate (**S2**) and cyclohexenyl acetate with different carbon nucleophiles led to ee values up to 98% and 93%, respectively. This catalytic system was also screened for the asymmetric hydrogenation of a large number of diverse 1,1-disubstituted and trisubstituted olefins, providing very high enantioselectivities (Scheme 2.33).[182]

The substrates were either minimally functionalized or α-substituted by a polar group. For an important number of substrates, ligand **L57a** proved to be the most effective ligand. Moreover, it was demonstrated that the sense of enantioselectivity is controlled by the configuration of the stereogenic center next to the phosphite moiety. Thus, the use of ligand **L58a** allowed the generation of the opposite enantiomer, with similar optical purity from most substrates.

2.3.4.4 BoQPhos and Aminophosphine Pyridine Ligands

In 2014, Qu and co-workers described BoQPhos (Bo Qu phosphine) ligands (**L60**, Figure 2.18), structurally analogous to the phosphine-based ligands POP, BI-DIME, and BIBOP.[183] These could readily be prepared by addition of a pyridyl moiety to *P*-stereogenic intermediates.

Methoxy-substituted **L60a** proved to be among the best ligands for the Ir-catalyzed hydrogenation of pyridinium salts.[184,185] This catalytic system allowed the preparation of piperidines bearing 2-alkyl and 2-aryl substituents of different nature, with enantioselectivities up to 86% and 98%, respectively (Scheme 2.34).

More recently, Fan *et al.* reported the preparation of aminophosphine-pyridine ligands **L61** (Figure 2.18).[186] These ligands were screened for the hydrogenation of benzazepines and benzodiazepines. Isopropyl-substituted ligand **L61a** proved to be the most active and enantioselective ligand, showing good diastereomeric ratios, with ee values of up to 99% for both types of substrate (Scheme 2.35).

2.3.4.5 Axially Chiral *P,N*-Pyridino Ligands

The advances in asymmetric catalysis, achieved by atropoisomeric BINAP (2,2'-Bis(diphenylphosphino)-1,1'-binaphthyl) ligand and ferrocene-based *P,N*-ligands,[5] prompted the synthesis of the first axially chiral *P,N*-ligand **L62** namely QUINAP

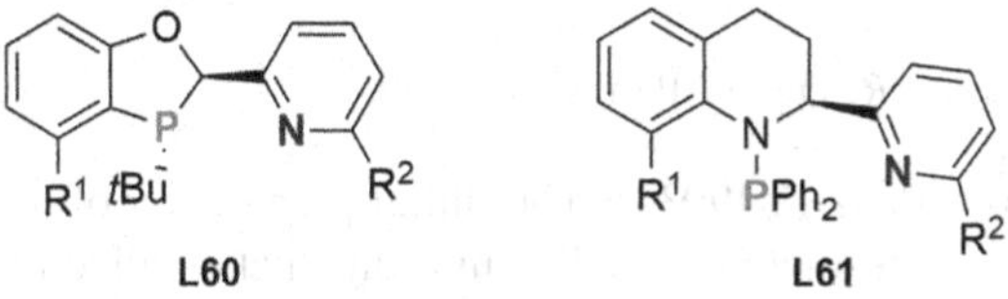

FIGURE 2.18 BoQPhos ligands (**L60**) and aminophosphine pyridine ligands (**L61**).

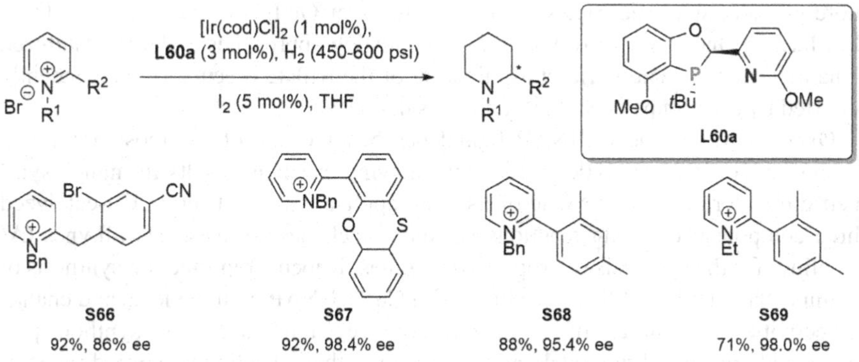

SCHEME 2.34 Highlighted results of the hydrogenation of pyridinium salts with BoQPhos **L60a**.

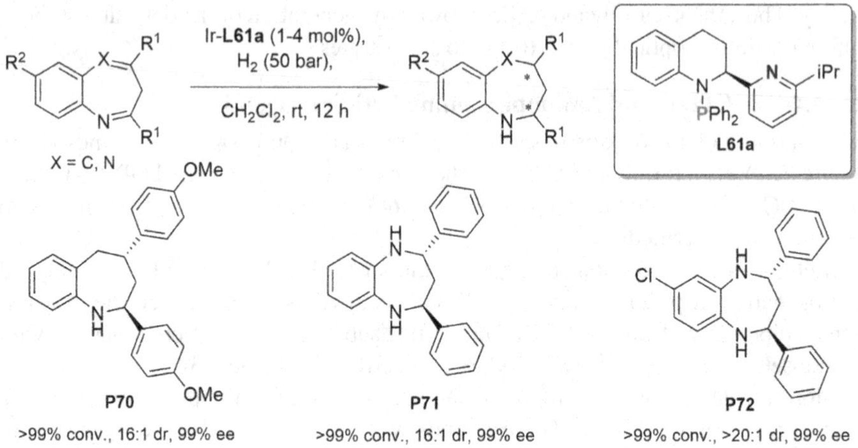

P70

>99% conv., 16:1 dr, 99% ee

P71

>99% conv., 16:1 dr, 99% ee

P72

>99% conv., >20:1 dr, 99% ee

SCHEME 2.35 Dihydrogenation of benzazepines and benzodiazepines with **L61a**.

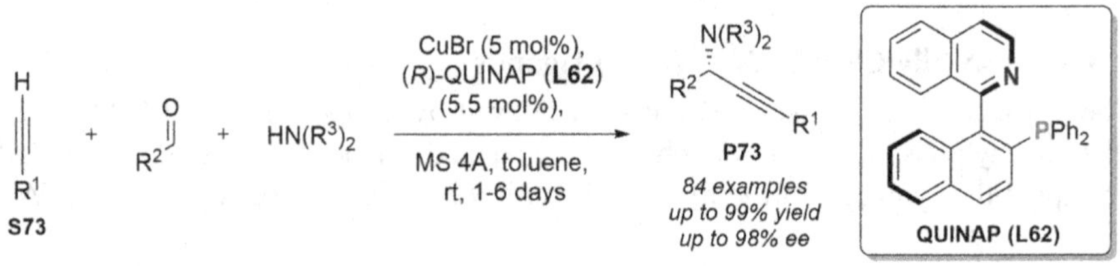

SCHEME 2.36 Use of (*R*)-QUINAP in A^3 coupling.

by Brown and co-workers in 1993.[187] The initial preparation of the ligand involved a final diastereomeric resolution of palladium salts that entailed two disadvantages: the need for a stochiometric quantity of palladium complex and the need for the introduction of the phosphine prior to the resolution step, which limited the route to ligand diversity. For this reason, over the years, many research groups have developed more efficient synthetic routes for the synthesis of QUINAP that overcome these drawbacks.[7] One of the most recent and interesting approaches is the Pd-catalyzed dynamic kinetic asymmetric phosphination of the triflate precursor, independently reported by the groups of Stoltz[188] and Lassaletta.[189]

Over the years, the QUINAP ligand has become one of the most prominent atropoisomeric *P,N*-catalytic systems by showing excellent results in many asymmetric transformations.[4,7,148] One of its main applications has been in Cu-catalyzed three-component coupling reactions between aldehydes, amines, and alkynes (A^3 coupling) for the synthesis of propargylic amines. Knochel reported the synthesis of an important number of these products with Cu/QUINAP in high yields and enantioselectivities (Scheme 2.36).[190–193] The use of three simple accessible synthons permits a wide range of functionalizations in each of them, leading to a broad range of propargylic amines, with different functional groups in either the α-amino carbon, the nitrogen atom, or the alkyne terminal.

In 2003, Schreiber reported the Ag-catalyzed [3+2] cycloaddition reaction between azomethine ylides and *tert*-butyl acrylate for the synthesis of pyrrolidines.[194] This transformation can generate up to four stereocenters in one step, with an endo:exo ratio of >20:1 and high enantioselectivities. In 2013, Reisman expanded this methodology to include the synthesis of pyrrolizidines by adding five equivalents of another dipolarophile and one equivalent of cinnamaldehyde after the first cycloaddition (Scheme 2.37).[195] Impressively, this reaction can generate up to six stereogenic centers in one step, with excellent enantioselectivity.

Initial studies by Brown in Rh-catalyzed hydroboration of vinylarenes showed the potential of QUINAP for this process, with ee values up to 94%.[196] In addition to the hydrolysis of the boronates for the preparation of primary alcohols, Brown also reported the amination reaction, expanding the method to the synthesis of different primary amines, with good to very high enantioselectivities (77–98% ee).[197,198] Independently, Fernandez and Brown improved the results and extended the scope of QUINAP for the hydroboration of vinylarenes.[199,200] QUINAP-derived ligands, with a quinazoline core named Quinazolinap (**L63**, Figure 2.19), were also used with an important range of vinylarenes, providing slightly higher enantioselectivities than QUINAP.[201,202] Interestingly, hydroboration of indene (**P78**) proceeded at very high enantioselectivities. The best results involving both ligands are summarized in Scheme 2.38.

Morken developed the Rh-catalyzed diboration of alkenes with dicatechol diboron, for which QUINAP performed better than BINAP.[203–205] An important number of olefins were dihydroxylated, to give the *syn*-diol product with high diastereo- and enantioselectivity, but at only low to good yields (Scheme 2.39). Unlike QUINAP-catalyzed hydroboration, the diboration does not need the presence of an aromatic

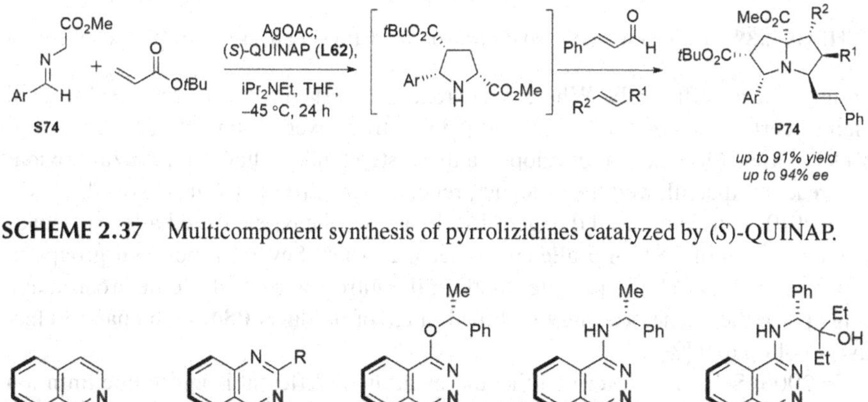

SCHEME 2.37 Multicomponent synthesis of pyrrolizidines catalyzed by (*S*)-QUINAP.

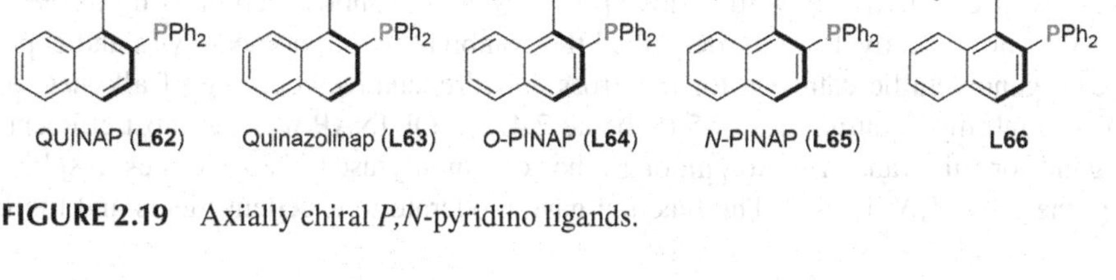

FIGURE 2.19 Axially chiral *P,N*-pyridino ligands.

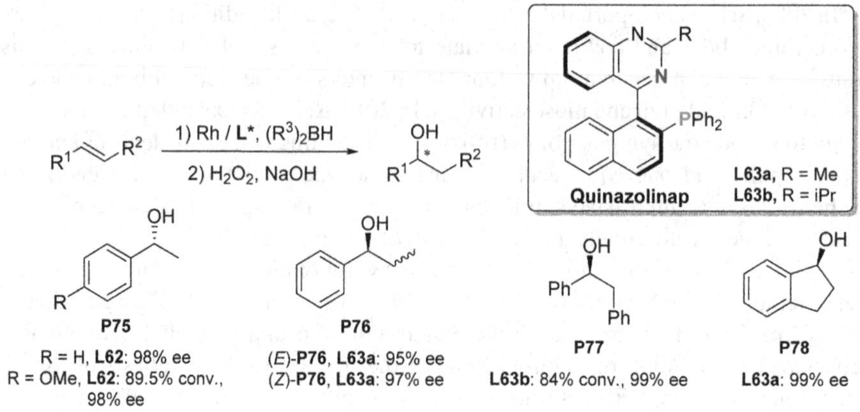

SCHEME 2.38 Selected results of the hydroboration of olefins with QUINAP and Quinazolinap. Unless otherwise specified, conversion is higher than 90%.

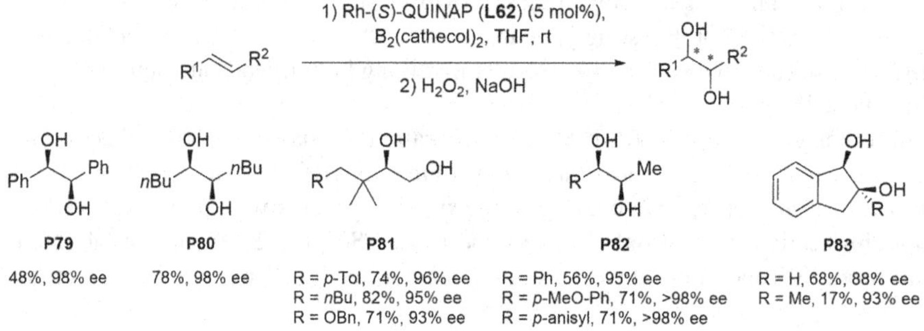

SCHEME 2.39 Rhodium-catalyzed diboration-oxidation of alkenes with (*S*)-QUINAP.

group to occur efficiently. Whereas the reaction worked for most *trans*-substituted olefins, certain mono- and *cis*-substituted substrates were obtained in lower enantioselectivities. Morken also developed a three-step tandem diboration/Suzuki/oxidation reaction that allowed the efficient preparation of different 1-aryl-2-ols.[204]

In 2010, Murakami used the QUINAP ligand for the Ni-catalyzed allene cycloaddition reaction of **S84** with allenes (Scheme 2.40).[206] Several functional groups of different nature could be tolerated both in the nitrogen atom of the heteroaromatic ring and in the allene, leading to a broad range of products **P84**, with enantioselectivities of up to 97% ee.

In 2006, Schreiber described the alkynylation of different isoquinoline iminium salts with Cu/QUINAP, with yields up to 95% and enantioselectivities up to 99% ee.[207] More recently, Li *et al.* described the combination of photocatalysis and copper organometallic catalysis for the cross-dehydrogenative-coupling of alkynes to *N*-aryl tetrahydroquinolines **S85** (Scheme 2.41).[208] QUINAP was the most efficient ligand for this transformation among other commonly-used bisoxazolines, bisphosphines, and *P,N*-ligands. The reaction gave moderate to excellent yields and high

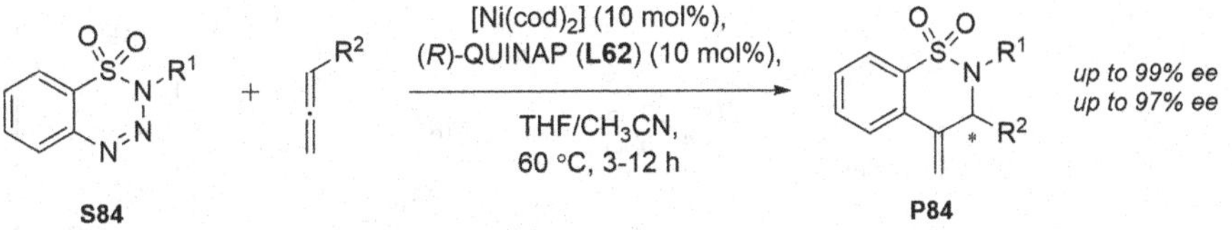

SCHEME 2.40   Application of (*R*)-QUINAP to the annulation of **S84** with allenes.

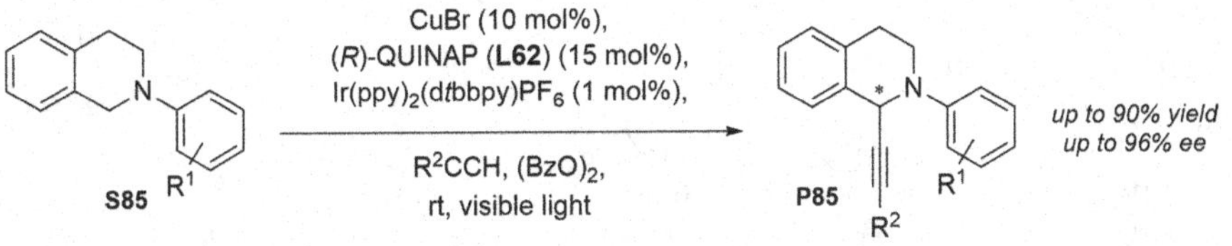

SCHEME 2.41   Cooperative catalysis for cross-dehydrogenative alkynylation with QUINAP.

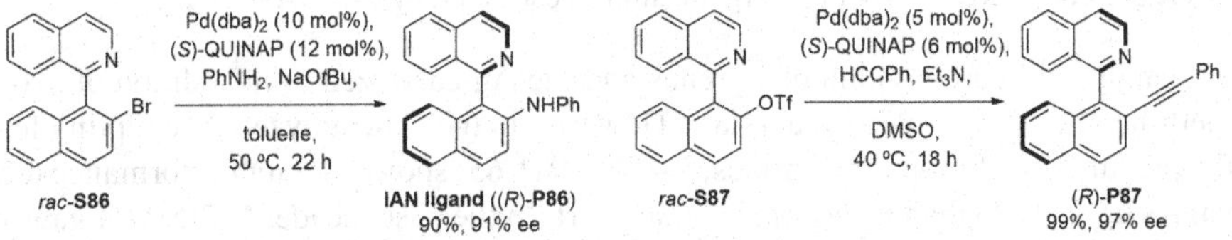

SCHEME 2.42   Pd-catalyzed dynamic kinetic asymmetric Buchwald-Hartwig amination and alkynylation with QUINAP.

ee values for a range of products, with different functional groups in the *N*-aryl and alkynyl moieties.

In 2016, Lassaletta and co-workers described the Pd-catalyzed dynamic kinetic asymmetric Buchwald–Hartwig amination and alkynylation of atropoisomeric heterobiaryls.[209,210] Ligand screening showed that (*S*)-QUINAP was the best choice among many other ligands for both reactions. Starting from racemic **S86**, the functionalized enantiopure biaryls were obtained at high yields and excellent enantioselectivities. A broad range of amino- and alkynyl-heterobiaryls could be synthesized, since both transformations tolerated the presence of several modifications in the biaryl backbone and the use of different aryl and alkyl amines or alkynes. The asymmetric amination was used for the synthesis of known ligands, such as IAN (2,2'-Bis(diphenylphosphino)-1,1'-binaphthyl), while the alkynylation process was used to synthesize new ligands, such as **P87** (Scheme 2.42).

At that time, the main disadvantage of QUINAP and Quinazolinap was the need for a stoichiometric amount of chiral palladium complex for its resolution. To overcome this drawback, Carreira *et al.* developed the analogous PINAP ligands (**L64–L66**; Figure 2.19), which include a chiral motif that allowed the easy separation of diastereomers by chromatography or crystallization without using chiral Pd salts.[211] The *O*-PINAP ligand (**L64**) gave ee values similar to QUINAP in both the

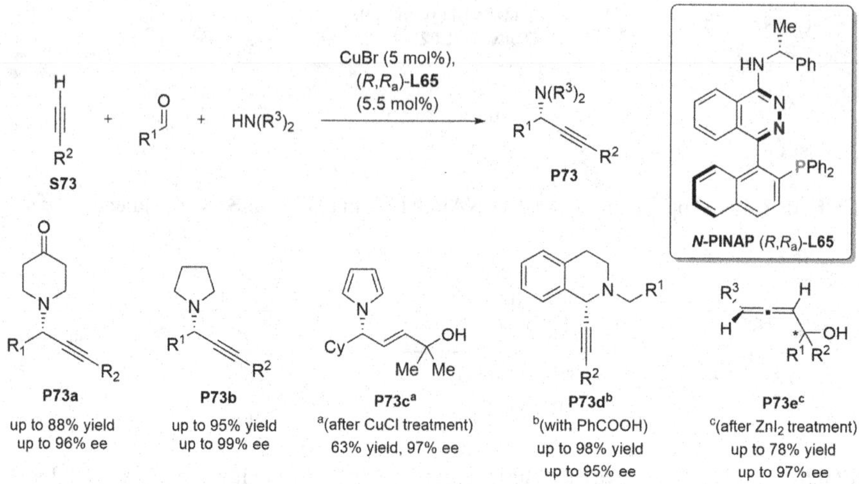

SCHEME 2.43 Synthesis of propargylic amines catalyzed by *N*-PINAP.

Rh-catalyzed hydroboration of styrenes and the Ag-catalyzed cycloaddition of azomethine ylides and *tert*-butyl acrylate. However, in the Cu-catalyzed A[3] coupling for the preparation of propargyl amines, *N*-PINAP (**L65**) showed better performance (ee values up to 99%) than its original counterpart. In the past decade, *N*-PINAP ligands have shown excellent results in different A[3] couplings and other alkynyl functionalization reactions.[148]

Carreira and co-workers also reported the preparation of propargylic amines bearing the more labile group 4-piperidone, in high yields and with enantioselectivities up to 96% (Scheme 2.43).[212] Ma and co-workers reported the A[3] coupling of pyrrolidine and different aromatic aldehydes, yielding the corresponding propargylic amines at high enantioselectivities, ranging from 91% to 99% ee (Scheme 2.43).[213] Ma has also developed a procedure for the synthesis of (*E*)-*N*-allyl pyrroles involving a Cu/PINAP A[3] coupling and subsequent [1,5]-hydride transfer catalyzed by CuCl (Scheme 2.43).[214] More recently, Oguri has applied the A[3] coupling, using Cu/PINAP as the catalytic system, in the synthesis of anti-malarial 6-aza-artemisinin compounds.[215]

Tetrahydroisoquinoline was also used as an amine source in A[3] couplings.[216] Unexpectedly, the α-alkynylated tetrahydroisoquinoline (**P73d**) was obtained instead of the corresponding propargylic amine (Scheme 2.43). The reaction took place with very good yields (up to 98%) and excellent enantioselectivities (up to 95% ee) for an important range of functionalized aromatic aldehydes and alkynes. This asymmetric redox-A[3] coupling strategy has been successfully used in the total synthesis of several natural products.[217,218,219] The Cu/PINAP three-component coupling was also useful for the synthesis of allenols by formation of the corresponding propargylic amine and posterior Zn-mediated deamination (Scheme 2.43).[220]

Carreira used the Cu/PINAP catalytic system in the alkynylation of Meldrum's acid derivatives.[221] However, ligand **L66** was the most efficient catalyst for this

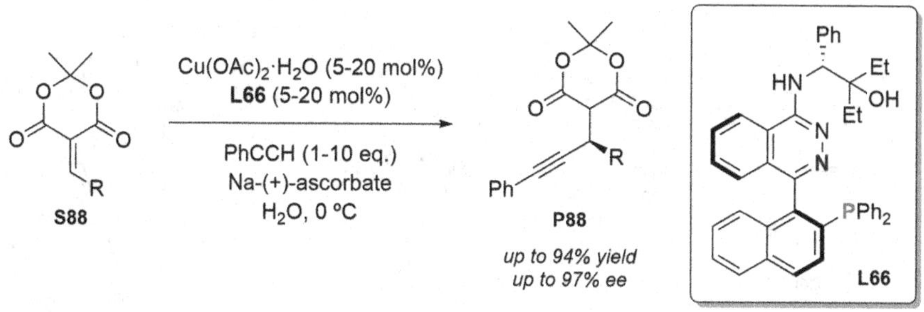

SCHEME 2.44 Copper-catalyzed alkynylation of Meldrum's acid derivatives with **L66**.

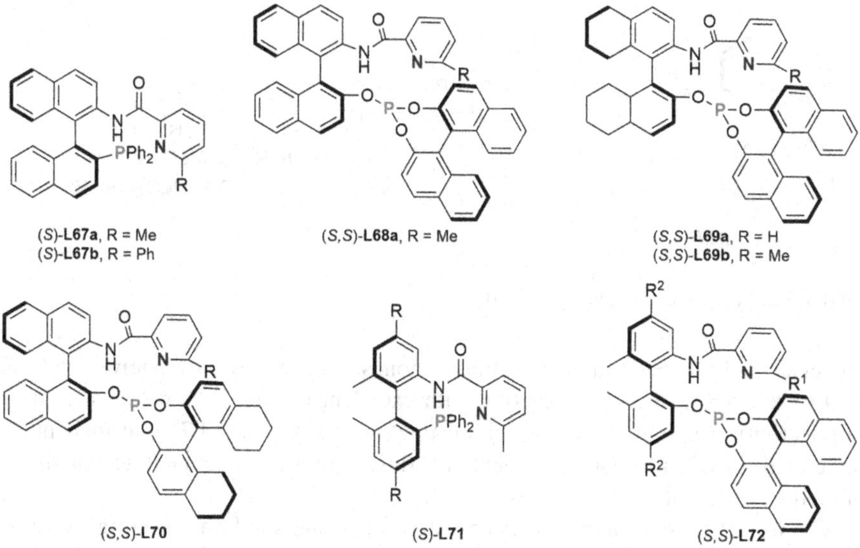

FIGURE 2.20 NOBIN-derived phosphine-pyridine ligands.

transformation, yielding products **P88** with high yields and ee values up to 97% (Scheme 2.44).

2.3.4.6 Phosphino-Amidopyridine Ligands

Back in 1999, Zhang reported **L67**, a new type of axially chiral bidentate phosphino-pyridine ligand, with a rigid amide linker (Figure 2.20).[222] Their synthesis could easily be achieved in four steps from commercial NOBIN (2-amino-2'-hydroxy-1,1'-binaphthyl). These ligands were applied in the Cu-catalyzed 1,4-addition of diethylzinc to different linear enones, achieving high yields and ee values up to 98%; for instance, the addition to 2-cyclohexen-1-one occurred at 92% ee.

Later on, Hu and co-workers described other ligands of this sort (**L68–L72**, Figure 2.20), containing modifications at the axially chiral backbone and chiral biaryl phosphites at the phosphorus atom. **L68** provided ee values up to 97% in the

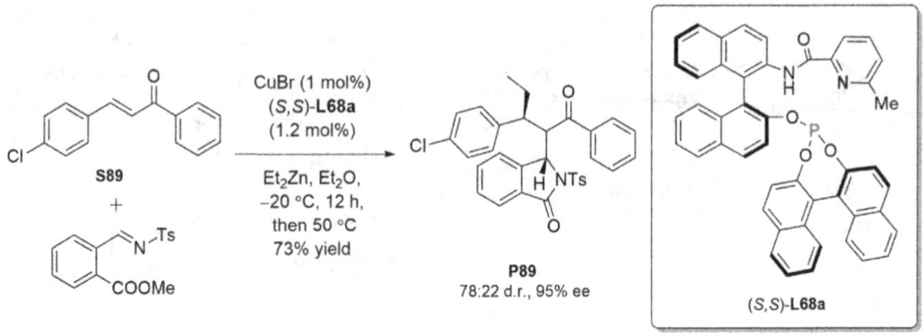
CuBr (1 mol%)
(S,S)-L68a
(1.2 mol%)
Et2Zn, Et2O,
–20 °C, 12 h,
then 50 °C
73% yield
S89
COOMe
Ts
P89
78:22 d.r., 95% ee
NTs
Me
(S,S)-L68a

SCHEME 2.45 Example of tandem conjugate addition-Mannich reaction.

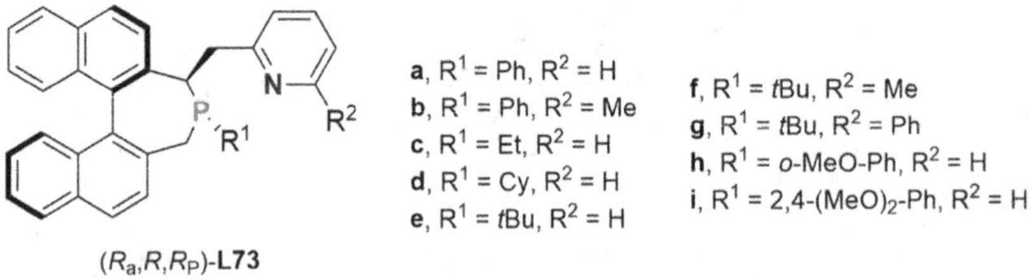
a, R1 = Ph, R2 = H
b, R1 = Ph, R2 = Me
c, R1 = Et, R2 = H
d, R1 = Cy, R2 = H
e, R1 = tBu, R2 = H
f, R1 = tBu, R2 = Me
g, R1 = tBu, R2 = Ph
h, R1 = o-MeO-Ph, R2 = H
i, R1 = 2,4-(MeO)2-Ph, R2 = H
(Ra,R,RP)-L73

FIGURE 2.21 Binepine-pyridyl ligands.

conjugate addition of diethylzinc to linear enones.[223] H_8-NOBIN-derivatives **L69** and **L70** were used in the same reaction with excellent results.[224] However, the enantioselectivity was only moderate with cyclic enones. Ligands **L71** and their phosphite analogs **L72** also gave excellent enantioselectivities in the conjugate addition of diethylzinc to enones (ee values up to 96%).[225–227]

The use of these ligands has been extended to the synthesis of chiral nitrogenated heterocycles. Ligands (*R*)-**L67a** and (*S,S*)-**L68a** were used in dual conjugate addition reactions.[228,229] Ligand (*S,S*)-**L68a** was used in different tandem conjugate addition-Mannich reactions for the synthesis of chiral isoindolinone derivatives (Scheme 2.45).[230] In general, these reactions proceeded at high yields, good diastereomeric ratios, and high enantioselectivities > 90% ee.

Ligand (*R*)-**L67b** (R = Ph) provided high yields and enantioselectivities (up to 98% ee) in the Cu-catalyzed addition of diethylzinc to different aldehydes.[231]

2.3.4.7 Binepine–Pyridyl Ligands

Mazet *et al.* have developed a series of ligands **L73** incorporating a 2-pyridyl moiety to binepine scaffolds (Figure 2.21).[232] This family was screened in the Pd-catalyzed intramolecular α-arylation of aldehydes, showing better performance than QUINAP, PINAP, or PHOX. The electron-richer ligand **L73i** showed the best results and was used with a range of α-branched aldehydes, mostly with good yields and excellent enantioselectivities (Scheme 2.46). This catalytic system also showed better performance than with other successful *P,P*- and *P,N*-ligands in the Heck reaction between

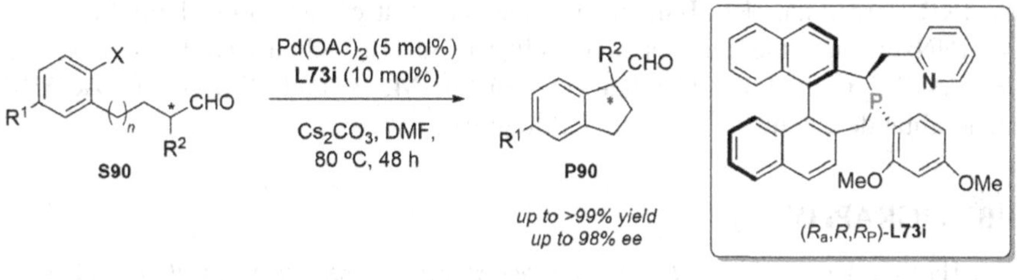

SCHEME 2.46   Results of the α-arylation of aldehydes with **L73i**.

2-substituted furans and aryl triflates for the synthesis of functionalized 2,5-dihydro-furans, with the ligand (R_a, R, R_P)-**L73e**, providing moderate yields, and enantioselectivities up to 94% ee for a wide range of substrates.[233]

2.4   CONCLUSIONS

Chiral *P,N*-ligands were initially developed to mimic the coordination pattern of the Crabtree catalyst. Whereas phosphino-oxazoline ligands are arguably the most successful and the best-known *P,N*-ligands, phosphino-pyridine ligands are the ones which more accurately match the original Crabtree system. This may explain the enormous success of this type of ligands in the Ir-catalyzed asymmetric hydrogenation of unfunctionalized alkenes. Despite this, a great effort has also been put into replacing the pyridine fragment with other nitrogenated heterocycles, such as oxazole, thiazole, or imidazole, which allows modification of the electronics of the metal, and thus the acidity of the corresponding iridium hydrides. The *N*-donor group has also been replaced by simple amines or imines, which permit other applications, such as allylic alkylation, aldol, or Mannich reactions to name but a few.

Overall, hundreds of ligands have been synthesized, tested, and successfully used in asymmetric catalysis. The general trend is to develop a scaffold with two or even three diversity points, that allow the preparation of a family of ligands, thus enabling ligand optimization. The chirality is generally located at the backbone. A common trend in ligand optimization is to change the *P*-donor fragment. This is, however, a weak link in the whole optimization process, since the number of commercially available dialkyl and diaryl phosphines is quite limited. A possible solution is to introduce chirality at this position with a phosphite with axial chirality or a *P*-stereogenic phosphine, which allows further improvement of the enantiomeric excess. In this regard, it would also be desirable to develop methods to increase diversity at the *N*-heterocyclic component. Most often, this is limited to different flat aromatics enclosing the *N*-donor site, and it is particularly likely that the introduction of chirality at this point would provide further improvements in *P,N*-ligands in the future.

2.5   ACKNOWLEDGMENTS

We acknowledge financial support from FEDER/Ministerio de Ciencia, Innovación y Universidades (MICINN)-Agencia Estatal de Investigación (CTQ2017-87840-P), and IRB Barcelona. IRB Barcelona is the recipient of institutional funding from MICINN through the Centres of Excellence Severo Ochoa award and from the CERCA Program of the Catalan government. M. B. and P. R. thank AGAUR (Generalitat de Catalunya) for PhD fellowships.

BIBLIOGRAPHY

1. Hartwig, J. *Organotransition Metal Chemistry: From Bonding to Catalysis*. University Science Books: Sausalito, 2010.
2. Chelucci, G.; Orrù, G.; Pinna, G. A. *Tetrahedron* 2003, *59* (48), 9471–9515.
3. Kamer, P. C. J.; van Leeuwen, P. W. N. M. *Phosphorus (III) Ligands in Homogeneous Catalysis*. John Wiley & Sons, Ltd.: Hoboken, NJ, 2012.
4. Carroll, M. P.; Guiry, P. J. *Chem. Soc. Rev.* 2014, *43* (3), 819–833.
5. Fernández, E.; Guiry, P. J.; Connole, K. P. T.; Brown, J. M. *J. Org. Chem.* 2014, *79* (12), 5391–5400.
6. Li, W.; Zhang, J. *Chem. Soc. Rev.* 2016, *45* (6), 1657–1677.
7. Rokade, B. V.; Guiry, P. J. *ACS Catal.* 2018, *8* (1), 624–643.
8. Crabtree, R. H.; Morris, G. E. *J. Organomet. Chem.* 1977, *135* (3), 395–403.
9. Crabtree, R. *Acc. Chem. Res.* 1979, *12* (9), 331–337.
10. Kagan, H. B.; Dang, T. P. *J. Am. Chem. Soc.* 1972, *94* (18), 6429–6433.
11. Pfaltz, A.; Drury, W. J. *Proc. Natl. Acad. Sci. U. S. A.* 2004, *101* (16), 5723–5726.
12. Morimoto, T.; Tachibana, K.; Achiwa, K. *Synlett* 1997.
13. Akermark, B.; Krakenberger, B.; Hansson, S.; Vitagliano, A. *Organometallics* 1987, *6* (3), 620–628.
14. Espinet, P.; Soulantica, K. Coord. *Chem. Rev.* 1999, 193–195, 499–556.
15. Hayashi, T.; Yamamoto, K.; Kumada, M. *Tetrahedron Lett.* 1974, *15* (49–50), 4405–4408.
16. Hayashi, T.; Tajika, M.; Tamao, K.; Kumada, M. *J. Am. Chem. Soc.* 1976, *98* (12), 3718–3719.
17. Jedlicka, B.; Kratky, C.; Weissensteiner, W.; Widhalm, M. *J. Chem. Soc. Chem. Commun.* 1993, 1329–1330.
18. Mosquera, Á.; Pena, M. A.; Sestelo, J. P.; Sarandeses, L. A. *Eur. J. Org. Chem.* 2013, *13*, 2555–2562.
19. Wang, M. C.; Liu, L. T.; Hua, Y. Z.; Zhang, J. S.; Shi, Y. Y.; Wang, D. K. *Tetrahedron: Asymmetry* 2005, *16* (15), 2531–2534.
20. Jin, M. J.; Takale, V. B.; Sarkar, M. S.; Kim, Y. M. *Chem. Commun.* 2006, *6* (6), 663–664.
21. Ahn, K. H.; Cho, C. W.; Park, J.; Sunwoo, L. *Tetrahedron: Asymmetry* 1997, *8* (8), 1179–1185.
22. Jia, H.; Liu, H.; Guo, Z.; Huang, J.; Guo, H. *Org. Lett.* 2017, *19* (19), 5236–5239.
23. Hayashi, T.; Fukushima, M.; Konishi, M.; Kumada, M. *Tetrahedron Lett.* 1980, *21* (1), 79–82.
24. von Matt, P.; Pfaltz, A. *Angew. Chem., Int. Ed.* 1993, *32* (4), 566–568.
25. Sprinz, J.; Helmchen, G. *Tetrahedron Lett.* 1993, *34* (11), 1769–1772.
26. Jumnah, R.; Williams, J. M. J.; Williams, A. C. *Tetrahedron Lett.* 1993, *34* (41), 6619–6622.
27. Anderson, J. C.; Cubbon, R. J.; Harling, J. D. *Tetrahedron: Asymmetry* 1999, *10* (15), 2829–2832.

28. Anderson, J. C.; Cubbon, R. J.; Harling, J. D. *Tetrahedron: Asymmetry* 2001, *12* (6), 923–935.
29. Vyskočil, Š.; Smrčina, M.; Hanuš, V.; Polášek, M.; Kočovský, P. *J. Org. Chem.* 1998, *63* (22), 7738–7748.
30. Hayashi, T. *J. Synth. Org. Chem. Japan* 1994, *52* (11), 900–911.
31. Fairlamb, I. J. S.; Lloyd-Jones, G. C.; Vyskočil, Š.; Kočovský, P. *Chem. Eur. J.* 2002, *8* (19), 4443–4453.
32. Vyskocil, S.; Smrcina, M.; Kocovsky, P. *Tetrahedron Lett.* 1998, *39*, 9289.
33. Kocovsky, P.; Vyskocil, S.; Cisarova, I.; Sejbal, J.; Tislerova, I.; Smrcina, M.; Lloyd-Jones, G. C.; Stephen, S. C.; Butts, C. P.; Murray, M.; Langer, V. *J. Am. Chem. Soc.* 1999, *121*, 7714.
34. Yin, J.; Buchwald, S. L. *J. Am. Chem. Soc.* 2000, *122* (48), 12051–12052.
35. Shen, X.; Jones, G. O.; Watson, D. A.; Bhayana, B.; Buchwald, S. L. *J. Am. Chem. Soc.* 2010, *132* (32), 11278–11287.
36. Wang, Y.; Guo, H.; Ding, K. *Tetrahedron: Asymmetry* 2000, *11* (20), 4153–4162.
37. Wang, Y.; Li, X.; Sun, J.; Ding, K. *Organometallics* 2003, *22* (9), 1856–1862.
38. Xie, J. B.; Xie, J. H.; Liu, X. Y.; Kong, W. L.; Li, S.; Zhou, Q. L. *J. Am. Chem. Soc.* 2010, *132* (13), 4538–4539.
39. Zhu, S. F.; Yu, Y. B.; Li, S.; Wang, L. X.; Zhou, Q. L. *Angew. Chem., Int. Ed.* 2012, *51* (35), 8872–8875.
40. Yu, Y. B.; Cheng, L.; Li, Y. P.; Fu, Y.; Zhu, S. F.; Zhou, Q. L. *Chem. Commun.* 2016, *52* (26), 4812–4815.
41. Li, S.; Zhang, J.; Li, H.; Feng, L.; Jiao, P. *J. Org. Chem.* 2019, *84* (15), 9460–9473.
42. Sladojevich, F.; Trabocchi, A.; Guarna, A.; Dixon, D. J. *J. Am. Chem. Soc.* 2011, *133* (6), 1710–1713.
43. Franchino, A.; Jakubec, P.; Dixon, D. J. *Org. Biomol. Chem.* 2015, *14* (1), 93–96.
44. Zhang, X.; Wang, X.; Gao, Y.; Xu, X. *Chem. Commun.* 2017, *53* (16), 2427–2430.
45. Shao, P. L.; Liao, J. Y.; Ho, Y. A.; Zhao, Y. *Angew. Chem., Int. Ed.* 2014, *53* (21), 5435–5439.
46. De La Campa, R.; Ortín, I.; Dixon, D. J. *Angew. Chem., Int. Ed.* 2015, *54* (16), 4895–4898.
47. De La Campa, R.; Manzano, R.; Calleja, P.; Ellis, S. R.; Dixon, D. J. *Org. Lett.* 2018, *20* (19), 6033–6036.
48. Ortín, I.; Dixon, D. J. *Angew. Chem., Int. Ed.* 2014, *53* (13), 3462–3465.
49. De La Campa, R.; Gammack Yamagata, A. D.; Ortín, I.; Franchino, A.; Thompson, A. L.; Odell, B.; Dixon, D. J. *Chem. Commun.* 2016, *52* (70), 10632–10635.
50. Franchino, A.; Chapman, J.; Funes-Ardoiz, I.; Paton, R. S.; Dixon, D. J. *Chem. Eur. J.* 2018, *24* (67), 17660–17664.
51. Liao, J. Y.; Shao, P. L.; Zhao, Y. *J. Am. Chem. Soc.* 2015, *137* (2), 628–631.
52. Zheng, S. C.; Wang, Q.; Zhu, J. *Angew. Chem., Int. Ed.* 2019, *58* (5), 1494–1498.
53. Cheng, H.; Zhang, R.; Yang, S.; Wang, M.; Zeng, X.; Xie, L.; Xie, C.; Wu, J.; Zhong, G. *Adv. Synth. Catal.* 2016, *358* (6), 970–976.
54. Peng, X. J.; Ho, Y. A.; Wang, Z. P.; Shao, P. L.; Zhao, Y.; He, Y. *Org. Chem. Front.* 2017, *4* (1), 81–85.
55. George, J.; Kim, H. Y.; Oh, K. *Org. Lett.* 2018, *20* (8), 2249–2252.
56. Wang, Z. P.; Xiang, S.; Shao, P. L.; He, Y. *J. Org. Chem.* 2018, *83* (18), 10995–11007.
57. Wang, Z. P.; Wu, Q.; Jiang, J.; Li, Z. R.; Peng, X. J.; Shao, P. L.; He, Y. *Org. Chem. Front.* 2018, *5* (1), 36–40.
58. Zhao, M. X.; Xiang, J.; Zhao, Z. Q.; Zhao, X. L.; Shi, M. *Org. Biomol. Chem.* 2020, *18* (8), 1637–1646.

59. Dong, X. Y.; Zhang, Y. F.; Ma, C. L.; Gu, Q. S.; Wang, F. L.; Li, Z. L.; Jiang, S. P.; Liu, X. Y. *Nat. Chem.* 2019, *11* (12), 1158–1166.

60. Zhang, Z. H.; Dong, X. Y.; Du, X. Y.; Gu, Q. S.; Li, Z. L.; Liu, X. Y. *Nat. Commun.* 2019, *10* (1), 1–10.

61. Hayashi, T.; Hayashi, C.; Uozumi, Y. *Tetrahedron: Asymmetry* 1995, *6* (10), 2503–2506.

62. Hu, X.; Dai, H.; Hu, X.; Chen, H.; Wang, J.; Bai, C.; Zheng, Z. *Tetrahedron: Asymmetry* 2002, *13* (15), 1687–1693.

63. Hu, X.; Bai, C.; Dai, H.; Chen, H.; Zheng, Z. *J. Mol. Catal. A Chem.* 2004, *218* (1), 107–112.

64. Hu, X.; Dai, H.; Bai, C.; Chen, H.; Zheng, Z. *Tetrahedron: Asymmetry* 2004, *15* (7), 1065–1068.

65. Yuan, H.; Zhou, Z.; Xiao, J.; Liang, L.; Dai, L. *Tetrahedron: Asymmetry* 2010, *21* (15), 1874–1884.

66. Dai, L.; Li, X.; Yuan, H.; Li, X.; Li, Z.; Xu, D.; Fei, F.; Liu, Y.; Zhang, J.; Zhou, Z. *Tetrahedron: Asymmetry* 2011, *22* (13), 1379–1389.

67. Noël, T.; Bert, K.; Van Der Eycken, E.; Van Der Eycken, J. *Eur. J. Org. Chem.* 2010, No. 21, 4056–4061.

68. Bert, K.; Noël, T.; Kimpe, W.; Goeman, J. L.; Van Der Eycken, J. *Org. Biomol. Chem.* 2012, *10* (42), 8539–8550.

69. T. Mino, W. Imiya, M. Y. *Synlett* 1997, 583.

70. Mino, Takashi, M. Shiotsuki, N. Yamamoto, T. Suenaga, M. Sakamoto, T. Fujita, M. Y. *J. Org. Chem.* 2001, *66* (5), 1795–1797.

71. Mino, T.; Ogawa, T.; Yamashita, M. *J. Organomet. Chem.* 2003, *665* (1–2), 122–126.

72. Fukuda, T.; Takehara, A.; Iwao, M. *Tetrahedron: Asymmetry* 2001, *12* (20), 2793–2799.

73. Jang, H. Y.; Seo, H.; Han, J. W.; Chung, Y. K. *Tetrahedron Lett.* 2000, *41* (26), 5083–5087.

74. Schenkel, L. B.; Ellman, J. A. *Org. Lett.* 2003, *5* (4), 545–548.

75. Schenkel, L. B.; Ellman, J. A. *J. Org. Chem.* 2004, 69 (6), 1800–1802.

76. Church, T. L.; Rasmussen, T.; Andersson, P. G. *Organometallics* 2010, *29* (24), 6769–6781.

77. Degrado, S. J.; Mizutani, H.; Hoveyda, A. H. *J. Am. Chem. Soc.* 2001, *123* (4), 755–756.

78. Hoveyda, A. *Handbook of Combinatorial Chemistry*; K. C. Nicolaou, R. Hanko, W. H., Ed.; Wiley-VCH: Weinheim, 2002; pp 991–1016.

79. Brown, M. K.; Hoveyda, A. H. *J. Am. Chem. Soc.* 2008, *130* (39), 12904–12906.

80. Mizutani, H.; Degrado, S. J.; Hoveyda, A. H. *J. Am. Chem. Soc.* 2002, *124* (5), 779–781.

81. Cesati, R. R.; De Armas, J.; Hoveyda, A. H. *J. Am. Chem. Soc.* 2004, *126* (1), 96–101.

82. Brown, M. K.; Degrado, S. J.; Hoveyda, A. H. *Angew. Chem., Int. Ed.* 2005, *44* (33), 5306–5310.

83. Luchaco-Cullis, C. A.; Hoveyda, A. H. *J. Am. Chem. Soc.* 2002, *124* (28), 8192–8193.

84. Weiss, M. E.; Carreira, E. M. *Angew. Chem., Int. Ed.* 2011, *50* (48), 11501–11505.

85. Wu, J.; Mampreian, D. M.; Hoveyda, A. H. *J. Am. Chem. Soc.* 2005, *127* (13), 4584–4585.

86. Zeng, X.; Gao, J. J.; Song, J. J.; Ma, S.; Desrosiers, J. N.; Mulder, J. A.; Rodriguez, S.; Herbage, M. A.; Haddad, N.; Qu, B.; Fandrick, K. R.; Grinberg, N.; Lee, H.; Wei, X.; Yee, N. K.; Senanayake, C. H. *Angew. Chem., Int. Ed.* 2014, *53* (45), 12153–12157.

87. Anderson, J. C.; Stepney, G. J.; Mills, M. R.; Horsfall, L. R.; Blake, A. J.; Lewis, W. *J. Org. Chem.* 2011, *76* (7), 1961–1971.

88. Degrado, S. J.; Mizutani, H.; Hoveyda, A. H. *J. Am. Chem. Soc.* 2002, *124* (45), 13362–13363.

89. Josephsohn, N. S.; Snapper, M. L.; Hoveyda, A. H. *J. Am. Chem. Soc.* 2003, *125* (14), 4018–4019.

90. Mandai, H.; Mandai, K.; Snapper, M. L.; Hoveyda, A. H. *J. Am. Chem. Soc.* 2008, *130* (52), 17961–17969.
91. Josephsohn, N. S.; Snapper, M. L.; Hoveyda, A. H. *J. Am. Chem. Soc.* 2004, *126* (12), 3734–3735.
92. Carswell, E. L.; Snapper, M. L.; Hoveyda, A. H. *Angew. Chem., Int. Ed.* 2006, *45* (43), 7230–7233.
93. Wieland, L. C.; Vieira, E. M.; Snapper, M. L.; Hoveyda, A. H. *J. Am. Chem. Soc.* 2009, *131* (2), 570–576.
94. Curti, C.; Battistini, L.; Ranieri, B.; Pelosi, G.; Rassu, G.; Casiraghi, G.; Zanardi, F. *J. Org. Chem.* 2011, *76* (7), 2248–2252.
95. Ranieri, B.; Curti, C.; Battistini, L.; Sartori, A.; Pinna, L.; Casiraghi, G.; Zanardi, F. *J. Org. Chem.* 2011, *76* (24), 10291–10298.
96. Wencel, J.; Rix, D.; Jennequin, T.; Labat, S.; Crévisy, C.; Mauduit, M. *Tetrahedron: Asymmetry* 2008, *19* (15), 1804–1809.
97. Moessner, C.; Bolm, C. *Angew. Chem., Int. Ed.* 2005, *44* (46), 7564–7567.
98. Lu, S. M.; Bolm, C. *Adv. Synth. Catal.* 2008, *350* (7–8), 1101–1105.
99. Lu, S. M.; Bolm, C. *Chem. Eur. J.* 2008, *14* (25), 7513–7516.
100. Biosca, M.; Pàmies, O.; Diéguez, M. *J. Org. Chem.* 2019, *84* (12), 8259–8266.
101. Zhou, Q. L. *Privileged Chiral Ligands and Catalysts.* Wiley-VCH Verlag GmbH & KGaA.: Weinheim, 2011.
102. Verendel, J. J.; Pàmies, O.; Diéguez, M.; Andersson, P. G. *Chem. Rev.* 2014, *114* (4), 2130–2169.
103. Brandt, P.; Hedberg, C.; Andersson, P. G. *Chem. Eur. J.* 2003, *9*, 339–347.
104. Källström, K.; Hedberg, C.; Brandt, P.; Bayer, A.; Andersson, P. G. *J. Am. Chem. Soc.* 2004, *126* (44), 14308–14309.
105. Hedberg, C.; Källström, K.; Brandt, P.; Hansen, L. K.; Andersson, P. G. *J. Am. Chem. Soc.* 2006, *128* (9), 2995–3001.
106. Engman, M.; Diesen, J. S.; Paptchikhine, A.; Andersson, P. G. *J. Am. Chem. Soc.* 2007, *129* (15), 4536–4537.
107. Cheruku, P.; Paptchikhine, A.; Church, T. L.; Andersson, P. G. *J. Am. Chem. Soc.* 2009, *131* (23), 8285–8289.
108. Tolstoy, P.; Engman, M.; Paptchikhine, A.; Bergquist, J.; Church, T. L.; Leung, A. W. M.; Andersson, P. G. *J. Am. Chem. Soc.* 2009, *131* (25), 8855–8860.
109. Cheruku, P.; Paptchikhine, A.; Ali, M.; Neudörfl, J. M.; Andersson, P. G. *Org. Biomol. Chem.* 2008, *6* (2), 366–373.
110. Mazuela, J.; Paptchikhine, A.; Tolstoy, P.; Pàmies, O.; Diéguez, M.; Andersson, P. G. *Chem. Eur. J.* 2010, *16* (2), 620–638.
111. Mazuela, J.; Paptchikhine, A.; Pàmies, O.; Andersson, P. G.; Diéguez, M. *Chem. Eur. J.* 2010, *16* (15), 4567–4576.
112. Li, J. Q.; Paptchikhine, A.; Govender, T.; Andersson, P. G. *Tetrahedron: Asymmetry* 2010, *21* (11–12), 1328–1333.
113. Paptchikhine, A.; Cheruku, P.; Engman, M.; Andersson, P. G. *Chem. Commun.* 2009, *40*, 5996–5998.
114. Peters, B. K.; Zhou, T.; Rujirawanich, J.; Cadu, A.; Singh, T.; Rabten, W.; Kerdphon, S.; Andersson, P. G. *J. Am. Chem. Soc.* 2014, *136* (47), 16557–16562.
115. Li, J. Q.; Liu, J.; Krajangsri, S.; Chumnanvej, N.; Singh, T.; Andersson, P. G. *ACS Catal.* 2016, *6* (12).
116. Ponra, S.; Yang, J.; Kerdphon, S.; Andersson, P. G. *Angew. Chem., Int. Ed.* 2019, *58* (27), 9282–9287.
117. Biosca, M.; Paptchikhine, A.; Pàmies, O.; Andersson, P. G.; Diéguez, M. *Chem. Eur. J.* 2015, *21* (8), 3455–3464.

118. Kaukoranta, P.; Engman, M.; Hedberg, C.; Bergquist, J.; Andersson, P. G. *Adv. Synth. Catal.* 2008, *350* (7–8), 1168–1176.

119. Verendel, J. J.; Li, J. Q.; Quan, X.; Peters, B.; Zhou, T.; Gautun, O. R.; Govender, T.; Andersson, P. G. *Chem. Eur. J.* 2012, *18* (21), 6507–6513.

120. Peters, B. K.; Liu, J.; Margarita, C.; Rabten, W.; Kerdphon, S.; Orebom, A.; Morsch, T.; Andersson, P. G. *J. Am. Chem. Soc.* 2016, *138* (36), 11930–11935.

121. Rabten, W.; Margarita, C.; Eriksson, L.; Andersson, P. G. *Chem. Eur. J.* 2018, *24* (7), 1681–1685.

122. Eicher, T.; Hauptmann, S.; Speicher, A. *The Chemistry of Heterocycles.* Wiley-VCH: Weinheim, 2003.

123. Liu, J.; Krajangsri, S.; Yang, J.; Li, J. Q.; Andersson, P. G. *Nat. Catal.* 2018, *1* (6), 438–443.

124. Álvarez-Yebra, R.; Rojo, P.; Riera, A.; Verdaguer, X. *Tetrahedron* 2019, *75* (32), 4358–4364.

125. Salomó, E.; Orgué, S.; Riera, A.; Verdaguer, X. *Angew. Chem., Int. Ed.* 2016, *55* (28), 7988–7992.

126. Busacca, C. A. U.S. Patent 6,316,620; European Patent 1218388, 2001.

127. Busacca, C. A.; Grossbach, D.; So, R. C.; O'Brien, E. M.; Spinelli, E. M. *Org. Lett.* 2003, *5* (4), 595–598.

128. Busacca, C. A.; Grossbach, D.; Campbell, S. J.; Dong, Y.; Eriksson, M. C.; Harris, R. E.; Jones, P. J.; Kim, J. Y.; Lorenz, J. C.; McKellop, K. B.; O'Brien, E. M.; Qiu, F.; Simpson, R. D.; Smith, L.; So, R. C.; Spinelli, E. M.; Vitous, J.; Zavattaro, C. *J. Org. Chem.* 2004, *69* (16), 5187–5195.

129. Menges, F.; Neuburger, M.; Pfaltz, A. *Org. Lett.* 2002, *4* (26), 4713–4716.

130. Guiu, E.; Claver, C.; Benet-Buchholz, J.; Castillón, S. *Tetrahedron: Asymmetry* 2004, *15* (21), 3365–3373.

131. Busacca, C. A.; Lorenz, J. C.; Grinberg, N.; Haddad, N.; Lee, H.; Li, Z.; Liang, M.; Reeves, D.; Saha, A.; Varsolona, R.; Senanayake, C. H. *Org. Lett.* 2008, *10* (2), 341–344.

132. Busacca, C. A.; Lorenz, J. C.; Saha, A. K.; Cheekoori, S.; Haddad, N.; Reeves, D.; Lee, H.; Li, Z.; Rodriguez, S.; Senanayake, C. H. *Catal. Sci. Technol.* 2012, *2* (10), 2083–2089.

133. Busacca, C. A.; Qu, B.; Grĕt, N.; Fandrick, K. R.; Saha, A. K.; Marsini, M.; Reeves, D.; Haddad, N.; Eriksson, M.; Wu, J. P.; Grinberg, N.; Lee, H.; Li, Z.; Lu, B.; Chen, D.; Hong, Y.; Ma, S.; Senanayake, C. H. *Adv. Synth. Catal.* 2013, *355* (8), 1455–1463.

134. Wang, A.; Bernasconi, M.; Pfaltz, A. *Adv. Synth. Catal.* 2017, *359* (15), 2523–2529.

135. Lorenz, J. C.; Busacca, C. A.; Feng, X. W.; Grinberg, N.; Haddad, N.; Johnson, J.; Kapadia, S.; Lee, H.; Saha, A.; Sarvestani, M.; Spinelli, E. M.; Varsolona, R.; Wei, X.; Zeng, X.; Senanayake, C. H. *J. Org. Chem.* 2010, *75* (4), 1155–1161.

136. Pfaltz, A. M. F. WO 2005021562 (Solvias AG, Switz.), 2005.

137. Baeza, A.; Pfaltz, A. *Chem. Eur. J.* 2010, *16* (13), 4003–4009.

138. Ganić, A.; Pfaltz, A. *Chem. Eur. J.* 2012, *18* (22), 6724–6728.

139. Cardoso, F. S. P.; Abboud, K. A.; Aponick, A. *J. Am. Chem. Soc.* 2013, *135* (39), 14548–14551.

140. Paioti, P. H. S.; Abboud, K. A.; Aponick, A. *J. Am. Chem. Soc.* 2016, *138* (7), 2150–2153.

141. Pappoppula, M.; Aponick, A. *Angew. Chem., Int. Ed.* 2015, *54* (52), 15827–15830.

142. Pappoppula, M.; Cardoso, F. S. P.; Garrett, B. O.; Aponick, A. *Angew. Chem., Int. Ed.* 2015, *54* (50), 15202–15206.

143. Mishra, S.; Liu, J.; Aponick, A. *J. Am. Chem. Soc.* 2017, *139* (9), 3352–3355.

144. DeRatt, L. G.; Pappoppula, M.; Aponick, A. *Angew. Chem., Int. Ed.* 2019, *58* (25), 8416–8420.

145. Rokade, B. V.; Guiry, P. J. *ACS Catal.* 2017, *7* (4), 2334–2338.
146. Paioti, P. H. S.; Abboud, K. A.; Aponick, A. *ACS Catal.* 2017, *7* (3), 2133–2138.
147. Lassaletta, J. M. *Atropisomerism and Axial Chirality.* World Scientific Publishing Europe Ltd.: London, UK, 2019.
148. Rokade, B. V.; Guiry, P. J. *J. Org. Chem.* 2019, *84* (9), 5763–5772.
149. Ito, K.; Kashiwagi, R.; Iwasaki, K.; Katsuki, T. *Synlett* 1999, *10*, 1563–1566.
150. Ito, K.; Kashiwagi, R.; Hayashi, S.; Uchida, T. *Synlett* 2001, *2*, 284–286.
151. Ito, K.; Akashi, S.; Saito, B.; Katsuki, T. *Synlett* 2003, *12*, 1809–1812.
152. Ito, K.; Ishii, A.; Kuroda, T.; Katsuki, T. *Synlett* 2003, *5*, 643–646.
153. Ito, K.; Imahayashi, Y.; Kuroda, T.; Eno, S.; Saito, B.; Katsuki, T. *Tetrahedron Lett.* 2004, *45* (39), 7277–7281.
154. Chelucci, G.; Saba, A.; Soccolini, F. *Tetrahedron* 2001, *57*, 9989–9996.
155. Malkov, A. V.; Bella, M.; Stará, I. G.; Kočovský, P. *Tetrahedron Lett.* 2001, *42* (16), 3045–3048.
156. Malkov, A. V.; Friscourt, F.; Bell, M.; Swarbrick, M. E.; Kočovský, P. *J. Org. Chem.* 2008, *73* (11), 3996–4003.
157. Drury, W. J.; Zimmermann, N.; Keenan, M.; Hayashi, M.; Kaiser, S.; Goddard, R.; Pfaltz, A. *Angew. Chem., Int. Ed.* 2004, *43* (1), 70–74.
158. Quan, X.; Parihar, V. S.; Bera, M.; Andersson, P. G. *Eur. J. Org. Chem.* 2014, *2014* (1), 140–146.
159. Meng, X.; Li, X.; Xu, D. *Tetrahedron: Asymmetry* 2009, *20* (12), 1402–1406.
160. Bunlaksananusorn, T.; Polborn, K.; Knochel, P. *Angew. Chem., Int. Ed.* 2003, *42* (33), 3941–3943.
161. Bunlaksananusorn, T.; Knochel, P. *J. Org. Chem.* 2004, *69* (14), 4595–4601.
162. Xie, J. H.; Zhu, S. F.; Zhou, Q. L. *Chem. Rev.* 2011, *111* (3), 1713–1760.
163. Kaiser, S.; Smidt, S. P.; Pfaltz, A. *Angew. Chem., Int. Ed.* 2006, *45* (31), 5194–5197.
164. Bianco, G. G.; Ferraz, H. M. C.; Costas, A. M.; Costa-Lotufo, L. V.; Pessoa, C.; De Moraes, M. O.; Schreins, M. G.; Pfaltz, A.; Silva, L. F. *J. Org. Chem.* 2009, *74* (6), 2561–2566.
165. Woodmansee, D. H.; Müllér, M. A.; Neuburger, M.; Pfaltz, A. *Chem. Sci.* 2010, *1* (1), 72–78.
166. Jessen, H. J.; Schumacher, A.; Schmid, F.; Pfaltz, A.; Gademann, K. *Org. Lett.* 2011, *13* (16), 4368–4370.
167. Schmid, F.; Bernasconi, M.; Jessen, H. J.; Pfaltz, A.; Gademann, K. *Synth.* 2014, *46* (7), 864–870.
168. Woodmansee, D. H.; Müller, M. A.; Tröndlin, L.; Hörmann, E.; Pfaltz, A. *Chem. Eur. J.* 2012, *18* (43), 13780–13786.
169. Tiefenbacher, K.; Tröndlin, L.; Mulzer, J.; Pfaltz, A. *Tetrahedron* 2010, *66* (33), 6508–6513.
170. Bernasconi, M.; Müller, M. A.; Pfaltz, A. *Angew. Chem., Int. Ed.* 2014, *53* (21), 5385–5388.
171. Olefins, A.; Bell, S.; Wu, B.; Kaiser, S.; Menges, F.; Netscher, T. *Science* 2006, *311*, 642–645.
172. Wang, A.; Fraga, R. P. A.; Hörmann, E.; Pfaltz, A. *Chem. - An Asian J.* 2011, *6* (2), 599–606.
173. Ratsch, F.; Schlundt, W.; Albat, D.; Zimmer, A.; Neudörfl, J. M.; Netscher, T.; Schmalz, H. G. *Chem. Eur. J.* 2019, *25* (19), 4941–4945.
174. Wang, A.; Wüstenberg, B.; Pfaltz, A. *Angew. Chem., Int. Ed.* 2008, *47* (12), 2298–2300.
175. Yoshinari, T.; Ohmori, K.; Schrems, M. G.; Pfaltz, A.; Suzuki, K. *Angew. Chem., Int. Ed.* 2010, *49* (5), 881–885.

176. Pischl, M. C.; Weise, C. F.; Haseloff, S.; Müller, M. A.; Pfaltz, A.; Schneider, C. *Chem. Eur. J.* 2014, *20* (52), 17360–17374.
177. Pauli, L.; Tannert, R.; Scheil, R.; Pfaltz, A. *Chem. Eur. J.* 2015, *21* (4), 1482–1487.
178. Baeza, A.; Pfaltz, A. *Chem. Eur. J.* 2010, *16* (7), 2036–2039.
179. Tosatti, P.; Pfaltz, A. *Angew. Chem., Int. Ed.* 2017, *56* (16), 4579–4582.
180. Schneekönig, J.; Liu, W.; Leischner, T.; Junge, K.; Schotes, C.; Beier, C.; Beller, M. *Org. Process Res. Dev.* 2020, *24* (3), 443–447.
181. Mazuela, J.; Pàmies, O.; Diéguez, M. *Chem. Eur. J.* 2013, *19* (7), 2416–2432.
182. Mazuela, J.; Pàmies, O.; Diéguez, M. *Adv. Synth. Catal.* 2013, *355* (13), 2569–2583.
183. Qu, B.; Samankumara, L. P.; Savoie, J.; Fandrick, D. R.; Haddad, N.; Wei, X.; Ma, S.; Lee, H.; Rodriguez, S.; Busacca, C. A.; Yee, N. K.; Song, J. J.; Senanayake, C. H. *J. Org. Chem.* 2014, *79* (3), 993–1000.
184. Qu, B.; Mangunuru, H. P. R.; Wei, X.; Fandrick, K. R.; Desrosiers, J. N.; Sieber, J. D.; Kurouski, D.; Haddad, N.; Samankumara, L. P.; Lee, H.; Savoie, J.; Ma, S.; Grinberg, N.; Sarvestani, M.; Yee, N. K.; Song, J. J.; Senanayake, C. H. *Org. Lett.* 2016, *18* (19), 4920–4923.
185. Qu, B.; Mangunuru, H. P. R.; Tcyrulnikov, S.; Rivalti, D.; Zatolochnaya, O. V.; Kurouski, D.; Radomkit, S.; Biswas, S.; Karyakarte, S.; Fandrick, K. R.; Sieber, J. D.; Rodriguez, S.; Desrosiers, J. N.; Haddad, N.; McKellop, K.; Pennino, S.; Lee, H.; Yee, N. K.; Song, J. J.; Kozlowski, M. C.; Senanayake, C. H. *Org. Lett.* 2018, *20* (5), 1333–1337.
186. Liu, Y.; Chen, F.; He, Y. M.; Li, C.; Fan, Q. H. *Org. Biomol. Chem.* 2019, *17* (20), 5099–5105.
187. Alcock, N. W.; Brown, J. M.; Hulmes, D. I. *Tetrahedron: Asymmetry* 1993, *4* (4), 743–756.
188. Bhat, V.; Wang, S.; Stoltz, B. M.; Virgil, S. C. *J. Am. Chem. Soc.* 2013, *135* (45), 16829–16832.
189. Ramírez-López, P.; Ros, A.; Estepa, B.; Fernández, R.; Fiser, B.; Gómez-Bengoa, E.; Lassaletta, J. M. *ACS Catal.* 2016, *6* (6), 3955–3964.
190. Gommermann, N.; Koradin, C.; Polborn, K.; Knochel, P. *Angew. Chem., Int. Ed.* 2003, *42* (46), 5763–5766.
191. Gommermann, N.; Knochel, P. *Synlett* 2005, No. 18, 2799–2801.
192. Gommermann, N.; Knochel, P. *Tetrahedron* 2005, *61* (48), 11418–11426.
193. Gommermann, N.; Knochel, P. *Chem. Eur. J.* 2006, *12* (16), 4380–4392.
194. Chen, C.; Li, X.; Schreiber, S. L. *J. Am. Chem. Soc.* 2003, *125* (34), 10174–10175.
195. Lim, A. D.; Codelli, J. A.; Reisman, S. E. *Chem. Sci.* 2013, *4* (2), 650–654.
196. Brown, J. M.; Hulmes, D. I.; Layzell, T. P. *J. Chem. Soc., Chem. Commun.* 1993, 1673–1674.
197. Fernandez, E.; Hooper, M. W.; Knight, F. I.; Brown, J. M. *Chem. Commun.* 1997, *21*, 173–174.
198. Knight, F. I.; Brown, J. M.; Lazzari, D.; Ricci, A.; Organica, C.; Blacker, A. J. *Tetrahedron* 1997, *53* (33), 11411–11424.
199. Segarra, A. M.; Daura-Oller, E.; Claver, C.; Poblet, J. M.; Bo, C.; Fernández, E. *Chem. Eur. J.* 2004, *10* (24), 6456–6467.
200. Black, A.; Brown, J. M.; Pichont, C. *Chem. Commun.* 2005, *1* (42), 5284–5286.
201. Connolly, D. J.; Lacey, P. M.; McCarthy, M.; Saunders, C. P.; Carroll, A. M.; Goddard, R.; Guiry, P. J. *J. Org. Chem.* 2004, *69* (20), 6572–6589.
202. Maxwell, A. C.; Flanagan, S. P.; Goddard, R.; Guiry, P. J. *Tetrahedron: Asymmetry* 2010, *21* (11–12), 1458–1473.
203. Morgan, J. B.; Miller, S. P.; Morken, J. P. *J. Am. Chem. Soc.* 2003, *125* (29), 8702–8703.
204. Miller, S. P.; Morgan, J. B.; Nepveux V, F. J.; Morken, J. P. *Org. Lett.* 2004, *6* (1), 131–133.

205. Trudeau, S.; Morgan, J. B.; Shrestha, M.; Morken, J. P. *J. Org. Chem.* 2005, *70* (23), 9538–9544.
206. Miura, T.; Yamauchi, M.; Kosaka, A.; Murakami, M. *Angew. Chem., Int. Ed.* 2010, *49* (29), 4955–4957.
207. Taylor, A. M.; Schreiber, S. L. *Org. Lett.* 2006, *8* (1), 143–146.
208. Perepichka, I.; Kundu, S.; Hearne, Z.; Li, C. J. *Org. Biomol. Chem.* 2015, *13* (2), 447–451.
209. Ramírez-López, P.; Ros, A.; Romero-Arenas, A.; Iglesias-Sigüenza, J.; Fernández, R.; Lassaletta, J. M. *J. Am. Chem. Soc.* 2016, *138* (37), 12053–12056.
210. Hornillos, V.; Ros, A.; Ramírez-López, P.; Iglesias-Sigüenza, J.; Fernández, R.; Lassaletta, J. M. *Chem. Commun.* 2016, *52* (98), 14121–14124.
211. Knöpfel, T. F.; Aschwanden, P.; Ichikawa, T.; Watanabe, T.; Carreira, E. M. *Angew. Chem., Int. Ed.* 2004, *43* (44), 5971–5973.
212. Aschwanden, P.; Stephenson, C. R. J.; Carreira, E. M. *Org. Lett.* 2006, *8* (11), 2437–2440.
213. Fan, W.; Ma, S. *Chem. Commun.* 2013, *49* (86), 10175–10177.
214. Fan, W.; Yuan, W.; Ma, S. *Nat. Commun.* 2014, *5* (May), 1–9.
215. Bonepally, K. R.; Hiruma, T.; Mizoguchi, H.; Ochiai, K.; Suzuki, S.; Oikawa, H.; Ishiyama, A.; Hokari, R.; Iwatsuki, M.; Otoguro, K.; Omura, S.; Oguri, H. *Org. Lett.* 2018, *20* (15), 4667–4671.
216. Lin, W.; Cao, T.; Fan, W.; Han, Y.; Kuang, J.; Luo, H.; Miao, B.; Tang, X.; Yu, Q.; Yuan, W.; Zhang, J.; Zhu, C.; Ma, S. *Angew. Chem., Int. Ed.* 2014, *53* (1), 277–281.
217. Lin, W.; Ma, S. *Org. Chem. Front.* 2014, *1* (4), 338–346.
218. Lin, W.; Ma, S. *Org. Chem. Front.* 2017, *4* (6), 958–966.
219. Zhou, S.; Tong, R. *Org. Lett.* 2017, *19* (7), 1594–1597.
220. Ye, J.; Li, S.; Chen, B.; Fan, W.; Kuang, J.; Liu, J.; Liu, Y.; Miao, B.; Wan, B.; Wang, Y.; Xie, X.; Yu, Q.; Yuan, W.; Ma, S. *Org. Lett.* 2012, *14* (5), 1346–1349.
221. Knöpfel, T. F.; Zarotti, P.; Ichikawa, T.; Carreira, E. M. *J. Am. Chem. Soc.* 2005, *127* (27), 9682–9683.
222. Hu, X.; Chen, H.; Zhang, X. *Angew. Chem., Int. Ed.* 1999, *38* (23), 3518–3521.
223. Hu, Y.; Liang, X.; Wang, J.; Zheng, Z.; Hu, X. *J. Org. Chem.* 2003, *68* (11), 4542–4545.
224. Hu, Y.; Liang, X.; Wang, J.; Zheng, Z.; Hu, X. *Tetrahedron: Asymmetry* 2003, *14* (24), 3907–3915.
225. Liang, Y.; Gao, S.; Wan, H.; Hu, Y.; Chen, H.; Zheng, Z.; Hu, X. *Tetrahedron: Asymmetry* 2003, *14*, 3211–3217.
226. Wan, H.; Hu, Y.; Liang, Y.; Gao, S.; Wang, J.; Zheng, Z.; Hu, X. *J. Org. Chem.* 2003, *68* (21), 8277–8280.
227. Luo, X.; Hu, Y.; Hu, X. *Tetrahedron: Asymmetry* 2005, *16* (6), 1227–1231.
228. Guo, S.; Xie, Y.; Hu, X.; Huang, H. *Org. Lett.* 2011, *13* (20), 5596–5599.
229. Wang, Q.; Li, S.; Hou, C. J.; Chu, T. T.; Hu, X. P. *Tetrahedron* 2019, *75* (29), 3943–3950.
230. Guo, S.; Xie, Y.; Hu, X.; Xia, C.; Huang, H. *Angew. Chem., Int. Ed.* 2010, *49* (15), 2728–2731.
231. Wang, Q.; Li, S.; Hou, C. J.; Chu, T. T.; Hu, X. P. *Appl. Organomet. Chem.* 2019, *33* (10), 1–8.
232. Nareddy, P.; Mantilli, L.; Guénée, L.; Mazet, C. *Angew. Chem., Int. Ed.* 2012, *51* (16), 3826–3831.
233. Borrajo-Calleja, G. M.; Bizet, V.; Bürgi, T.; Mazet, C. *Chem. Sci.* 2015, *6* (8), 4807–4811.

3 Chiral Bidentate Heterodonor P,S/O Ligands

Jèssica Margalef and Miquel A. Pericàs

CONTENTS

3.1 INTRODUCTION

Asymmetric metal-catalysis is one of the most important synthetic strategies for preparing enantioenriched compounds, occupying a central position in areas ranging from medicinal chemistry to chiral materials. Asymmetric metal-catalysis has benefited from the enormous diversity of chiral ligands that are available nowadays, which are able to modulate the reactivity of the metal center and to control the stereochemical outcome of the reactions they mediate. In recent years, ligand design has progressively evolved from C_2-homodonor to heterodonor species. The latter class of ligands facilitates the introduction of electronic differentiation, which is important in controlling substrate coordination and reactivity, *via* the different *trans* effects of the two dissimilar donor groups. The field of heterodonor ligands was initially dominated by P,N-ligands, in particular by phosphine-oxazolines, and, more recently, by phosphite-oxazolines. Chiral P,O and P,S heterodonor ligands have traditionally played a less important role than P,N ligands. In the case of P,O ligands, this can be attributed to its hemilabile character, due to the presence of both a hard (O) and a soft (P) base on the same metal center. On the other hand, the less frequent use of P,S ligands for asymmetric catalysis can be explained by the formation of difficult-to-control diastereomeric mixtures of active species, since most of the sulfur centers present in those ligands become stereogenic following coordination to the metal. Nevertheless, the number of successful applications of P,O and P,S ligands has increased over the past few years, which may lead to a renaissance of the use of these ligands in asymmetric catalysis.

3.2 APPLICATION OF HETERODONOR P,O LIGANDS IN ASYMMETRIC CATALYSIS

The hemilability of the P,O ligands facilitates several transformations at the metal center, such as oxidative addition, ligand exchange, isomerization, etc., which often expedite catalytic activity. Nevertheless, this hemilability can cause a detrimental effect on enantioselectivity, since the ligand can be coordinated in a monodentate fashion in the enantio-discriminating transition state. MOP (2-(diphenylphosphin o)-2'-alkoxy-1,1'-binaphthyl) ligands constitute one of the earliest examples in which a P,O ligand acts as monodentate ligand (Figure 3.1).[1,2] Despite this, MOP ligands have been used successfully in many asymmetric transformations, such as the asymmetric Pd-catalyzed substitution and reduction of allylic esters,[1c,e–g] the Rh-catalyzed addition of boronic acids to ketones, and the hydrosilylation of 1-alkenes.[1a,b,h] In all these cases, the ligand acts as a chiral monophosphine with a hanging ether group, which is able to establish a secondary interaction with the incoming nucleophile/ reagent. Such secondary interactions, as demonstrated earlier by Hayashi and coworkers,[3] are crucial to achieving high levels of enantioselectivity.

Heterodonor phosphine-phosphine oxide ligands have been shown to be effective bidentate ligands for several asymmetric transformations.[4] The early work of Faller and coworkers described the use of the BINAP(O) (2'-(diphenylphosphino)- [1,1'-binaphthalen]-2-yl)diphenylphosphine oxide) ligand in the highly enantioselective Ru-catalyzed Diels-Alder reactions (Scheme 3.1a).[5] More recently, the groups of Oestreich[6] and Hou[7] independently demonstrated the effectiveness of BINAP(O) in the Pd-catalyzed asymmetric Heck reaction (Scheme 3.1b). They reported that the use of the BINAP(O) ligand, instead of BINAP, markedly changed the regio- and enantioselectivity in the arylation of cyclic alkenes. In particular, the arylation of 2,3-dihydrofuran behaved perfectly regiodivergently; whereas the use of BINAP favored the formation of the thermodynamically more stable 2-aryl-2,3-dihydrofuran, the use of BINAP(O) led to the preferential formation of 2-aryl-2,5-dihydrofuran. In addition, little alkene migration was observed when BINAP(O) was used. Zhou's group used BINAP(O) for a "domino reaction" that allowed the synthesis of fused carbo- and heterocycles, involving an asymmetric intermolecular Heck reaction, followed by a diastereoselective cyclization (Scheme 3.1b).[7a] Pd-BINAP(O) has also been successfully used in the kinetic resolution of 2-substituted-dihydrofurans *via* a Heck reaction, producing optically enriched *trans*-2,5-disubstituted-dihydrofurans and 2-substituted dihydrofurans in high yields and at high enantiomeric excess (ee) values (*S* factor of up to 70; Scheme 3.1b).[7b]

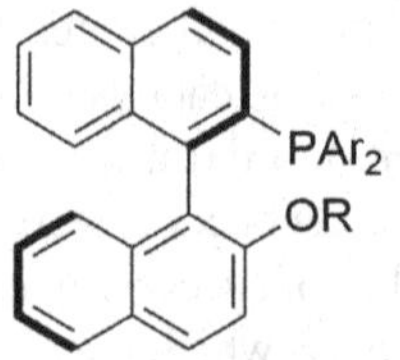

FIGURE 3.1 (*R*)-MOP-type ligands.

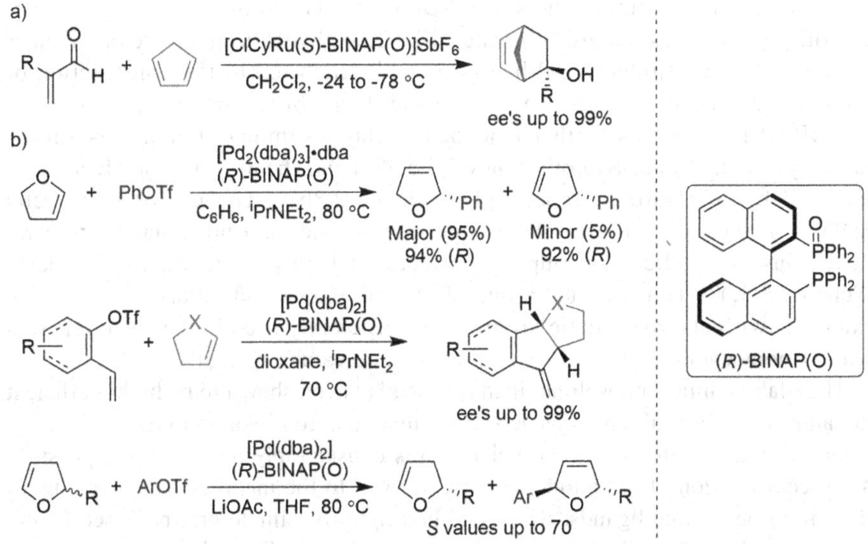

SCHEME 3.1 Representative examples of the use of the BINAP(O) ligand in the asymmetric a) Diels-Alder and b) Heck reactions.

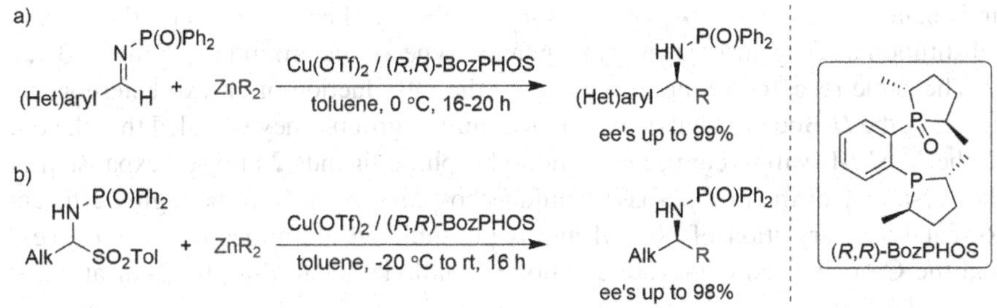

SCHEME 3.2 Synthesis of α-chiral amines using bis(phosphine) monoxide BozPHOS ligand.

Another interesting example can be found in the work of Charette's group that disclosed the use of BozPHOS (1,2-Bis[(2S,5S)-2,5-dimethylphospholano]benzene monooxide), a monoxide version of the Me-Duphos (1,2-bis[(2R,5R)-2,5-dimethylphospholano]benzene), in the Cu-catalyzed 1,2-addition of diorganozinc reagents to *N*-phosphinoylarylaldimines (up to 99% ee; Scheme 3.2a).[8] However, the application of such a strategy can be hampered by the accessibility and stability of the *N*-phosphinoylalkylaldimines. To address this drawback, the same group demonstrated that the reaction also proceeds well using the sulfinic adduct of *N*-phosphinoylimines (Scheme 3.2b).[9]

The latest bis(phosphine) monoxide design, with an outstanding applicability, can be found described in the work of Zhou and coworkers, which revealed the use of spirocyclic SDP(O) ligands in several intermolecular Heck reactions

(Scheme 3.3). Interestingly, the use of SDP(O) ligands allowed not only the successful arylation of standard substrates, like 2,3-dihydrofuran or cyclopentene,[10] but also of 5-substituted-2,3-dihydrofurans,[11] which led to the construction of a chiral quaternary carbon center (Scheme 3.3a). More interestingly, the use of SDP(O) ligands was further extended to the desymmetrization of 4-substituted-cyclopent-1-enes and other bicyclic olefins *via* the asymmetric Heck reaction and hydroarylation, respectively (Scheme 3.3b).[12] The use of spirocyclic SDP(O) ligands also allowed the unique asymmetric intermolecular Heck reaction, using aryl halides as coupling partners, instead of aryl and vinyl triflates (Scheme 3.3c).[13] Thus, a wide range of aryl bromides and chlorides, including examples with heteroaromatic groups, were efficiently used in the Heck reaction with various cyclic olefins (35 examples with ee values typically >95%).

Hemilabile amido-phosphine ligands have also been shown to be highly efficient in catalysis of several asymmetric transformations. Tomioka and coworkers early demonstrated that the reaction efficiency was considerably reliant on the possibility of coordination of the amide carbonyl oxygen to the metal center.[14] Among all the amido-phosphine ligands, we should highlight the simple proline-based ligand **1**, developed by Tomioka and coworkers (Scheme 3.4). This ligand proved to be highly efficient in the Rh-catalyzed asymmetric 1,4-addition of arylboronic acids to cycloalkenones (ee values up to 97%)[14] as well as in the Cu-catalyzed conjugate addition of dialkylzinc reagents to nitroalkenes (up to 80% ee)[15] and in the highly regio- and enantioselective (regioselectivities up to >99% and ee values up to 91%) allylic substitution of Grignard reagents to cinnamyl-type allylic bromides[16] (Scheme 3.4a).

The same research group also explored the introduction of an extra stereogenic center at the *N*-Boc (*tert*-butyloxycarbonyl) amido group. They revealed that the use of the *N*-Boc-L-valine-connected amido-phosphine ligands **2** allowed expansion of the versatility of the proline-based amido-phosphino ligands to the highly efficient Rh-catalyzed arylation of *N*-tosyl and *N*-phosphinoyl aldimines (up to 99% ee)[17] and the Cu-catalyzed conjugate addition of dialkylzincs to β-aryl-α,β-unsaturated

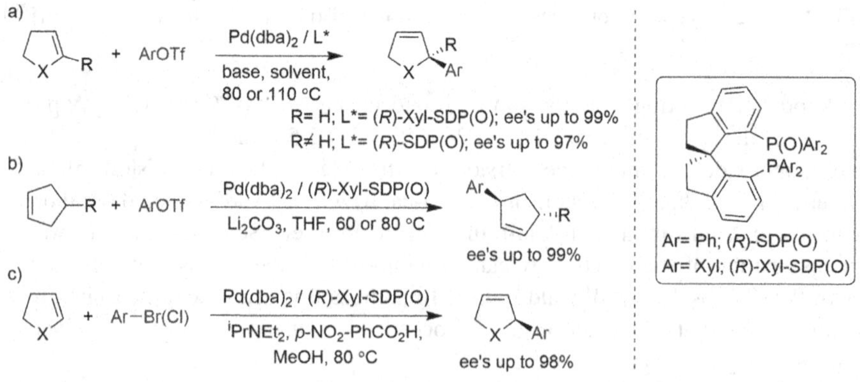

SCHEME 3.3 Application of bis(phosphine) monoxide SDP(O) ligands in the asymmetric Pd-catalyzed Heck reaction.

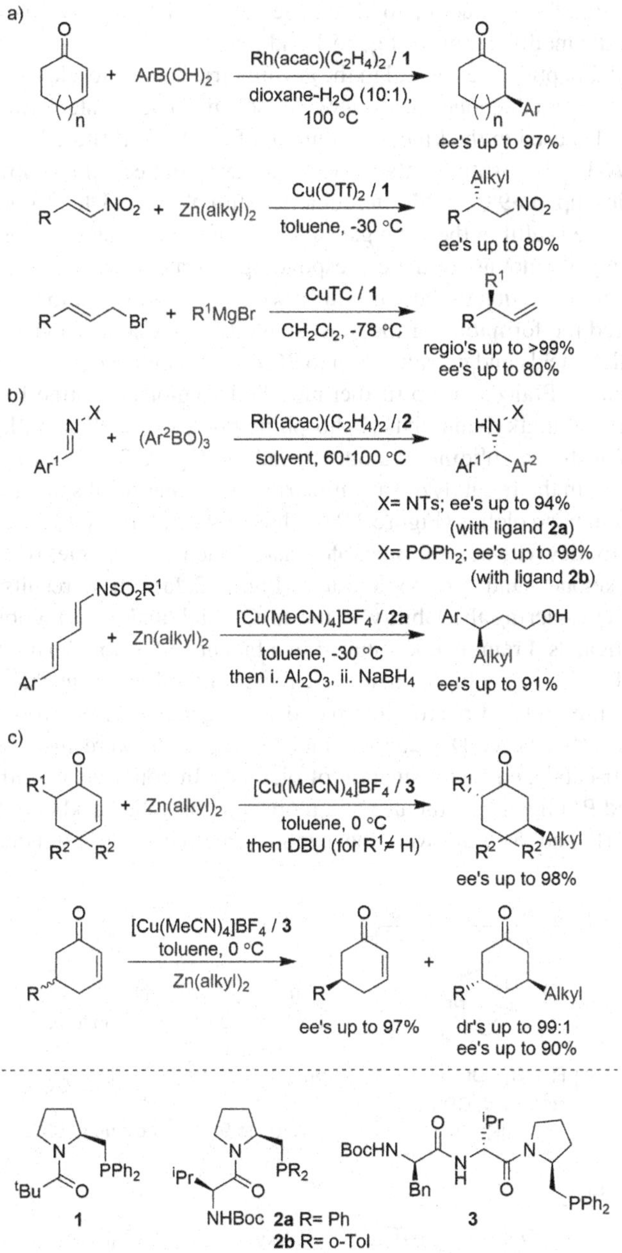

SCHEME 3.4 Application of proline-based amido-phosphine ligands **1–3** in asymmetric catalysis.

sulfonylaldimines[18] (ee values up to 91%; Scheme 3.4b). Inspired by this success, they studied peptidic modifications of ligand **1** and found that ligand **3**, involving a small D-Phe-D-Val dipeptide, was useful in the asymmetric conjugate addition of organozinc reagents to cycloalkenones (up to 98% ee; Scheme 3.4c).[19] Interestingly, the Cu/**3** catalyst was also used in the kinetic resolution of (*rac*)-5-substituted cycloalkenones to yield *trans*-3,5-disubstituted alkanones, with excellent ee values (up to 90%) and *trans/cis* ratios (up to >98/2 dr (diastereomeric ratio); Scheme 3.4c).[20] Later, Cu/**3** was also shown to be useful in the conjugate addition of 6-substituted cyclohexenones to give almost equal amounts of the corresponding *cis*- and *trans*-disubstituted cyclohexanones. Epimerization of these mixtures with 1,8-diazabicyclo[5.4.0]undec-7-ene (DBU) favored the formation of the most stable *trans*-cyclohexenones as the major product in high yields and ee values (up to 96% ee; Scheme 3.4c).[21]

More recently, Pfaltz's group further modified Tomioka's proline-based ligand **1** to include several dialkyl and dialkyl phosphino groups, as well as bulky amide and urea groups at the pyrrolidine N-atom (ligands **4**; Figure 3.2).[22,23] Ligands **4** were efficiently used in the Ir-catalyzed asymmetric hydrogenation of some types of minimally functionalized olefins (Figure 3.2a). Thus, Ir/**4** catalytic systems achieved high enantioselectivities with several substrate classes, such as *trans*-methyl stilbene, α,β-unsaturated ketones, and carboxylic esters (Figure 3.2a). These results were rather unexpected, considering the lability of the Ir–O bond, and unambiguously demonstrated that ligands **4** remain coordinated in a bidentate fashion during the catalytic cycle, probably due to the highly acidic character of the Ir^III/Ir^V intermediates.

More recently, related pyrrolidine-based P,O ligands derived from inexpensive carbohydrates (D-ribose, D-mannose, and D-arabinose) were also prepared and used in the Ir-catalyzed hydrogenation of olefins.[24] In comparison with the related proline-based P,O ligands **4**, the use of a more rigid bicyclic backbone derived from D-mannose (ligand **5**) improved the enantioselectivity and extended the range

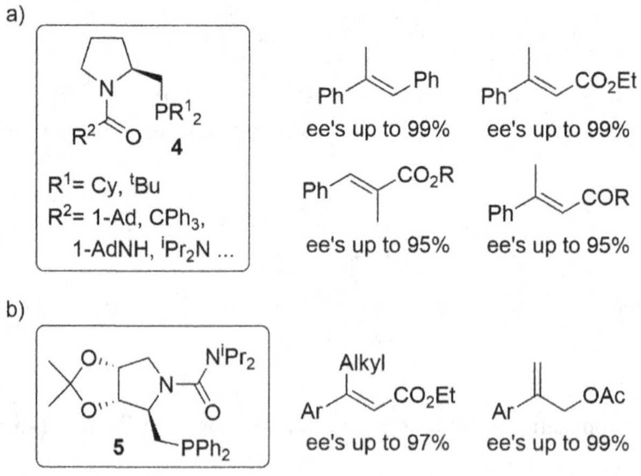

FIGURE 3.2 Ir-catalyzed asymmetric hydrogenation of minimally functionalized olefins with a) proline-based P,O ligands **4** and b) D-mannose pyrrolidine-based ligand **5**.

of substrates that could be reduced, including 1,1-disubstituted allylic acetates (Figure 3.2b). As a further advantage, and unlike ligand **4**, ligand **5** does not require the presence of less stable, bulkier phosphine substituents for optimal performance.

Another relevant class of amido-phosphine P,O ligands are the non-biaryl, atropoisomeric ligands **6** and **7** (Scheme 3.5). Ligand **6** was successfully used in the Pd-catalyzed allylic substitution reaction of 1,3-diphenylallyl acetate, using dimethyl malonate as the nucleophile (ee values up to 95%; Scheme 3.5a).[25] The Pd/**6** catalyst was also used for the asymmetric Heck reaction of 2,3-dihydrofuran with aryl triflates, albeit with only moderate enantioselectivities (up to 55% ee).[26] More recently, Cia and Xu groups developed the novel amido-phosphino ligand **7**, which proved to be highly effective in asymmetric Ag-catalyzed [3+2] cycloaddition reactions. Thus, catalyst Ag/**7** mediated the [3+2] cycloaddition of aldiminoesters with nitroalkenes to yield optically enriched nitrosubstituted pyrrolidines with excellent diastereo- and enantioselectivities (up to >99:1 dr and up to 99% ee; Scheme 3.5b).[27] The same catalyst was also used in the preparation of imide-containing pyrrolidines by reaction of iminoesters with maleimides (dr values up to >98:2 and ee values up to 99%; Scheme 3.5b).[28]

Sulfonamide phosphines **8–11** represent another type of heterodonor P,O ligands, that have recently shown their potential in asymmetric catalysis (Figure 3.3).[29,30]

So far, only ligands **8** have proved to be extremely useful in several asymmetric transformation, such as the Cu-catalyzed [3+2] cycloaddition of azomethine ylides with a range of β-trifluoromethyl β,β-disubstituted enones and α-trifluoromethyl α,β-unsaturated esters,[29a–b] the Pd-catalyzed Suzuki reaction for the synthesis of axially chiral phosphonates and phosphine oxides,[29c] the Pd-catalyzed intramolecular Heck reaction of allyl aryl ethers,[29d] and the Pd-catalyzed intermolecular Heck reaction of aryl triflates and alkynes[29e] (Scheme 3.6). Ligands **9–11** have also shown their

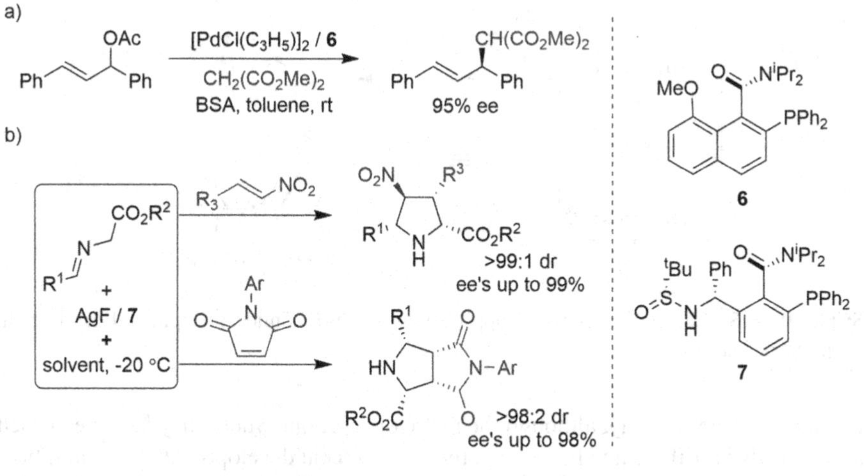

SCHEME 3.5 Application of non-biaryl atropoisomeric amido-phosphine ligands **6** and **7** in enantioselective a) Pd-catalyzed allylic substitution and b) Ag-catalyzed [3+2] cycloaddition reactions.

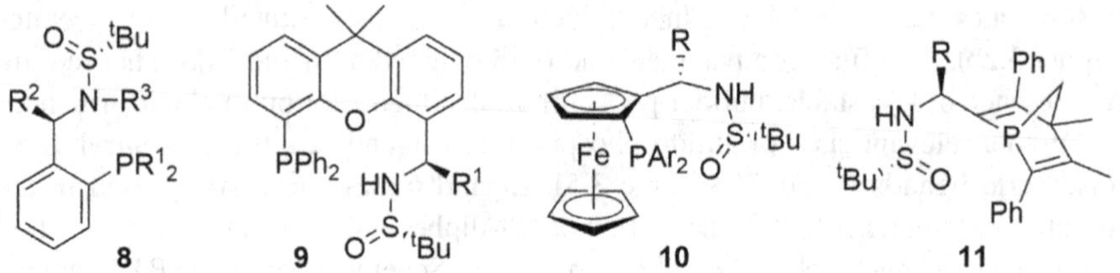

FIGURE 3.3 Chiral sulfonamide-phosphine ligands **8–11**.

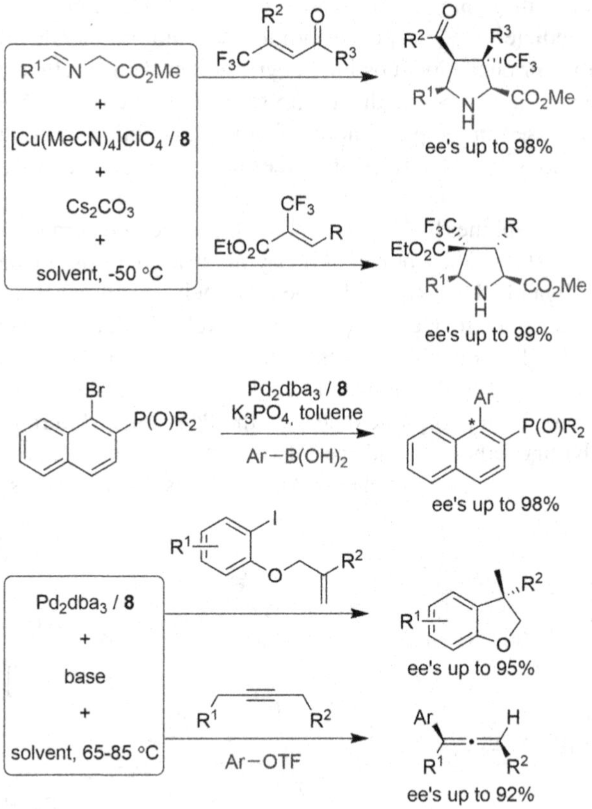

SCHEME 3.6 Selected successful applications of sulfonamide-phosphine ligands **8** in asymmetric catalysis.

usefulness more recently, although the range of reactions where they have been used successfully is still limited because of their very recent development. The Xantphos-inspired (ligand **9** has been used in the arylation of sulfenate anions (up to 99% ee),[30a] while ligands **10** and **11** have been shown to be useful in the boroacylation of 1,1-disubstituted allenes[30b] and 1,3-dipolar cycloadditions[30c–d], respectively.

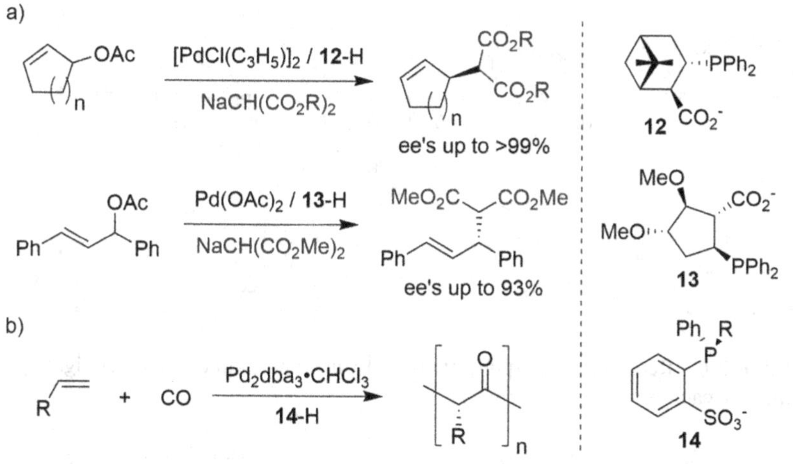

a)

ee's up to >99%

ee's up to 93%

b)

SCHEME 3.7 Representative successful applications of anionic P,O ligands **12–14**.

Finally, it should be mentioned that the use of anionic P,O ligands is far less developed than that of its neutral counterparts, in spite of the early successful application of phosphine-carboxylate ligands **12** and **13** (Scheme 3.7a).[31] These ligands provided high enantioselectivities in the Pd-catalyzed allylic substitution reactions (ee values up to >99%; Scheme 3.7a). More recently, the use of phosphine-sulfonate ligands **14** enabled the asymmetric copolymerization of polar vinyl monomers with carbon monoxide to yield highly head-to-tail isotactic γ-polyketone polymers (Scheme 3.7b).[32]

3.3 APPLICATION OF HETERODONOR P,S LIGANDS IN ASYMMETRIC CATALYSIS

P-thioether ligands are predominant among the P,S ligands.[33] Their use in asymmetric catalysis, however, suffers from an important drawback, namely the control of the diastereomeric mixtures of organometallic species arising from the sulfur atom, which becomes a stereogenic center upon coordination. In recent years, the number of successful applications and the versatility of the catalytic systems based on P,S ligands have grown considerably.[34] This is mainly because chemists have learned how to control the coordination of the sulfur thioether group with the chirality of the ligand backbone. One of the first successful exploitations of such control can be found in the development of phosphinite-thioether ligands **15** (Scheme 3.8).[35] The Evans group demonstrated that, by optimizing the different ligand parameters at this simple backbone, it is possible to control the thioether coordination to the metal. As a result, ligands **15** were the first family of P-thioether ligands to achieve excellent results with several asymmetric transformations, such as the Rh-catalyzed hydrosilylation and hydrogenation reactions,[35c,36] as well as in the Pd-catalyzed allylic substitution reactions[35a,b] (Scheme 3.8). However, the substrate/reagents range was rather

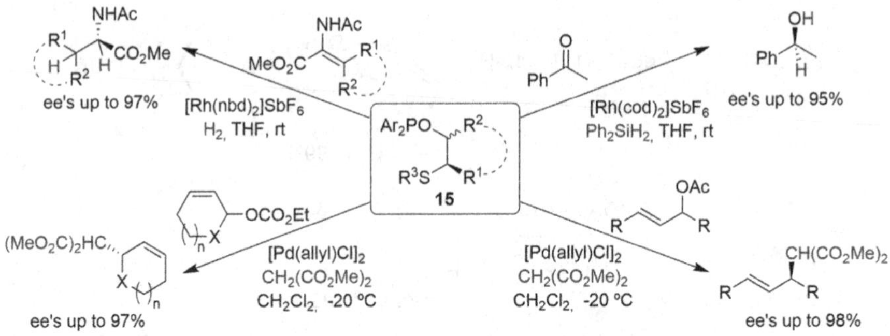

SCHEME 3.8 Representative applications of Evan's phosphinite-thioether ligands **15** in asymmetric catalysis.

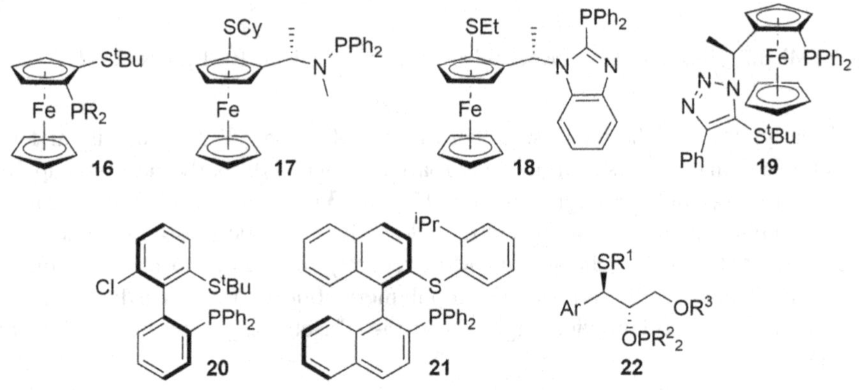

FIGURE 3.4 Phosphine/phosphinite-thioether ligands **16–22**.

limited. It should be noted that a careful optimization of the reaction conditions was also needed to exert a better control of the sulfur coordination. In the case of the allylic substitution reactions, for example, the reaction needs to be performed at low temperature (−20 °C) to achieve such excellent enantioselectivities.

Later, other families of phosphine/phosphinite-thioether ligands were successfully applied in catalysis, including those with planar (ligands **16–19**; Figure 3.4), axial (ligands **20–21**; Figure 3.4), and central chirality (ligands **22**; Figure 3.4). With the exception of ligands **20–21**, these ligands have allowed expansion of the range of asymmetric reactions and/or the substrate/reagents scope which could be used successfully.

Ferrocenyl-based Fesulphos ligands **16**, developed by Carretero's group, have become particularly useful in many asymmetric transformations, achieving high enantioselectivities in combination with either Pd and Cu catalyst precursors.[37] Thus, catalysts Pd/**16** have shown high enantioselectivities in the allylic substitution of the model substrate *rac*-1,3-diphenylallyl acetate, using malonates or *N*-nucleophiles,[37a,b] and in the ring opening of heterobicyclic alkenes, using diorganylzinc reagents[37a,c,f]

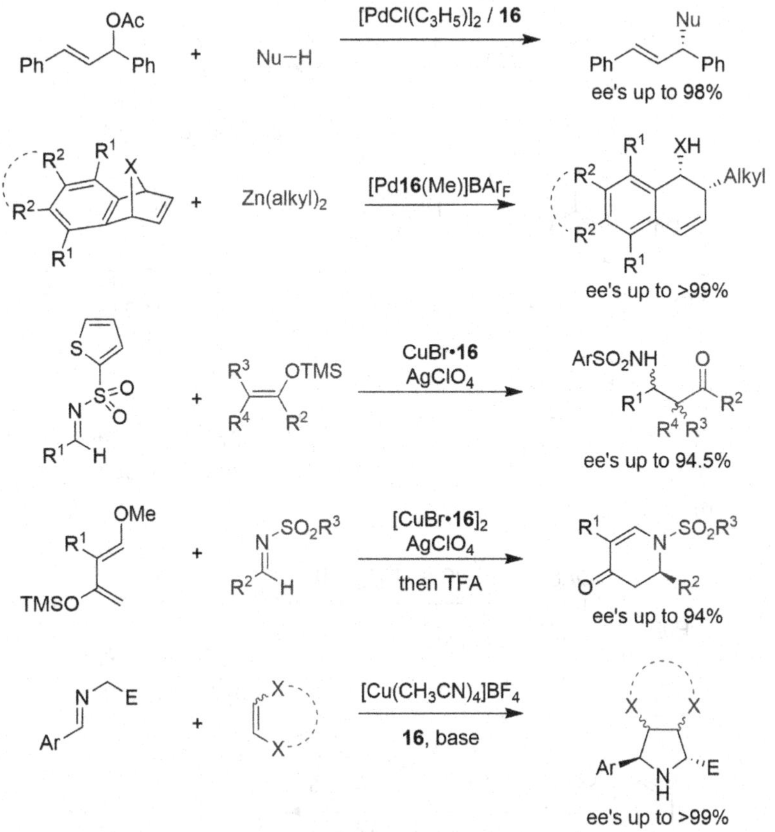

SCHEME 3.9 Representative applications of ferrocenyl-based Fesulphos ligand **16** in asymmetric catalysis.

(ee values up to 98% and >99%, respectively; Scheme 3.9). On the other hand, the catalysts Cu/**16** achieved excellent enantiocontrol in Mannich-type reactions of *N*-sulfonylimines with several electrophiles,[37h,l] (aza)-Diels-Alder reactions of electron-rich alkenes with aldimines,[37d,g] and 1,3-cycloaddition reactions of azomethine ylides with a wide range of activated alkenes[37e,i,k,m,o-r] (up to >99% ee; Scheme 3.9).

Later on, Chan's group developed the ligands **17**[38] and **18**[39] (Figure 3.4), which combine both planar and central chirality, and demonstrated their versatility in Pd-catalyzed allylic substitution reactions. The use of Pd/**17** not only provided high ee values in the allylic substitution of *rac*-1,3-diphenylallyl acetate with dimethyl malonate and a range of amines (up to 98% ee; Scheme 3.10a),[38a] but also with a range of aliphatic alcohols (ee values up to 98%; Scheme 3.10a)[38b]. The latter represents one of the first successful examples of the allylic etherification reactions with non-aromatic alcohols. More recently, the use of ligand **18,** containing a benzimidazole unit, extended the range of nucleophiles of Pd/**17** to include less-studied nucleophiles, such as indoles (ee values up to 96%; Scheme 3.10b).[39a] Pd/**18** also

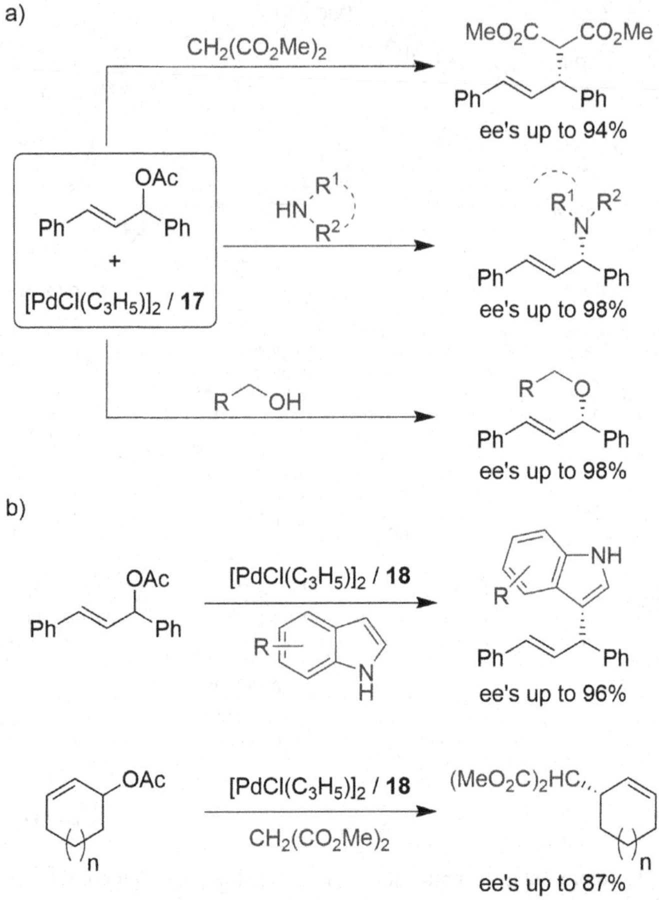

a)

b)

SCHEME 3.10 Representative applications in Pd-catalyzed allylic substitution reactions, using ferrocenyl-based ligands a) **17** and b) **18**.

proved to be able to expand the range of substrates to include some cyclic substrates (n = 0–2; Scheme 3.10b).[39b] In this way, the allylic alkylation of *rac*-cyclic allylic acetates with dimethyl malonate achieved ee values of up to 87% (Scheme 3.10b).

Fukuzawa's group developed ThioClick-ferrophos ligand **19** (Figure 3.4) and successfully applied it in the Ag-catalyzed Mannich reactions of *N*-tosylimines with a glycine Schiff base.[40a] The reaction proceeded with moderate diastereoselectivities and high enantioselectivities (up to 7:3 dr and up to 98% ee). Ag/**19** efficiently catalyzed the 1,3-dipolar cycloadditions of azomethine ylides with several activated alkenes, achieving enantioselectivities similar to those achieved using catalyst Cu/ Fesulphos **16** (up to 99% ee).[40b–g] Ferrocenyl-based ligand **19** was also successfully used in the Ag-catalyzed conjugate addition of a range of Michael acceptors to different imino esters, oxazoline esters and related substrates (ee values up to >99%; Scheme 3.11).[40i–l] Interestingly, the same authors also demonstrated that, by

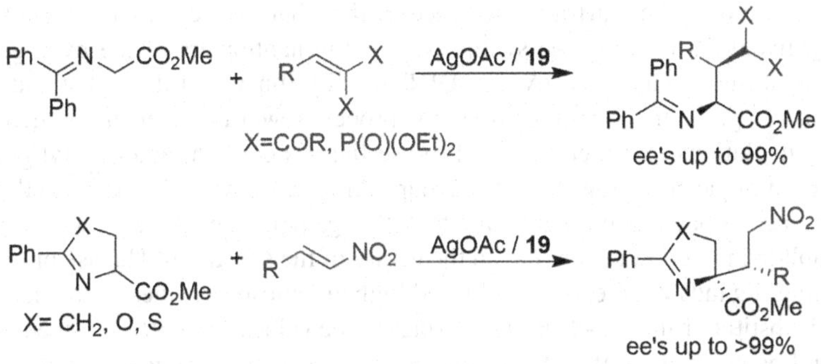

SCHEME 3.11 Representative examples of Ag-catalyzed conjugate addition with ferrocenyl-based P,S ligand 19.

switching the positions of the thioether and the phosphine moieties, high enantioselectivities could also be achieved in the Pd-catalyzed allylic substitution reactions (up to 90% ee).[40m]

Chiral biaryl ligands 20 and 21 (Figure 3.4), with axial chirality, have also been used in asymmetric catalysis. However, their reaction scope is rather narrow, compared with their P,P and P,O counterparts, such as BINAP and MOP ligands. Thus, ligands 20 and 21 achieve excellent enantioselectivities only in the Pd-catalyzed allylic alkylation of *rac*-1,3-diphenylallyl acetate with a range of indoles (ee values up to 95%).[41]

Arylglycidol-based phosphinite-thioether ligands 22 (Figure 3.4) have been found to be useful in allylic substitution and hydrogenation reactions.[42] Thus, Pd/22 catalytic systems achieved high enantioselectivities similar to those obtained by Evans ligands 15 in the allylic substitution of di- and trisubstituted linear allylic acetates with a range of malonate-type nucleophiles and amines, working under milder reaction conditions (e.g., at room temperature), as well as extending the nucleophile scope to the less-studied aliphatic alcohols.[42a] Ligands 22 were also used in the Rh-catalyzed hydrogenation of dehydroamino acids, albeit with less success than with the related ligands 15.[42b] Interestingly, ligands 22 have proved to be highly efficient in the asymmetric Ir-catalyzed hydrogenation of non-functionalized olefins,[42c] in spite of the known difficulties of enantiocontrol associated with substrates lacking metal-coordinating functionalities.[43] Most of the successful catalysts in the reduction of non-functionalized olefins used in the past were too specific for a certain double bond geometry and substitution pattern of the olefin. For example, the most successful cases have been reported for the trisubstituted (*E*)-non-functionalized alkenes and, to a lesser extent, for (*Z*)-trisubstituted and 1,1'-disubstituted alkenes. On the positive side, the new phosphinite–thioether ligands 22 could hydrogenate not only (*E*)-trisubstituted but also a variety of 1,1'-disubstituted alkenes, with results comparable to the best ones reported in the literature with Ir–P,N catalysts. An important practical advantage offered by ligands 22 lies in the fact that these ligands are synthesized from readily available arylglycidols in only three steps.[42c] In addition, both enantiomeric series of these P,S ligands are equally available (by simple selection of the tartrate ester used in the Sharpless epoxidation, leading to the arylglycidol), allowing the synthesis of both enantiomers of the hydrogenated products at high enantioselectivities. DFT (density-functional theory) studies were necessary to guide the ligand optimization process toward high enantioselectivities. These calculations indicated the need to use ligands containing a mesityl group at the carbon adjacent to the thioether group (Ar = 2,4,6-Me$_3$-C$_6$H$_2$), as well as the convenience of having bulky aromatic thioether groups (such as 2,6-dimethylphenyl or 1-naphthyl moieties, depending on the nature of the substrate). The use of mesityl-containing ligands **23**, therefore, achieved high enantioselectivities for a wide range of *E*-trisubstituted and 1,1'-disubstituted olefins (ee values up to >99%; Scheme 3.12). It should be pointed out that the achievement of excellent enantioselectivities could also be extended to olefins containing relevant poorly coordinative polar groups, like α,β-disubstituted enones, β,β'-disubstituted unsaturated esters, vinyl boronates, and olefins with trifluoromethyl substituents. The effective hydrogenation of such a wide range of olefins is of great significance, since their reduced products are key structural chiral units found in many high-value chemicals (e.g. α- and β-chiral ketones are ubiquitous in natural products, fragrances, agrochemicals, and drugs, while chiral borane compounds are useful building blocks for synthesis, because the C–B bond can readily be converted to C–O, C–N, and C–C bonds, with retention of the chirality).[43] Remarkably, the catalytic systems could also be recycled up to three times by using the environmentally benign propylene carbonate, by simple extraction with hexane. DFT calculations also indicated that the reaction proceeds *via* an Ir(III)/Ir(V) catalytic system, in which the enantioselectivity-determining step is the migration of a hydride to the coordinated alkene. In addition, the analysis of the transition states allowed the development of a quadrant model system, that facilitates rationalization of the catalytic results.

Another strategy to control the preferential formation of a single diastereomer upon coordination of the thioether group consists of introducing a biaryl phosphite or phosphoramidite group into the ligand design. The most successful phosphite/ phosphoramidite-thioether ligand designs are presented in Figure 3.5.

The use of the ribofuranoside phosphite-thioether ligand **24** represented the first application of P,S ligands in the asymmetric hydrogenation of non-functionalized

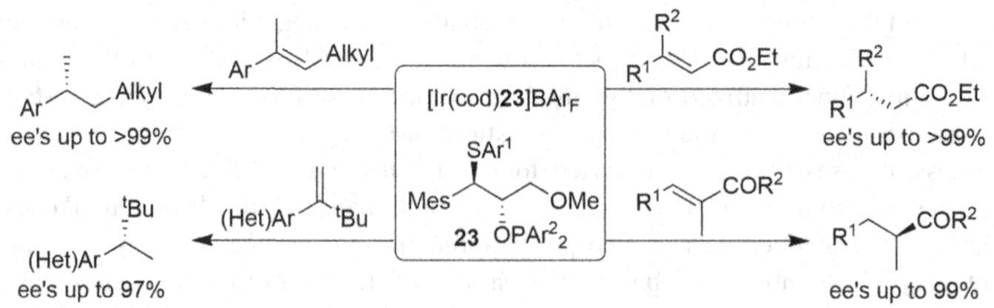

SCHEME 3.12   Representative examples of Ir-catalyzed asymmetric hydrogenation of non-functionalized olefins and olefins with poorly coordinative groups, using mesityl-containing ligands **23**.

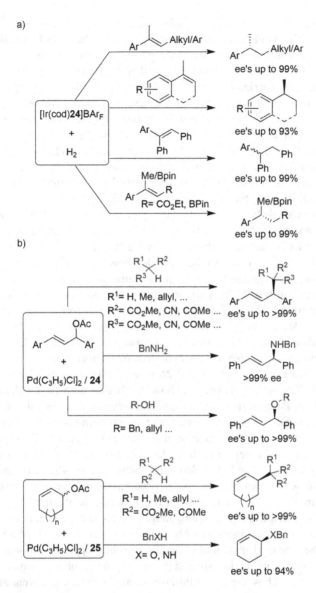

SCHEME 3.13 Representative examples of the use of furanoside ligands **24** and **25** in the a) Ir-catalyzed asymmetric hydrogenation of non-functionalized olefins and b) Pd-catalyzed allylic substitution reactions.

olefins (up to 99% ee; Scheme 3.13a).[44] Catalyst Ir/**24** proved to be highly efficient in the hydrogenation of not only methyl stilbene-type *E*-trisubstituted olefins, but also *Z*-trisubstituted olefins and triaryl-substituted olefins, providing a feasible entry point to synthetic pathways for valuable compounds containing diarylmethine chiral centers. This catalytic system also provided high enantioselectivities in the hydrogenation of many 1,1'-disubstituted olefins. Interestingly, for this substrate class, both enantiomers of the hydrogenated product can be readily attained by changing the configuration of the biaryl phosphite moiety in the ligand structure. Another interesting feature of this ligand is that the furanoside backbone is able to control the tropoisomerization of conformationally flexible biaryl phosphite moieties. So, for most of the substrates studied, similarly high enantioselectivities have been achieved with the cheap achiral bulky biphenyl phosphite moiety present in ligand **24**. Ligand **24** was also shown to be extremely useful in the Pd-catalyzed allylic substitution of hindered 1,3-disubstituted allylic acetates with a range of *C*-nucleophiles (such as α-substituted malonates, diketones, cyano esters, etc.) and with a range of *N*- and *O*-nucleophiles (up to >99% ee; Scheme 3.13b).[45] To achieve high enantioselectivities for unhindered linear and cyclic substrates, the use of the related xylofuranoside ligand **25** was required (ee values up to >99%; Scheme 3.13b). This feature was rationalized with the aid of NMR studies on the Pd-allyl intermediates and of DFT calculations of the transition states, using cyclohex-2-en-1-yl acetate as the model substrate. These studies revealed how the configuration at C-3 of the furanoside backbone affected the size of the chiral pocket in the catalytic species. Thus, by using catalyst Pd/**25**, only one of the two possible *syn/syn* diastereomer Pd-1,3-cyclohexenyl-allyl intermediates is predominantly formed (dr value >20:1).[45b]

Later, the development of the indene-based phosphite-thioether ligand **26** (Figure 3.5) demonstrated that a simpler backbone than the furanoside one present in ligands **24** and **25** is also able to achieve high enantioselectivities in the Pd-catalyzed allylic substitution reactions.[46] Ligand **26** is synthesized in only three steps from cheap indene. The simple indene backbone facilitated both DFT calculations and NMR studies of Pd-allyl intermediates, which were used to optimize the thioether and phosphite substituents in the search for the optimum catalyst. As a result, catalyst Pd/**26** is one of the few catalytic systems able to provide excellent enantioselectivities (typically >95% ee) for both hindered and unhindered allylic acetates with a broad range of *C*-, *N*-, and *O*-nucleophiles (Scheme 3.14; 40 compounds in total). In comparison with previous, furanoside-based P,S ligands, which have emerged as some of the most successful catalysts for this process, the P,S ligand **26** achieved a greater activity and a wider nucleophile range (i.e., including the addition of pyrroles and a broader range of amines). Notably, the excellent performance of ligand **26** was preserved when the environmentally friendly solvent propylene carbonate was used. Mechanistic investigations shed some light onto the wide substrate range, which was exceptionally rare. Thus, the use of Pd/**26** not only favors the preferential formation of one of the possible Pd-allyl intermediates, but also speeds up the nucleophilic attack toward the terminal allylic carbon atom *trans* to the phosphite moiety of most stable Pd-allyl intermediates. In addition, the authors took advantage of the great diversity of the allylic substitution products arising from the introduction of malonates containing allyl and propargylic groups, for the synthesis of a range of chiral functionalized carbo- and heterocycles, as well as polycarbocycles. The former compounds were prepared by means of ring-closing metathesis, whereas the latter were prepared *via* the Pauson-Khand reaction (Scheme 3.15).

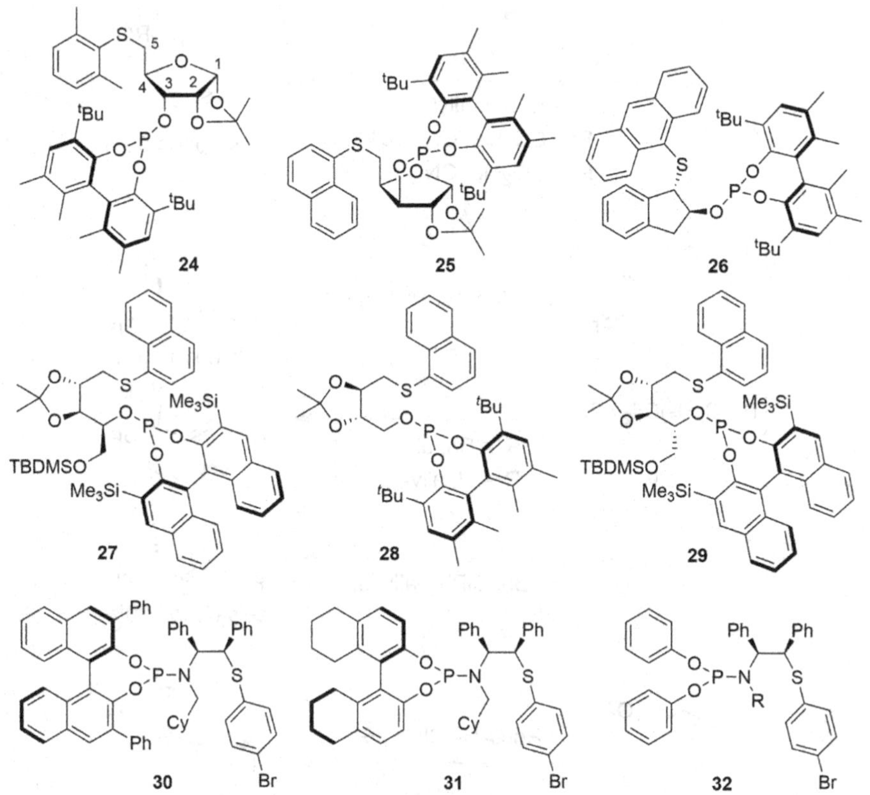

FIGURE 3.5 Phosphite/phosphoramidite-thioether ligands **24–32**.

More recently, a large family of TADDOL (α,α,α',α'-tetraaryl-2,2-disubstituted 1,3-dioxolane-4,5-dimethanol)-type phosphite-thioether ligands was synthesized from carbohydrates (D-mannitol and L-tartaric acid). Screening of this library with respect to several asymmetric transformations led to the discovery of ligands **27–29** (Figure 3.5).[47] Ligand **27** proved to be highly efficient in the Ir-catalyzed asymmetric hydrogenation of *E*-trisubstituted non-functionalized olefins. Catalyst Ir/**27** achieved high enantioselectivities (ee values up to 95%) in the hydrogenation of *trans*-methyl stilbene-type substrates, as well as β,β'-disubstituted unsaturated esters, α,β-disubstituted enones, and lactones and lactams bearing an exocyclic double bond (up to 95% ee).[47b] It should be pointed out that, for most of these substrates, the selenoether version of the ligands provided somewhat higher enantioselectivities than did the thioether analogs.[47b] On the other hand, the use of catalyst Ir/**28** was necessary to maximize enantioselectivity in the hydrogenation of 1,1'-disubstituted olefins (ee values up to 99%).[47b] The use of ligand **29**, with the configuration at both carbons adjacent to the phosphite group and at the phosphite moiety opposite to that of ligand **27**, provided excellent enantioselectivities in the Ir-catalyzed hydrogenation of the

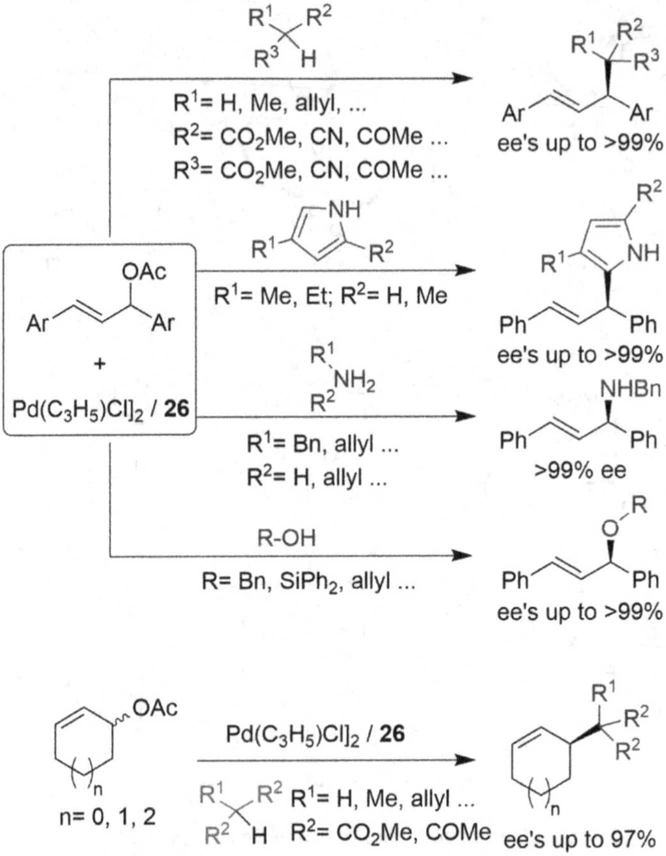

SCHEME 3.14 Pd-catalyzed allylic substitution of disubstituted linear and cyclic substrates with C-, N-, and O-nucleophiles using indene-based phosphite-thioether ligand **26**.

challenging cyclic β-enamides (up to 99% ee; Scheme 3.16a). Their hydrogenation products, such as 2-aminotetralines and 3-aminochromanes, are structural fragments found in biologically active natural products and therapeutic agents.[48] Despite this, only a few examples of successful hydrogenation of these substrates had previously been reported, mainly with Rh-/Ru-catalysts, and even then with a limited substrate range, except for two recent reports wit Ir/P–N catalysts.[49] Also noteworthy, the switch from Ir/**29** to Rh/**29** led to the formation of the opposite enantiomer of the corresponding 2-aminotetralines and 3-aminochromanes at similarly high enantiose-lectivities (up to 95% ee; Scheme 3.16a).[47a] This observation represents one of the rare examples of enantio-switchable metal-catalyzed transformation. This allows, for instance, access to both enantiomers of the precursors for the synthesis of rotigotine (a non-ergoline dopamine agonist indicated for the treatment of Parkinson's disease and restless legs syndrome)[48a] and alnespirone (a selective 5-HT1A receptor full agonist of the azapirone chemical class).[48d] In addition, the use of Rh/**27** and Rh/**29** catalysts

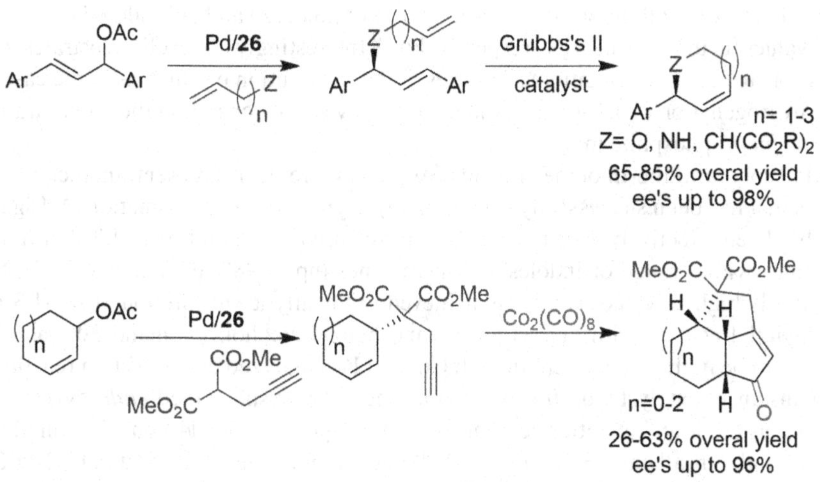

SCHEME 3.15 Preparation of functionalized carbo- and heterocycles *via* ring-closing metathesis as well as polycarbocyles *via* Pauson–Khand reactions.

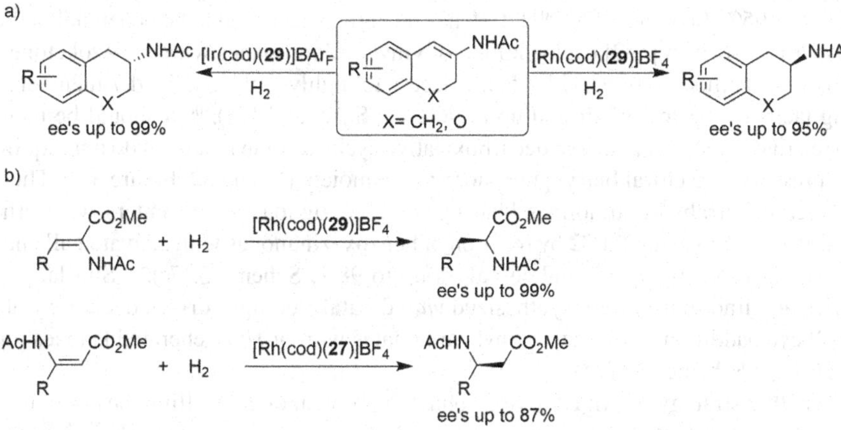

SCHEME 3.16 Application of phosphite-thioether ligands **27** and **29** in the asymmetric hydrogenation of a) cyclic β-enamides using Rh- and Ir-catalysts and b) dehydroamino acid derivatives.

proved to be useful in the asymmetric hydrogenation of functionalized olefins, such as dehydroamino acid derivatives (Scheme 3.16b).[47b] This again is a very unusual feature, since the hydrogenation of both functionalized and non-functionalized olefins follows very different catalytic cycles, and each type of substrate has been shown to require a particular catalyst type (Rh-/PP-catalysts for functionalized and Ir/PN-catalysts for non-functionalized olefins) for optimal results.[43]

Interestingly, the use of ligand **29** and its derivatives containing different silylated protecting groups (e.g., TBDPS or TIPS) also achieved high enantioselectivities in the Pd-catalyzed allylic substitution of a range of hindered and unhindered substrates (ee values up to 99% and 91%, respectively).[47c] Interestingly, for cyclic substrates, it is possible to select the enantiomeric series of the substitution product, as in the case of the hydrogenation of 1,1'-disubstituted olefins, by simply changing the configuration of the biaryl phosphite moiety.

Phosphoramidite-thioether ligands 30–32 (Figure 3.5) represent another ligand class that has been successfully used in many asymmetric transformations.[50] Ligand 30 has been effectively used in the Pd-catalyzed allylic substitution of 1,3-diarylallyl acetates with a range of indoles and hydrazones (up to 98% ee; Scheme 3.17a).[50a,b] Similarly high ee values were also achieved in the allylic substitution of rac-1,3-di-phenylallyl acetate with benzyl amine and benzyl alcohol.[50a] Ligand 30 was also shown to be highly competent in both Cu- and Pd-catalyzed cycloaddition reactions. For instance, catalyst Cu/30 generated a range of polysubstituted endo pyrroles in high diastereo- and enantioselectivities via 1,3-dipolar cycloaddition of azomethine ylides and nitroalkenes (Scheme 3.17b).[50c] Interestingly, the use of related H_8-BINOL (1,1'-bi-2-naphthol)-derived ligand 31 (Figure 3.5) led to the formation of the exo pyrroles (Scheme 3.17b).[50c] Catalyst Pd/30 was successfully used in inverse-electron demand decarboxylative [4+2] cycloaddition reactions. Thus, highly functionalized dihydroquinol-2-ones were produced with excellent selectivities (dr > 20:1 and ee values up to 95%; Scheme 3.17c).[50d] Pd/30 has recently been found to be beneficial in the visible-light-driven [5+2] cycloaddition of vinyl cyclopropanes with α-diazoketones. This new methodology provides facile access to highly functionalized 7-membered ring lactones (up to 16:1 dr and up to 96% ee; Scheme 3.17d).[50e] It should be mentioned that some Pd-catalyzed decarboxylative cycloaddition reactions do not require the presence of a chiral biaryl phosphoramidite moiety (ligand 32; Figure 3.5). Thus, a range of tetrahydroquinolines bearing three contiguous stereocenters were efficiently prepared using Pd/32 by reaction of benzoxazinanones with activated alkenes (dr values typically >95:5 and ee values up to 98%; Scheme 3.17e).[50f] Similarly, a series of quinolinones were synthesized via Pd-catalyzed light-driven decarboxylate [4+2] cycloaddition of tosylated vinyl carbamates with in-situ-generated ketenes (up to 96% ee; Scheme 3.17f).[50g]

Another strategy by which to overcome the problem of controlling the configuration of the S-thioether group upon coordination to the metal center is based on the replacement of the thioether group by a chiral sulfoxide. Figure 3.6 shows the most successful P-sulfoxide ligands developed to date.

The use of ligands 33 constituted the first fruitful application of P-sulfoxide ligands in many asymmetric transformations.[51] Thus, excellent enantioselectivities (up to 98% ee) were obtained in the Rh/33-catalyzed asymmetric 1,4-addition of arylboronic acids to a wide range of electron-deficient olefins (Scheme 3.18a).[51a] The introduction of a second sulfoxide moiety at the other ortho position of the phosphine group (ligand 34) allowed the construction of chiral γ,γ-diaryl-substituted carbonyl compounds via the same transformation, which led to the synthesis of bioactive compounds such as sertraline (up to >99% ee; Scheme 3.18b).[52] Ligands 33 also allowed the first Cu-catalyzed formation of α-aryl-β-borylstannanes by means

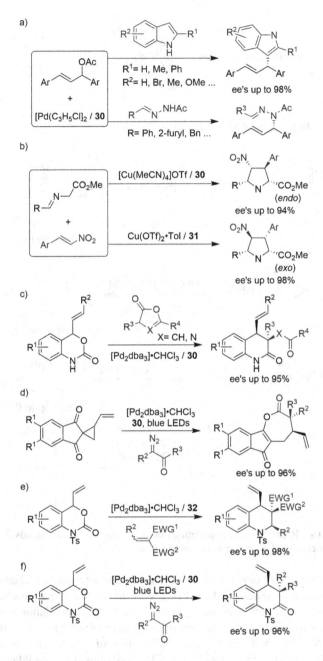

SCHEME 3.17 Representative applications of phosphoramidite-thioether ligands **30–32** in asymmetric catalysis.

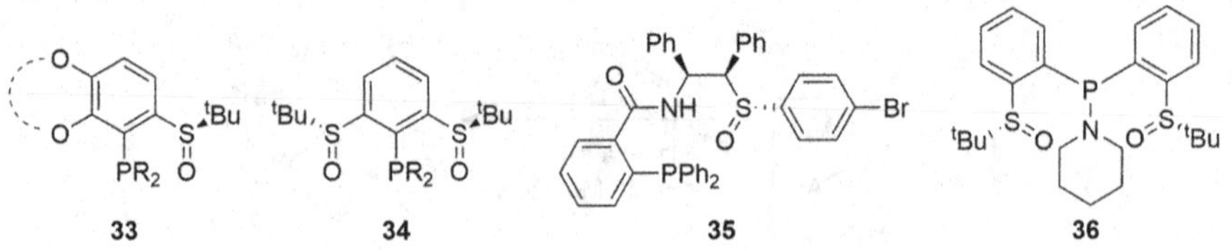

FIGURE 3.6 Heterodonor P-sulfoxide ligands **33–36**.

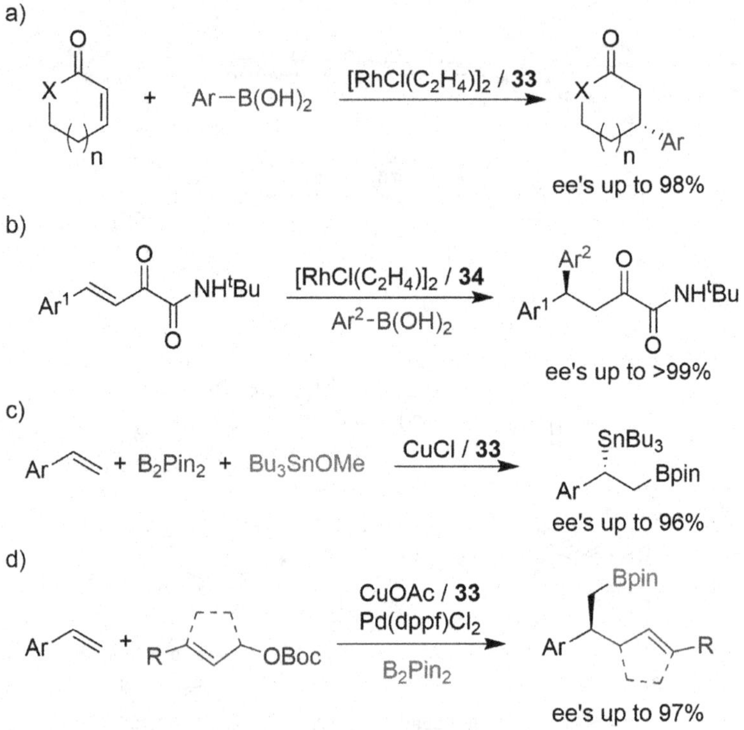

SCHEME 3.18 Representative applications of phosphine-sulfoxide ligands **33** and **34** in several metal-catalyzed asymmetric transformations.

of a three-component borylstannation of aryl-substituted alkenes (Scheme 3.18c).[51b] Such transformation relies on the efficiency of controlling the stereochemistry of the BCu addition, as well as its ability to facilitate the transmetalation of enantioenriched alkyl-Cu species, with retention of the configuration. More recently, a cooperative Cu-/Pd-catalyzed asymmetric allylboration of alkenes has been reported (Scheme 3.18d).[51c] Thus, the use of CuOAc/**33,** in combination with Pd(dppf)Cl₂ catalysts, allowed the three-component reaction of styrenes, B₂(pin)₂ (Bis(pinacolato) diboron), and allyl carbonates at high enantioselectivities (up to 97% ee).

Phosphine-sulfoxide ligand **35** (Figure 3.6) proved to have a very broad nucleophile substrate range in the Pd-catalyzed allylic substitution of 1,3-diarylallyl

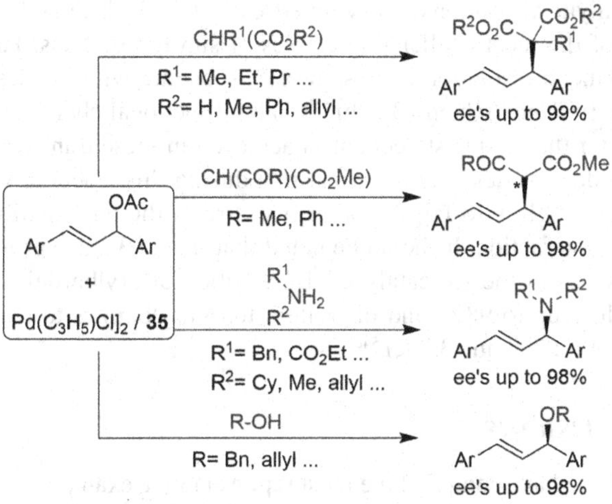

SCHEME 3.19 Pd-catalyzed allylic substitution of 1,3-diarylallyl acetates using phosphine-sulfoxide ligand **35**.

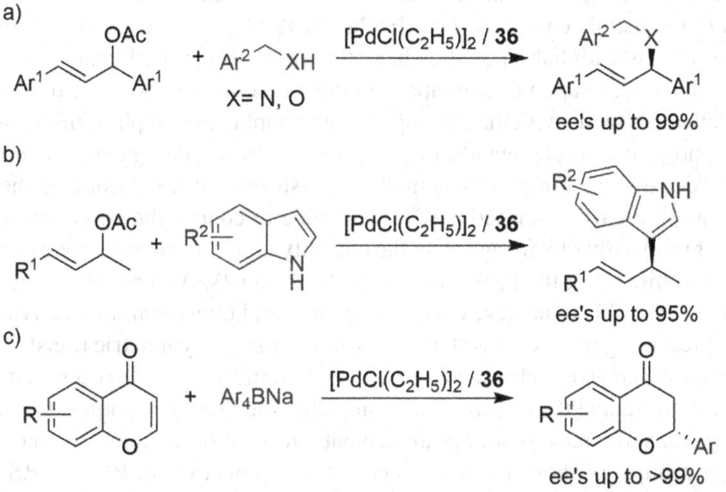

SCHEME 3.20 Representative applications of phosphine-bis(sulfoxide) ligand **36** in several metal-catalyzed asymmetric transformations.

acetates.[53] As a result, excellent enantioselectivities were achieved using various malonates, including examples with different functionalities at the α-position, as well as ketoesters, amines, alcohols, and indoles (ee values up to 99%; Scheme 3.19).

More recently, phosphine-bis(sulfoxide) ligand **36** (Figure 3.6) also proved to be highly effective at the Pd-catalyzed allylic etherification and amination of 1,3-diarylallyl acetates with benzylic alcohols and amines (ee values up to 99%; Scheme 3.20a).[54a] Even more interesting were the excellent results achieved in the

Pd-catalyzed allylic alkylation of unsymmetrically 1,3-disubstituted allylic acetates, which is one of the most challenging classes of substrates. Thus, Pd/**36** catalyzed the dynamic kinetic resolution of this class of substrate with indoles (yields up to 84% and up to 95% ee; Scheme 3.20b).[54b] The bifunctional character of this ligand is responsible for the unique stereocontrol achieved in these transformations. Thus, the two sulfoxide moieties play a distinct role during this reaction; whereas one of them coordinates tightly to Pd, the other one directs the nucleophilic attack *via* a hydrogen-bond interaction. It should be noted that ligand **36** also provided excellent enantioselectivities in the Rh-catalyzed 1,4-addition of arylboronic acids to cyclic enones (ee values up to 98%) and of sodium tetra-arylborates to chromenones (ee values up to >99%; Scheme 3.20c).[54c]

3.4   CONCLUSIONS

In this chapter, we have compiled the most representative examples of the use of P,O and P,S ligands in asymmetric metal catalysis. Traditionally, the use of such ligand classes in asymmetric catalysis has been hampered by the hemilabile character of the O-donor group in P,O ligands and, in the case of P,S ligands, by the formation of diastereomeric mixtures of organometallic species following coordination of the S atom to the metal center. The hemilability in P,O ligands can be controlled either by using metals with harder acidic characters or by designing ligands able to form a stable chelate ring upon coordination. In this latter context, we have described the recent advances made with the development and application of phosphine-phosphin-oxides, phosphine-amide ligands, or phosphine-sulfonamide ligands, among others, to achieve a diverse range of asymmetric transformations. Concerning the use of P,S ligands, we have described how it is possible to control the coordination of the S-donor group, either by using solely the chirality at the ligand backbone or by combining the chirality at the backbone and at the P-moiety. As a result, the number of ligands that are able to achieve comparable (or even better) enantioselectivities than those representing the previous state-of-the-art in many asymmetric transformations has increased considerably in the past decade. Another, more recent, successful strategy to overcome the drawback of controlling the S-coordination is based on the use of chiral sulfoxides as structural elements, instead of the most commonly used thioether groups. In short, the latest successful designs in both P,O and P,S ligands have increased the potential of these heterodonor ligands to achieve a diverse range of catalytic transformations, constituting, in some cases, the current state-of-the-art. In addition, most of them are highly modular, stable in air and water, and easy to synthesize. The combination of all these features makes these less-studied ligand classes particularly appealing for further research in asymmetric catalysis.

3.5   ACKNOWLEDGEMENTS

We gratefully acknowledge financial support from the Spanish Ministry of Economy and Competitiveness (CTQ2015-69136-R and PID2019-1092336RB-I00, CTQ2016-74878-P, PID2019-104904GB-I00, and Severo Ochoa Excellence Accreditation

2014-2018, SEV-2013-0319), European Regional Development Fund (AEI/ FEDER, UE), the Catalan Government (2017SGR1472 and 2017SGR1139), CERCA Programme/Generalitat de Catalunya, and the "La Caixa" Foundation.

REFERENCES

1. For representative applications of MOP ligands, see: a) Y. Uozumi, T. Hayashi, *J. Am. Chem. Soc.* **1991**, *113*, 9887–9888; b) Y. Matsumoto, M. Naito, Y. Uozumi, T. Hayashi, *J. Chem. Soc. Chem. Commun.* **1993**, 1468–1469; c) T. Hayashi, H. Iwamura, M. Naito, Y. Matsumoto, Y. Uozumi, M. Miki, K. Yanagi, *J. Am. Chem. Soc.* **1994**, *116*, 775–776; d) K. Kitayama, Y. Uozumi, T. Hayashi, *J. Chem. Soc. Chem. Commun.* **1995**, 1533–1534; e) T. Hayashi, M. Kawatsura, H. Iwamura, Y. Yamaura, Y. Uozumi, *Chem. Commun.* **1996**, 1767–1768; f) T. Hayashi, M. Kawatsura, Y. Uozumi, *Chem. Commun.* **1997**, 561–562; g) T. Hayashi, M. Kawatsura, Y. Uozumi, *J. Am. Chem. Soc.* **1998**, *120*, 1681–1687; h) R. Shintani, M. Inoue, T. Hayashi, *Angew. Chem. Int. Ed.* **2006**, *45*, 3353–3356.
2. For related MOP-type ligands that coordinates as monodentated ligand, see for instance: a) W. Fu, M. Nie, A. Wang, Z. Cao, W. Tang, *Angew. Chem. Int. Ed.* **2015**, *54*, 2520–2524; b) M. Nie, W. Fu, Z. Cao, W. Tang, *Org. Chem. Front.* **2015**, *2*, 1322–1325; c) N. Hu, G. Zhao, Y. Zhang, X. Liu, G. Li, W. Tang, *J. Am. Chem. Soc.* **2015**, *137*, 6746–6749.
3. (a) T. Hayashi, *Pure Appl. Chem.* **1988**, *60*, 7–12. (b) T. Hayashi, A. Yamamoto, Y. Ito, E. Nishioka, H. Miura, K. Yanagi, *J. Am. Chem. Soc.* **1989**, *111*, 6301–6311.
4. V. V. Grushin, *Chem. Rev.* **2004**, *104*, 1629–1662.
5. J. W. Faller, B. J. Grimmond, D. G. D'Alliessi, *J. Am. Chem. Soc.* **2001**, *123*, 2525–2529.
6. T. H. Wöste, M, Oestreich, *Chem. Eur. J.* **2011**, *17*, 11914–11918.
7. a) J. Hu, H. Hirao, Y. Li, J. Zhou, *Angew. Chem. Int. Ed.* **2013**, *52*, 8676–8680; b) H. Li, S.-l. Wan, C.-H. Ding, X.-L. Hou, *RSC Adv.* **2015**, *5*, 75411–75414.
8. A. A. Boezio, J. Pytkowicz, A. Côté, A. B. Charette, *J. Am. Chem. Soc.* **2003**, *125*, 14260–14261.
9. A. Côté, A. A. Boezio, A. B. Charette, *PNAS* **2004**, *101*, 5405–5410.
10. J. Hu, Y. Lu, Y. Lia, J. Zhou, *Chem. Commun.* **2013**, *49*, 9425–9427.
11. Q.-S. Zhang, S.-L. Wan, D. Chen, C.-H. Ding, X.-L. Hou, *Chem. Commun.* **2015**, *51*, 12235–12238.
12. S. Liu, J. Zhou, *Chem. Commun.* **2013**, *49*, 11758–11760.
13. C. Wu, J. Zhou, *J. Am. Chem. Soc.* **2014**, *136*, 650–652.
14. M. Kuriyama, K. Nagai, K. Yamada, Y. Miwa, T. Taga, K. Tomioka, *J. Am. Chem. Soc.* **2002**, *124*, 8932–8939.
15. F. Valleix, K. Nagai, T. Soeta, M. Kuriyama, K. Yamada. K. Tomioka, *Tetrahedron* **2005**, *61*, 7420–7424.
16. K. B. Selim, K. Yamada, K. Tomioka, *Chem. Commun.* **2008**, 5140–5142.
17. a) M. Kuriyama, T. Soeta, X. Hao, Q. Chen, K. Tomioka, *J. Am. Chem. Soc.* **2004**, *126*, 8128–8129; b) X. Hao, M. Kuriyama, Q. Chen, Y. Yamamoto, K. Yamada, K. Tomioka, *Org. Lett.* **2009**, *11*, 4470–4473.
18. T. Soeta, M. Kuriyama, K. Tomioka, *J. Org. Chem.* **2005**, *70*, 297–300.
19. T. Soeta, K. Selim, M. Kuriyama, K. Tomioka, *Adv. Synth. Catal.* **2007**, *349*, 629–635.
20. T. Soeta, K. Selim, M. Kuriyama, K. Tomioka, *Tetrahedron* **2007**, *63*, 6573–6576.
21. K. Selim, T. Soeta, K. Yamada, K. Tomioka, *Chem. Asian J.* **2008**, *3*, 342–350.
22. D. Rageot, D. H. Woodmansee, B. Pugin, A. Pfaltz, *Angew. Chem. Int. Ed.* **2011**, *50*, 9598–9601.

23. For other P,O-ligands containing an urea or a carbamate group see for instance: a) J. Meeuwissen, R. J. Detz, A. J. Sandee, B. de Bruina, J. N. H. Reek, *Dalton Trans.* **2010**, *39*, 1929–1931; b) J. Meeuwissen, R. Detz, A. J. Sandee, B. de Bruin, M. A. Siegler, A. L. Spek, J. N. H. Reek, *Eur. J. Inorg. Chem.* **2010**, 2992–2997; c) Á. M. Pálvölgyi, M. Schnürch, K. Bica-Schröder, *Tetrahedron* **2020**, *76*, 131246–131253.

24. P. Elías-Rodríguez, C. Borràs, A. T. Carmona, J. Faiges, I. Robina, O. Pàmies, M. Diéguez, *ChemCatChem* **2018**, *10*, 5414–5424.

25. W.-M. Dai, K. K. Y. Yeung, J.-T. Liu, Y. Zhang, I. D. Williams, *Org. Lett.* **2002**, *4*, 1615–1618.

26. W.-M. Dai, K. K. Y. Yeung, Y. Wang, *Tetrahedron* **2004**, *60*, 4425–4430.

27. X.-F. Bai, T. Song, Z. Xu, C.-G. Xia, W.-S. Huang, L.-W. Xu, *Angew. Chem. Int. Ed.* **2015**, *54*, 5255–5259.

28. X.-F. Bai, J. Zhang, C.-G. Xia, J.-X. Xu, L.-W. Xu, *Tetrahedron* **2016**, *72*, 2690–2699.

29. b) Z.-M. Zhang, B. Xu, S. Xu, H.-H. Wu, J. Zhang, *Angew. Chem. Int. Ed.* **2016**, *55*, 6324–6328; b) B. Xu, Z.-M. Zhang, S. Xu, B. Liu, Y. Xiao, J. Zhang, *ACS Catal.* **2017**, *7*, 210–214; c) Y. Zhang, Y. Li, B. Pan, H. Xu, H. Liang, X. Jiang, B. Liu, M.-K. Tse, L. Qiu, *Chemistry Select* **2019**, *4*, 5122–5125; d) Z.-M. Zhang, B. Xu, Y. Qian, L. Wu, Y. Wu, L. Zhou, Y. Liu, J. Zhang, *Angew. Chem. Int. Ed.* **2018**, *57*, 10373–10377; e) C. Zhu, H. Chu, G. Li, S. Ma, J. Zhang, *J. Am. Chem. Soc.* **2019**, *141*, 19246–19251.

30. a) L. Wang, M. Chen, P. Zhang, W. Li, J. Zhang, *J. Am. Chem. Soc.* **2018**, *140*, 3467–3473; b) J. Han, W. Zhou, P.-C. Zhang, H. Wang, R. Zhang, H.-H. Wu, J. Zhang, *ACS Catal.* **2019**, *9*, 6890–6895; c) Z. Gan, M. Zhi, R. Han, E.-Q. Li, Z. Duan, F. Mathey, *Org. Lett.* **2019**, *21*, 2782–2785; d) M. Zhi, Z. Gan, R. Ma, H. Cui, E.-Q. Li, Z. Duan, F. Mathey, *Org. Lett.* **2019**, *21*, 3210–3213; e) H. Cui, K. Li, Y. Wang, M. Song, C. Wang, D. Wei, E.-Q. Li, Z. Duan, F. Mathey, *Org. Biomol. Chem.* **2020**, *18*, 3740–3746.

31. a) G. Knühl, P. Sennhenn, G. Helmchen, *J. Chem. Soc. Chem. Commun.* **1995**, 1845–1846; b) H. Inoue, Y. Nagaoka, K. Tomioka, *J. Org. Chem.* **2002**, *67*, 5864–5867.

32. A. Nakamura, T. Kageyama, H. Goto, B. P. Carrow, S. Ito, K. Nozaki, *J. Am. Chem. Soc.* **2012**, *134*, 12366–12369.

33. For reviews on the use of P-S ligands in asymmetric catalysis, see: a) F. Loi Lam, F. Yee Kwong, A. S. C. Chan, *Chem. Commun.* **2010**, *46*, 4646–4667; b) J. C. Carretero, J. Adrio, M. Rodríguez Rivero, Chiral Ferrocene in Asymmetric Catalysis, ed. L.-X. Dai, X.-L. Hou, *Sulfur– and Selenium–Containing Ferrocenyl Ligands in Chiral Ferrocenes in Asymmetric Catalysis*, Wiley-VCH, Weiheim, **2010**, 257–282; c) R. G. Arrayás, J. C. Carretero, *Chem. Commun.* **2011**, *47*, 2207–2211; d) J. Margalef, O. Pàmies, M. A. Pericàs, M. Diéguez, *Chem. Commun.* **2020**, *56*, 10795–10808.

34. For early examples on the use of P-thioether ligands, see: a) A. Albinati, P. S. Pregosin, K. Wick, *Organometallics* **1996**, *15*, 2419–2421; b) S. Gladiali, A. Dore, D. Fabbri, *Tetrahedron: Asymmetry* **1994**, *5*, 1143–1146; c) S. Gladiali, S. Medici, G. Pirri, S. Pulacchini S. Fabbri, *Can. J. Chem.* **2001**, *79*, 670–678; d) J. Kang, S. H. Yu, J. I. Kim, H. G. Cho, *Bull. Korean Chem. Soc.* **1995**, *16*, 439–443.

35. a) D. A. Evans, K. R. Campos, J. S. Tedrow, F. E. Michael, M. R. Gagné, *J. Org. Chem.* **1999**, *64*, 2994–2995; b) D. A. Evans, K. R. Campos, J. S. Tedrow, F. E. Michael, M. R. Gagné, *J. Am. Chem. Soc.* **2000**, *122*, 7905–7920; c) D. A. Evans, F. E. Michael, J. S. Tedrow, K. R. Campos, *J. Am. Chem. Soc.* **2003**, *125*, 3534–3543.

36. C. Borràs, M. Biosca, O. Pàmies, M. Diéguez, *Organometallics* **2005**, *34*, 5321–5334.

37. a) J. Priego, O. G. Mancheño, S. Cabrera, R. G. Arrayás, T. Llamas, J. C. Carretero, *Chem. Commun.* **2002**, 2512–2513; b) O. G. Mancheño, J. Priego, S. Cabrera, R. G. Arrayás, T. Llamas, J. C. Carretero, *J. Org. Chem.* **2003**, *68*, 3679–3686; c) S. Cabrera, R. G. Arrayás, J. C. Carretero, *Angew. Chem. Int. Ed.* **2004**, *43*, 3944–3947; d) O. G. Mancheño, R. G. Arrayás, J. C. Carretero, *J. Am. Chem. Soc.* **2004**, *126*, 456–457;

e) S. Cabrera, R. G. Arrayás, J. C. Carretero, *J. Am. Chem. Soc.* **2005**, *127*, 16394–16395; f) S. Cabrera, R. G. Arrayás, I. Alonso, J. C. Carretero, *J. Am. Chem. Soc.* **2005**, *127*, 17938–17947; g) S. Cabrera, O. G. Mancheño, R. G. Arrayás, I. Alonso, P. Mauleón, J. C. Carretero, *Pure Appl. Chem.* **2006**, *78*, 257–265; h) A. S. González, R. G. Arrayás, J. C. Carretero, *Org. Lett.* **2006**, *8*, 2977–2980; i) S. Cabrera, R. G. Arrayás, B. Martín-Matute, F. P. Cossío, J. C. Carretero, *Tetrahedron* **2007**, *63*, 6587–6602; j) B. M. Matute, S. I. Pereira, E. Peña-Cabrera, J. Adrio, A. M. S. Silva, J. C. Carretero, *Adv. Synth. Catal.* **2007**, *349*, 1714–1724; k) A. López-Pérez, J. Adrio, J. C. Carretero, *J. Am. Chem. Soc.* **2008**, *130*, 10084–10085; l) A. S. González, R. G. Arrayás, M. R. Rivero, J. C. Carretero, *Org. Lett.* **2008**, *10*, 4335–4337; m) J. Hernández-Toribio, R. G. Arrayás, B. Martín-Matute, J. C. Carretero, *Org. Lett.* **2009**, *11*, 393–396; n) E. Hernando, R. G. Arrayás, J. C. Carretero, *Chem. Commun.* **2012**, *48*, 9622–9624; o) J. Adrio and J. C. Carretero, Chem. Commun. **2014**, *50*, 12434–12446; p) A. Pascual-Escudero; A de Cózar, F. P. Cossío, J. Adrio, J. C. Carretero, *Angew. Chem. Int. Ed.* **2016**, *55*, 15334–15338; q) A. Molina, A. Pascual-Escudero, J. Adrio, J. C. Carretero, *J. Org. Chem.* **2017**, *82*, 11238–11246; r) J. Corpas, A. Ponce, J. Adrio, J. C. Carretero, *Org. Lett.* **2018**, *20*, 3179–3182.

38. a) F. L. Lam, T. T. L. Au-Yeung, H. Y. Cheung, S. H. L. Kok, W. S. Lam, K. Y. Wongaand, A. S. C. Chan, *Tetrahedron: Asymmetry* **2006**, *17*, 497–499; b) F. Loi Lam, T. Tin-Lok Au-Yeung, F. Yee Kwong, Z. Zhou K. Yin Wong, A. S. C. Chan, *Angew. Chem. Int. Ed.* **2008**, *47*, 1280–1283.

39. a) H. Y. Cheung, W.-Y. Yu, F. L. Lam, T.-T. Au-Yeung, Z.-Y. Zhou, T. H. Chan, A. S. Chan, *Org. Lett.* **2007**, *9*, 4295–4298; b) H. Y. Cheung, W.-Y. Yu, T. T. L. Au-Yeung, Z. Zhou, A. S. C. Chan, *Adv. Synth. Catal.* **2009**, *351*, 1412–1422.

40. a) K. Imae, K. Simizu, K. Ogata, S. Fukuzawa, *J. Org. Chem.* **2011**, *76*, 3604–3608; b) I. Oura, K. Shimizu, K. Ogata, S. Fukuzawa, *Org. Lett.* **2010**, *12*, 1752–1755; c) K. Shimizu, K. Ogata, S. Fukuzawa, *Tetrahedron Lett.* **2010**, *51*, 5068–5070; d) K. Imae, T. Konno, K. Ogata, S. Fukuzawa, *Org. Lett.* **2012**, *14*, 4410–4413; e) S. Watanabe, A. Tada, Y. Tokoro, S.. Fukuzawa, *Tetrahedron Lett.* **2014**, *55*, 1306–1309; f) A. Tada, S. Watanabe, M. Kimura, Y. Tokoro, S. Fukuzawa, *Tetrahedron Lett.* **2014**, *55*, 6224–6226; g) M. Kimura, Y. Matsuda, A. Koizumi, C. Tokumitsu, Y. Tokoro, S. Fukuzawa, *Tetrahedron* **2016**, *72*, 2666–2670; h) T. Konno, S. Watanabe, T. Takahashi, Y. Tokoro, S. Fukuzawa, *Org. Lett.* **2013**, *15*, 4418–4421; i) M. Kimura, A. Tada, Y. Tokoro, S. Fukuzawa, *Tetrahedron Lett.* **2015**, *56*, 2251–2253; j) A. Koizumi, Y. Matsuda, R. Haraguchi, S. Fukuzawa, *Tetrahedron Asymmetry* **2017**, *28*, 428–432; k) A. Koizumi, M. Harada, R. Haraguchi, S. Fukuzawa, *J. Org. Chem.* **2017**, *82*, 8927–8932; l) S. Kato, Y. Suzuki, K. Suzuki, R. Haraguchi, S. Fukuzawa, *J. Org. Chem.* **2018**, *83*, 13965–13972. m) M. Kato, T. Nakamura, K. Ogata, S. Fukuzawa, *Eur. J. Org. Chem.* **2009**, 5232–5238.

41. a) A. Berthelot-Bréhier, A. Panossian, F. Colobert, F. R. Leroux, *Org. Chem. Front.* **2015**, *2*, 634–644; b) T. Hoshi, K. Sasaki, S. Sato, Y. Ishii, T. Suzuki, H. Hagiwara, *Org. Lett.* **2011**, *13*, 932–935.

42. a) X. Caldentey, M. A. Pericàs, *J. Org. Chem.* **2010**, *75*, 2628–2644; b) X. Caldentey, X: C. Cambeiro, M. A. Pericàs, *Tetrahedron* **2011**, *67*, 4161–4168; c) J. Margalef, X. Caldentey, E. A. Karlsson, M. Coll, J. Mazuela, O. Pàmies, M. Diéguez, M. A. Pericàs, *Chem. Eur. J.* **2014**, *20*, 12201–12214.

43. For reviews on the asymmetric hydrogenation of unfunctionalized olefins, see: a) X. Cui, K. Burgess, *Chem. Rev.* **2005**, *105*, 3272–3296; b) S. J. Roseblade, A. Pfaltz, *Acc. Chem. Res.* **2007**, *40*, 1402–1411; c) O. Pàmies, P. G. Andersson, M. Diéguez, *Chem. Eur. J.* **2010**, *16*, 14232–14240; d) D. H. Woodmansee, A. Pfaltz, *Chem. Commun.* **2011**, *47*, 7912–7916; e) Y. Zhu, K. Burgess, *Acc. Chem. Res.* **2012**, *45*, 1623–1636; f)

J. J. Verendel, O. Pàmies, M. Diéguez, P. G. Andersson, *Chem. Rev.* **2014**, *114*, 2130–2169; g) S. Gruber, A. Pfaltz, A. *Angew. Chem. Int. Ed.* **2014**, *53*, 1896–1900; h) C. Margarita, P. G. Andersson, *J. Am. Chem. Soc.* **2017**, *139*, 1346–1356.

44. a) M. Coll, O. Pàmies, M. Diéguez, *Chem. Commun.* **2011**, *47*, 9215–9217; b) M. Coll, O. Pàmies, M. Diéguez, *Adv. Synth. Catal.* **2013**, *355*, 143–160.

45. a) M. Coll, O. Pàmies, M. Diéguez, *Org. Lett.* **2014**, *16*, 1892–1895; b) J. Margalef, M. Coll, P.-O. Norrby, O. Pàmies, M. Diéguez, *Organometallics* **2016**, *35*, 3323–3335.

46. M. Biosca, J. Margalef, X. Caldentey, M. Besora, C. Rodríguez-Escrich, J. Saltó, X. C. Cambeiro, F. Maseras, O. Pàmies, M. Diéguez, M. A. Pericàs, *ACS Catal.* **2018**, *8*, 3587–3601.

47. a) J. Margalef, O. Pàmies, M. Diéguez, *Chem. Eur. J.* **2017**, *23*, 813–822; b) J. Margalef, C. Borràs, S. Alegre, E. Alberico, O. Pàmies, M. Diéguez, *ChemCatChem* **2019**, *11*, 2142–2168; c) J. Margalef, C. Borràs, S. Alegre, O. Pàmies, M. Diéguez, *Dalton Trans.* **2019**, *48*, 12632–12643.

48. a) D. Q. Pharm, A. Nogid, *Clin. Ther.* **2008**, *30*, 813–824 (Rotigotine); b) J. I. Osende, D. Shimbo, V. Fuster, M. Dubar, J. J. Badimon, *J. Thromb. Haemost.* **2004**, *2*, 492–498 (Terutroban); c) S. B. Ross, S.-O. Thorberg, E. Jerning, N. Mohell, C. Stenfors, C. Wallsten, I. G. Milchert, G. A. Ojteg, *CNS Drug Rev.* **1999**, *5*, 213–232 (Robalzotan); d) B. Astier, L. Lambás Señas, F. Soulière, P. Schmitt, N. Urbain, N. Rentero, L. Bert, L. Denoroy, B. Renaud, M. Lesourd, C. Muñoz, G. Chouvet, *Eur. J. Pharmacol.* **2003**, *459*, 17–26 (Alnespirone).

49. a) E. Salom, S. Orgué, A. Riera, X. Verdaguer, *Angew. Chem. Int. Ed.* **2016**, *55*, 7988–7992; b) M. Magre, O. Pàmies, M. Diéguez, *ACS Catal.* **2016**, *6*, 5186–5190.

50. a) B. Feng, X.-Y. Pu, Z.-C. Liu, W.-J. Xiao, J.-R. Chen, *Org. Chem. Front.* **2016**, *3*, 1246–1249; b) B. Lu, B. Feng, H. Ye, J.-R. Chen, W.-J. Xiao, *Org. Lett.* **2018**, *20*, 3473–3476; c) B. Feng, J.-R. Chen, Y.-F. Yang, B. Lu, W.-J. Xiao, *Chem. Eur. J.* **2018**, *24*, 1714–1719; d) Y.-N. Wang, Q. Xiong, L.-Q. Lu, Q.-L. Zhang, Y. Wang, Y. Lan, W. J. Xiao, *Angew. Chem. Int. Ed.* **2019**, *58*, 11013–11017; e) M.-M. Li, Q. Xiong, B.-L. Qu, Y.-Q. Xiao, Y. Lan, L.-Q. Lu, W.-J. Xiao, *Angew. Chem. Int. Ed.* **2020**, *56*, 10795–10808; f) Y. Wei, L. Q. Lu, T. R. Li, B. Feng, Q. Wang, W. J. Xiao, H. Alper, *Angew. Chem. Int. Ed.* **2016**, *55*, 2200–2204; g) M. M. Li, Y. Wei, J. Liu, H. W. Chen, L. Q. Lu, W. J. Xiao, *J. Am. Chem. Soc.* **2017**, *139*, 14707–14713.

51. a) F. Lang, D. Li, J. Chen, J. Chen, L. Li, L. Cun, J. Zhu, J. Deng, J. Liao, *Adv. Synth. Catal.* **2010**, *352*, 843–846; b) T. Jia, P. Cao, D. Wang, Y. Lou, J. Liao, *Chem. Eur. J.* **2015**, *21*, 4918–4922; c) T. Jia, P. Cao, B. Wang, Y. Lou, X. Yin, M. Wang, J. Liao, *J. Am. Chem. Soc.* **2015**, *137*, 13760–13763.

52. J. Wang, M. Wang, P. Cao, L. Jiang, G. Chen, J. Liao, *Angew. Chem. Int. Ed.* **2014**, *53*, 6673–6677.

53. a) H.-G. Cheng, B. Feng, L.-Y. Chen, W. Guo, X.-Y. Yu, L.-Q. Lu, J.-R. Chen, W.-J. Xiao, *Chem. Commun.* **2014**, *50*, 2873–2875; b) B. Feng, H.-G. Cheng, J.-R. Chen, Q.-H. Deng, L.-Q. Lu, W.-J. Xiao, *Chem. Commun.* **2014**, *50*, 9550–9553; c) L.-Y. Chen, X.-Y. Yu, J.-R. Chen, B. Feng, H. Zhang, Y.-H. Qi, W.-J. Xiao, *Org. Lett.* **2015**, *17*, 1381–1384.

54. a) L. Du, P. Cao, J. Liao, *Acta Chim. Sinica* **2013**, *71*, 1239–1242; b) L. Du, P. Cao, J. Xing, Y. Lou, L. Jiang, L. Li, J. Liao, *Angew. Chem. Int. Ed.* **2013**, *52*, 4207–4211; c) J. Chen, J. Chen, F. Lang, X. Zhang, L. Cun, J. Zhu, J. Deng, J. Liao, *J. Am. Chem. Soc.* **2010**, *132*, 4552–4553.

4 Chiral Bidentate Heterodonor P-P′ ligands

Antonio Pizzano

CONTENTS

4.1 HISTORICAL PERSPECTIVE

The origin and initial development stages of asymmetric catalysis have largely been associated with the use of C_2 symmetric diphosphines. Seminal studies by Kagan and Knowles on the performance of Rh catalysts based on DIOP[1] and DIPAMP[2] ligands, respectively (Figure 4.1), in the asymmetric hydrogenation of olefins demonstrated the possibility of inducing a high enantioselectivity in a metal- catalyzed reaction, well exploited in the industrial synthesis of L-DOPA.[3] From these fundamental contributions, the study of other catalysts based on C_2 symmetric diphosphines greatly developed the field of asymmetric hydrogenation. For instance, ligands of the BINAP

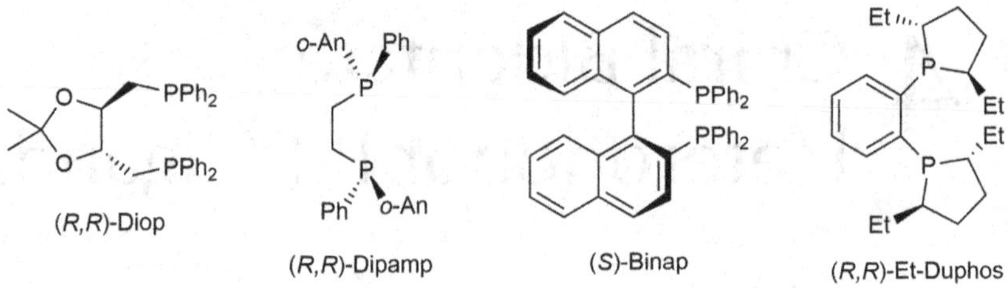

FIGURE 4.1 Structures of prominent C_2 symmetric diphosphines.

family provided outstanding results in the ruthenium- catalyzed hydrogenation of ketones.[4] Likewise, the application of rhodium DUPHOS catalysts in the asymmetric hydrogenation of olefins expanded the synthetic scope of this reaction.[5] Due to the success of these ligands, research on catalysts based on phosphorus ligands largely placed the focus on C_2 symmetric diphosphines.

In this context, a breakthrough occurred in 1993 with the discovery by Takaya and Nozaki of a Rh catalyst for the asymmetric hydroformylation of olefins based on a chelating ligand, which contained a phosphine and a phosphite fragment named BINAPHOS.[6] Non-asymmetric olefin hydroformylation is a process of paramount industrial importance,[7] already matured when research into asymmetric catalysis launched, so it is not surprising that the discovery of an effective catalyst for asymmetric hydroformylation was an early goal for asymmetric catalysis. However, this reaction posed significant difficulties. Attempts using Rh catalysts with C_2 symmetric diphosphine ligands (otherwise effective in olefin hydrogenation) did not achieve satisfactory results.[8] Moreover, a Pt(II)/SnCl$_2$ system using these ligands achieved good enantioselectivity in some cases, but low regioselectivity and catalyst activity were observed, as well as undesirable substrate hydrogenation.[9] In this scenario, the appearance of the BINAPHOS catalyst brought an unprecedented high enantioselectivity along with a relatively good regioselectivity and a high catalytic activity in the asymmetric hydroformylation of diverse olefins (e. g., styrene; Scheme 4.1).

BINAPHOS ligand attracted interest with respect to other phosphine-phosphites for the asymmetric hydroformylation reaction[10] and, concomitantly, for the performance of catalysts based on these ligands in alternative enantioselective reactions. As a result, the application of phosphine-phosphites to allyl substitution[11] and Rh hydrogenation[12] were then reported. In the latter reaction, catalysts showed outstanding levels of catalytic activity and enantioselectivity, similar to those observed with diphosphine catalysts. Moreover, fueled by the outstanding performances of chiral phosphoramidites,[13] the incorporation of phosphine-phosphoramidites was a natural extension of the field. Another landmark in this area can be considered to be the discovery of the JOSIPHOS ferrocenyl diphosphine ligand by Togni and coworkers.[14] Although both coordinating fragments in this ligand are phosphine ones, very versatile synthetic procedures enable the introduction of two unequal phosphine fragments.[15] These ligands, as well as phosphine-phosphites[16] and phosphine-phosphoramidites[17] typically possess a tunable modular structure, which facilitates a systematic and independent optimization of both P fragments and therefore a very

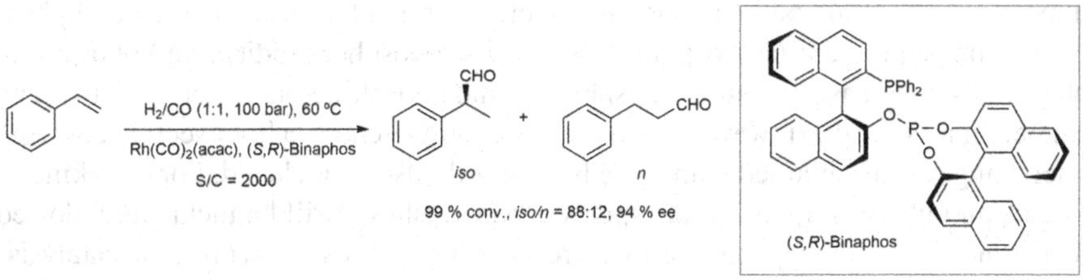

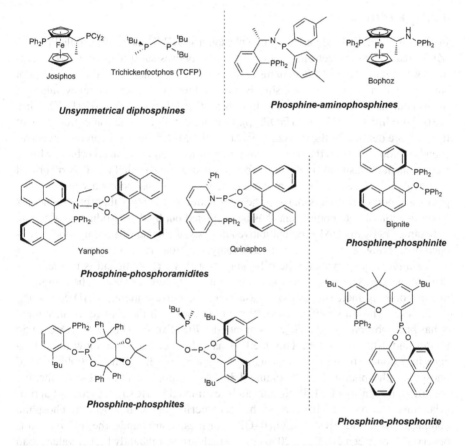

SCHEME 4.1 Hydroformylation of styrene with the Binaphos catalyst.

FIGURE 4.2 Representative examples of heterobidentate P-P' ligands with common names given, where appropriate.

precise adjustment of the catalyst structure.[18] Accordingly, it is not surprising that research on heterobidentate P-P' ligands has grown impressively, and many state-of-the-art applications of asymmetric catalysis are based on these kinds of ligands.

This chapter will cover bidentate ligands possessing one phosphine fragment and an unequal P(III) coordinating function: phosphine, aminophosphine, phosphinite, phosphonite, phosphoramidite, and phosphite (Figure 4.2). Regarding relative

importance, it should be mentioned that more attention has been devoted to diphosphines, phosphine-aminophosphines, phosphine-phosphoramidites and phosphine-
phosphites, whereas phosphine-phosphinites and phosphine-phosphonites have been
studied in less detail. However, for the sake of completeness, a brief section covering
interesting results obtained with these ligands will also be included. For each kind of
ligand, an outline of the general aspects of the synthesis will be included, followed
by the more relevant applications of corresponding ligands in asymmetric catalysis.

4.2   GENERAL FEATURES OF HETERODONOR P-P′ LIGANDS

4.2.1   ELECTRONIC PROPERTIES

In general, the overall donor ability of a phosphorus(III) ligand L of formula P(X)(Y)
(Z) decreases as the electronegativity of the phosphorus substituents increases. This
trend was quantified by Tolman in his classic study of the analysis of the IR spectra of
$Ni(L)(CO)_3$ complexes.[19] Thus, a shift is observed toward higher frequency values of
the A_1 band as the donor ability of L diminishes (e.g., 2069 (PPh_3), 2075 ($P(OPh)Ph_2$),
2080 ($P(OPh)_2Ph$), 2085 cm^{-1} ($P(OPh)_3$)). In more detail, the electronic properties of
ligand L are defined by the two components of the M–L bond. A σ component corresponds to donation from the phosphorus lone pair to an empty d metal orbital, while a
π component corresponds to donation from a filled metal d orbital to a P–X σ* orbital
(Figure 4.3a).[20] Thus, while phosphines are good σ donors but poor π acceptors, phosphites are poorer σ donors but significantly better π acceptors than phosphines. A
quantification of both components has been carried out by the introduction of χ_d and
π_p parameters by the QALE (Quantitative Analysis of Ligand Effects) method, developed by Giering, which has been used to analyze a broad range of ligands.[21]

A simple way to evaluate the σ basicity of a P(X)(Y)(Z) ligand is by measurement of the $^1J(P,Se)$ of the corresponding selenide, P(=Se)(X)(Y)(Z). The magnitude
of this coupling increases as the σ basicity of the corresponding P(III) compound
decreases.[22] Thus, a significant variation in $^1J(P,Se)$ with the change of substituent
X has been observed for $P(=Se)X_3$ compounds: 705 (Et), 735 (Ph), 784 (NMe_2), 954
(OMe) and 1025 Hz (OPh). This method has also been used to show the difference in σ-donor strength of coordinating fragments of phosphine-phosphites[23] and
phosphine-phosphoramidites[24] (Figure 4.3b). As a complement to these results, the
overall donor ability of a P-P′ ligand can be estimated by IR spectroscopy of appropriate metal carbonyls. This analysis has been performed with phosphine-phosphite
complexes of formula Rh(Cl)(CO)(P–OP). For these compounds, the υ(CO) band is
observed [23] between 2046 and 2056 cm^{-1}, which are significantly higher values than
those reported for the corresponding complexes of diphosphines $R_2PCH_2CH_2PR_2$
(2010 cm^{-1} (R = Ph), 1990 cm^{-1} (R = ^iPr)),[25] demonstrating the lower overall donor
ability of the phosphine-phosphite ligands.

4.2.2 C_2 VS C_1 LIGAND SYMMETRY

From a fundamental viewpoint, the use of a C_2 symmetric bidentate ligand
reduced by half the number of possible reaction intermediates,[26] with an inherent

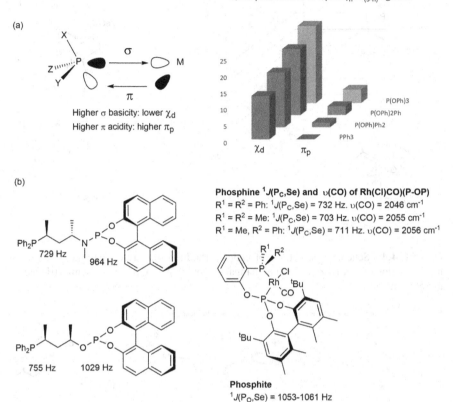

FIGURE 4.3 Schematic drawing of components of a metal-phosphorus ligand bond. (a) Values of $^1J(P,Se)$ (Hz) of diselenides of ligands shown and (b) $\upsilon(CO)$ values for Rh(Cl)(CO)(phosphine-phosphite) complexes.

simplification of the catalytic system. Let us consider, at this point, the structure of rhodium cationic dihydrides, which are key intermediates in the hydrogenation of chelating olefins, such as methyl Z-α-acetamidocinnamate (MAC, Figure 4.4).[27] These intermediates have an octahedral geometry, and mechanistic studies indicate that the catalytically relevant ones are those with the carbonyl oxygen and one of the hydride ligands located in mutually *trans* coordination positions. For the case of a C_2 diphosphine, only one *pro-R* and one *pro-S* intermediates are possible (Figure 4.4a). In the case of a bidentate ligand with unequivalent P and P' fragments (Figure 4.4b), there are four possible dihydrides. However, as olefins and hydrides have different electronic and steric features, there should be energy differences between the two *pro-R* diastereomers (as well as between the two *pro-S* ones).

The study of the structure in solution of [Rh(MAC)(P–OP)]⁺ complexes (P–OP = phosphine-phosphite ligand, Figure 4.5) has shown that only isomers with the olefin located in a coordination position *trans* to the phosphine are observed.[23] This preference has been attributed to the π-accepting properties of the phosphite, which

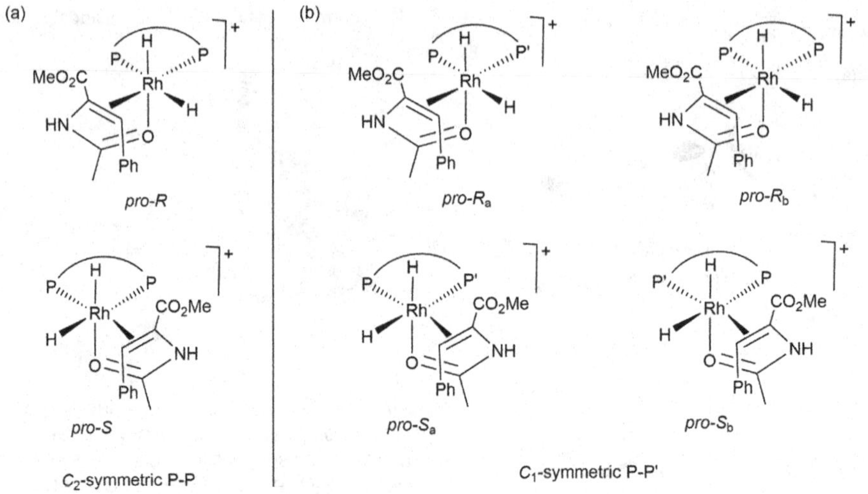

FIGURE 4.4 Schematic structures of cationic rhodium dihydrides, intermediates in the hydrogenation of methyl Z-α-acetamidocinnamate bearing (a) C_2 or (b) C_1 symmetric bidentate P-P' ligands.

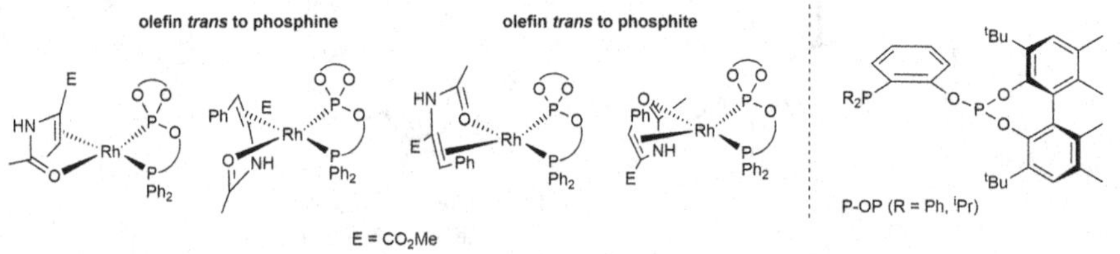

FIGURE 4.5 Structures of isomers of [Rh(P–OP)(MAC)]$^+$ complexes. Complex charge has been omitted for clarity.

disfavors structures where the latter and the π-acceptor olefin are located in mutually *trans* coordination positions. If this effect were transferred to the dihydrides, the participation of only one *pro-R* and one *pro-S* dihydride would be expected, simplifying the catalytic system. In this scenario, the P fragment adjacent to the olefin gains a prominent role in the induction of chirality. This is the case of the phosphite fragment in the enantioselective hydrogenation of several types of olefins, with Rh phosphine-phosphite catalysts.[16b]

4.3 SYNTHESIS AND CATALYTIC APPLICATIONS OF BIDENTATE HETERODONOR P-P' LIGANDS

4.3.1 P-P' DIPHOSPHINES

4.3.1.1 Synthesis and Structure

In this section two groups of diphosphines will be considered. The first corresponds to JOSIPHOS ligands[15] and to more elaborate structures based on the

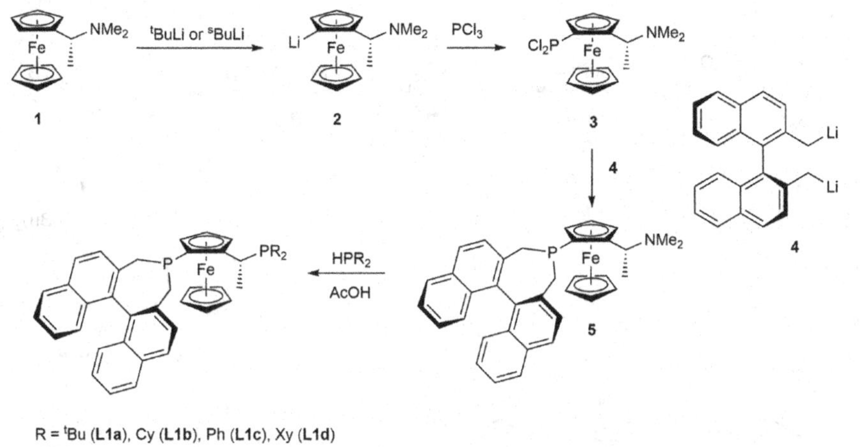

R = ^{t}Bu (**L1a**), Cy (**L1b**), Ph (**L1c**), Xy (**L1d**)

SCHEME 4.2 Synthesis of ferrocenyl diphosphines **L1a–L1d**.

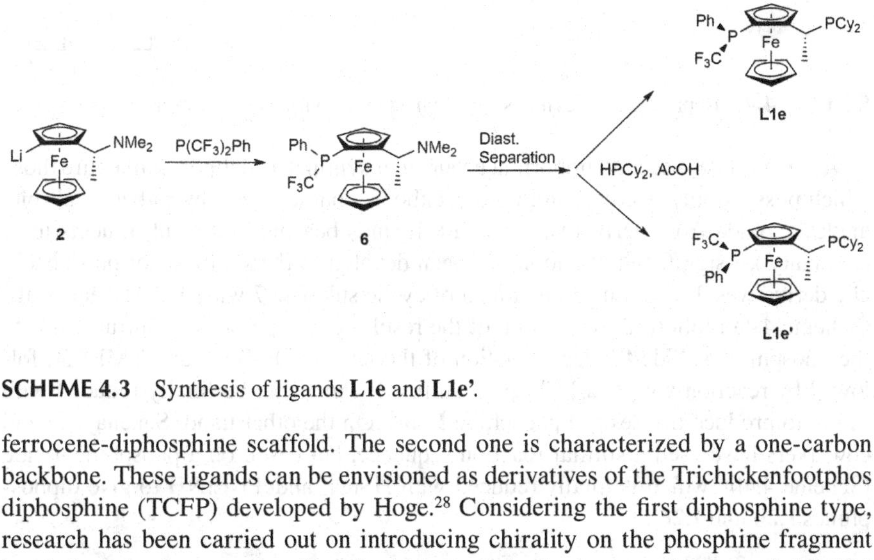

SCHEME 4.3 Synthesis of ligands **L1e** and **L1e'**.

ferrocene-diphosphine scaffold. The second one is characterized by a one-carbon backbone. These ligands can be envisioned as derivatives of the Trichickenfootphos diphosphine (TCFP) developed by Hoge.[28] Considering the first diphosphine type, research has been carried out on introducing chirality on the phosphine fragment bonded to the ferrocene moiety. In this research, the key starting material is Ugi amine (**1**, Scheme 4.2), which can be lithiated with ^{t}BuLi or ^{s}BuLi to give compound **2**. Reaction of the latter with PCl$_3$ yields the corresponding dichlorophoshine amine **3**, which is then reacted with the dilithiated reagent **4** to give amine phosphine **5**. The final reaction, with a secondary phosphine in acetic acid, gives the expected nucleophilic substitution with retention of the configuration, leading to corresponding binaphane diphosphines **L1a–L1d**.[29] On the other hand, Togni and coworkers have reported the ability of a phosphorus-bound CF$_3$ fragment to act as a leaving group in a nucleophilic substitution (Scheme 4.3). Thus, reaction of **2** with PhP(CF$_3$)$_2$ leads to phosphine-amine **6** as a mixture of two diastereomers, which can be separated by flash chromatography. Further reaction with a secondary phosphine, as mentioned above, generates the P-stereogenic diphosphines **L1e** and **L1e'**.[30]

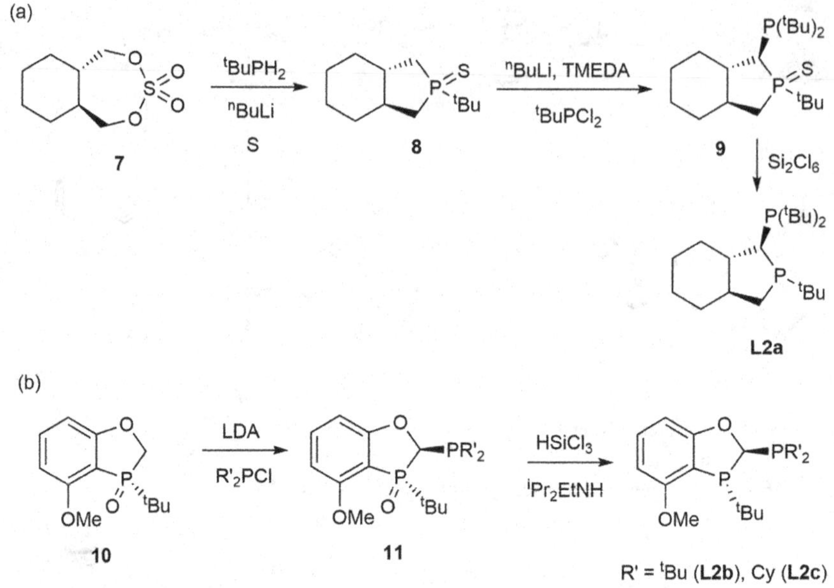

(a)

(b)

R' = ᵗBu (**L2b**), Cy (**L2c**)

SCHEME 4.4 Representative synthesis of diphosphines with a one-carbon backbone.

As mentioned, a second important group of unsymmetrical diphosphines are those which possess only one carbon in the backbone, characterized by rather small bite angles.[31] Aside from derivatives of TCFP ligands bearing bulky substituents (e.g., adamantyl),[32] significant attention has been devoted to the synthesis of phosphacyclic derivatives. For instance, treatment of cyclic sulphate **7** with ᵗBuPH$_2$ and ⁿBuLi (Scheme 4.4a), followed by reaction of the resulting phosphine with sulfur, leads to the phosphine–sulfide **8**. Deprotonation of the latter with ᵗBuLi and TMEDA, followed by reaction with ᵗBu$_2$PCl, gives sulfide **9** which can be finally reduced with Si$_2$Cl$_6$ to produce the desired phosphine **L2a**.[33] On the other hand, Senenayake and coworkers have used a similar reaction sequence, but based on a phosphine–oxide (Scheme 4.4b), which is finally reduced with HSiCl$_3$ and ᶦPr$_2$EtNH to give diphosphines **L2b** and **L2c**.[34]

4.3.1.2 Catalytic applications

Among the P-P' diphosphines, JOSIPHOS ligands belong to the selected class of privileged ligands and occupy a prominent place.[15] Due to their modular structure and easy tunability, along with the commercial availability of a wide library of ligands, they have been used extensively. As a consequence, good results on challenging transformations continue to appear in the literature.

In the field of Rh-catalyzed asymmetric hydrogenation, catalysts bearing P-P' diphosphines play a very important role. Aside from the application to the hydrogenation of benchmark substrates, like α-,β-dehydroamino acids or itaconates,[32–35] these catalysts have provided good results in more challenging reactions, such as in the asymmetric hydrogenation of tetrasubstituted olefins (Scheme 4.5). Examples

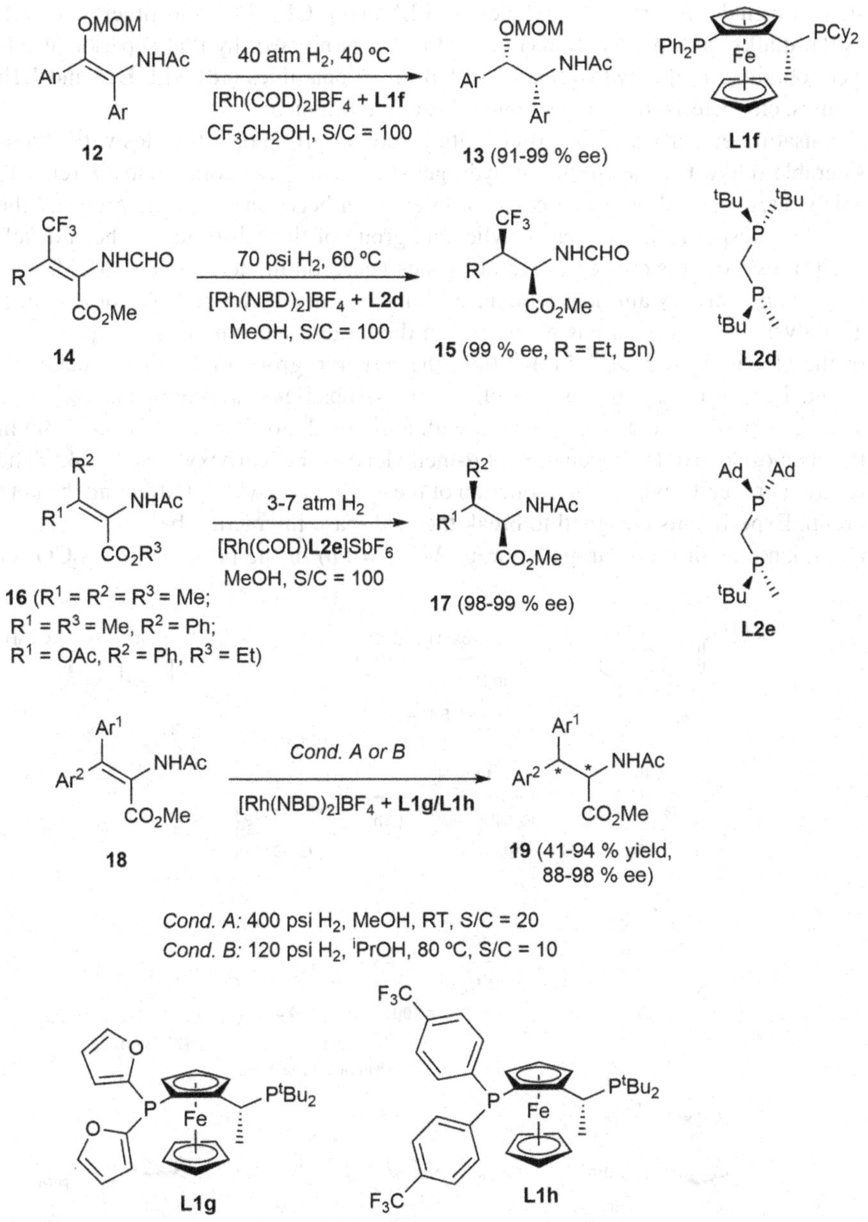

SCHEME 4.5 Hydrogenation of tetrasubstituted olefins with Rh catalysts based on P-P' diphosphines.

studied include substrates with β-alcoxy (**12**)[36] and β-CF$_3$ (**14**)[37] substituents, as well as β,β-dialkyl and β-alkyl-β-aryl ones (**16**).[38] It is noteworthy that superior results were obtained in the hydrogenation of β,β-diaryl enamides (**18**) with **L1g** and **L1h** ligands, characterized by a low donor phosphine fragment.[39]

Unsaturated carboxylic acids constitute another group of substrates with a considerable interest for asymmetric hydrogenation. For these compounds, a remarkable strategy, based on a non-covalent interaction between an amino group of the chiral diphosphine and the carboxylic acid group of the substrate, has been developed. Based on this concept, several ligands based on the ferrocene scaffold, possessing an auxiliary amine group, have been examined (Scheme 4.6). For instance, a catalyst based on **L3a** has been used in the large-scale synthesis of a precursor of the aliskiren drug **21**.[40] In addition, the research group of X. Zhang has used ligand **L3b** in the hydrogenation of several α-substituted acrylic acids **22**.[41] The same group has used a related ligand with a highly donor P^{t}Bu$_2$ fragment (**L3c**) in the chemoselective hydrogenation of α-methylene-γ-keto carboxylic acids **24**.[42] The reaction proceeds, with clean reduction of the olefin bond without affecting the keto group. Experiments designed to break the acid–base interaction between substrate and ligand, as in the hydrogenation of **24** (R = Ph) in the presence of Cs$_2$CO$_3$ or

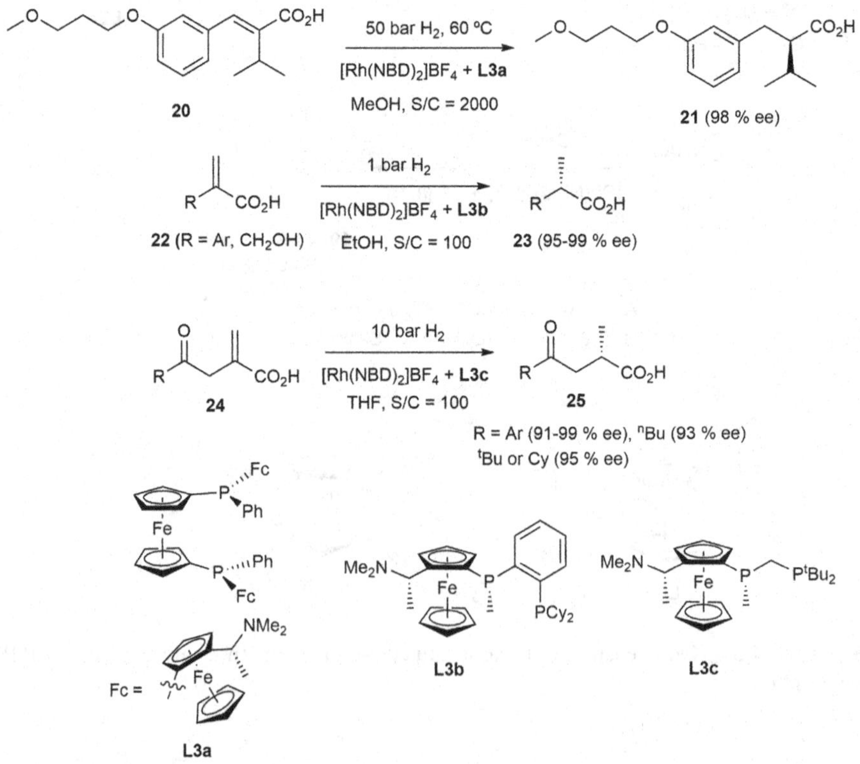

SCHEME 4.6 Hydrogenation of unsaturated carboxylic acids.

NEt_3, result in a dramatic drop in enantioselectivity, down to 20 and 8 % enantiomeric excess (ee), respectively.

In the field of Rh-catalyzed asymmetric hydrogenation, another challenging goal is chemoselective hydrogenation. Relevant examples of this topic are the hydrogenation of cyclic keto enamides **26** (Scheme 4.7)[43] and nitro olefins **28** (Scheme 4.8),[44] for which the optimized catalytic system only shows the reduction of the olefin bond. Also of interest, the optimization of the catalytic system included the examination of a wide variety of diphosphines, including C_2 symmetric ones, whereas, in both types, the best results were provided by a C_1 symmetric ligand, *ent*-**L2d** and **L1i**, respectively.

By the use of suitable ligands, Rh catalysts may also be capable of reducing C=N and C=O bonds enantioselectively. For instance, the research group of W. Zhang has reported the chemoselective and enantioselective hydrogenation of alkynyl hydrazones **30** to give chiral propargyl hydrazines **31** with good enantioselectivities (Scheme 4.9).[45] On the other hand, Togni and coworkers have used a JOSIPHOS ligand **L1j** in the hydrogenation of fluoroalkyl ketones **32** (Scheme 4.10).[46] High enantioselectivities were obtained in this reaction from substrates bearing an aryl substituent. The catalyst exhibited a greater reactivity for trifluoromethyl ketones than for methyl ones. The catalytic system also provided satisfactory results in the case of perfluoroethyl and perfluoropropyl substrates. Another interesting example was the hydrogenation of pyridinium bromides **34** (Scheme 4.11).[47] In this reaction a very favorable effect of the addition of NEt_3 on conversion was observed.

The use of P-P' diphosphines has also provided remarkable results in Ir-catalyzed hydrogenations. For instance, the groups of A. Togni[48] and S. Zhang[49] have studied the hydrogenation of challenging 1-aryl dihydroisoquinolines **36** with catalysts based on JOSIPHOS-type ligands. These reactions required an acid additive in order to proceed. In particular, catalysts based on **L1a**[29] provided exceedingly

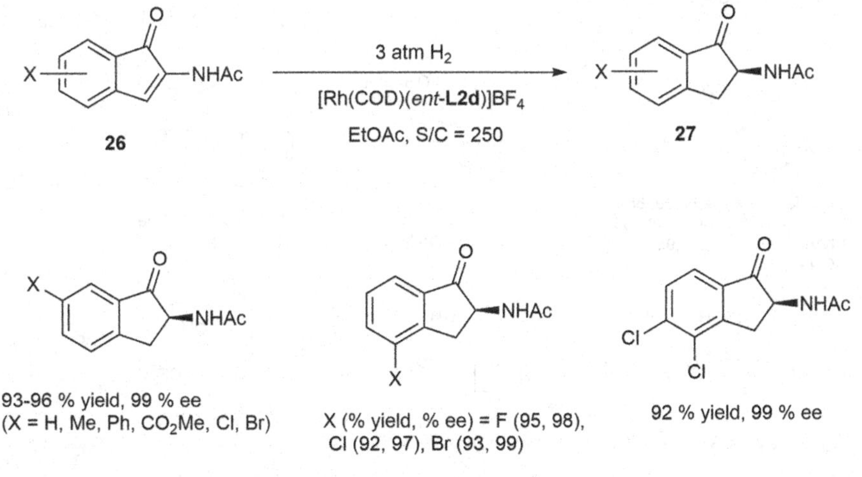

SCHEME 4.7 Asymmetric hydrogenation of cyclic α-dehydroaminoketones **26**.

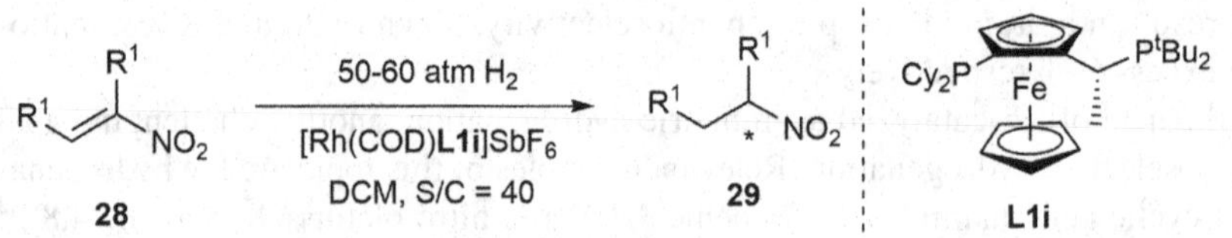

Representative results (% yield, % ee):
R¹ = Me; R² = Ph (92, 90), 4-MeO-Ph (62, 82), 3,4,-(MeO)₂-Ph (95, 77)
4-F (90, 26), 4-Br (91, 86), 1-Naphthyl (86, 85), 2-thienyl (88, 92), 3-pyridyl (50, 81).
R¹ = Et; R² = Ph (41, 88), 4-MeO-Ph (55, 88).

SCHEME 4.8 Asymmetric hydrogenation of nitroalkenes **28**.

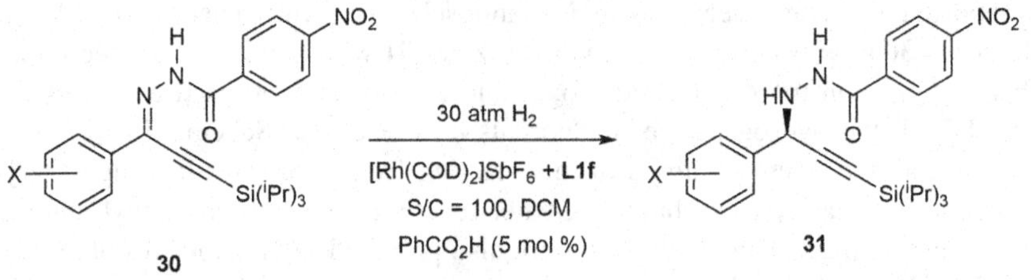

Representative results (% yield, % ee):
X = H (90, 90), 4-Me (92, 85), 4-MeO (88, 95),
4-F (90, 91), 4-tBu (85, 94), 3-Br (87, 90),

SCHEME 4.9 Asymmetric hydrogenation of alkynyl hydrazones **30**.

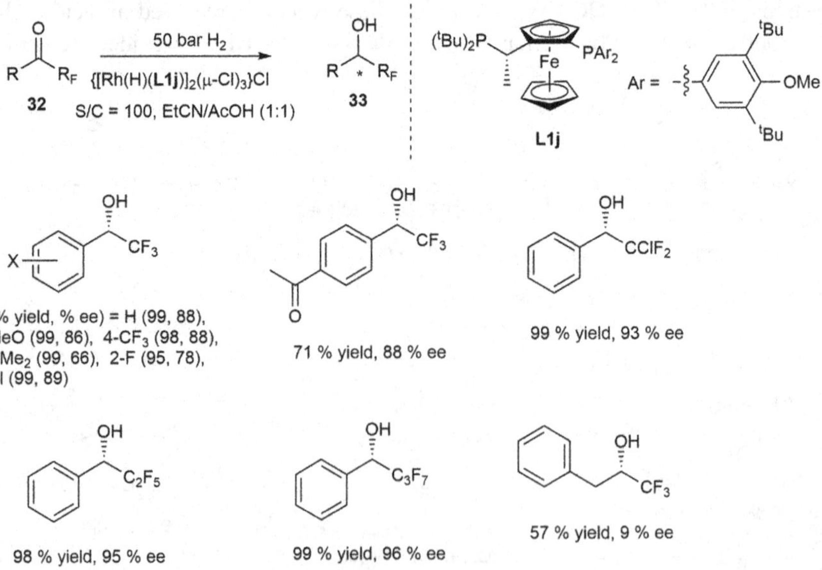

X (% yield, % ee) = H (99, 88),
4-MeO (99, 86), 4-CF₃ (98, 88),
4-NMe₂ (99, 66), 2-F (95, 78),
3-Cl (99, 89)

71 % yield, 88 % ee

99 % yield, 93 % ee

98 % yield, 95 % ee

99 % yield, 96 % ee

57 % yield, 9 % ee

SCHEME 4.10 Asymmetric hydrogenation of fluoroalkyl ketones **32**.

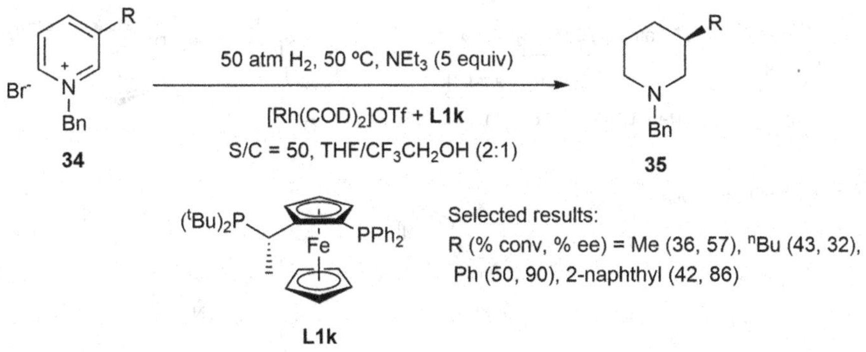

SCHEME 4.11 Asymmetric hydrogenation of pyridinium bromides **34**.

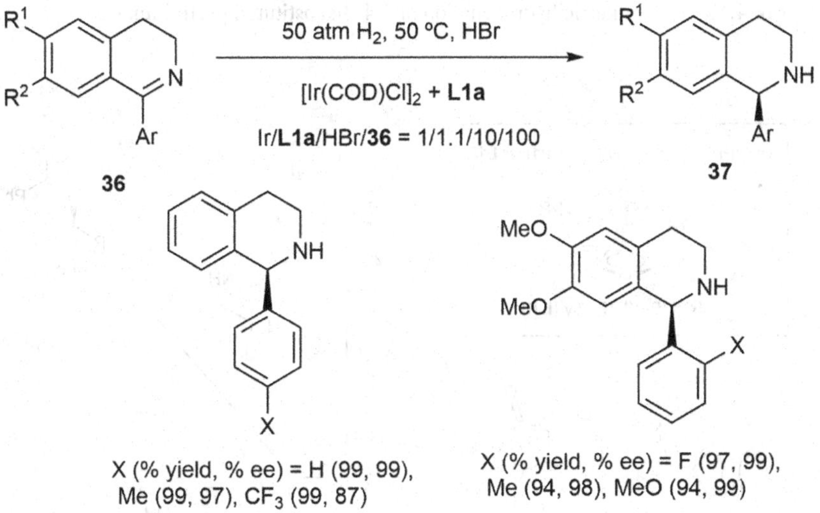

X (% yield, % ee) = H (99, 99),
Me (99, 97), CF$_3$ (99, 87)

X (% yield, % ee) = F (97, 99),
Me (94, 98), MeO (94, 99)

SCHEME 4.12 Asymmetric hydrogenation of 1-aryl dihydroisoquinolines **36**.

high enantioselectivities on a wide variety of substrates (Scheme 4.12), even in the case of sterically hindered ones containing *ortho*-substituted aryl fragments. There are several appropriate precursors for the synthesis of biologically active chiral tetrahydroisoquinolines among products **37**. Another application of interest regards the hydrogenation of pyrimidines **38** with a catalyst based on ligand *ent*-**L1k** (Scheme 4.13).[50] In this study a strong influence of additives has been detected. Among those additives investigated, the best results were obtained with Yb(OTf)$_3$. To rationalize this effect, substrate activation by Yb coordination to the N-atom located at position 3 has been proposed.

On the other hand, Breit has used Rh catalysts, based on JOSIPHOS ligands **L1i** and **L1l,** on a wide range of couplings between allenes and different nucleophilic reagents (Scheme 4.14a), leading to a plethora of chiral allylic building blocks

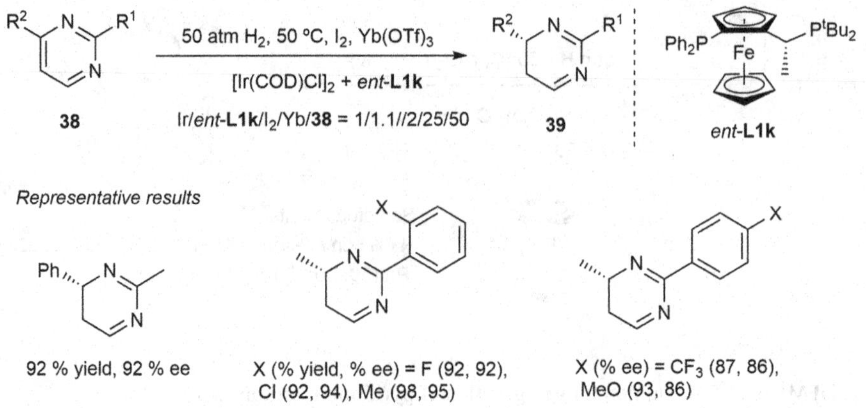

SCHEME 4.13 Asymmetric hydrogenation of 2,4-disubstituted pyrimidines **38**.

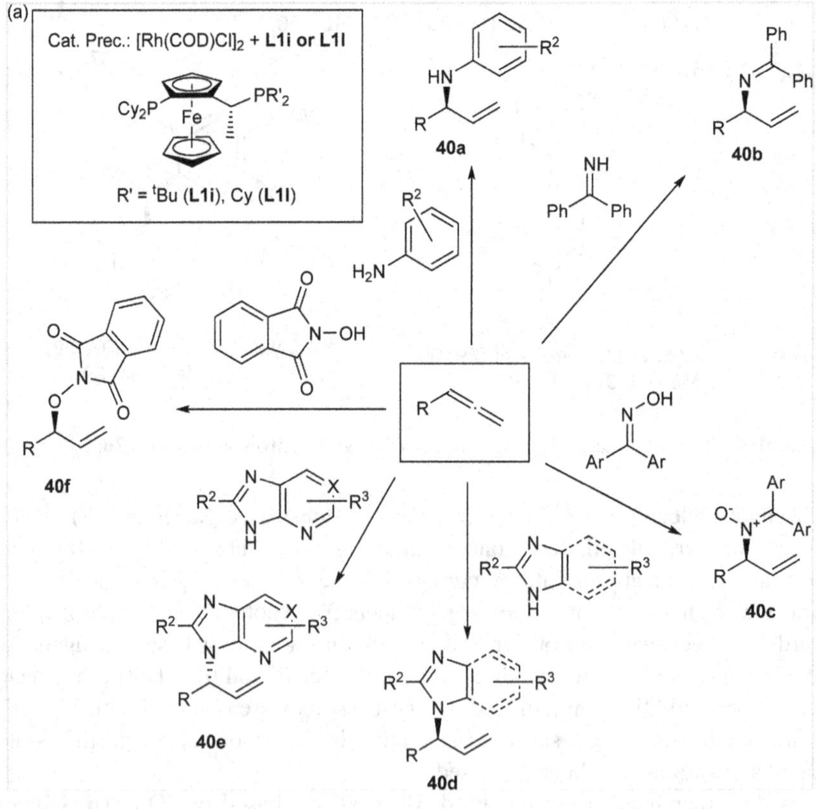

SCHEME 4.14A Types of enantioselective additions of nucleophiles to allenes catalyzed by Rh complexes based on Josiphos ligands.

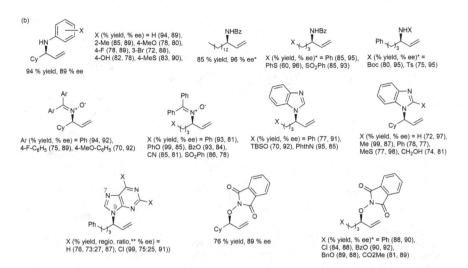

SCHEME 4.14B Representative results. *Compounds obtained after hydrolysis and acylation of imines **40b**; **Regio-isomeric ratio of N^9 and N^7 compounds.

(Scheme 4.14b). For instance, diverse allyl amines **40a** could be generated by the addition of anilines.[51] On the other hand, benzophenone imine was used to prepare imines **40b** which can be readily converted into the corresponding primary amines.[52] Likewise, these catalysts are also able to effect the addition of imidazole and purine derivatives to give corresponding products **40d** and **40e**,[53,54] as well as of diaryl oximes to obtain nitrones **40c** and N-hydroxyphthalimide to produce compounds **40f**. It should also be noted that acid additives, like pyridinium p-toluenesulfonate, p-methoxy benzoic acid, or L-tartaric acid, were used in the synthesis of **40b**,[52] **40c**[55] and **40f**,[56] respectively. In this regard, mechanistic investigations propose an oxidative addition of carboxylic acids to the Rh(I) precursor which leads to a Rh(III) hydride that causes the hydrometalation of the allene.[57]

Significant advances have also been made in the area of Pd-catalyzed couplings using JOSIPHOS-type ligands. For instance, Kim and Cho have reported a Suzuki–Miyaura coupling between (diborylmethyl)silanes and aryl iodides to generate the corresponding silylboronate esters **41** with moderate to good enantioselectivities (80–99 % ee, Scheme 4.15).[58] For this process, catalyst optimization identified a ligand with donor properties dissimilar to those of the phosphine fragments (**L1m**). Compounds **41** are suitable precursors for the preparation of diverse chiral benzylic silanes. On the other hand, Lassaletta and Fernández have described a Heck reaction between racemic heteroaryl sulfonates and vinyl ethers or enamides which proceeds with a simultaneous dynamic kinetic resolution, leading to heterobiaryls **43** with high levels of enantioselectivity (Scheme 4.16).[59] Following an analogous strategy, the same authors have described a very convenient procedure for the synthesis of a family of Quinaphos ligands **44** with good to excellent enantioselectivities, by a coupling between heterobiaryl sulfonates **42** and silylphosphines (Scheme 4.17).[60] Finally, the group of Walsh has described a rather general arylation of racemic

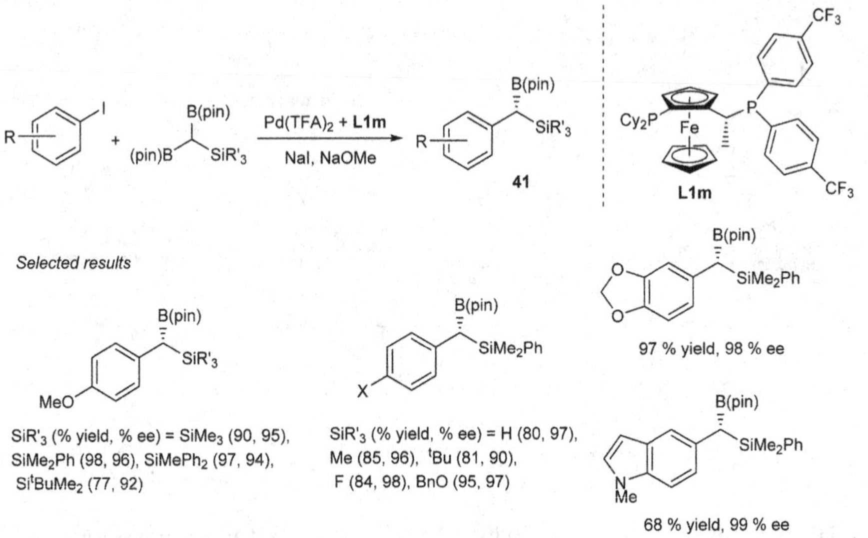

SCHEME 4.15 Pd-catalyzed Suzuki coupling between (diborylmethyl)silanes and aryl halides.

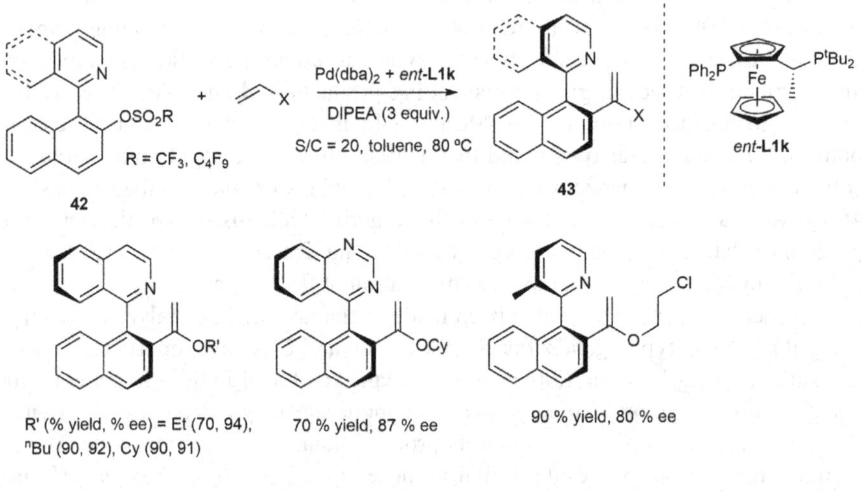

SCHEME 4.16 Dynamic kinetic enantioselective Heck reaction for the synthesis of heterobiaryls **43**.

benzylic sulfoxides **45** to produce diaryl sulfoxides **46,** with high levels of enantioselectivity.[61] Mechanistic investigations of this transformation propose an aryl attack to a coordinated sulfenate, generated by debenzylation of the substrate (Scheme 4.18).

Moreover, significant advances have been possible with the use of JOSIPHOS ligands in Cu-catalyzed transformations. For instance, the research group of Oestreich has described the addition of silicon Grignard reagents to olefins **47**, activated by a

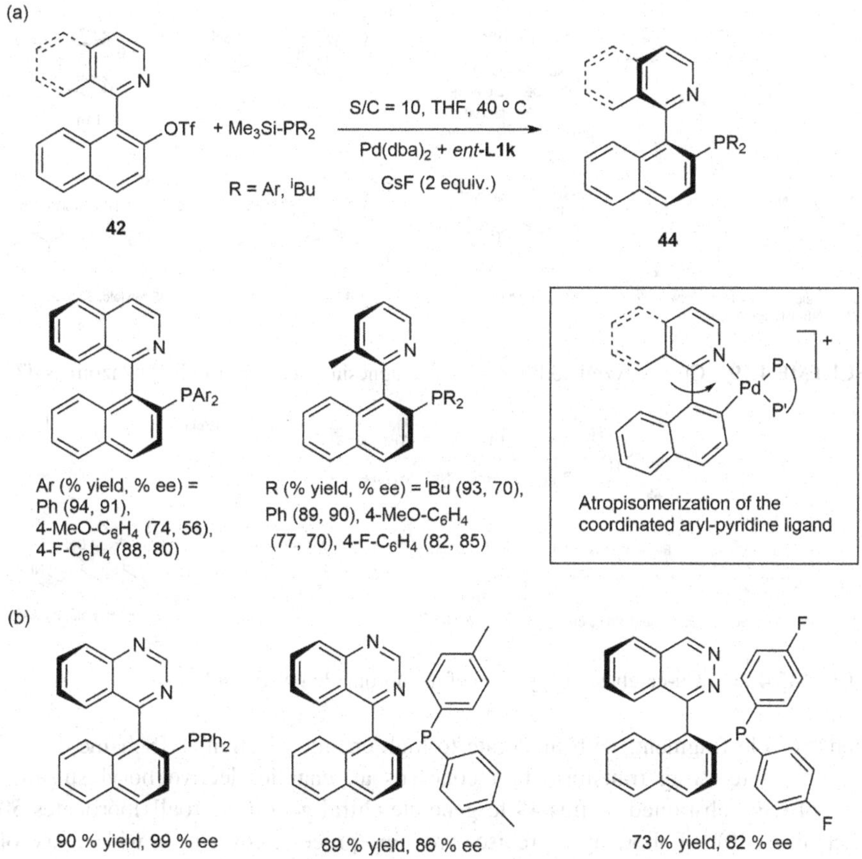

(a)

42 + Me₃Si-PR₂

R = Ar, ¹Bu

$\xrightarrow[\text{Pd(dba)}_2 + ent\text{-L1k}]{\text{S/C} = 10, \text{THF}, 40\,^\circ\text{C}}$

CsF (2 equiv.)

44

Ar (% yield, % ee) =
Ph (94, 91),
4-MeO-C₆H₄ (74, 56),
4-F-C₆H₄ (88, 80)

R (% yield, % ee) = ¹Bu (93, 70),
Ph (89, 90), 4-MeO-C₆H₄
(77, 70), 4-F-C₆H₄ (82, 85)

Atropisomerization of the
coordinated aryl-pyridine ligand

(b)

90 % yield, 99 % ee   89 % yield, 86 % ee   73 % yield, 82 % ee

SCHEME 4.17 (a) Dynamic kinetic asymmetric C–P coupling for the synthesis of P–N ligands **44**. (b) Additional examples of P,N ligands prepared by this methodology.

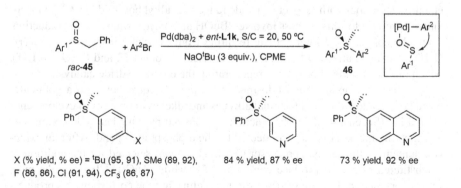

rac-**45** + Ar²Br

$\xrightarrow[\text{NaO}^t\text{Bu (3 equiv.), CPME}]{\text{Pd(dba)}_2 + ent\text{-L1k, S/C} = 20, 50\,^\circ\text{C}}$

46

X (% yield, % ee) = ¹Bu (95, 91), SMe (89, 92),   84 % yield, 87 % ee   73 % yield, 92 % ee
F (86, 86), Cl (91, 94), CF₃ (86, 87)

SCHEME 4.18 Pd-catalyzed arylation of racemic sulfoxides **45**.

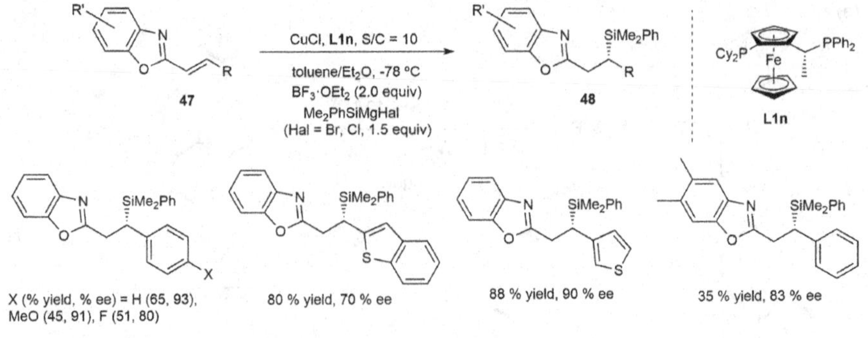

SCHEME 4.19 Cu-catalyzed addition of silyl magnesium reagents to vinyl oxazolines **47**.

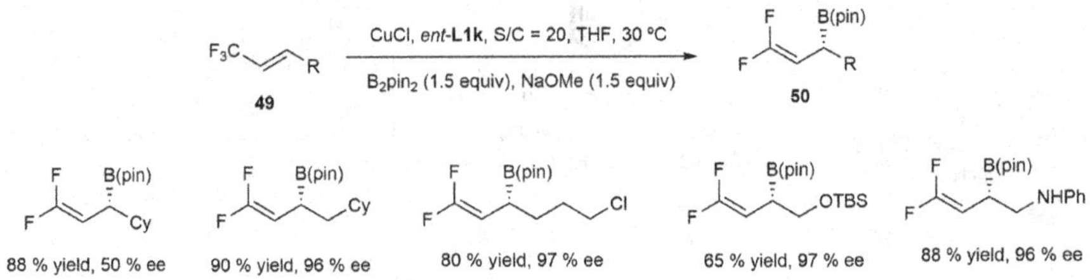

SCHEME 4.20 Cu-catalyzed borylation of trifluoromethyl olefins **49**.

benzoxazole fragment, with moderate to high enantioselectivities (Scheme 4.19).[62] Another interesting transformation concerns an enantioselective boryl substitution of CF$_3$-substituted olefins **49** to generate chiral *gem*-difluoroallylboronates **50** (Scheme 4.20).[63] Following a precise screening process, covering a wide range of JOSIPHOS ligands, *ent*-**L1k** achieved the best results. The enantioselectivity of the reaction diminished abruptly with the increase in steric hindrance caused by the R substituent, as shown in the result obtained with the cyclohexyl substrate (R = Cy). Finally, a Cu- catalyzed reduction of alkenyl boramides **51** has been described recently by Park and Yun (Scheme 4.21).[64] The process provides high to excellent enantioselectivities with respect to a wide range of β-alkyl boramides **52**. The mechanism proposed for this process involves ^{t}BuOH and a hydrosilane for the reduction. Thus, the copper alkyl intermediate resulting from the olefin insertion into Cu(H) (**L1f**) complex is attacked by the alcohol to release product **52** and Cu(O^{t}Bu)(**L1f**), which reacts with the hydrosilane, regenerating the copper hydride catalyst.

A remarkable application of Josiphos-type ligands corresponds to a gold-catalyzed [2+2] cycloaddition of terminal alkynes and alkenes to give chiral cyclobutenes **54** and **54'** (Scheme 4.22).[65] The reaction is catalyzed by dinuclear gold complexes, in which the metal atoms are connected by the diphosphine ligand. After an extensive screening, the catalyst precursor (AuCl)$_2$(**L1o**) was selected for the addition of disubstituted olefins to produce compounds **54**, while (AuCl)$_2$(**L1p**) was the most satisfactory one in the case of trisubstituted olefins and the corresponding products **54'**. A complementary study by density functional theory (DFT) methods indicated

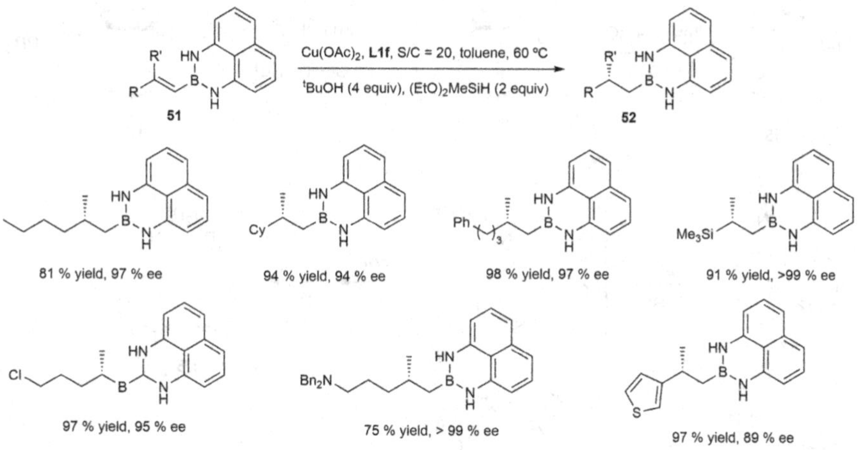

SCHEME 4.21 Cu-catalyzed reduction of alkenyl boramides **51**.

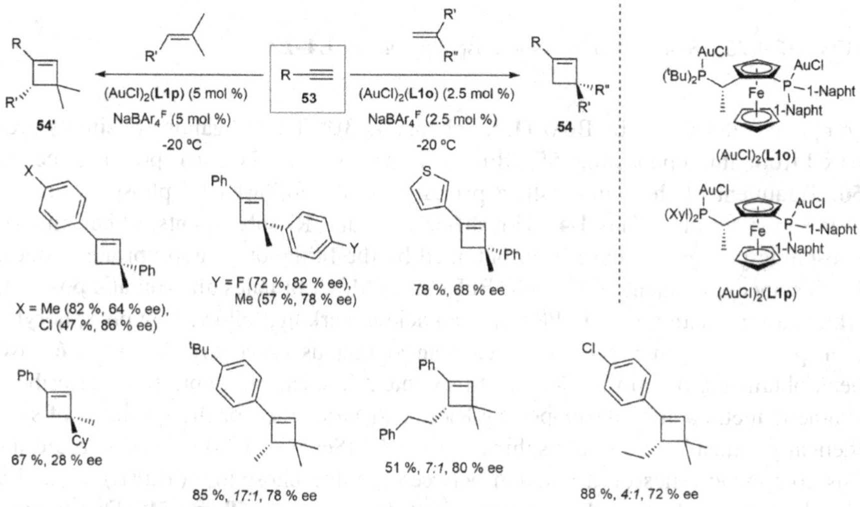

SCHEME 4.22 Synthesis of cyclobutenes **54** and **54'** by gold-catalyzed [2+2] cycloaddition (regio-isomer ratio of compounds **54'** in italics).

that only one metal center participates in the cycloaddition reaction, with the second one being necessary for the attainment of high enantioselectivity.

4.3.2 Phosphine-Aminophosphines

4.3.2.1 Synthesis

The next class of ligands covered in this chapter corresponds to phosphine-aminophosphines. The interest in this class of ligands was stimulated by Bophoz

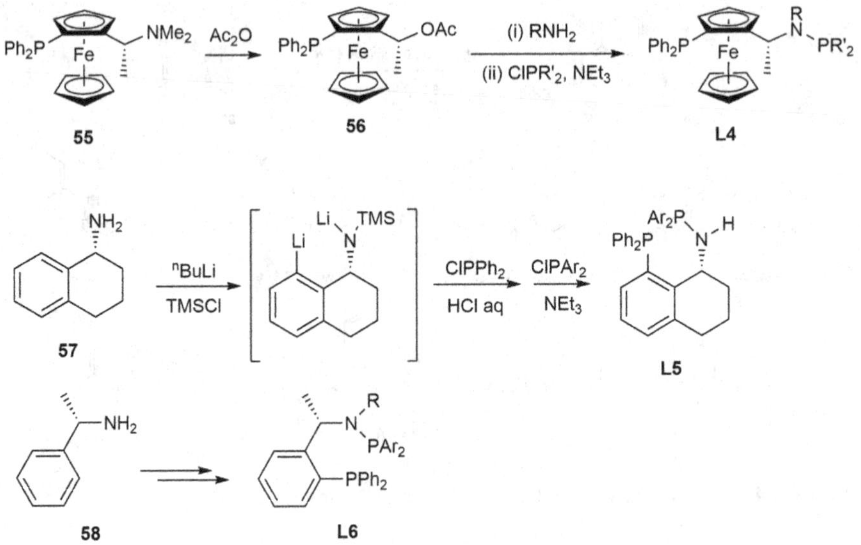

SCHEME 4.23   Synthesis of amino-phosphine ligands **L4–L6**.

compounds developed by Boaz (**L4**, Scheme 4.23).[66] These ligands are simply prepared from aminophosphine **55**, which is transformed to the corresponding acetate
56. Treatment of the latter with a primary amine, followed by phosphorylation,
produces a set of ligands **L4,** with different R and R' substituents. Other amino-
phosphine-phosphines have been obtained by the lithiation of appropriate amines.
For instance, treatment of **57** with [n]BuLi and TMSCl lithiates an aromatic position,
which, after reaction with ClPPh$_2$ and an acidic work-up, followed by phosphorylation, produces ligands **L5**.[67] Following an analogous procedure, ligands **L6** have
been obtained from amine **58**.[68] In this context it seems appropriate to describe a
synthetic methodology developed by Riera and Verdaguer for the synthesis of synthetically valuable amino-phosphine borane **62** (Scheme 4.24). The first route for
this compound relies on a reaction between a chlorophosphine ([t]Bu)(R)PCl and a
chiral amine,[69] which leads to a mixture of diastereomers **59** and **59'**. Due to easy
racemization of the starting chlorophosphines, this step involves a dynamic kinetic
resolution, although diastereomeric ratios observed are not very high (from 1:1 to
6:1). Nevertheless, diastereomers **59** and **59'** can conveniently be separated by either
crystallization or chromatography. Subsequent key reductive cleavage, using Li
in NH$_3$, led to compound **62**. Alternatively, these authors have developed a very
elegant synthetic route, starting from *cis*-1-amino-2-indanol,[70] following a similar
approach to that developed by Jugé for the synthesis of *P*-stereogenic phosphines.[71]
From compound **60**, a nucleophilic ring opening with a Grignard reagent or an alkyl
lithium reagent gives aminophosphine borane **61**. Reductive cleavage then leads to
compound **62**. The latter has been used in the preparation of an amino derivative of
the TCFP diphosphine named MAXPHOS (**L7**). To that end, **62** was reacted with

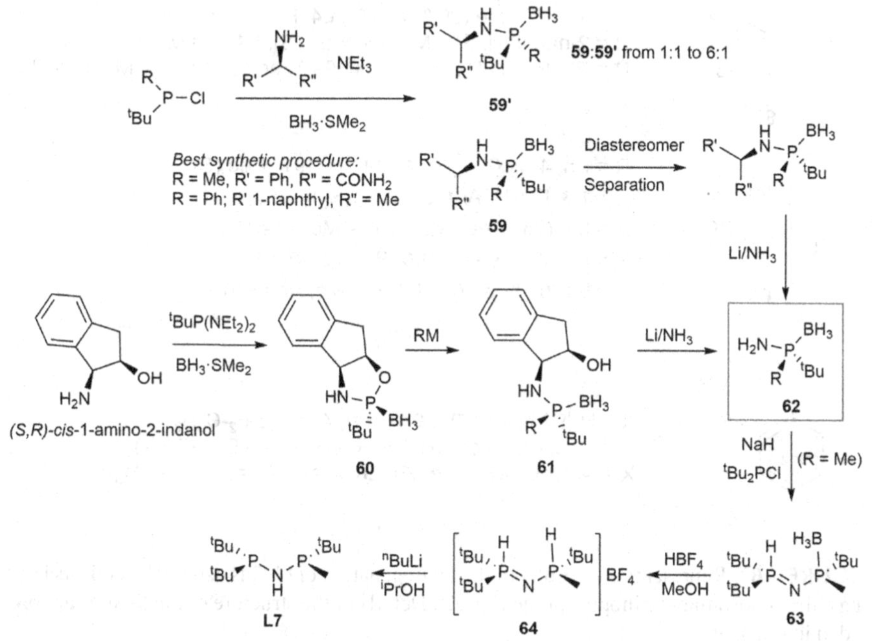

SCHEME 4.24 Synthetic routes to **62** and the Maxphos ligand (**L7**).

^{t}Bu$_2$PCl to give compound **63**. Deboronation of the latter, using an acidic treatment, led to the phosphonium salt **64**, which, upon deprotonation, led to the desired **L7**.

4.3.2.2 Catalytic Applications

The majority of applications of phosphine-aminophosphine ligands regard Rh-catalyzed asymmetric hydrogenation reactions of chelating olefins, such as dehydroamino acids or acyl enamides (Figure 4.6). For instance, Boaz has applied ligand **L4** (R = Me, Ph) in the hydrogenation of several α-dehydroamino acids **65**, with very high enantioselectivities.[66] An extremely high catalytic activity was observed in the hydrogenation of α-N-acetyl cinnamic acid. On the other hand, Hu and coworkers have described the application of ligands **L6** in the hydrogenation of several β-acylamino acrylates **66** and aryl enamides **67**, with generally high enantioselectivities.[72]

In addition, the hydrogenation of challenging unsaturated phosphonates **68** has been successfully described with Rh catalysts, based on phosphine-aminophosphine ligands (Scheme 4.25).[73] A catalyst based on ligands of type **L5** provides outstanding enantioselectivities for β-alkyl substituents, as well as for more demanding β-aryl ones.

Another interesting application in this context is the asymmetric hydrogenation of α,β-unsaturated esters bearing a γ-phthalimido substituent (**70**, Scheme 4.26), with a Rh catalyst based on a P-stereogenic phosphine-aminophosphine **L8**.[74] High

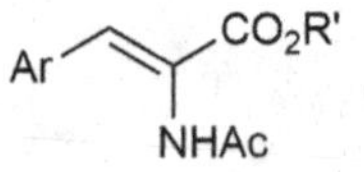

65

R = Ph, R' = Me (99 % ee (S); **L4**, R = Me, R' = Ph)
R = 2-naphthyl, R' = Me (98 % ee (S); **L4**, R = Me, R' = Ph)
R = Ph, R' = H (TOF = 49900; 99 % ee (S); **L4**, R = Me, R' = Ph)

66

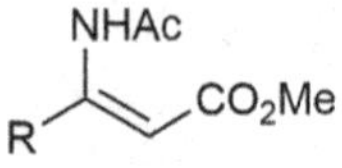

R = Ph, 4-MeO-C$_6$H$_5$, 4-Cl-C$_6$H$_5$ (91 % ee (S),
   L6: R = H, Ar = 3,5-F$_2$-C$_6$H$_3$)
R = Me (98 % ee (R); **L6**: R = Me, Ar = Ph)
R = Et (95 % ee (R); **L6**: R = Me, Ar = Ph)
R = ^{i}Pr (91 % ee (R); **L6**: R = Me, Ar = Ph)

67

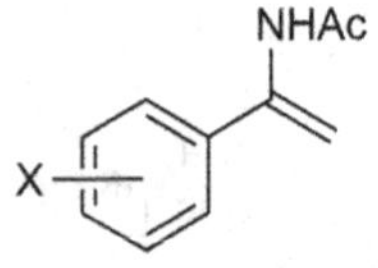

X = H (94 % ee (R); **L6**: R = H, Ar = 3,5-F$_2$-C$_6$H$_3$)
X = 4-Cl (95 % ee (R); **L6**: R = H, Ar = 3,5-F$_2$-C$_6$H$_3$)
X = 4-MeO (93 % ee (R); **L6**: R = H, Ar = 3,5-F$_2$-C$_6$H$_3$)

FIGURE 4.6 Representative results in the hydrogenation of olefins using Rh catalysts bearing chiral phosphine-aminophosphine ligands. Details of the structure of the ligands are provided in brackets.

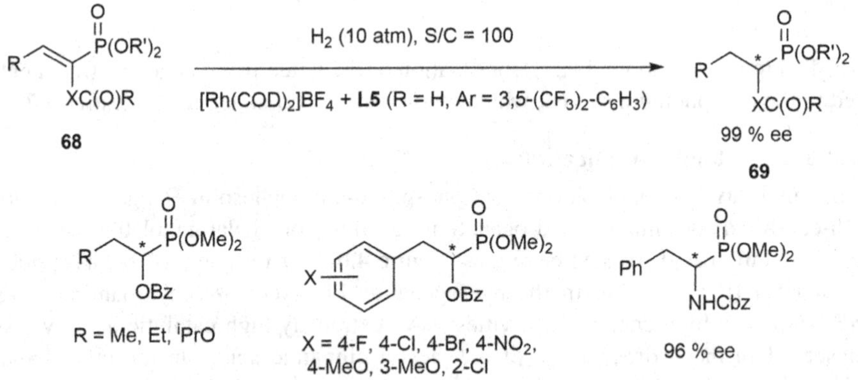

68

H$_2$ (10 atm), S/C = 100
[Rh(COD)$_2$]BF$_4$ + **L5** (R = H, Ar = 3,5-(CF$_3$)$_2$-C$_6$H$_3$)

69

99 % ee

R = Me, Et, ^{i}PrO

X = 4-F, 4-Cl, 4-Br, 4-NO$_2$,
4-MeO, 3-MeO, 2-Cl

96 % ee

SCHEME 4.25 Hydrogenation of phosphonates **68** with Rh catalysts based on ligand **L5**.

enantioselectivities were obtained for the hydrogenation of substrates bearing different aryl or heteroaryl substituents at position β. This reaction provides a convenient synthesis route for R enantiomers of the pharmaceutical compounds baclofen and rolipram.

To end this section, it is appropriate to mention an application of a phosphine-aminophosphine ligand **L4** in the synthesis of chiral biaryls. The process is based on a Pd-catalyzed coupling (Scheme 4.27), which produces compounds **73** with good to high enantioselectivity.[75] Products **73** are rather versatile compounds for the

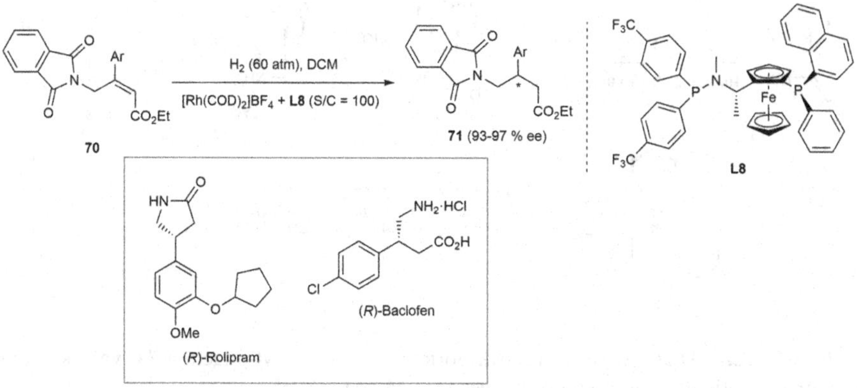

SCHEME 4.26 Hydrogenation of esters **70** with Rh catalysts based on ligand **L8**.

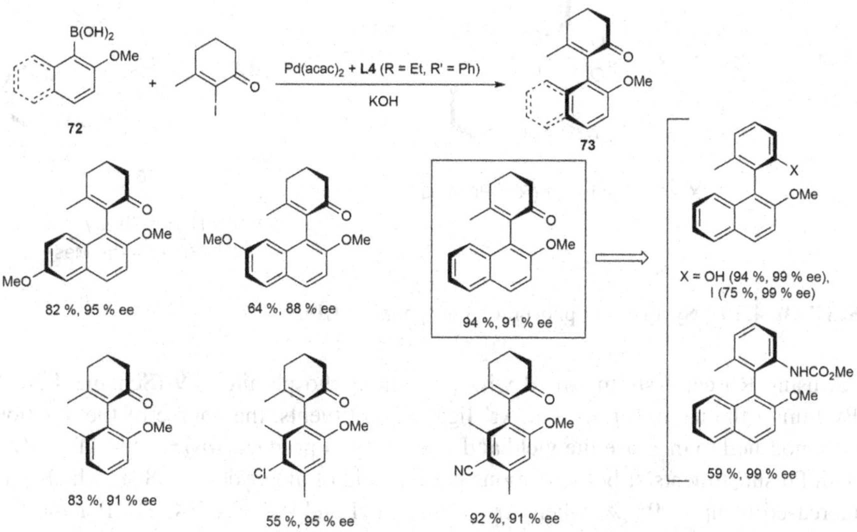

SCHEME 4.27 Synthesis of axially chiral compounds **73** by a Pd-catalyzed coupling.

preparation of a variety of axially chiral, biaryl derivatives, as exemplified in the case of the 2-methoxy naphthalene compound.

4.3.3 PHOSPHINE-PHOSPHONITES AND PHOSPHINE-PHOSPHINITES

Chiral phosphine-phosphinites and phosphine-phosphonites are two groups of ligands which have deserved less attention than the other types of ligands covered in this chapter. However, some relevant results from the use of chiral phosphine-phosphinites and phosphine-phosphonites deserve mention. For instance, Takaya and Nozaki have described the hydroformylation of β-lactam **74** to give aldehydes

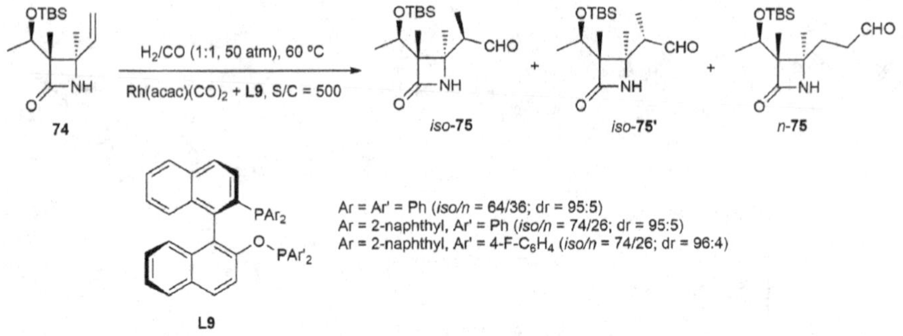

SCHEME 4.28   Diastereoselective hydroformylation of 4-vinyl-β-lactam **74**. Values of diastereomeric ratio (dr) refer to ratios between *iso*-**75** and *iso*-**75'**.

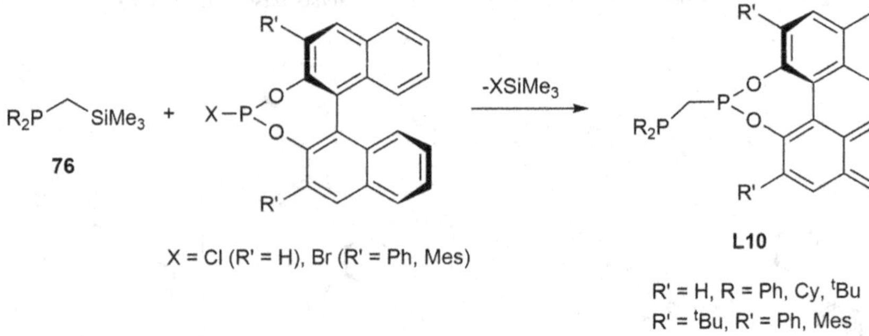

SCHEME 4.29   Synthesis of phosphine-phosphonites **L10**.

75, using Rh catalysts modified with phosphine-phosphinites **L9** (Scheme 4.28).[76] By tuning the nature of Ar and Ar' ligand substituents, the course of the reaction was modified to increase the yield and selectivity of product *iso*-**75**. Thus, a catalyst with Ph substituents at both positions gave a yield of aldehydes of 58 %, which was increased to up to 95 %, when Ar = 2-naphthyl and Ar' = 4-F-C_6H_4. For the latter, good regioselectivity (74:26) and diastereoselectivity (96:4) were also observed. *Iso*-**75** is a convenient starting material for the synthesis of 1β-methylcarbapenem antibiotics.[77]

On the other hand, Pringle and coworkers have described a rather simple synthesis of phosphine-phosphonites **L10** (Scheme 4.29).[78] Thus, the direct reaction between silylphosphines **76** and chloro- or bromophosphites derived from BINOL leads to these ligands and the release of $ClSiMe_3$ or $BrSiMe_3$. Ligands **L10** have been examined in the Rh-catalyzed hydrogenation of benchmark olefins, like dimethyl itaconate and methyl Z-α-acetamidocinnamate, with enantioselectivities up to 99 % ee.

Another interesting application of phosphine-phosphonites was described by the group of Reek. These authors have described the synthesis of ligands **L11** from bromo phosphine **77** (Scheme 4.30).[79] Reaction of the latter with ^tBuLi and $(Et_2N)_2PCl$ leads to the corresponding phosphine-diaminophosphine **78**, which reacted with a variety

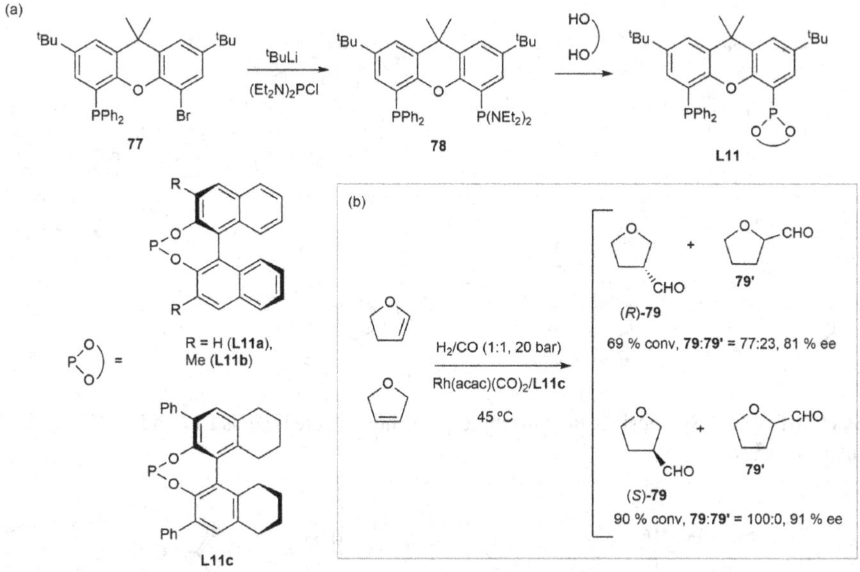

SCHEME 4.30 (a) Synthesis of phosphine-phosphonite ligands **L11**. (b) Application of **L11** to the asymmetric hydroformylation of dihydrofurans.

of BINOL derivatives to give the desired ligands **L11** in high yields. These ligands have been used in the asymmetric hydroformylation of dihydrofurans with good levels of both regio- and enantioselectivity.

4.3.4 Phosphine-Phosphoramidites

4.3.4.1 Synthesis

The synthesis route of phosphine-phosphoramidites is very similar to that of phosphine-aminophosphines, except that, in the introduction of the second phosphorus functionality, a chlorophosphite (instead of a chlorophosphine) reagent is used. For illustrative purposes, the synthesis of some selected phosphine-phosphoramidite ligands is briefly described below. For instance, Leitner and coworkers have described a convenient synthesis of ligands **L12**, starting from 2-fluoro-1-bromobenzene (Scheme 4.31a).[80] First, Pd-catalyzed amination produces the corresponding fluoroamine **80**, which, upon reaction with KPPh$_2$, produces aminophosphine **81**, which can be used for the introduction of diverse phosphoramidite fragments by treatment with ^{n}BuLi and an appropriate chlorophosphite. Alternatively, the group of X. Zhang has described the synthesis of a family of ligands **L13** called YANPHOS (Scheme 4.31b).[81] Starting from known aminophosphine **82** and following a methodology described in the literature,[82] a set of aminophosphines **83** was prepared and then phosphorylated to give **L13**. Another interesting phosphine-phosphoramidite ligand is called INDOLPHOS (**L14**, Scheme 4.32).[83] The synthesis of this ligand starts with protection of the NH fragment of indole **84**, by lithiation, and reaction

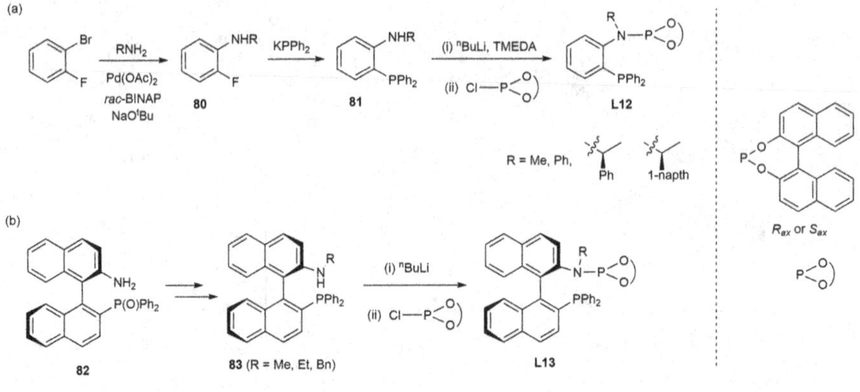

(a)

(b)

SCHEME 4.31 Synthesis of (a) phosphine-phosphoramidite ligands **L12** and (b) YANPHOS ligands **L13** (b).

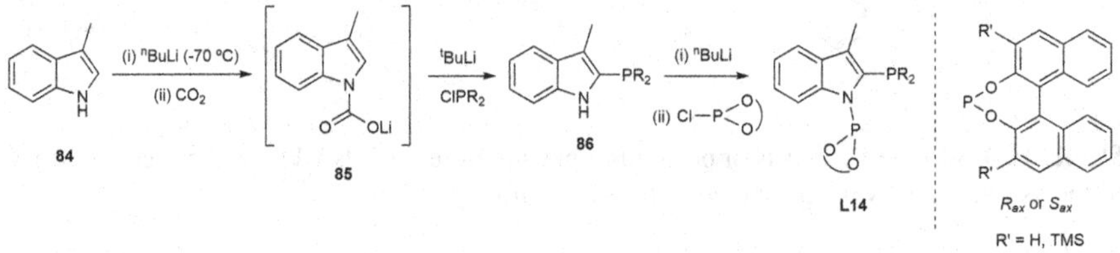

SCHEME 4.32 Synthesis of Indolphos ligands **L14**.

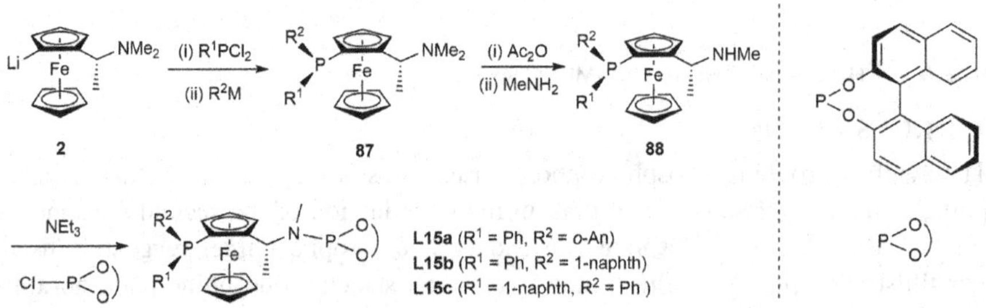

L15a (R^1 = Ph, R^2 = o-An)
L15b (R^1 = Ph, R^2 = 1-naphth)
L15c (R^1 = 1-naphth, R^2 = Ph)

SCHEME 4.33 Synthesis of phosphine-phosphoramidites **L15a–L15c**.

with CO_2 to give **85**. Subsequent reaction with ^{t}BuLi and a chlorophosphine leads to aminophosphine **86**, which can finally be phosphorylated to give **L14**.

Another interesting class of phosphoramidites are the ferrocenyl ligands **L15**, structurally related to Bophoz ligands (**L4**). Chen has described a rather powerful route for the preparation of a set of ligands **L15**, characterized by a *P*-stereogenic phosphine fragment (Scheme 4.33).[84] Reaction of chiral reagent **2** with a dichloro-phosphine and an alkylating or arylating metal reagent produces aminophosphines **87,** with a very high diastereoselectivity. Subsequent substitution of the NMe$_2$ fragment, to give **88**, followed by phosphorylation, finally produces the desired ligands.

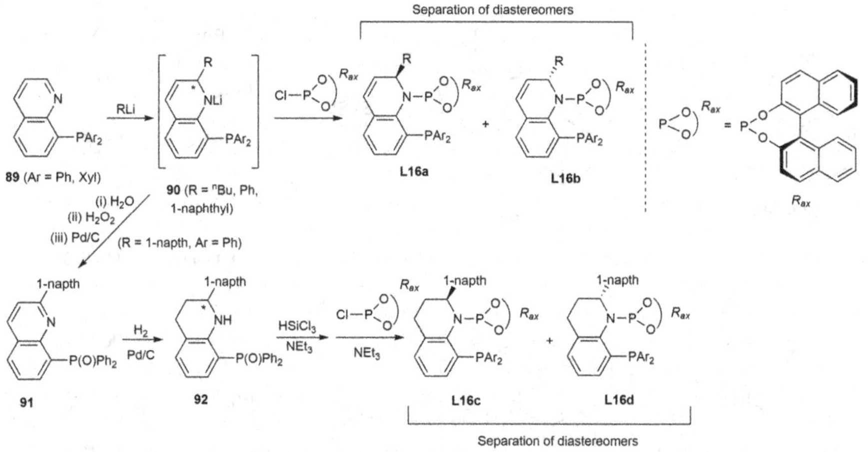

SCHEME 4.34 Synthesis of phosphine-phosphoramidites **L16**.

In addition, the group of Leitner has reported the synthesis of a family of ligands named Quinaphos (**L16**), starting from phosphino-quinoline **89**, and following two alternative routes (Scheme 4.34).[85] The first route relies on the reaction of **89** with an organolithium reagent, followed by treatment with the chlorophosphite of *R*-BINOL to produce a mixture of diastereomers **L16a** and **L16b**, which can be separated by column chromatography or crystallization, depending on the R substituent. The corresponding enantiomers have been generated by using the chlorophosphite of *S*-BINOL. Alternatively, a second, more elaborate route was required to prepare dihidro-Quinaphos ligands **L16c** and **L16d**, due to the lack of reactivity of amine derivatives of **90** towards hydrogenation. Satisfactory results were otherwise obtained by hydrogenation of phosphine oxide **91** to the corresponding tetrahydroquinoline **92,** followed by reduction of the phosphine oxide group and phosphorylation.

To end this section, it is appropriate to bring to the reader's attention a family of phosphine-phosphoradiamidite ligands (**L17**, Scheme 4.35). Compared with previous examples, they contain two N substituents and one O substituents. Diastereomeric ligands **L17** named BettiPhos have been prepared from amino alcohol **93** or aminophosphine **95** by selecting the order of addition and the configuration of these reagents.[86]

4.3.4.2 Catalytic applications

Rh-catalyzed asymmetric hydrogenation of common olefins, such as dimethyl itaconate,[87] α-dehydroamino acids[88] or aryl-enamides[89] has been used as a diagnostic tool to validate the design of phosphine-phosphoramidite ligands. In addition, some less-explored substrates have been examined with Rh catalysts bearing this kind of ligands. For instance, Zheng and coworkers have described an extremely enantioselective catalyst, based on ligand **L18a**, for the hydrogenation of unsaturated phosphonates **68**, suitable for β-aryl, β-alkyl and β-alkoxo derivatives (Scheme 4.36a).[90] On the other hand, Du and Hu have described the hydrogenation of phosphine oxides **96**.

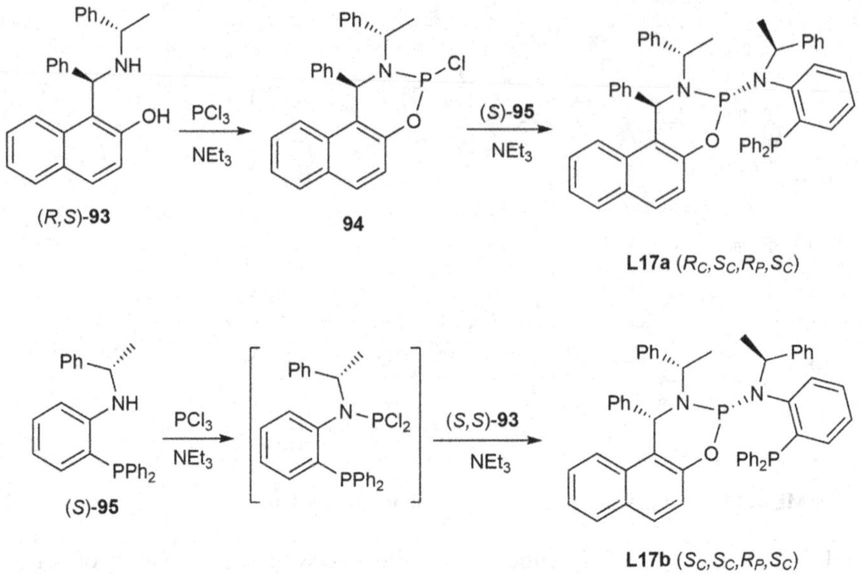

SCHEME 4.35 Synthesis of phosphine-phosphoradiamidite BETTIPHOS ligands **L17**.

This reaction requires relatively high pressures of hydrogen (Scheme 4.36b).[91] As well, an important influence of solvent was observed when catalyst loading was lowered, with trifluoroethanol achieving the best results. Under the optimized conditions, enantioselectivities are very high and the reaction provides access to a wide variety of β-acylamino phosphine oxides **97**. The latter compounds can be very conveniently converted by chiral β-aminophosphine ligands, as exemplified by the preparation of compounds **98**. On the other hand, a preferable approach to the synthesis of α-hydroxyphosphonates is the direct hydrogenation of α-ketophosphonates **99** (Scheme 4.37). In this regard, ligand **L19b** acts as a good catalyst for aryl-substituted substrates, with enantioselectivities between 80 and 87 % ee.[92]

Another interesting application of Rh-catalyzed asymmetric hydrogenation is the preparation of synthetically useful Roche esters **102** (Scheme 4.38). Zheng and coworkers studied the performance of several Rh catalysts, based on phosphine-phosphoramidites, indicating that the one based on **L18a** is the most satisfactory one.[93] Thus, the hydrogenation of unsaturated esters **101** led to the desired products **102,** with enantioselectivities ranging between 79 and 97 % ee. In this context, application of the catalyst based on the INDOLPHOS ligand (**L14**, R = ^iPr, R' = H, S_{ax}) achieved an excellent enantioselectivity of 98 % ee in the hydrogenation of the methyl ester, although the reaction was performed at −40 °C.[83]

A remarkable application of phosphine-phosphoramidites involves the asymmetric hydrogenation of 1-alkyl vinyl esters **103** (Scheme 4.39). The asymmetric hydrogenation of enol esters by Rh catalysts is a well-known reaction, but the substrates covered until recently were mostly limited to activated compounds bearing an electron-withdrawing olefin substituent (CN, CF$_3$, CO$_2$R or aryl).[94] The nature of this

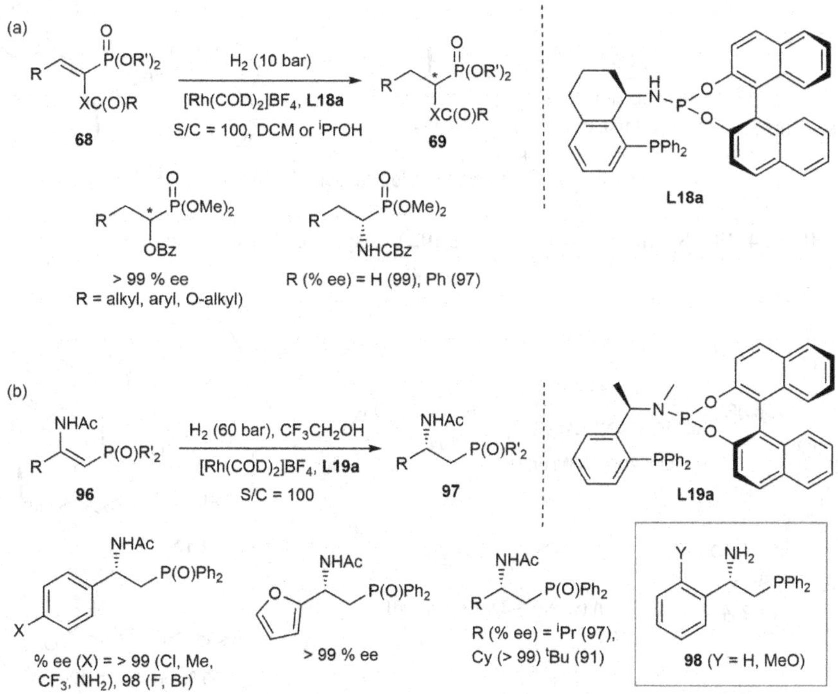

(a)

(b)

SCHEME 4.36 Hydrogenation of unsaturated phosphonates **68** and phosphine oxides **96** with Rh catalysts based on phosphine-phosphoramidite ligands.

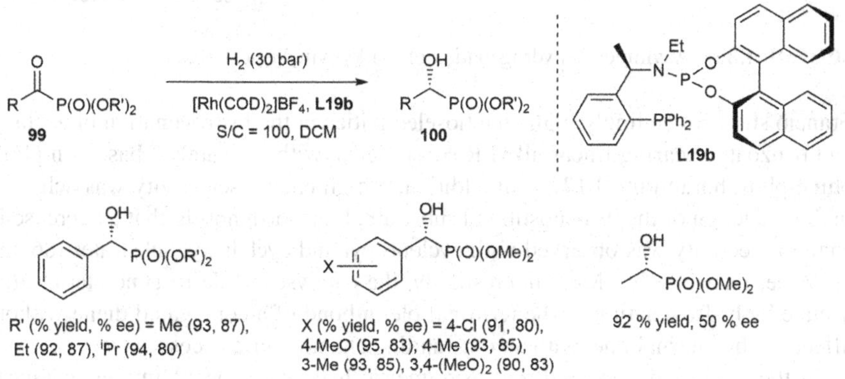

SCHEME 4.37 Hydrogenation of α-ketophosphonates **99**.

substituent may strongly affect the regioselectivity of the olefin insertion step and hence the enantioselectivity and even product configuration.[95] Moreover, due to an easy deprotection of hydrogenated products, this reaction permits a very convenient route for the synthesis of 2-alkanols, though rather difficult to obtain in high enantioselectivity by ketone asymmetric hydrogenation. Results reported by Leitner and

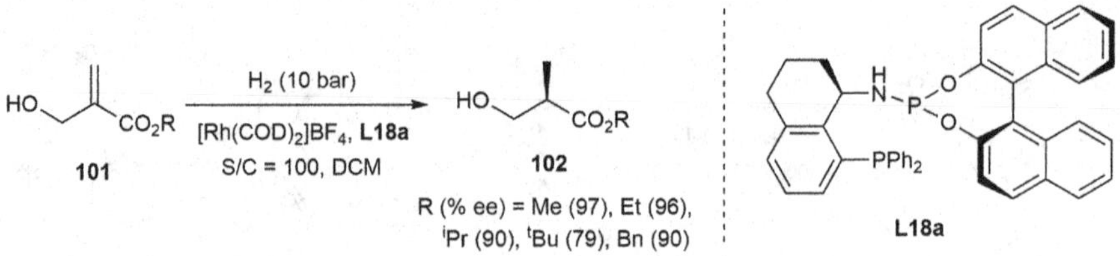

SCHEME 4.38 Synthesis of Roche esters **102** by asymmetric hydrogenation.

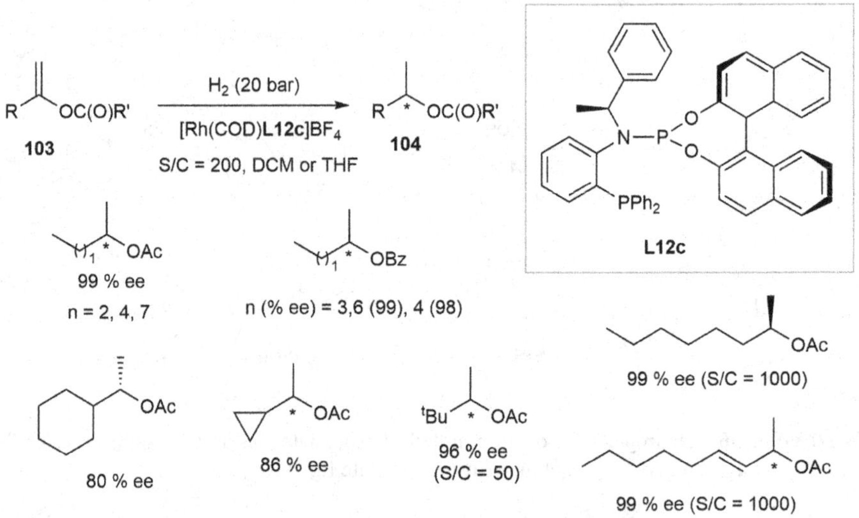

SCHEME 4.39 Asymmetric hydrogenation of 1-alkyl vinyl esters **103**.

Franciò show exceedingly high enantioselectivities in the hydrogenation of acetates and benzoates bearing linear alkyl R substituents, with the catalyst based on phosphine-phosphoramidite **L12c**.[96] In addition, a high enantioselectivity was achieved in the reduction of the tBu-substituted substrate. It is worth noting that a decrease in enantioselectivity was observed with cyclopropyl and cyclohexyl substrates (86 and 80 % ee, respectively). Most interestingly, the catalyst is able to generate an allyl acetate by hydrogenation of the terminal olefin bond of a conjugated diene without affecting the internal one, with an outstanding enantiomeric excess.

In the field of the Ir-catalyzed asymmetric hydrogenation of imines, catalysts based on phosphine-phosphoramidites have also produced remarkable results. For instance, Hu studied the reduction of a set of N-aryl imines **106**, with the catalyst based on **L19c** ligand (Scheme 4.40a). This system required the presence of KI for a satisfactory performance. Generally, high enantioselectivities were observed in amines **107**, with the exception of sterically encumbered substrates bearing ortho-substituted aryl rings.[97] This drawback was solved by the same authors using a catalyst bearing the ligand **L19d** (Scheme 4.40b),[98] leading to outstanding

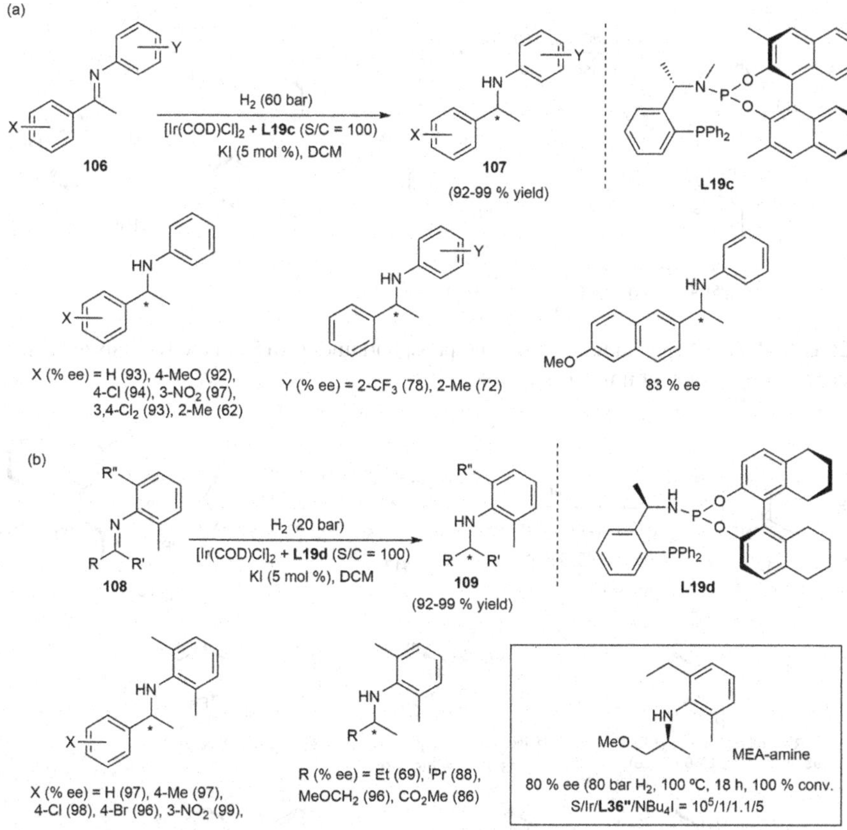

SCHEME 4.40 Application of phosphine-phosphoramidites in the Ir-catalyzed asymmetric hydrogenation of imines.

enantioselectivities in the hydrogenation of sterically hindered imines **108**. Thus, values between 95 and 99 % ee were obtained for imines derived from methyl aryl ketones. In addition, respectable values of enantioselectivities were obtained for challenging imines derived from 2-butanone (69 % ee) and 3-methyl-2-butanone (88 % ee). Most remarkably, the catalytic system is capable of providing a practical synthesis for the monoethanolamine-based amine (MEA-amine), a precursor of the metholachlor herbicide, giving full conversion and 80 % ee for a reaction performed at a substrate to catalyst (S/C) ratio of 100,000.

A very interesting type of substrate for hydrogenation are imino esters **110**, as the reaction products can be converted into α-amino acids (Scheme 4.41),[99] particularly those with α-aryl substituents, which are not accessible by Rh-catalyzed olefin hydrogenation. In a reaction catalyzed by an Ir complex bearing the ligand **L15d**, generally high enantioselectivities were obtained in the synthesis of aryl glycine derivatives. This reaction is very sensitive to the additives used and negligible

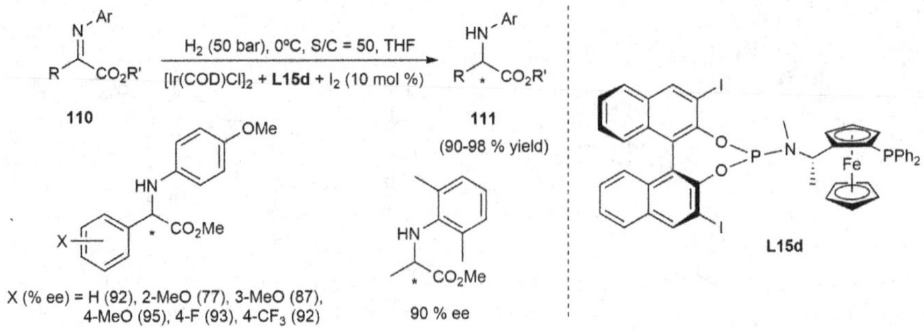

SCHEME 4.41 Application of phosphine-phosphoramidite **L15d** in the Ir-catalyzed asymmetric hydrogenation of imino esters **110**.

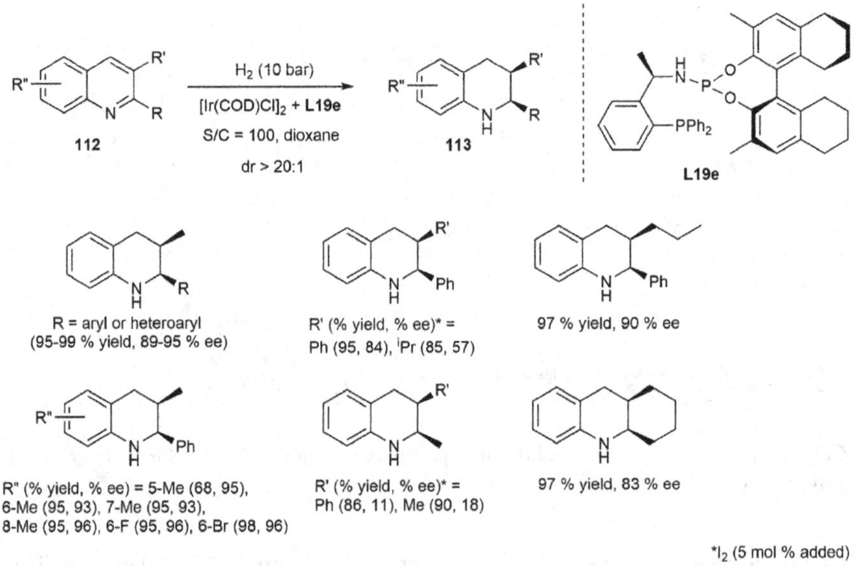

SCHEME 4.42 Hydrogenation of 2,3-disubstituted quinolines with an Ir/phosphine-phosphoramidite catalyst.

conversions were observed in the absence of I_2. A high enantioselectivity (90 % ee) was also observed in the case of the N-aryl alanine derivatives.

The hydrogenation of 2,3-disubstituted quinolines **112** (Scheme 4.42) is a difficult reaction to achieve with high levels of both diastereoselectivity and enantioselectivity. To deal with this challenge, Hu and coworkers studied the performance of Ir catalysts, based on phosphine-phosphoramidites, in this reaction.[100] After a catalyst screening, a catalyst bearing ligand **L19e** achieved the best results. This catalyst reduces a wide range of substituted quinolines under relatively mild conditions, with a high diastereoselectivity. The enantioselectivity, however, depends on the substitution pattern of the substrate. Thus, the system is best suited for quinolines with an aryl and a *n*-alkyl substituents at positions 2 and 3, respectively, resulting in enantioselectivities of between 89 and 95 % ee. For these substrates, no additive is required, whereas, in some other examples, I_2 was added as additive. Increasing the steric effects, by including a Ph or a ^{i}Pr substituent at position 3 leads to a decrease in enantioselectivity, whereas substrates with an alkyl substituent at position 2 led to significantly lower enantioselectivities.

On the other hand, an interesting transformation, related to that depicted in Scheme 4.21, is the Cu-catalyzed reduction of unsaturated esters **114,** using a silane and an alcohol (Scheme 4.43).[101] Depending on the nature of the X and R substituents, enantioselectivities in the range between 76 and 99 % ee were observed.

A reaction of considerable synthetic interest is enantioselective hydroformylation. This is, however, a rather challenging transformation, as the generation of the desired branched aldehyde with high optical purity requires the control of chemo-, regio-, and enantioselectivity of the process. Significant advances in this field have been reported with catalysts based on phosphine-phosphoramidite ligands. For instance, the research group of X. Zhang has studied the performance of YANPHOS ligands **L13** (Scheme 4.44) in the hydroformylation of benchmark olefins. These ligands are structurally similar to Binaphos phosphine-phosphites (Scheme 4.1), although the presence of the NR fragment reduces the size of the chiral pocket generated by the chelating ligand. This structural modification has a positive effect on enantioselectivity.[102] Furthermore, the reactivity profile is somewhat different, with best results being obtained under lower pressures of the CO/H$_2$ (1:1) mixture, typically at 20 bar. Under these conditions, very high enantioselectivities, around 98 % ee, were obtained for a set of vinyl arenes with a catalyst based on **L13b**. Likewise, vinyl esters were hydroformylated with high enantioselectivities (93–98 % ee). This catalyst is characterized by a phosphite BINOL fragment, which showed better results than other biaryl fragments. A major drawback of the performance of this catalyst is a relatively low level of regioselectivity (also observed in the case of BINAPHOS), with *iso/n* ratios between 87:13 and 91:9 for vinyl arenes and of 93:7 for vinyl acetate. In this regard, remarkable results have been obtained with the catalyst based on the BETTIPHOS ligand **L17b** (Scheme 4.35), which has shown *iso/n* ratios up to 1,000 in the hydroformylation of vinyl esters (Scheme 4.46).[103] This catalyst also achieves relatively high enantioselectivities, with values between 91 and 95 % ee.

Another interesting application of the YANPHOS ligand is the hydroformylation of *N*-, *S*-, or *O*-containing heteroaryl olefins **118** (Scheme 4.46).[104] In line with the results described before in the hydroformylation of vinyl arenes, high enantioselectivities (91–96 % ee) and moderate regioselectivities (*iso/n* ratios between 4 and 10) were observed in the synthesis of a wide range of aldehydes **119**.

Taking advantage of the rich reactivity of chiral aldehydes generated in the asymmetric hydroformylation reaction,[105] X. Zhang and coworkers have described several procedures based on the further transformation of these aldehydes. For instance, hydroformylation of allyl amines **120** (Scheme 4.47) lead to aldehydes suitable for an intramolecular attack of the amine to give hemiaminals **121**.[106] As no water elimination is achieved, this process constitutes an interrupted intramolecular hydroaminomethylation. Compounds **121** are stable enough to be transformed in a subsequent

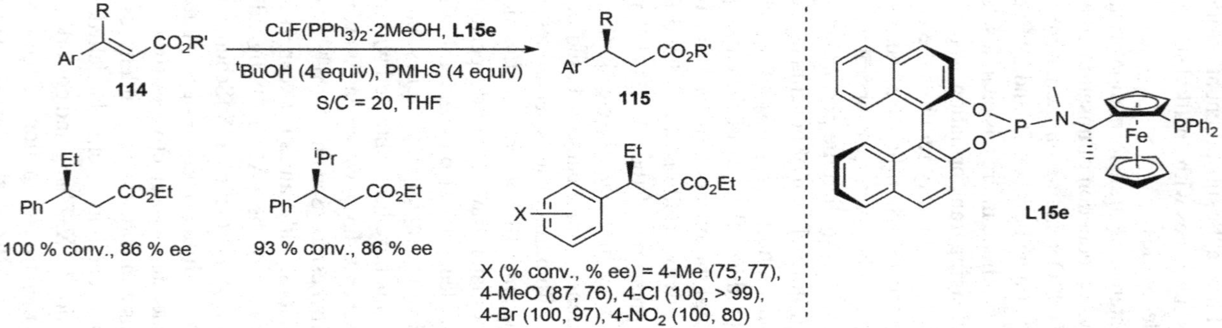

SCHEME 4.43 Reduction of unsaturated esters **114** by a copper catalyst based on ligand **L15e**.

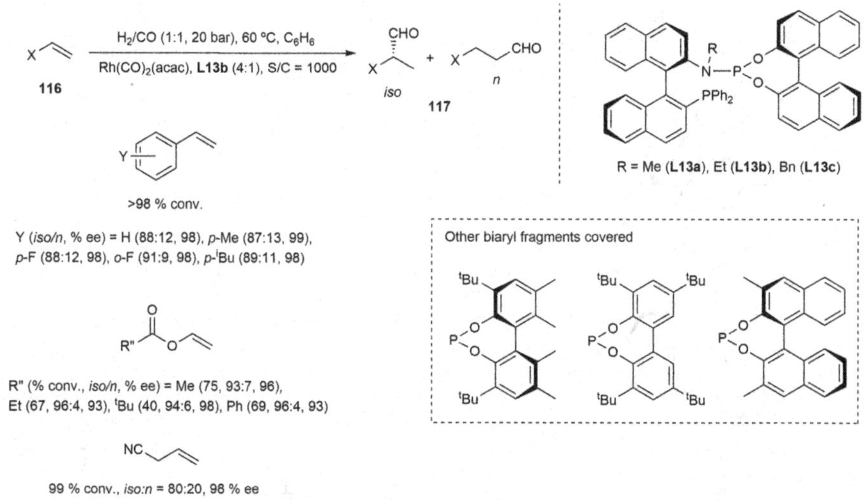

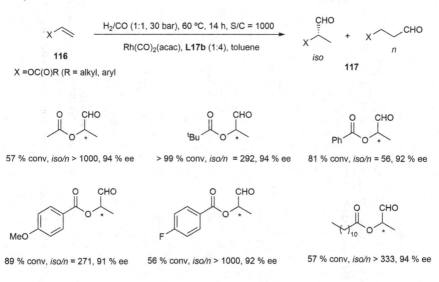

SCHEME 4.44 Results on the asymmetric hydroformylation of olefins **116** by Rh-YANPHOS catalysts.

SCHEME 4.45 Synthesis of α-acyloxy-aldehydes obtained by hydroformylation of vinyl esters using a Rh catalyst based on the BettiPhos ligand **L17b**.

oxidation with pydidinium chlorochromate (PCC) and sodium acetate to produce pyrrolidinones **122**. Similarly, compounds **121** can be reduced by HSiEt$_3$ in the presence of BF$_3$·Et$_2$O, to achieve pyrrolidines **123**. In both types of transformation, high yields and high enantioselectivities were observed. The formation of five-membered rings results from a high preference for the γ-formyl isomer. However, for a substrate bearing a n-propyl R substituent, regioselectivity is not as pronounced and a lower

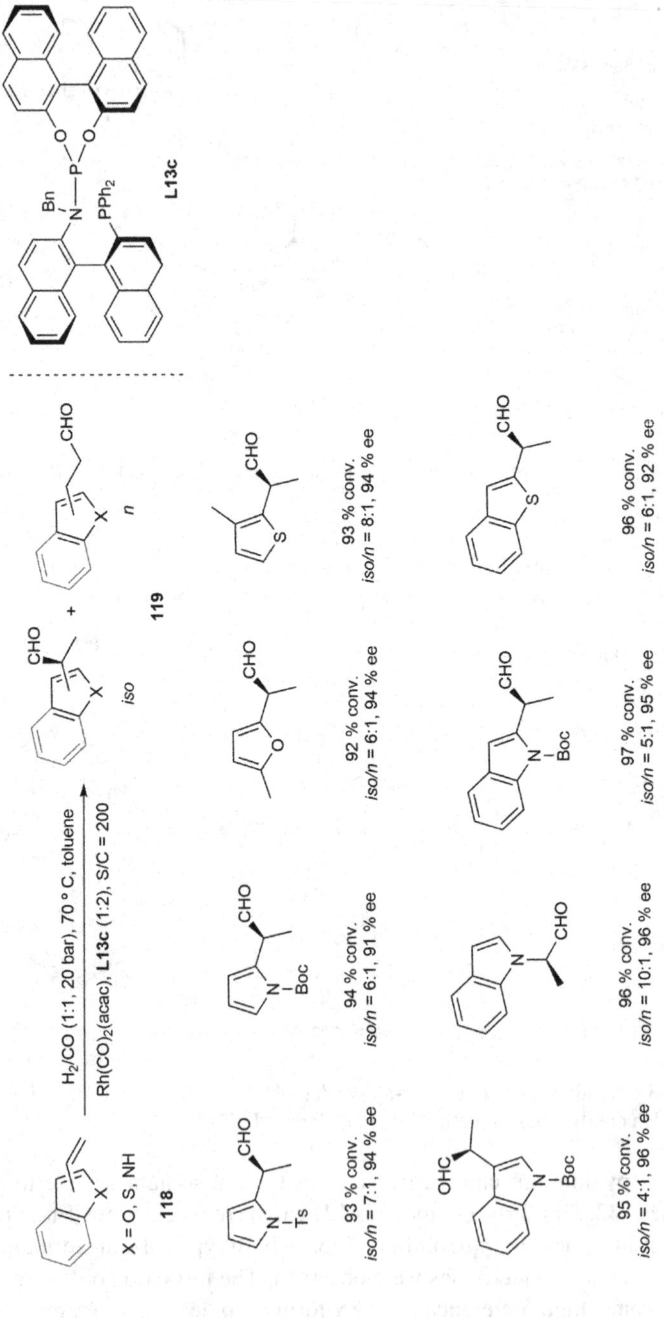

SCHEME 4.46 Asymmetric hydroformylation of vinyl heteroarenes **118** by a Rh-YANPHOS catalyst.

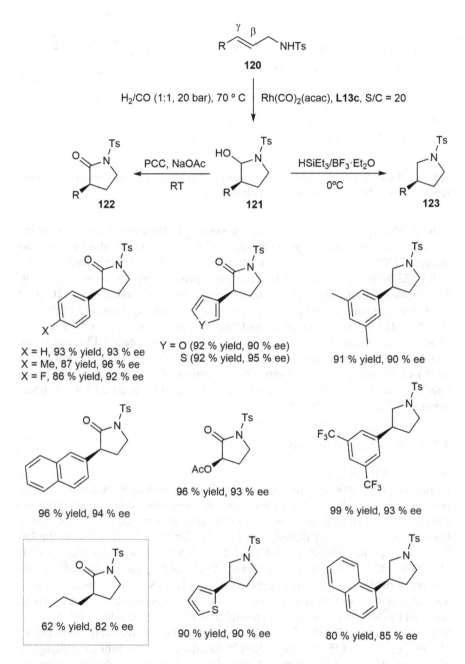

SCHEME 4.47 Synthesis of hemiaminals **121** by an intramolecular asymmetric interrupted hydroaminomethylation. Application in the synthesis of pyrrolidinones **122** and pyrrolidines **123**.

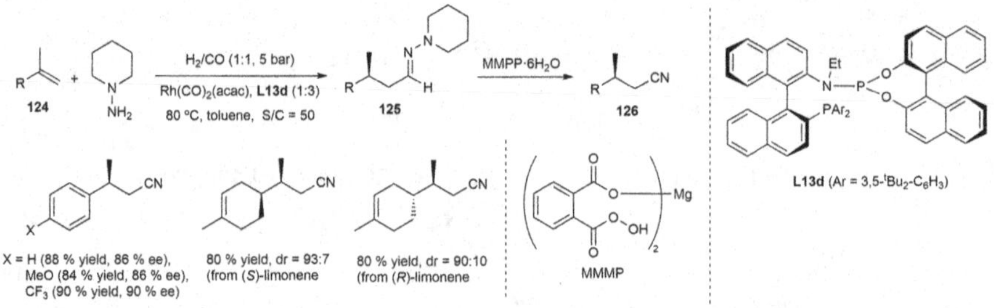

SCHEME 4.48 Synthesis of chiral nitriles **126** by an asymmetric hydroformylation/condensation process followed by oxidation of hydrazones **125**.

yield of the corresponding pyrrolidinone is observed. Another remarkable contribution from the same research group considers the synthesis of chiral nitriles without the use of HCN (Scheme 4.48).[107] The procedure is based on the formation of hydrazones **125** by the reaction of hydroformylation products with 1-amino-pyperidine. These hydrazones are subsequently transformed into the desired nitriles **126** by an aza-Cope elimination, produced by magnesium monoperoxyphthalate hexahydrate (MMMP·6H$_2$O). It should be mentioned that the authors have explored the reaction of olefins with different substitution patterns. Among them, ligand **L13d**, which possesses an *S* axial configuration in both naphthyl fragments, provides the best results for 1,1-disubstituted olefins. In turn, for monosubstituted olefins and 1,2-disubstituted ones, the best results were obtained with a catalyst based on 1,2-bis(2,5-diphenylphospholane)ethane (Ph-BPE).

4.3.5 Phosphine-Phosphites

4.3.5.1 Synthesis

Phosphine-phosphites are typically prepared by a reaction between a hydroxyphosphine and a chlorophosphite. Of these reagents, the latter are very easily prepared from a diol, whereas hydroxyphosphines usually require more synthetic effort. For instance, synthesis of BINAPHOS ligands **L20** depends on access to the key hydroxyphosphine **129** (Scheme 4.49).[108] **L29** is prepared in two steps from ditriflate of BINOL **127**, followed by a Pd- catalyzed P–C coupling and the reduction of the phosphine oxide **128**. Likewise, access to ligands **L21** (Scheme 4.50a) depends on a suitable synthesis of phenol phosphines **131**. These reagents can conveniently be obtained by demethylation of anisyl phosphines **130**.[23] On the other hand, Schmalz and coworkers have described a very versatile procedure for the synthesis of phenol phosphine boranes **135** (Scheme 4.50b),[109] based on a bromo-lithium exchange in phosphinite boranes **134**, which is followed by a migration of the fosfino-borane fragment. From **135**, a family of phosphine-phosphites **L22**, based on (*R*,*R*)-TADDOL, has been prepared. In addition, suitable procedures for the synthesis of ligands, bearing only one carbon in the backbone, have also been described. Thus, the reaction between diarylphospholane borane **136** and formaldehyde produces the

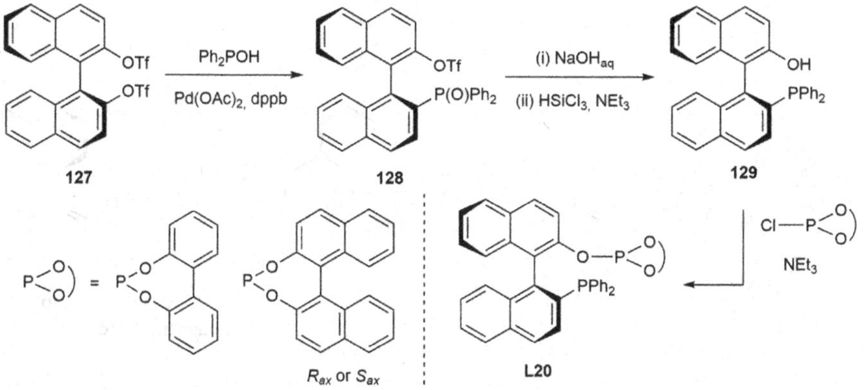

SCHEME 4.49 Synthesis of ligands **L20**.

corresponding borane-protected hydroxyphosphine **137** (Scheme 4.51a).[110] Similarly, enantioselective reduction of acetyl phosphine **138** leads, after borane protection, to chiral hydroxy phosphine-borane **139** (Scheme 4.51b).[111] On the other hand, reduction by LiAlH$_4$ of acetate phosphine **140** leads to the corresponding hydroxy ferrocenyl phosphine **141** (Scheme 4.52).[112] In this context, it should be mentioned that, in many of these procedures, the phosphine is deboronated at the end of the synthetic route by reaction with 1,4-diazabicyclo[2.2.2]octane (DABCO).

4.3.5.2 Catalytic applications

As mentioned in the section of phosphine-phosphoramidites (Section 4.3.4), many phosphine-phosphites have been tested in the Rh-catalyzed asymmetric hydrogenation of benchmark olefins such as methyl Z-α-acetamidocinnamate, methyl α-acetamidoacrylate or dimethyl itaconate.[12,23,113–115] In addition, broad-range catalysts, based on this kind of ligand, have also been described. For instance, a catalyst based on **L26** produces a variety of β-aryl-α-amino acids in an enantiopure form (Scheme 4.53).[116] In this area, a topic of interest is the introduction of N-protecting groups, which are synthetically more useful than N-acetyl. In this regard, Vidal-Ferran and coworkers have described effective catalysts based on ligands **L27** for the reduction of this type of substrate with high enantioselectivities (Scheme 4.54).[117] Interestingly, this catalytic system is also effective for a substrate bearing a β–methoxy substituent.

On the other hand, Rh phosphine-phosphite catalysts have been examined in the hydrogenation of phosphonate enamides **96** (Scheme 4.36b), although the results did not outperform those obtained with phosphine-phosphoramidite ligands described above.[118] Similarly, enantioselectivities up to 93 % ee were observed in the hydrogenation of enamides from β-tetralones, using catalysts based on ligands **L21**,[119] although these results did not reach the outstanding performance shown by Ir catalysts bearing P,N ligands.[120]

As mentioned above, the hydrogenation of enol esters **103** is a very valuable approach for the synthesis of aliphatic alcohols. In this regard, ligands **L28** provide

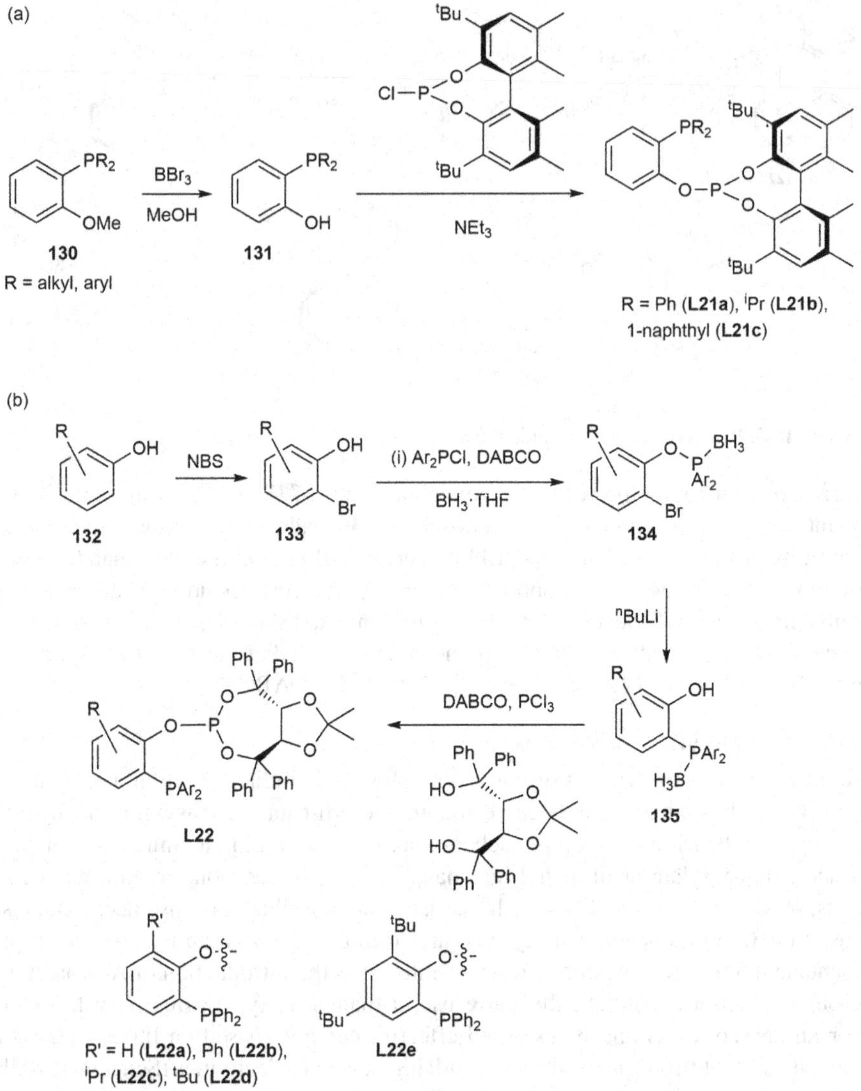

(a)

130
R = alkyl, aryl

131

R = Ph (**L21a**), ^{i}Pr (**L21b**),
1-naphthyl (**L21c**)

(b)

132

133

134

L22

135

R' = H (**L22a**), Ph (**L22b**),
^{i}Pr (**L22c**), ^{t}Bu (**L22d**)

L22e

SCHEME 4.50   Synthesis of phosphine-phosphite ligands **L21** and **L22**.

a powerful catalytic system, suitable for the hydrogenation of a wide range of these substrates with high enantioselectivity (Scheme 4.55).[121] As described before, substrates bearing a cycloalkyl R substituent constitute a particularly difficult subset of substrates **103** to hydrogenate with a high enantioselectivity. By tuning the phosphine fragment in ligands **L28**, enantioselectivities greater than 90 % ee were obtained for these substrates. The reaction is also suitable for the synthesis of fluoro-aryl and bromo-aryl, as well as N-phthalimido derivatives. Interestingly, an enantioreversal was observed in the configuration of products of hydrogenation of ^{t}Bu

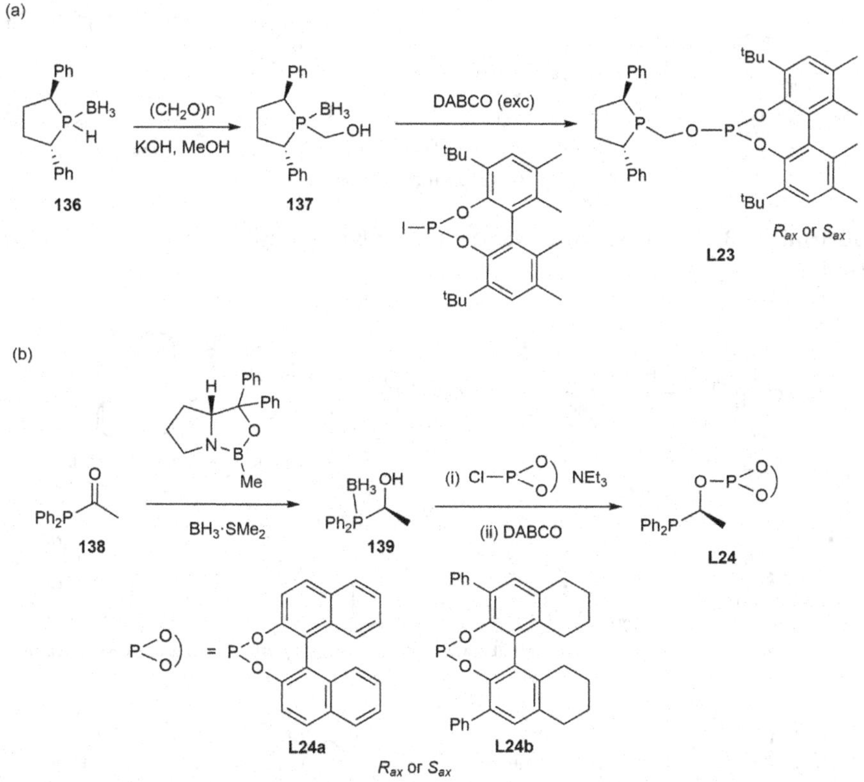

(a)

SCHEME 4.51   Synthesis of small bite angle ligands (a) L23 and (b) L24.

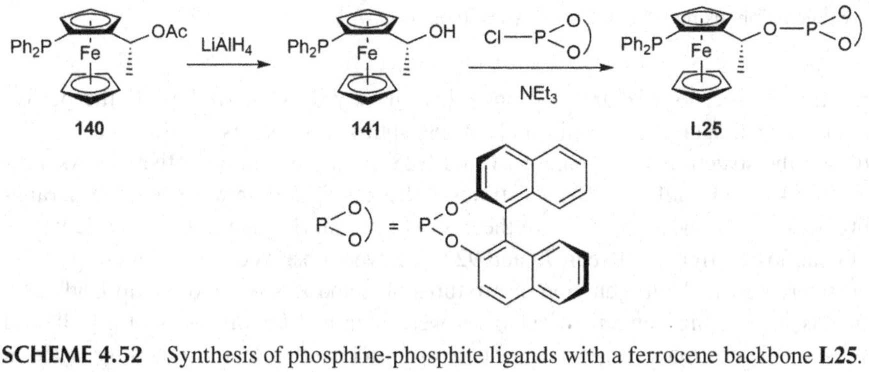

SCHEME 4.52   Synthesis of phosphine-phosphite ligands with a ferrocene backbone L25.

and Ph substrates, analogous to that described in the case of structurally analogous N-acetyl enamides.[122]

The hydrogenation of α,β-disubstituted enol esters 144 (Scheme 4.56) poses difficulties additional to those associated with the hydrogenation of esters 103, due to higher olefin substitution and the influence of the olefin configuration of 144 in the

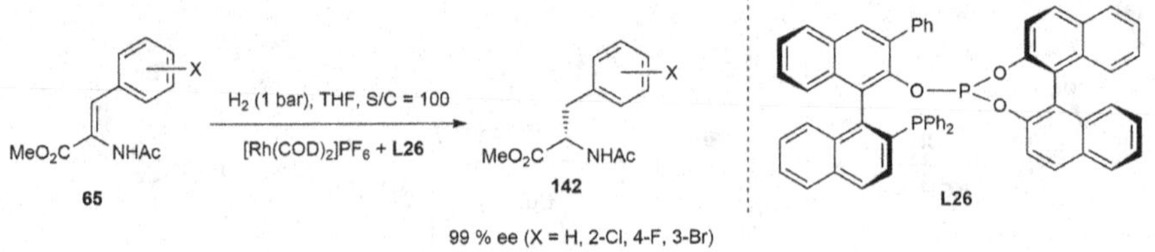

SCHEME 4.53 Enantioselective hydrogenation of β-aryl dehydroamino acids by Rh-**L26** catalyst.

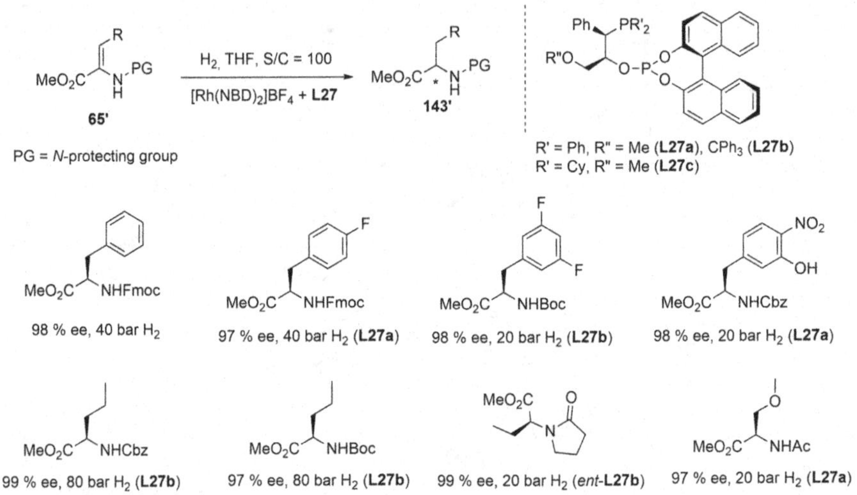

SCHEME 4.54 Selected applications of the hydrogenation of α-dehydroamino acids by hydrogenation, using Rh catalysts bearing ligands **L27**.

reaction. Cadierno, Pizzano and coworkers have published studies on the performance of Rh catalysts, based on chiral phosphine-phosphites, in this reaction. By tuning the structures of ligands **L21** and **L28**, highly enantioselective catalysts for α,β-dialkyl and α-alkyl-β-aryl substrates with Z configuration were developed, ranging between 91 and 99 % ee.[123] In the case of α,β-diaryl substrates, moderate values of enantioselectivity, between 79 and 92 % ee, were observed. An aspect of particular interest is the hydrogenation of mixtures of E and Z isomers of compounds **144**. In this regard, high enantioselectivities were obtained for mixtures of α,β-dialkyl substituted substrates, while a significant drop in enantioselectivity was observed in the case of α-alkyl-β-aryl ones.

Another class of interesting enol esters for asymmetric hydrogenation are β-acyloxy phosphonates **146** (Scheme 4.57). In the hydrogenation of these substrates by Rh catalysts based on ligands **L21** and **L28** (Scheme 4.57a),[124] the formation of achiral secondary products lacking the benzoate group (**147'**) was observed, in addition to desired products **147**. The ratio between the two types of products depends on

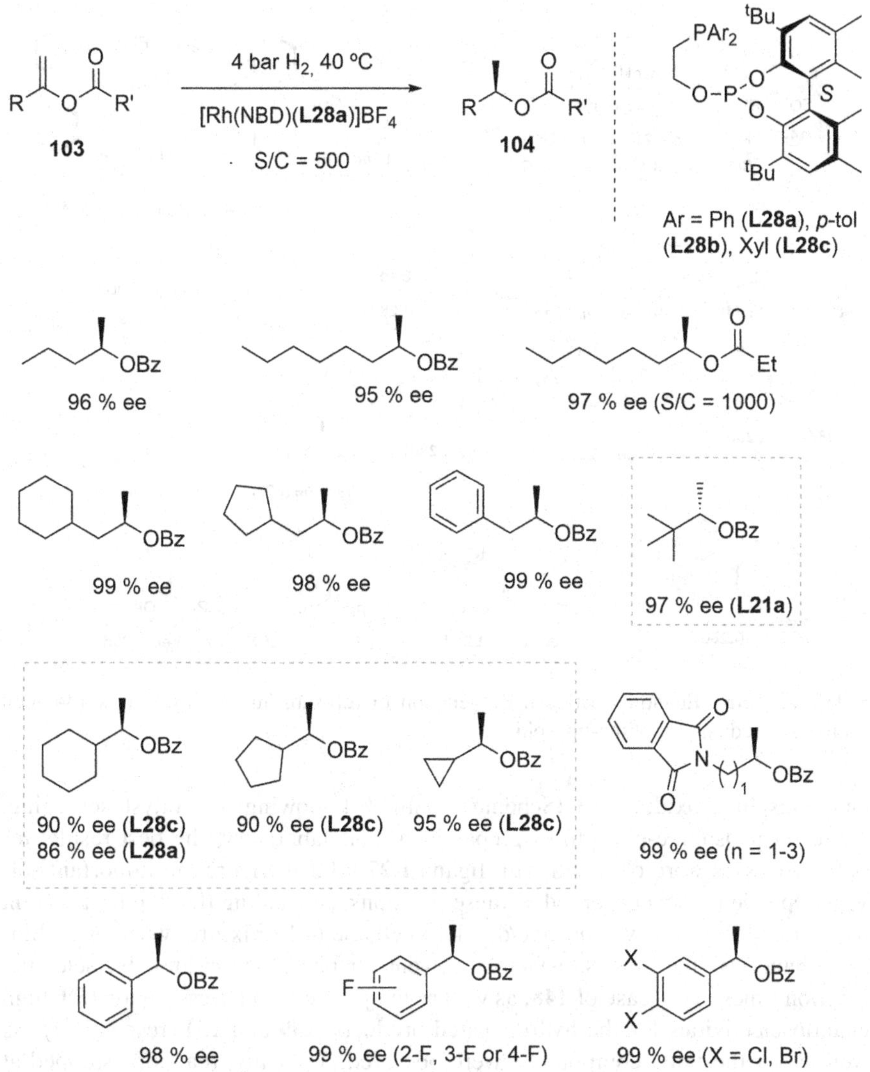

SCHEME 4.55 Enantioselective hydrogenation of 1-substituted vinyl esters **103** with catalysts based on phosphine-phosphites **L28**.

the phosphine-phosphite ligand used, with the formation of **147** being most favorable in the case of **L21b**. With the corresponding catalyst, very high enantioselectivities, of between 95 and 99 % ee, have been reported. For the formation of products **147′**, a benzoate- elimination step has been proposed (Scheme 4.57b).[118]

The use of olefin asymmetric hydrogenation reactions in kinetic resolution processes is another topic of interest. In this regard, Vidal-Ferran and coworkers have applied this methodology to racemic sulfoxides **148** (Scheme 4.58a)[125]

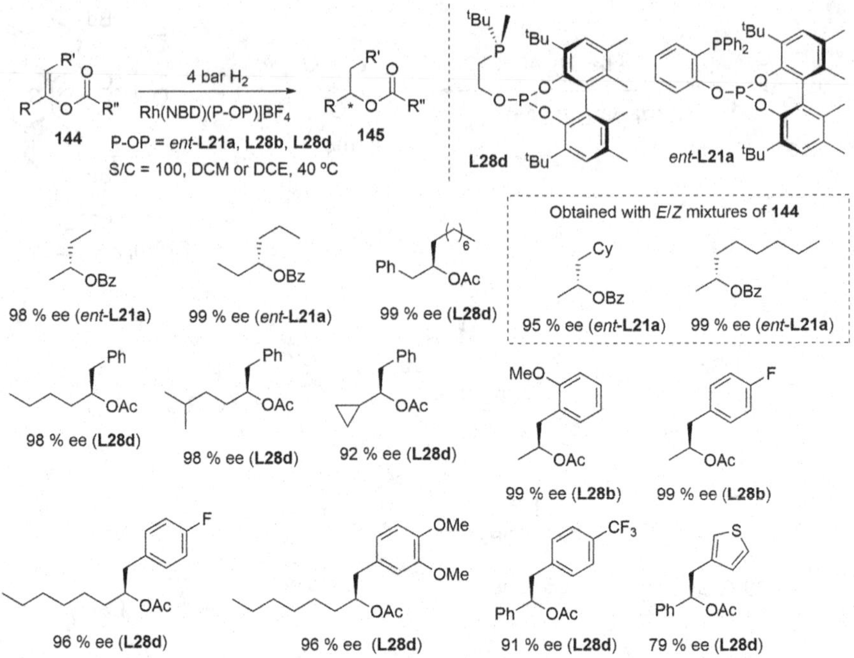

SCHEME 4.56 Enantioselective hydrogenation of α,β-substituted vinyl esters **144** with catalysts based on phosphine-phosphites.

and phosphine-oxides **150** (Scheme 4.58b).[126] Following a catalyst screening, based on phosphine-phosphites prepared in their laboratory, the best results for both processes were obtained with ligand **L27d**. Furthermore, an important solvent dependence was observed in these reactions, concluding that the best solvent for sulfoxides was a cyclohexane/dichloromethane (4:1) mixture, whereas dichloromethane was the most satisfactory solvent for phosphine oxides. By selecting reaction times in the case of **148**, as well as temperature and pressure for **150**, high enantioselectivities for the hydrogenated products (**149** and **151**, respectively) as well as for the vinyl compounds, were observed. Typically, reactions stopped at ca. 40 % conversion are appropriate for the reduced products, while conversion values of approximately 60 % are best suited to recover the starting material with a high enantioselectivity. Vinyl derivatives have attracted considerable interest due to further transformation, as shown in the preparation of a bis-phospholane derivative (Scheme 4.58c).

The family of ligands **L27** have also generated remarkable results in the area of the hydrogenation of heterocycles. More specifically, iridium catalysts, based on these ligands, achieve high enantioselectivities in the hydrogenation of the C=N bond of a wide range of heterocycles (**152–154**, Scheme 4.59).[127] Some of these substrates, particularly benzoxazinones and benzothiazinones **153**, were rather unreactive and high pressures were required to achieve satisfactory conversions.

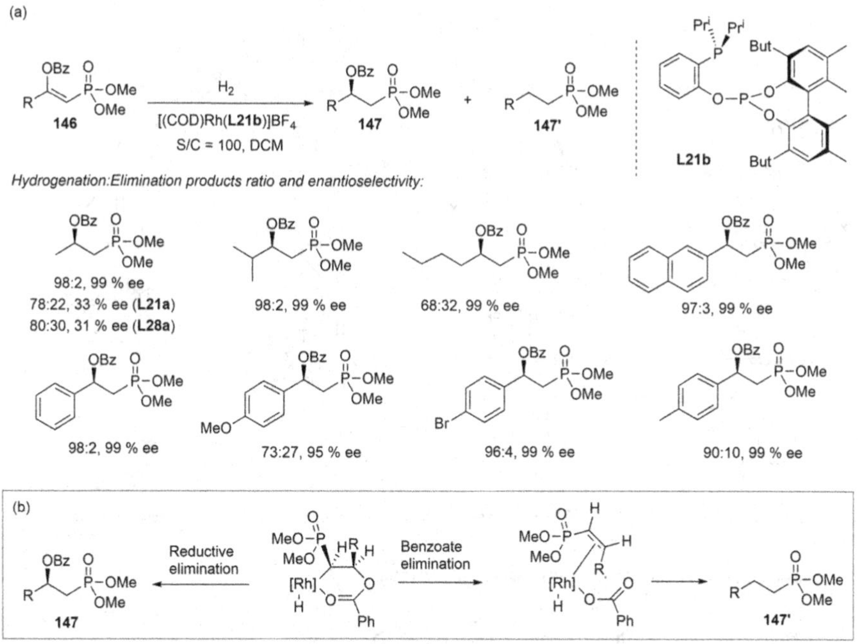

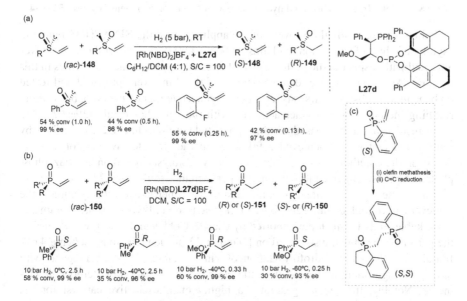

SCHEME 4.57 (a) Enantioselective hydrogenation of β-benzoyloxyphosphonates **146**. (b) Proposed routes for the formation of compounds **147** and **147'**.

SCHEME 4.58 Kinetic resolution of (a) racemic sulfoxides **148** and (b) phosphine-oxides **150** by catalytic olefin hydrogenation. (c) Synthesis of a bis-phospholane derivative.

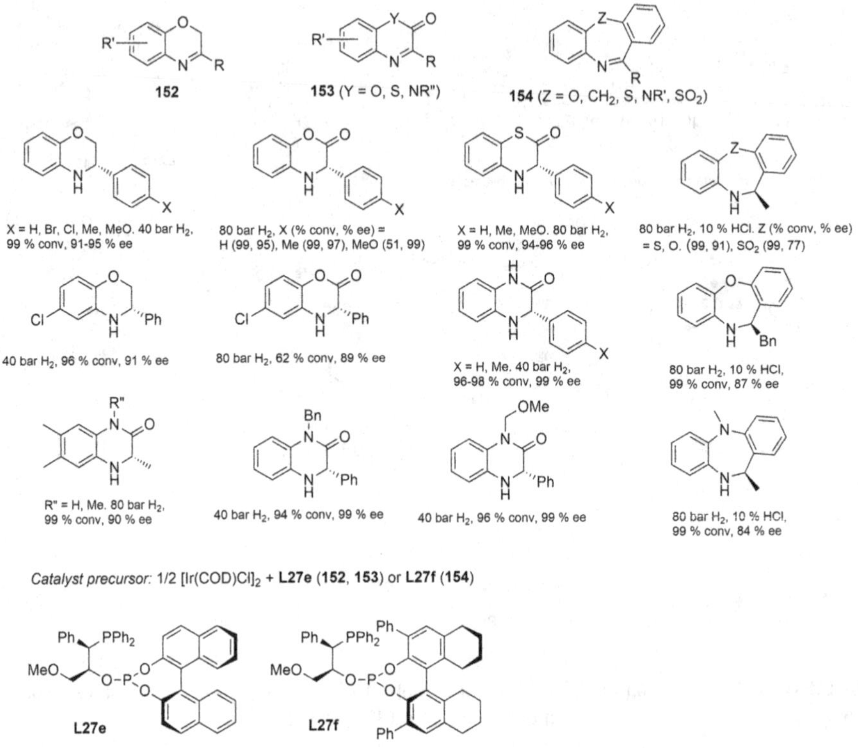

152

153 (Y = O, S, NR")

154 (Z = O, CH$_2$, S, NR', SO$_2$)

X = H, Br, Cl, Me, MeO. 40 bar H$_2$,
99 % conv, 91-95 % ee

80 bar H$_2$, X (% conv, % ee) =
H (99, 95), Me (99, 97), MeO (51, 99)

X = H, Me, MeO. 80 bar H$_2$,
99 % conv, 94-96 % ee

80 bar H$_2$, 10 % HCl, Z (% conv, % ee)
= S, O. (99, 91), SO$_2$ (99, 77)

40 bar H$_2$, 96 % conv, 91 % ee

80 bar H$_2$, 62 % conv, 89 % ee

X = H, Me. 40 bar H$_2$,
96-98 % ee

80 bar H$_2$, 10 % HCl,
99 % conv, 87 % ee

R" = H, Me. 80 bar H$_2$,
99 % conv, 90 % ee

40 bar H$_2$, 94 % conv, 99 % ee

40 bar H$_2$, 96 % conv, 99 % ee

80 bar H$_2$, 10 % HCl,
99 % conv, 84 % ee

Catalyst precursor: 1/2 [Ir(COD)Cl]$_2$ + **L27e** (**152**, **153**) or **L27f** (**154**)

L27e

L27f

SCHEME 4.59 Iridium-catalyzed hydrogenation of C=N bonds of heterocycles **152–154**.

Following the seminal discovery of the application of the BINAPHOS ligand in asymmetric hydroformylation of olefins (Scheme 4.60a),[128] considerable interest was attracted with respect to the performance of alternative phosphine-phosphites in this reaction. As a consequence, diverse ligands of this kind were prepared and tested on benchmark substrates, like styrene and vinyl acetate (Scheme 4.60b), but without reaching the enantioselectivities achieved with the Binaphos catalyst.[111,129] Among these results, a catalyst based on diazaphospholane-phosphite **L29**, described by Landis and coworkers, achieved a 90 % ee in the hydroformylation of styrene. However, these authors have described alternative catalysts, based on bis-diazaphospholane ligands, which achieve a superior performance to that of **L29**.[129c,130]

One of the biggest challenges in the area of asymmetric hydroformylation is the achievement of high enantioselectivity in the reaction of alkyl-substituted olefins. A remarkable study in this topic, reported by Clarke, Cobley and coworkers, describes the performance of catalysts based on ligands **L23** in such reactions (Scheme 4.61). Initial studies on the hydroformylation of vinyl acetate identified a ligand with S,S,S_{ax} configuration as the one with the *matched* combination of stereogenic elements. Notably, this ligand generates a highly enantioselective catalyst for the hydroformylation of several alkyl-substituted olefins.[110] For instance, 93 % ee and a favorable regioselectivity in favor of the branched isomer (3:1) was observed in the

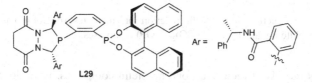

H2/CO
Rh(CO)2(acac)
P-OP
116
CHO
iso
CHO
n
117
PPh2
(R,S)-L20
(a) Results with Binaphos ligand (R,S)-L20 (Typical reaction conditions: H2/CO 1:1, 100 bar, 60 °C, > 95 % conv)
Y (iso/n, % ee)
H (88:12, 94)
MeO (87:13, 88)
Cl (87:13, 93)
iBu (88:12, 92)
iso/n = 24:76, 82 % ee
iso/n = 86:14, 96 % ee
iso/n = 89:11, 85 % ee
AcO
iso/n = 86:14, 92 % ee
Boc
α:β = 67:33
97 % ee (α), 71 % ee (β)
76 % ee
(b) Results with other P-OP ligands
Ph
L22d (20 bar, 50 °C): 60 % conv., iso/n = 98:2, 81 % ee
L21c (20 bar, 50 °C): 48 % conv., iso/n = 98:2, 71 % ee
L29 (H2/CO: 1:3, 11 bar, 40 °C): 48 % conv., iso/n = 95:5, 90 % ee
AcO
(S,S,Sax)-L23 (10 bar, 60 °C): 99 % conv., iso/n = 99:1, 83 % ee
(R,Rax)-L24b (10 bar, 40 °C): 99 % conv., iso/n = 99:1, 74 % ee
Ar
Ar
L29
Ar =
NH
Ph

SCHEME 4.60 (a) Selected results on the application of the BINAPHOS catalyst on the asymmetric hydroformylation of olefins. (b) Results obtained with other catalysts based on phosphine-phosphite ligands in the hydroformylation of styrene and vinyl acetate.

hydroformylation of 1-hexene. Similarly, remarkable levels of regio- and enantioselectivity were observed in the hydroformylation of benzylic substrates.

Mechanistic studies of the asymmetric hydroformylation reaction with phosphine-phosphite catalysts propose olefin insertion into Rh(H)(CO)(olefin)(P–OP) intermediates (Figure 4.7a) as the regio- and enantio-determining step.[131] Therefore, knowledge of the structure of the hydrido-olefin intermediate is of considerable interest. A key aspect in this regard corresponds to the coordination mode of the P–OP ligand in this intermediate. To that end, $Rh(H)(CO)_2$(P–OP) complexes have been

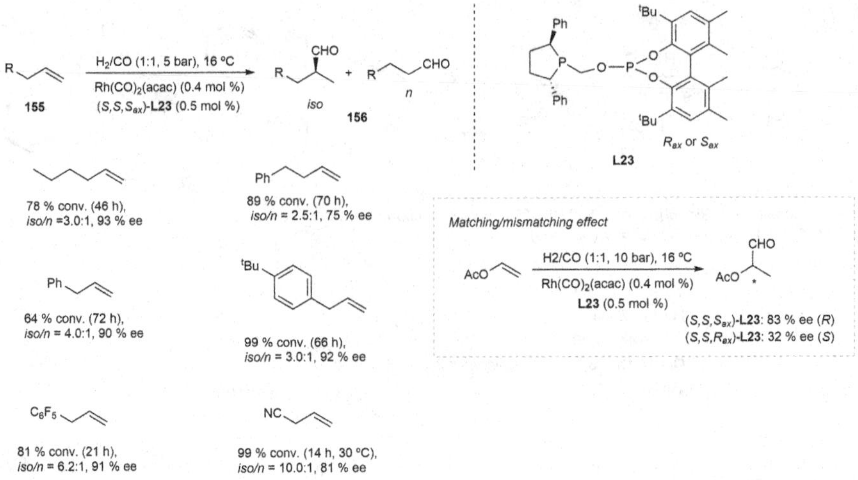

SCHEME 4.61 Hydroformylation of olefins **155** with Rh catalysts based on BOBPHOS ligand **L23**.

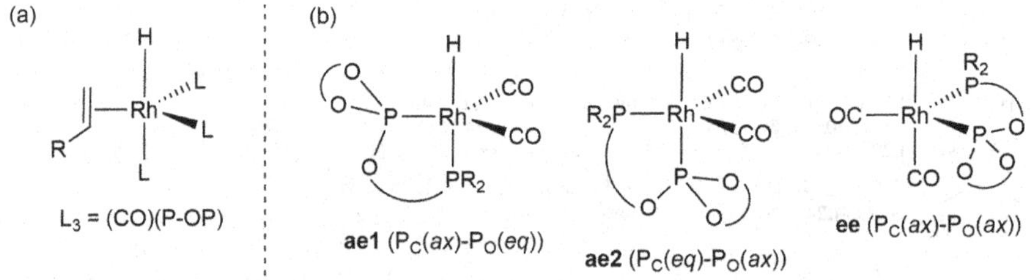

FIGURE 4.7 (a) Structure of key olefin reaction intermediates. (b) Coordination mode of phosphine-phosphite ligands in Rh(H)(CO)$_2$(P–OP) complexes.

considered as models of the corresponding olefin complexes. Hydrido-dicarbonyls can be generated by the reaction of Rh(acac)(CO)$_2$ and the P–OP ligand under pressure of the CO/H$_2$ mixture. Three isomers are possible for these compounds (Figure 4.7b), depending on the coordination mode of the P–OP ligand, namely two equatorial-axial (**ae1** and **ae2**) and one diequatorial (**ee**), characterized by different bite angles of the P–OP ligand (ca. 90°, 90°, and 120°, respectively). Most P–OP ligands considered here have bite angles near 90° and therefore prefer an axial-equatorial coordination. Moreover, considering the electronic properties of phosphine and phosphite fragments, an structure of type **ae1** should be preferred.[132] However, the difference between the two structures is not high and the preference of one over another or an equilibrium between both of them may be observed.[16b] For the prominent BINAPHOS ligand, the coordination mode of type **ae2** is strongly preferred and was the only isomer observed in initial studies.[128a] However, detailed NMR experiments have more recently detected the presence of the minor **ae1** isomer.[133] It seems

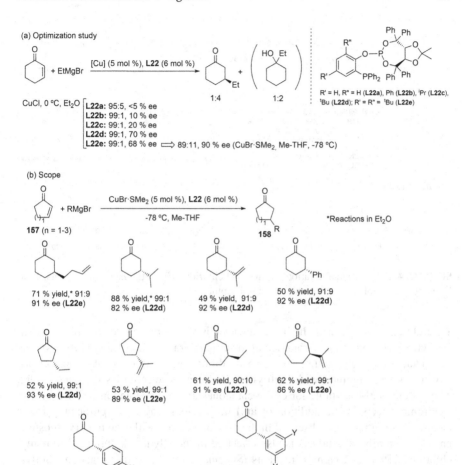

SCHEME 4.62 (a) Optimization study and (b) scope of conjugate addition of Grignard reagents to cycloalkenones **157**, catalyzed by Cu catalysts based on ligands **L22**.

that the coordination mode of the P–OP ligand in the catalytically relevant species depends on the particular catalytic system. In the hydroformylation of olefins **155**, with a catalyst based on (S,S,S_{ax})-**L23**,[134] a close agreement between computational results and experimental observations is obtained if an **ae1** coordination mode is considered.

Inspired by the work of Feringa and coworkers in the Cu-catalyzed conjugate additions of Grignard reagents,[135] the group of Schmalz has developed a particularly effective system for the addition of these reagents to cyclohexenone, based on phosphine-phosphite ligands developed in their laboratory.[136] Initial studies on the addition of EtMgBr to cycloxexenone (Scheme 4.62a) pointed to catalysts based on ligands **L22d** and **L22e** as the most satisfactory ones for the formation of the desired

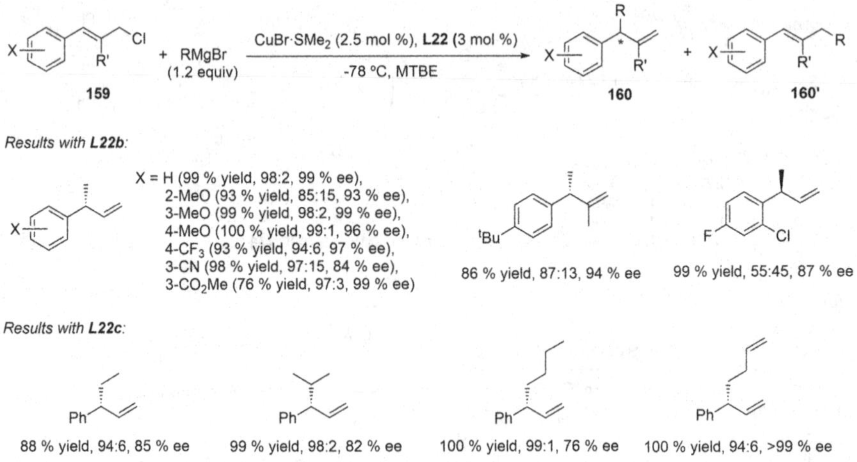

Results with **L22b**:

X = H (99 % yield, 98:2, 99 % ee),
2-MeO (93 % yield, 85:15, 93 % ee),
3-MeO (99 % yield, 98:2, 99 % ee),
4-MeO (100 % yield, 99:1, 96 % ee),
4-CF$_3$ (93 % yield, 94:6, 97 % ee),
3-CN (98 % yield, 97:15, 84 % ee),
3-CO$_2$Me (76 % yield, 97:3, 99 % ee)

86 % yield, 87:13, 94 % ee

99 % yield, 55:45, 87 % ee

Results with **L22c**:

88 % yield, 94:6, 85 % ee

99 % yield, 98:2, 82 % ee

100 % yield, 99:1, 76 % ee

100 % yield, 94:6, >99 % ee

SCHEME 4.63 Allylic alkylation of cinnamyl chlorides **159** with Grignard reagents, catalyzed by Cu complexes bearing phosphine-phosphite ligands.

1,4-addition product, with high enantioselectivity. Moreover, an improvement in catalytic performance was obtained at lower temperatures and using CuBr·SMe$_2$ as the metal source. Under these reaction conditions, a family of 3-substituted cyclohexanones was obtained with high yields and enantioselectivities (Scheme 4.62b). The system is also suitable for cyclopentenones and cycloheptenones.[137] Regarding electronic effects in the addition of aryl magnesium reagents, significantly lower enantioselectivities were observed in the case of electron-withdrawing aryl substituents. This family of catalysts is also effective in the allylic alkylation of cinnamyl chlorides **159** with Grignard reagents (Scheme 4.63).[138] For this reaction, catalyst optimization studies identified those based on **L22b** and **L22c** ligands as being the most satisfactory ones, achieving high levels of both regio- and enantioselectivity, particularly in the reactions with MeMgBr additives (84–99 % ee).

The ability of Ni complexes bearing phosphine-phosphites to effect the asymmetric hydrocyanation of olefins was first demonstrated by Nozaki in the case of norbornene, in a moderately enantioselective reaction (48 % ee),[139] while Rajanbabu and Casalnuovo described a highly enantioselective hydrocyanation of 2-vinyl-6-methoxy-napththalene by a Ni diphosphinite catalyst, although lower enantioselectivities were achieved with other vinyl arenes.[140] Against this background, Schmalz and coworkers have described the use of Ni catalysts, including ligands **L22**, in the hydrocyanation of aryl and heteroaryl olefins (Scheme 4.64). Among ligands examined, the xylyl derivative **L22f** achieved the best results. A detailed analysis of reaction conditions, including the nature of the CN source, has led to the identification of appropriate conditions for either the use of trimethylsilyl cyanide (TMS-CN, conditions A)[141] or HCN (conditions B).[142] In the first case, HCN is generated by reaction with MeOH. Of great importance under these circumstances is the concentration of HCN in the solution, as the catalyst decomposes in the absence of this reagent,

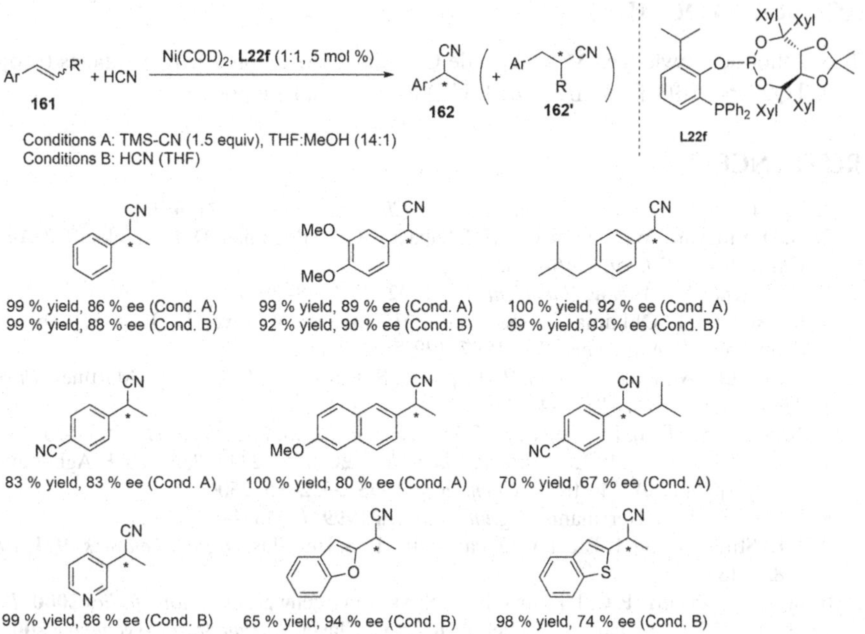

SCHEME 4.64 Ni-catalyzed hydrocyanation of aryl and heteroaryl olefins **161**.

whereas an excess of it leads to the undesired formation of $Ni(CN)_2(\mathbf{L22f})$. The scope of the reaction covers a wide range of aryl and heteroaryl products **162**, with enantioselectivities in the 80–90 % ee range. Interestingly, an exquisite regioselectivity was observed, as products **162'** were not detected. Considering the ease of conversion of these nitriles into the corresponding carboxylic acids, this catalytic system provides a very convenient route to production of anti-inflammatory chiral 2-aryl propionic acids.

4.4 CONCLUDING REMARKS

In this chapter, a wide range of chelating ligands containing a phosphine and another P(III) function has been presented. In most of the cases, ligands have a modular structure, and their synthetic methods allow a fine tuning of substituents in critical parts of the ligand and, therefore, of the catalyst. Due to this precise catalyst optimization, impressive results for a plethora of asymmetric catalytic transformations have been reported, surpassing in many cases the performance shown by catalysts based on the ubiquitous C_2-symmetric diphosphine ligands. Considering, as well, the wide range of electronic properties that P-P' ligands can exhibit with an appropriate modification of the second P functionality, the range of applications of this kind of ligands appears to have little limit. Finally, as many of the studies described have been released in recent years, it can be expected that this area of asymmetric catalysts will retain its considerable importance into the near future.

ACKNOWLEDGMENT

The author acknowledges Ministerio de Ciencia, Innovación y Universidades (project CTQ2016-75193-P; AEI/FEDER, UE) for financial support.

REFERENCES

1. T. P. Dang, H. B. Kagan, *J. Chem. Soc. D Chem. Commun.* **1971**, 481.
2. B. D. Vineyard, W. S. Knowles, M. J. Sabacky, G. L. Bachman, D. J. Weinkauff, *J. Am. Chem. Soc.* **1977**, *99*, 5946–5952.
3. W. S. Knowles, *Angew. Chem. Int. Ed.* **2002**, *41*, 1998–2007.
4. R. Noyori, T. Ohkuma, M. Kitamura, H. Takaya, N. Sayo, H. Kumobayashi, S. Akutagawa, *J. Am. Chem. Soc.* **1987**, *109*, 5856–5858.
5. M. J. Burk, M. F. Gross, T. G. P. Harper, C. S. Kalberg, J. R. Lee, J. P. Martinez, *Pure Appl. Chem.* **1996**, *68*, 37–44.
6. N. Sakai, S. Mano, K. Nozaki, H. Takaya, *J. Am. Chem. Soc.* **1993**, *115*, 7033–7034.
7. (a) A. C. Brezny, C. R. Landis, *Acc. Chem. Res.* **2018**, *51*, 2344–2354. (b) F. Agbossou, J.-F. Carpentier, A. Mortreux, *Chem. Rev.* **1995**, *95*, 2485–2506.
8. D. Gleich, W. A. Herrmann, *Organometallics* **1999**, *18*, 4354–4361.
9. J. K. Stilled, H. Su, P. Brechot, G. Parrinello, L. S. Hegedus, *Organometallics* **1991**, *10*, 1183–1189.
10. (a) S. Deerenberg, P. C. J. Kamer, P. W. N. M. Van Leeuwen, *Organometallics* **2000**, *19*, 2065–2072. (b) O. Pàmies, A. Net, A. Ruiz, C. Claver, *Tetrahedron: Asymmetry* **2002**, *12*, 3441–3445.
11. (a) S. Deerenberg, H. S. Schrekker, G. P. F. van Strijdonck, P. C. J. Kamer, P. W. N. M. van Leeuwen, J. Fraanje, K. Goubitz, *J. Org. Chem.* **2000**, *65*, 4810–4817. (b) O. Pàmies, G. P. F. van Strijdonck, M. Diéguez, S. Deerenberg, G. Net, A. Ruiz, C. Claver, P. C. J. Kamer, P. W. N. M. van Leeuwen, *J. Org. Chem.* **2001**, *66*, 8867–8871.
12. (a) O. Pamies, M. Dieguez, G. Net, A. Ruiz, C. Claver, *Chem. Commun.* **2000**, 2383–2384. (b) S. Deerenberg, O. Pàmies, M. Diéguez, C. Claver, P. C. J. Kamer, P. W. N. M. Van Leeuwen, *J. Org. Chem.* **2001**, *66*, 7626–7631. (c) A. Suarez, A. Pizzano, *Tetrahedron-Asymmetry* **2001**, *12*, 2501–2504.
13. J. F. Teichert, B. L. Feringa, *Angew. Chem. Int. Ed.* **2010**, *49*, 2486–2528.
14. A. Togni, C. Breutel, A. Schnyder, F. Spindler, H. Landert, A. Tijani, *J. Am. Chem. Soc.* **1994**, *116*, 4062–4066.
15. H. U. Blaser, B. Pugin, F. Spindler, E. Mejía, A. Togni, Josiphos Ligands: From Discovery to Technical Applications, in *Privileged Chiral Ligands and Catalysts.* Edited by Qi-Lin Zhou, WILEY-VCH, **2011**.
16. (a) H. Fernández-Pérez, P. Etayo, A. Panossian, A. Vidal-Ferran, *Chem. Rev.* **2011**, *111*, 2119–2176. (b) A. Pizzano, *Chem. Rec.* **2016**, *16*, 2599.
17. X. S. Chen, C. J. Hou, X. P. Hu, *Synth. Commun.* **2016**, *46*, 917–941.
18. A. Pfaltz, W. J. Drury, *Proc. Natl. Acad. Sci. U. S. A.* **2004**, *101*, 5723–5726.
19. C. A. Tolman, *Chem. Rev.* **1977**, *77*, 313–348.
20. (a) G. A. Ardizzoia, S. Brenna, *Phys. Chem. Chem. Phys.* **2017**, *19*, 5971–5978. (b) M. Rahman, H. Y. Liu, K. Eriks, A. Prock, W. P. Giering, *Organometallics* **1989**, *8*, 1–7.
21. (a) M. R. Wilson, D. C. Woska, A. Prock, W. P. Giering, *Organometallics* **1993**, *12*, 1742–1752. (b) M. R. Wilson, A. Prock, W. P. Giering, A. L. Fernandez, C. M. Haar, S. P. Nolan, B. M. Foxman, *Organometallics* **2002**, *21*, 2758–2763.
22. (a) B. David, S. C. Polytechnic, P. Street, S. S. Wb, *J. Chem. Soc., Dalton Trans.* **1982**, 51–54. (b) S. M. Socol, J. G. Verkade, *Inorg. Chem.* **1984**, *23*, 3487–3493. (c) T. S. Barnard, M. R. Mason, *Organometallics* **2001**, *20*, 206–214.

23. A. Suarez, M. A. Mendez-Rojas, A. Pizzano, *Organometallics* **2002**, *21*, 4611–4621.
24. S. Balogh, G. Farkas, I. Tóth, J. Bakos, *Tetrahedron Asymmetry* **2015**, *26*, 666–673.
25. (a) A. R. Sanger, *J. Chem. Soc., Dalton Trans.* **1977**, 121. (b) A. Del Zotto, L. Costella, A. Mezzetti, P. J. Rigo, *J. Organomet. Chem.* **1991**, *414*, 109.
26. J. K. Whitesell, *Chem. Rev.* **1989**, *89*, 1581–1590.
27. (a) I. D. Gridnev, T. Imamoto, *Acc. Chem. Res.* **2004**, *37*, 633–644. (b) I. D. Gridnev, T. Imamoto, *Chem. Commun.* **2009**, 7447–7464.
28. G. Hoge, H.-P. Wu, W. S. Kissel, D. A. Pflum, D. J. Greene, J. Bao, *J. Am. Chem. Soc.* **2004**, *126*, 5966–5967.
29. H. Nie, Y. Zhu, X. Hu, Z. Wei, L. Yao, G. Zhou, P. Wang, R. Jiang, S. Zhang, *Org. Lett.* **2019**, *21*, 8641–8645.
30. R. Schwenk, A. Togni, *Dalton Trans.* **2015**, *44*, 19566–19575.
31. S. M. Mansell, *Dalton Trans.* **2017**, *46*, 15157–15174.
32. Y. Sawatsugawa, K. Tamura, N. Sano, T. Imamoto, *Org. Lett.* **2019**, *21*, 8874–8878.
33. K. Huang, X. Zhang, T. J. Emge, G. Hou, B. Cao, X. Zhang, *Chem. Commun.* **2010**, *46*, 8555–8557.
34. W. Tang, A. G. Capacci, A. White, S. Ma, S. Rodriguez, B. Qu, J. Savoie, N. D. Patel, X. Wei, N. Haddad, N. Grinberg, N. K. Yee, D. Krishnamurthy, C. H. Senanayake, *Org. Lett.* **2010**, *12*, 1104–1107.
35. (a) H. P. Wu, G. Hoge, *Org. Lett.* **2004**, *6*, 3645–3647. (b) I. D. Gridnev, Y. Liu, T. Imamoto, *ACS Catal.* **2014**, *4*, 203–219. (c) E. Cristóbal-Lecina, P. Etayo, S. Doran, M. Revés, P. Martín-Gago, A. Grabulosa, A. R. Costantino, A. Vidal-Ferran, A. Riera, X. Verdaguer, *Adv. Synth. Catal.* **2014**, *356*, 795–804. (d) M. Birch, S. Challenger, J. P. Crochard, D. Fradet, H. Jackman, A. Luan, E. Madigan, N. McDowall, K. Meldrum, C. M. Gordon, et al., *Org. Process Res. Dev.* **2011**, *15*, 1172–1177.
36. J. Meng, M. Gao, H. Lv, X. Zhang, *Org. Lett.* **2015**, *17*, 1842–1845.
37. C. Benhaim, L. Bouchard, G. Pelletier, J. Sellstedt, L. Kristofova, S. Daigneault, *Org. Lett.* **2010**, *12*, 2008–2011.
38. Y. Sawatsugawa, K. Tamura, N. Sano, T. Imamoto, *Org. Lett.* **2019**, *21*, 8874–8878.
39. C. Molinaro, J. P. Scott, M. Shevlin, C. Wise, A. Ménard, A. Gibb, E. M. Junker, D. Lieberman, *J. Am. Chem. Soc.* **2015**, *137*, 999–1006.
40. W. Chen, P. J. McCormack, K. Mohammed, W. Mbafor, S. M. Roberts, J. Whittall, *Angew. Chem. Int. Ed.* **2007**, *46*, 4141–4144.
41. C. Chen, H. Wang, Z. Zhang, S. Jin, S. Wen, J. Ji, L. W. Chung, X. Q. Dong, X. Zhang, *Chem. Sci.* **2016**, *7*, 6669–6673.
42. C. Chen, S. Wen, M. Geng, S. Jin, Z. Zhang, X. Q. Dong, X. Zhang, *Chem. Commun.* **2017**, *53*, 9785–9788.
43. Z. Zhang, Q. Hu, Y. Wang, J. Chen, W. Zhang, *Org. Lett.* **2015**, *17*, 5380–5383.
44. S. Li, K. Huang, X. Zhang, *Chem. Commun.* **2014**, *50*, 8878–8881.
45. D. Fan, Y. Hu, F. Jiang, Z. Zhang, W. Zhang, *Adv. Synth. Catal.* **2018**, *360*, 2228–2232.
46. F. Brüning, H. Nagae, D. Käch, K. Mashima, A. Togni, *Chem. Eur. J.* **2019**, *25*, 10818–10822.
47. M. Renom-Carrasco, P. Gajewski, L. Pignataro, J. G. de Vries, U. Piarulli, C. Gennari, L. Lefort, *Chem. Eur. J.* **2016**, *22*, 9528–9532.
48. R. Schwenk, A. Togni, *Dalton Trans.* **2015**, *44*, 19566–19575.
49. H. Nie, Y. Zhu, X. Hu, Z. Wei, L. Yao, G. Zhou, P. Wang, R. Jiang, S. Zhang, *Org. Lett.* **2019**, *21*, 8641–8645.
50. R. Kuwano, Y. Hashiguchi, R. Ikeda, K. Ishizuka, *Angew. Chem. Int. Ed.* **2015**, *54*, 2393–2396.
51. M. L. Cooke, K. Xu, B. Breit, *Angew. Chem. Int. Ed.* **2012**, *51*, 10876–10879.
52. K. Xu, Y.-H. Wang, V. Khakyzadeh, B. Breit, *Chem. Sci.* **2016**, *7*, 3313–3316.

53. K. Xu, N. Thieme, B. Breit, *Angew. Chem. Int. Ed.* **2014**, *53*, 2162–2165.
54. N. Thieme, B. Breit, *Angew. Chem. Int. Ed.* **2017**, *56*, 1520–1524.
55. Y. H. Wang, B. Breit, *Chem. Commun.* **2019**, *55*, 7619–7622.
56. Z. Liu, B. Breit, *Org. Lett.* **2018**, *20*, 300–303.
57. U. Gellrich, A. Meißner, A. Steffani, M. Kähny, H. J. Drexler, D. Heller, D. A. Plattner, B. Breit, *J. Am. Chem. Soc.* **2014**, *136*, 1097–1104.
58. J. Kim, S. H. Cho, *ACS Catal.* **2019**, *9*, 230–235.
59. J. A. Carmona, V. Hornillos, P. Ramírez-López, A. Ros, J. Iglesias-Sigüenza, E. Gómez-Bengoa, R. Fernández, J. M. Lassaletta, *J. Am. Chem. Soc.* **2018**, *140*, 11067–11075.
60. P. Ramírez-López, A. Ros, B. Estepa, R. Fernández, B. Fiser, E. Gómez-Bengoa, J. M. Lassaletta, *ACS Catal.* **2016**, *6*, 3955–3964.
61. T. Jia, M. Zhang, S. P. McCollom, A. Bellomo, S. Montel, J. Mao, S. D. Dreher, C. J. Welch, E. L. Regalado, R. T. Williamson, et al., *J. Am. Chem. Soc.* **2017**, *139*, 8337–8345.
62. W. Mao, W. Xue, E. Irran, M. Oestreich, *Angew. Chem. Int. Ed.* **2019**, *58*, 10723–10726.
63. R. Kojima, S. Akiyama, H. Ito, *Angew. Chem. Int. Ed.* **2018**, *57*, 7196–7199.
64. Y. Park, J. Yun, *Org. Lett.* **2019**, *21*, 8779–8782.
65. C. García-Morales, B. Ranieri, I. Escofet, L. López-Suarez, C. Obradors, A. I. Konovalov, A. M. Echavarren, *J. Am. Chem. Soc.* **2017**, *139*, 13628–13631.
66. N. W. Boaz, E. B. Mackenzie, S. D. Debenham, S. E. Large, J. A. Ponasik, *J. Org. Chem.* **2005**, *70*, 1872–1880.
67. M. Qiu, X. P. Hu, J. Di Huang, D. Y. Wang, J. Deng, S. B. Yu, Z. C. Duan, Z. Zheng, *Adv. Synth. Catal.* **2008**, *350*, 2683–2689.
68. D. Y. Wang, J. Di Huang, X. P. Hu, J. Deng, S. B. Yu, Z. C. Duan, Z. Zheng, *J. Org. Chem.* **2008**, *73*, 2011–2014.
69. M. Revés, C. Ferrer, T. León, S. Doran, P. Etayo, A. Vidal-Ferran, A. Riera, X. Verdaguer, *Angew. Chem. Int. Ed.* **2010**, *49*, 9452–9455.
70. T. León, A. Riera, X. Verdaguer, *J. Am. Chem. Soc.* **2011**, *133*, 5740–5743.
71. M. Dutartre, J. Bayardon, S. Juge, *Chem. Soc. Rev.* **2016**, *45*, 5771–5794.
72. X. M. Zhou, J. Di Huang, L. Bin Luo, C. L. Zhang, Z. Zheng, X. P. Hu, *Tetrahedron Asymmetry* **2010**, *21*, 420–424.
73. M. Qiu, X.-P. Hu, J.-D. Huang, D.-Y. Wang, J. Deng, S.-B. Yu, Z.-C. Duan, Z. Zheng, *Adv. Synth. Catal.* **2008**, *350*, 2683–2689.
74. J. Deng, Z. C. Duan, J. Di Huang, X. P. Hu, D. Y. Wang, S. B. Yu, X. F. Xu, Z. Zheng, *Org. Lett.* **2007**, *9*, 4825–4828.
75. C. Pan, Z. Zhu, M. Zhang, Z. Gu, *Angew. Chem. Int. Ed.* **2017**, *56*, 4777–4781.
76. K. Nozaki, W. G. Li, T. Horiuchi, H. Takaya, T. Saito, A. Yoshida, K. Matsumura, Y. Kato, T. Imai, T. Miura, et al., *J. Org. Chem.* **1996**, *61*, 7658–7659.
77. D. H. Shih, L. Cama, B. G. Christensen, *Tetrahedron Lett.* **1985**, *26*, 587–590.
78. E. Louise Hazeland, A. M. Chapman, P. G. Pringle, H. A. Sparkes, *Chem. Commun.* **2015**, *51*, 10206–10209.
79. S. H. Chikkali, R. Bellini, G. Berthon-Gelloz, J. I. van der Vlugt, B. de Bruin, J. N. H. Reek, *Chem. Commun.* **2010**, *46*, 1244–1246.
80. M. Eggenstein, A. Thomas, J. Theuerkauf, G. Franciò, W. Leitner, *Adv. Synth. Catal.* **2009**, *351*, 725–732.
81. X. Zhang, B. Cao, Y. Yan, S. Yu, B. Ji, X. Zhang, *Chem. Eur. J.* **2010**, *16*, 871–877.
82. K. Sumi, T. Ikariya, R. Noyori, *Can. J. Chem.* **2000**, *78*, 697.
83. J. Wassenaar, M. Kuil, M. Lutz, A. L. Spek, J. N. H. Reek, *Chem. Eur. J.* **2010**, *16*, 6509–6517.
84. W. Chen, W. Mbafor, S. M. Roberts, J. Whittall, *J. Am. Chem. Soc.* **2006**, *128*, 3922–3923.

85. T. Pullmann, B. Engendahl, Z. Zhang, M. Hölscher, A. Zanotti-Gerosa, A. Dyke, G. Franciò, W. Leitner, *Chem. Eur. J.* **2010**, *16*, 7517–7526.
86. C. Schmitz, K. Holthusen, W. Leitner, G. Franciò, *ACS Catal.* **2016**, *6*, 1584–1589.
87. (a) S. Balogh, G. Farkas, I. Tóth, J. Bakos, *Tetrahedron Asymmetry* **2015**, *26*, 666–673. (b) C. Schmitz, K. Holthusen, W. Leitner, G. Franciò, *Eur. J. Org. Chem.* **2017**, *2017*, 4111–4116. (c) K. A. Vallianatou, I. D. Kostas, J. Holz, A. Börner, *Tetrahedron Lett.* **2006**, *47*, 7947–7950.
88. (a) Q. H. Zeng, X. P. Hu, Z. C. Duan, X. M. Liang, Z. Zheng, *Tetrahedron Asymmetry* **2005**, *16*, 1233–1238. (b) M. Qiu, X. P. Hu, D. Y. Wang, J. Deng, J. Di Huang, S. B. Yu, Z. C. Duan, Z. Zheng, *Adv. Synth. Catal.* **2008**, *350*, 1413–1418.
89. (a) X. P. Hu, Z. Zheng, *Org. Lett.* **2004**, *6*, 3585–3588. (b) J. Di Huang, X. P. Hu, Z. C. Duan, Q. H. Zeng, S. B. Yu, J. Deng, D. Y. Wang, Z. Zheng, *Org. Lett.* **2006**, *8*, 4367–4370.
90. D.-Y. Wang, X.-P. Hu, J.-D. Huang, J. Deng, S.-B. Yu, Z.-C. Duan, X.-F. Xu, Z. Zheng, *Angew. Chem. Int. Ed.* **2007**, *46*, 7810–7813.
91. H. Q. Du, X. P. Hu, *Org. Lett.* **2019**, *21*, 8921–8924.
92. Q. Li, C. J. Hou, Y. J. Liu, R. F. Yang, X. P. Hu, *Tetrahedron Asymmetry* **2015**, *26*, 617–622.
93. M. Qiu, D. Y. Wang, X. P. Hu, J. Di Huang, S. B. Yu, J. Deng, Z. C. Duan, Z. Zheng, *Tetrahedron Asymmetry* **2009**, *20*, 210–213.
94. M. J. Burk, *J. Am. Chem. Soc.* **1991**, *113*, 8518–8519.
95. S. Feldgus, C. R. Landis, *Organometallics* **2001**, *20*, 2374–2386.
96. T. M. Konrad, P. Schmitz, W. Leitner, G. Franciò, *Chem. Eur. J.* **2013**, *19*, 13299–13303.
97. Q. Li, C. J. Hou, X. N. Liu, D. Z. Huang, Y. J. Liu, R. F. Yang, X. P. Hu, *RSC Adv.* **2015**, *5*, 13702–13708.
98. C. J. Hou, Y. H. Wang, Z. Zheng, J. Xu, X. P. Hu, *Org. Lett.* **2012**, *14*, 3554–3557.
99. X. H. Hu, X. P. Hu, *Adv. Synth. Catal.* **2019**, *361*, 5063–5068.
100. X. H. Hu, X. P. Hu, *Org. Lett.* **2019**, *21*, 10003–10006.
101. C. J. Hou, W. L. Guo, X. P. Hu, J. Deng, Z. Zheng, *Tetrahedron Asymmetry* **2011**, *22*, 195–199.
102. Y. Yan, X. Zhang, *J. Am. Chem. Soc.* **2006**, *128*, 7198–7202.
103. C. Schmitz, K. Holthusen, W. Leitner, G. Franciò, *ACS Catal.* **2016**, *6*, 1584–1589.
104. B. Wei, C. Chen, C. You, H. Lv, X. Zhang, *Org. Chem. Front.* **2017**, *4*, 288–291.
105. P. Eilbracht, L. Bärfacker, C. Buss, C. Hollmann, B. E. Kitsos-Rzychon, C. L. Kranemann, T. Rische, R. Roggenbuck, a Schmidt, *Chem. Rev.* **1999**, *99*, 3329–3366.
106. C. Chen, S. Jin, Z. Zhang, B. Wei, H. Wang, K. Zhang, H. Lv, X. Q. Dong, X. Zhang, *J. Am. Chem. Soc.* **2016**, *138*, 9017–9020.
107. X. Li, C. You, J. Yang, S. Li, D. Zhang, H. Lv, X. Zhang, *Angew. Chem. Int. Ed.* **2019**, *58*, 10928–10931.
108. Y. Uozumi, A. Tanahashi, S.-Y. Lee, T. Hayashi, *J. Org. Chem.* **1993**, *58*, 1945–1948.
109. J. Velder, T. Robert, I. Weidner, J.-M. Neudörfl, J. Lex, H.-G. Schmalz, *Adv. Synth. Catal.* **2008**, *350*, 1309–1315.
110. G. M. Noonan, J. A. Fuentes, C. J. Cobley, M. L. Clarke, *Angew. Chem. Int. Ed.* **2012**, *51*, 2477–2480.
111. H. Fernández-Pérez, J. Benet-Buchholz, A. Vidal-Ferran, *Org. Lett.* **2013**, *15*, 3634–3637.
112. X. Jia, X. Li, W. S. Lam, S. H. Kok, L. Xu, G. Lu, C.-H. Yeung, A. S. Chan, *Tetrahedron: Asymmetry* **2004**, *15*, 2273–2278.
113. I. Arribas, S. Vargas, M. Rubio, A. Suarez, C. Domene, E. Alvarez, A. Pizzano, *Organometallics* **2010**, *29*, 5791–5804.

114. H. Fernández-Pérez, S. M. A. Donald, I. J. Munslow, J. Benet-Buchholz, F. Maseras, A. Vidal-Ferran, *Chem. Eur. J.* **2010**, *16*, 6495–6508.
115. G. Farkas, S. Balogh, Á. Szöllősy, L. Ürge, F. Darvas, J. Bakos, *Tetrahedron: Asymmetry* **2011**, *22*, 2104–2109.
116. Y. Yan, Y. Chi, X. Zhang, *Tetrahedron: Asymmetry* **2004**, *15*, 2173–2175.
117. (a) P. Etayo, J. L. Núñez-Rico, H. Fernández-Pérez, A. Vidal-Ferran, *Chem. Eur. J.* **2011**, *17*, 13978–13982. (b) H. Fernández-Pérez, M. A. Pericàs, A. Vidal-Ferran, *Adv. Synth. Catal.* **2008**, *350*, 1984–1990. (c) J. L. Núñez-Rico, P. Etayo, H. Fernández-Pérez, A. Vidal-Ferran, *Adv. Synth. Catal.* **2012**, *354*, 3025–3035.
118. M. Á. Chávez, S. Vargas, A. Suárez, E. Álvarez, A. Pizzano, *Adv. Synth. Catal.* **2011**, *353*, 2775–2794.
119. I. Arribas, M. Rubio, P. Kleman, A. Pizzano, *J. Org. Chem.* **2013**, *78*, 3997–4005.
120. (a) M. Magre, O. Pàmies, M. Diéguez, *ACS Catal.* **2016**, *6*, 5186–5190. (b) E. Salomó, S. Orgué, A. Riera, X. Verdaguer, *Angew. Chem. Int. Ed.* **2016**, *55*, 7988–7992.
121. P. Kleman, P. J. González-Liste, S. E. García-Garrido, V. Cadierno, A. Pizzano, *ACS Catal.* **2014**, *4*, 4398–4408.
122. (a) M. J. Burk, G. Casy, N. B. Johnson, *J. Org. Chem.* **1998**, *63*, 6084–6085. (b) I. D. Gridnev, N. Higashi, T. Imamoto, *J. Am. Chem. Soc.* **2000**, *122*, 10486–10487.
123. (a) P. J. González-Liste, F. León, I. Arribas, M. Rubio, S. E. García-Garrido, V. Cadierno, A. Pizzano, *ACS Catal.* **2016**, 3056–3060. (b) F. León, P. J. González-Liste, S. E. García-Garrido, I. Arribas, M. Rubio, V. Cadierno, A. Pizzano, *J. Org. Chem.* **2017**, *82*, 5852–5867.
124. S. Vargas, A. Suárez, E. Álvarez, A. Pizzano, *Chem. Eur. J.* **2008**, *14*, 9856–9859.
125. J. R. Lao, H. Fernández-Pérez, A. Vidal-Ferran, *Org. Lett.* **2015**, *17*, 4114–4117.
126. H. Fernández-Pérez, A. Vidal-Ferran, *Org. Lett.* **2019**, *21*, 7019–7023.
127. (a) J. L. Núñez-Rico, A. Vidal-Ferran, *Org. Lett.* **2013**, *15*, 2066–2069. (b) B. Balakrishna, A. Bauzá, A. Frontera, A. Vidal-Ferran, *Chem. Eur. J.* **2016**, *22*, 10607–10613.
128. (a) K. Nozaki, N. Sakai, T. Nanno, T. Higashijima, S. Mano, T. Horiuchi, H. Takaya, *J. Am. Chem. Soc.* **1997**, *119*, 4413–4423. (b) T. Horiuchi, T. Ohta, E. Shirakawa, K. Nozaki, H. Takaya, *J. Org. Chem.* **1997**, *62*, 4285–4292. (c) T. Horiuchi, T. Ohta, K. Nozaki, H. Takaya, *Chem. Commun.* **1996**, 155–156.
129. (a) M. Rubio, A. Suarez, E. Alvarez, C. Bianchini, W. Oberhauser, M. Peruzzini, A. Pizzano, *Organometallics* **2007**, *26*, 6428–6436. (b) T. Robert, Z. Abiri, J. Wassenaar, A. J. Sandee, S. Romanski, J.-M. Neudörfl, H.-G. Schmalz, J. N. H. Reek, *Organometallics* **2010**, *29*, 478–483. (c) A. L. Watkins, B. G. Hashiguchi, C. R. Landis, *Org. Lett.* **2008**, *10*, 4553–4556.
130. J. Klosin, C. R. Landis, *Acc. Chem. Res.* **2007**, *40*, 1251–1259.
131. T. Horiuchi, E. Shirakawa, K. Nozaki, H. Takaya, *Organometallics* **1997**, *16*, 2981–2986.
132. C. P. Casey, E. L. Paulsen, E. W. Beuttenmueller, B. R. Proft, B. A. Matter, D. R. Powell, *J. Am. Chem. Soc.* **1999**, *121*, 63–70.
133. D. A. Castillo Molina, C. P. Casey, I. Müller, K. Nozaki, C. Jäkel, *Organometallics* **2010**, *29*, 3362–3367.
134. P. Dingwall, J. A. Fuentes, L. Crawford, A. M. Z. Slawin, M. Bühl, M. L. Clarke, *J. Am. Chem. Soc.* **2017**, *139*, 15921–15932.
135. F. López, A. J. Minnaard, B. L. Feringa, *Acc. Chem. Res.* **2007**, *40*, 179–188.
136. (a) T. Robert, J. Velder, H.-G. Schmalz, *Angew. Chem. Int. Ed.* **2008**, *47*, 7718–7721. (b) Q. Naeemi, M. Dindaroğlu, D. P. Kranz, J. Velder, H.-G. Schmalz, *European J. Org. Chem.* **2012**, *2012*, 1179–1185.
137. Q. Naeemi, T. Robert, D. P. Kranz, J. Velder, H.-G. Schmalz, *Tetrahedron: Asymmetry* **2011**, *22*, 887–892.

138. W. Lölsberg, S. Ye, H.-G. Schmalz, *Adv. Synth. Catal.* **2010**, *352*, 2023–2031.
139. T. Horiuchi, E. Shirakawa, K. Nozaki, H. Takaya, *Tetrahedron: Asymmetry* **1997**, *8*, 57–63.
140. T. V. RajanBabu, A. L. Casalnuovo, *J. Am. Chem. Soc.* **1996**, *118*, 6325–6326.
141. A. Falk, A.-L. Göderz, H.-G. Schmalz, *Angew. Chem. Int. Ed.* **2013**, *52*, 1576–1580.
142. A. Falk, A. Cavalieri, G. S. Nichol, D. Vogt, H.-G. Schmalz, *Adv. Synth. Catal.* **2015**, *357*, 3317–3320.

5 Chiral Tridentate-Based Ligands

Uchchhal Bandyopadhyay, Basker Sundararaju,
Rinaldo Poli, and Eric Manoury

CONTENTS

5.1 INTRODUCTION

Although rarely described in the literature only twenty years ago, a very large number of chiral tridentate ligands has now been developed and used in numerous catalyzed reactions. Amongst all these ligands, we have decided to present in this chapter only the most recent and efficient ones, covering a large range of the possible coordination patterns. Indeed, because of space limitations, many interesting ligands and valuable catalytic systems could not be included here, but we hope that this chapter will provide a useful overview of the available chiral tridentate ligands for use in asymmetric catalysis.

5.2 NNN LIGANDS

The first successful family of chiral tridentate ligands belongs to the Pybox (pyridinebis(oxazoline); Figure 5.1) family,[1–3] introduced in 1989 by Nishiyama *et al.*,[4] which has been successfully used in various reactions, such as the Mukaiyama aldol reaction, in particular with Cu(II) catalysts, with main contributions from Evans and

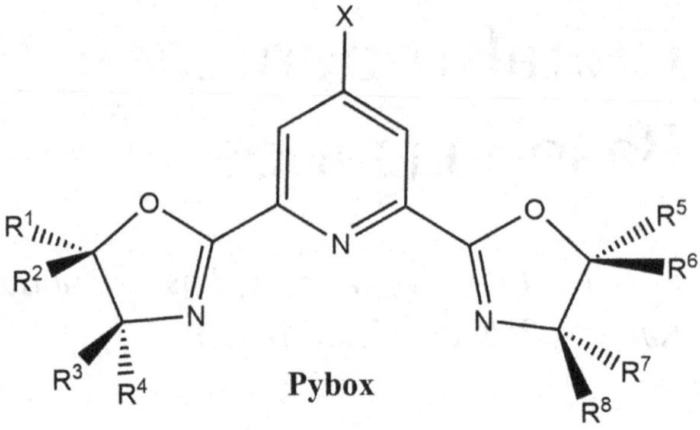

FIGURE 5.1 Pybox ligands.

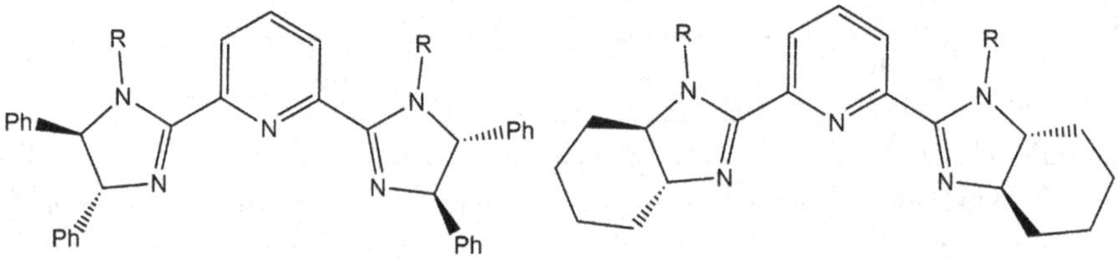

Pybim, R = H or -C(O)-R', R' beeing an alkyl, an aryl or an alkoxy group)

FIGURE 5.2 Pybim ligands.

coworkers[5] the hydrosilylation of ketones, mainly in combination with Rh(III); [4, 6, 7] the cyclopropanation with Ru(II); [8] the allylic oxidation with Cu(II); [9, 10] the Diels–Alder reaction with Sc(III)[11–13]; and several others. Pybox ligands are now well established and privileged ligands, regularly used to develop new catalytic systems.[14–22]

The Pybox structure has inspired the design of other chiral tridentate ligands, such as by retaining the central pyridine part and replacing one or both oxazoline rings with other groups. In 2005, Beller *et al.* developed a new family of ligands, called Pybim (pyridinebis(imidazoline)) from various chiral 1,2-diamines (Figure 5.2).[23] Ruthenium complexes of Pybim were tested in the asymmetric epoxidation of olefins by hydrogen peroxide; the chemoselectivities of the epoxides ranged from low to excellent with moderate enantioselectivities (epoxide up to 71% enantiomeric excess (ee) for (*trans*)-stilbene with ligands, from (*trans*)-stilbenediamine (Figure 5.2; left structure with R=1-naphtyloxy or (+)-menthyloxy).[23] Pybim ligands were also used in the zinc-catalyzed hydrosilylation of acetophenone, although with low enantioselectivities (16–24% ee).[24]

Fernandez and Lassaletta developed new pyridinebis(hydrazones) (Figure 5.3) and used them in the asymmetric Diels-Alder reaction of *N*-crotonyloxazolidin-2-one with cyclopentadiene.[25] Good diastereoselectivities (endo: exo ratio up to 92/8) and enantioselectivities (ee up to 74%) were obtained using ligand **5.3.c** and Cu(OTf)$_2$.

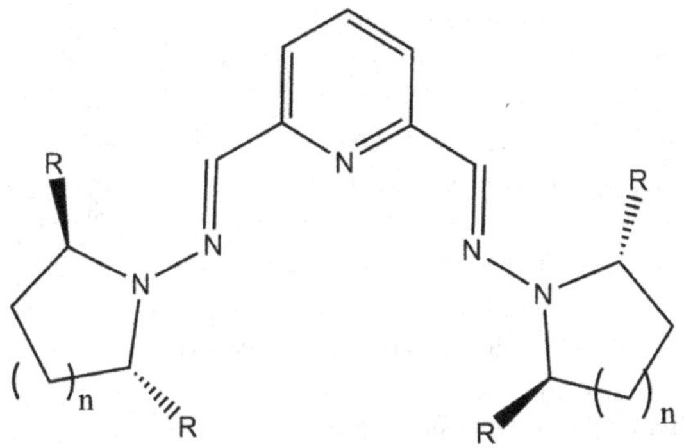

5.3.a: R = Ph, n = 1; 5.3.b: R = Ph, n = 2; 5.3.c: R = Me, n = 1

FIGURE 5.3   Pyridine bis(hydrazone) ligands.

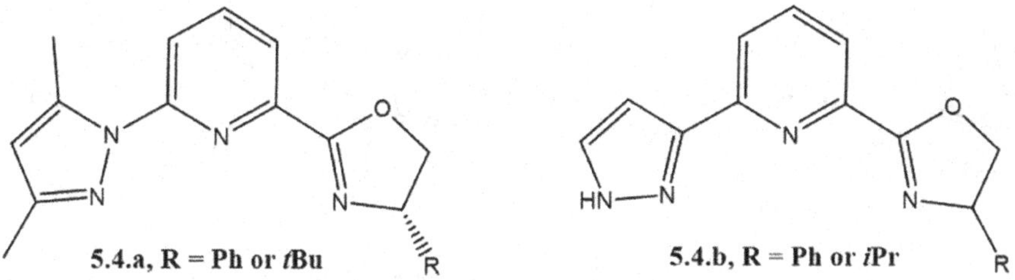

5.4.a, R = Ph or *t*Bu   **5.4.b, R = Ph or *i*Pr**

FIGURE 5.4   Pyrazole-pyridine-oxazoline ligands.

Pyrazoline-pyridine-oxazolines have been developed by Yu and coworkers for the ruthenium-catalyzed asymmetric transfer hydrogenation of ketones (Figure 5.4).[26] High catalytic activities and enantioselectivities were obtained with ruthenium(II) complexes of ligands **5.4.b** in the asymmetric hydrogenation of various aryl methyl ketones (ee of the alcohols = 67–98% for R = *i*Pr).

The same team also developed imidazole-pyridine-oxazolines (Figure 5.5). Ruthenium(II) complexes of these ligands were used in the asymmetric transfer hydrogenation of aryl and alkyl methyl ketones, the N-tosyl (NTs)-bearing ligands yielding the most active and enantioselective catalysts (ee from 73 to 99.9% for R=*i*Pr).[27, 28]

In 2014, Huang and coworkers developed new iminopyridine-oxazolines (IPO; Scheme 5.1). Their cobalt(II) complexes were used as catalysts in the first cobalt-catalyzed asymmetric hydroboration of alkenes with HBpin (pinacolborane) with complete regioselectivity and very high enantioselectivities, especially for 1-aryl-1'-alkyl-alkene and complex **5.5** (Scheme 5.1).[29] The same research group used iron(II) complexes of IPO ligands for the asymmetric alkene hydrosilylation. Excellent results were obtained for aryl methyl ketones and a crowded IPO ligand (R = *t*Bu, R' = CHPh₂), with yields in the 63–99% range and ee in the 64–93% range for almost all substrates.[30] Huang *et al.* also used Mn(II) complexes of IPO ligands

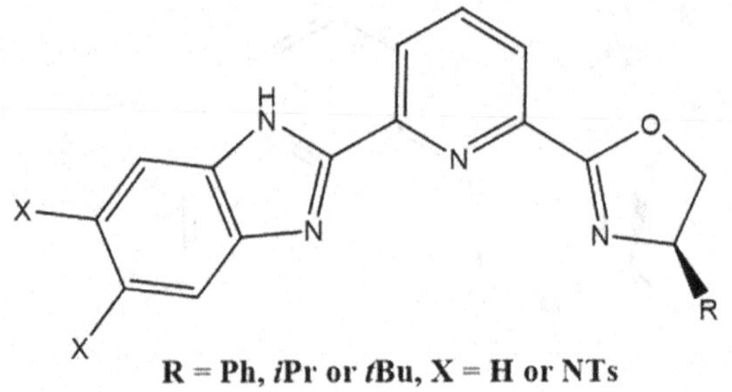

R = Ph, *i*Pr or *t*Bu, X = H or NTs

FIGURE 5.5 Benzimidazole-pyridine-oxazoline ligands.

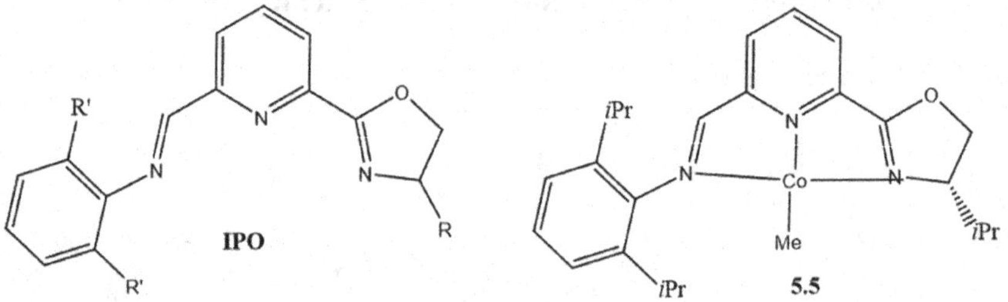

SCHEME 5.1 Imininopyridine-oxazoline (IPO) ligands. Asymmetric hydroboration of 1,1'-disubstituted alkene.

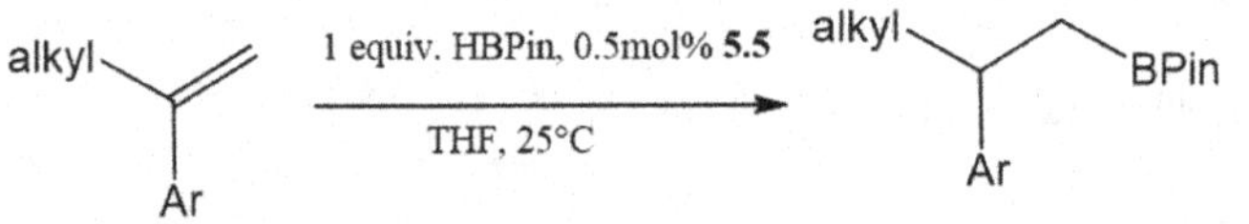

in the asymmetric alkene hydrosilylation, again with good results for aryl methyl ketones and the same crowded IPO ligand (R = *t*Bu, R' = CHPh₂): 68–99% yields and 79–92% ee.[31]

Independently from Huang's group, Lu *et al.* synthesized IPO ligands *via* another synthetic pathway and used them in the iron-catalyzed asymmetric hydroboration (ee in the 87–92% range)[32] and hydrosilylation (ee in the 78–99% range)[33] of 1,1'-disubstituted alkenes.

In addition to IPO ligands, Lu *et al.* also developed oxazoline-aminoisopropyl-pyridines (OAP; Figure 5.6). The $CoCl_2$(OAP) complexes proved to be efficient catalysts for the asymmetric hydroboration of 1-alkylstyrenes (yields in the 35–81% range, and ee in the 65–95% range, with the ligand with R^1=2,6-(Me)₂C₅H₃, R^2=*t*Bu, and R^3=H), with the opposite product configuration to that obtained with the corresponding IPO.[34]

To the best of our knowledge, Zhang *et al.* described the first analog of Pybox where the central pyridine was replaced by another nitrogen ligand (Ambox;

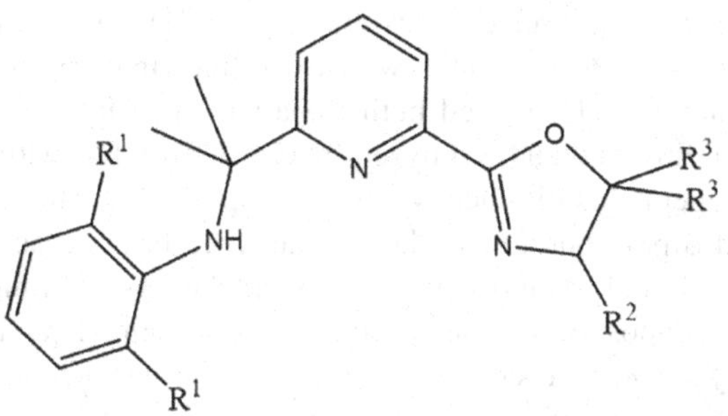

FIGURE 5.6 Oxazoline-aminoisopropylpyridine (OAP) ligands.

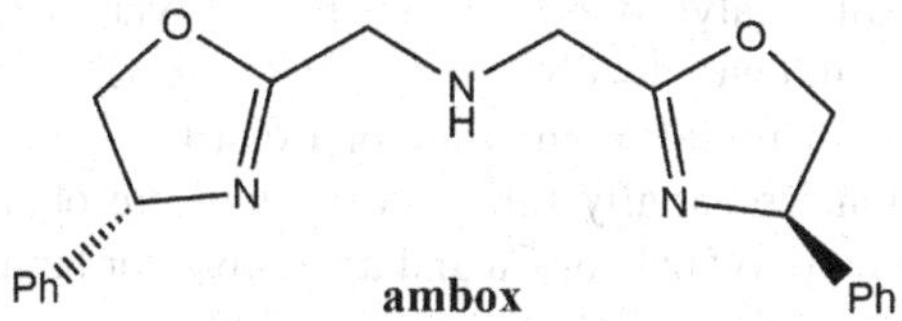

FIGURE 5.7 Ambox ligands.

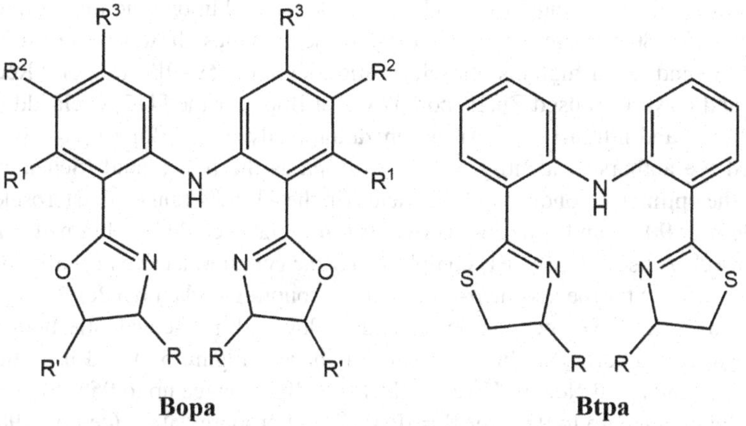

FIGURE 5.8 Bopa ligands and thiazoline analogs.

Figure 5.7). Ruthenium(II) complexes of Ambox were used in the asymmetric transfer hydrogenation of aryl alkyl ketones. For most substrates, the reaction was complete within 10 min with 1 mol% of catalyst with high enantioselectivities (ee in the 90–98% range).[35]

In 2002, Guiry and coworkers described a new Pybox analog with another linker between the two oxazoline groups: bis(oxazolynylphenyl)amine (Bopa; Figure 5.8). Bopa are potential anionic NNN-tridentate ligands after deprotonation. The authors used Bopa ligands in the acetophenone asymmetric transfer hydrogenation with high yields but low enantioselectivities (ee up to 18%).[36] Du, Xu, and coworkers synthesized analogs of Bopa ligands with oxazoline rings replaced by thiazoline rings (Btpa; Figure 5.8). They used both Bopa and Btpa in the Henry reaction of α-ketoesters with nitromethane, catalyzed by $Cu(OTf)_2/NEt_3$, with yields generally over 50% and moderate enatioselectivities (ee = up to 82%, similar to those for the related Bopa and Btpa)[37] but also in the Zn-catalyzed Henry reaction (with $ZnEt_2$), also with high yields and enantioselectivities (ee in the 13–85% range but with product configurations opposite to those obtained with the same ligands and the copper catalyst).[38] Nishiyama *et al.* used Bopa ligands in the iron(II)-catalyzed asymmetric hydrosilylation of ketones with high yields and enantioselectivities (up to 95% with the Bopa ligand with R=$CHPh_2$, R'=H, R^1= R^2= R^3=H), which were highly dependent on the iron source ($Fe(OAc)_2$ or $FeCl_3/Zn$).[39–41] Bopa ligands were also successfully used in the cobalt-catalyzed asymmetric hydrosilylation of ketones (ee in the 38–98% range with ligand with R=Ph, R'=H, R^1= R^2= R^3=H).[40] Ward and coworkers used calcium[42] and lanthanide[43] complexes of Bopa for the asymmetric intramolecular hydroamination; essentially full conversions were obtained with moderate enantioselectivities (ee up to 50% for Ca and up to 46% for lanthanides).

Hu and coworkers used Bopa Fe(III) complexes for the asymmetric cross-coupling reaction of alkyl halides with aryl Grignard reagents, with low enantioselectivities (ee up to 19%) (Scheme 5.2a).[44] Xiao *et al.* used Bopa zinc complexes for the synthesis of chromans by a tandem Friedel-Crafts/Michael addition reaction with moderate yields (54–80% range), very high diastereoselectivities (diastereomer excess (de) over 90%) and rather high enantioselectivities (ee in the 59–91% range).[45] Recently, Rovis and coworkers used Zn(II) complexes of Bopa for the [4+2] cycloaddition of 1-azadienes and nitroalkenes. An optimization study of the Bopa ligand structure allowed the authors to identify a difluoro-Bopa as the best ligand (Scheme 5.2b). Under the optimized conditions, high yields (in the 43–87% range), diastereoselectivities (de over 90%), and enantioselectivities (up to 92%) could be obtained.[46] Zhang and coworkers used Bopa Cu(I) complexes for the enantioselective aryl alkynylation of alkenes[47] and for the alkyne/α-bromoamide coupling (Scheme 5.2c).[48]

Nakada *et al.* developed a Bopa ligand analog with *o,o'*-covalently bound phenyl rings, called carbazole bis(oxazolines) (Cbzbox; Figure 5.9) and used them in chromium-catalyzed Nozaki-Hiyama aldehyde allylation (ee up to 95% for R = *i*Pr), methallylation (ee up to 95% for R = *i*Pr)[49, 50] and propargylation (ee up to 98% for R = *t*Bu).[51]

Gade and coworkers described a new family of chiral, potentially anionic NNN tridentate ligands after deprotonation, namely the bis(oxazolinylmethylidene)isoindolines (Boxmi; Figure 5.10).[52]

Boxmi complexes were used in the enantioselective nickel-catalyzed oxindole fluorination (Scheme 5.3a),[52] the chromium-catalyzed Nozaki-Hiyama-Kishi reaction (Scheme 5.3b),[52] and the copper-catalyzed β-ketoester alkylation (Scheme 5.3c).[53] Boxmi copper complexes proved to be inefficient catalysts in the asymmetric oxindole alkylation, whereas the corresponding Zn complexes were highly successful (Scheme 5.3d).[54] The same zinc complexes were also used in the asymmetric oxindole cyanation, with 4-acetylphenyl cyanate as cyanide source (Scheme 5.3e).[55]

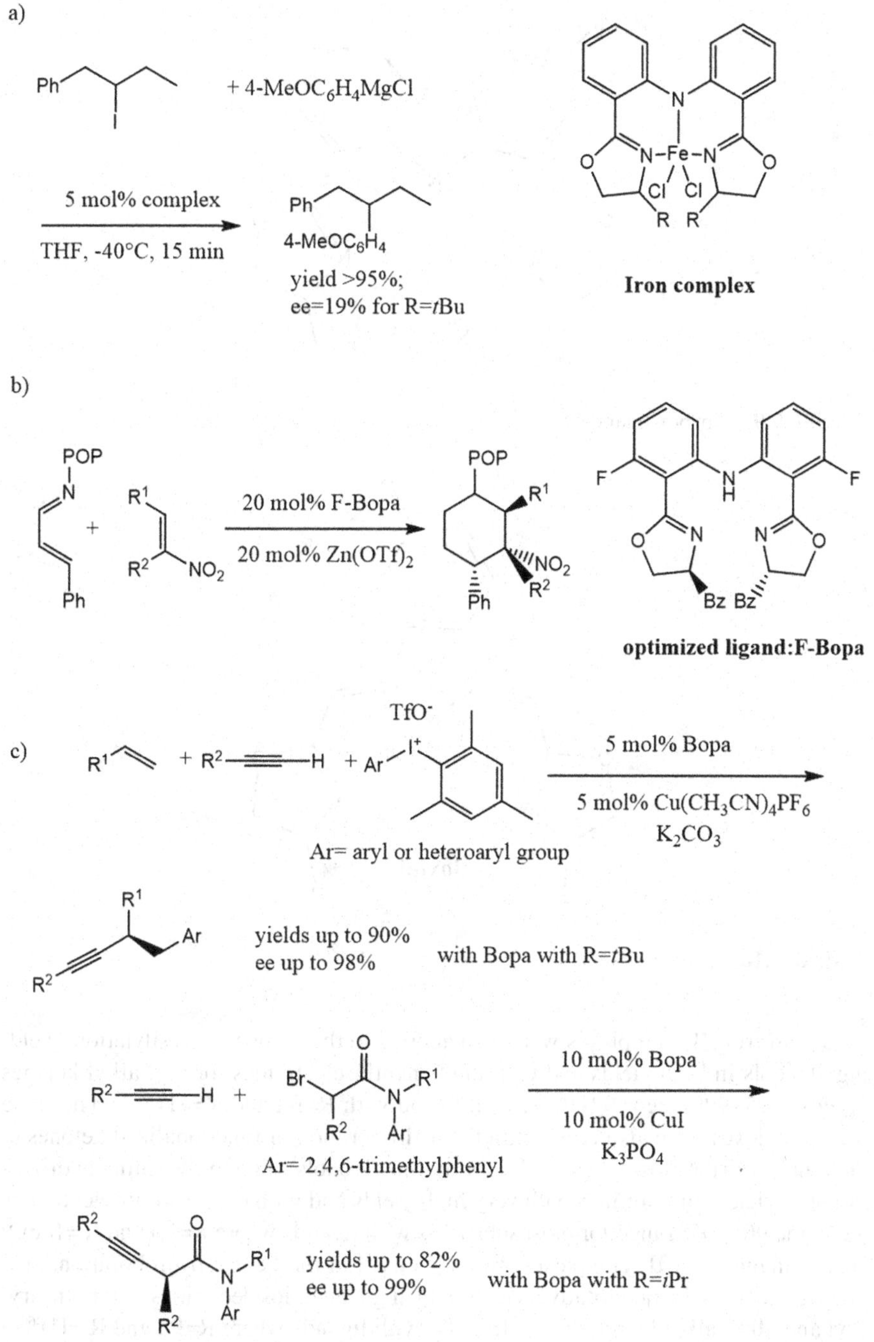

a)

Ph [structure] + 4-MeOC$_6$H$_4$MgCl

5 mol% complex
⟶
THF, -40°C, 15 min

Ph [structure]
4-MeOC$_6$H$_4$
yield >95%;
ee=19% for R=tBu

Iron complex

b)

POP [structure] + R^1 [structure] R^2 NO$_2$

20 mol% F-Bopa
⟶
20 mol% Zn(OTf)$_2$

POP [structure] R^1
///NO$_2$
R^2
Ph

optimized ligand:F-Bopa

c)

R^1 [structure] + R^2—≡—H + Ar—I$^+$ [structure] TfO$^-$

Ar= aryl or heteroaryl group

5 mol% Bopa
⟶
5 mol% Cu(CH$_3$CN)$_4$PF$_6$
K$_2$CO$_3$

R^1 [structure] Ar
R^2

yields up to 90%
ee up to 98%   with Bopa with R=tBu

R^2—≡—H + Br [structure] R^1
R^2 Ar

Ar= 2,4,6-trimethylphenyl

10 mol% Bopa
⟶
10 mol% CuI
K$_3$PO$_4$

R^2 [structure] R^1
R^2 Ar

yields up to 82%
ee up to 99%   with Bopa with R=iPr

SCHEME 5.2   Some asymmetric reactions with Bopa ligands.

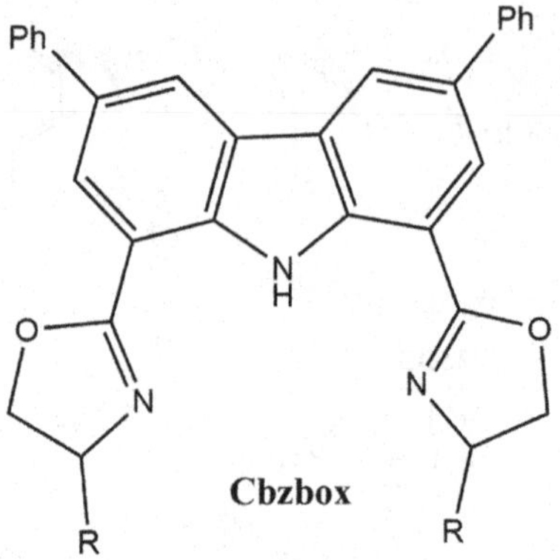

FIGURE 5.9 Cbzbox ligands.

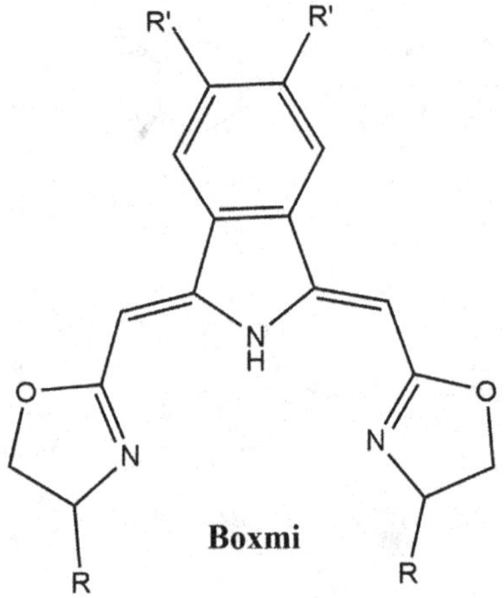

FIGURE 5.10 Boxmi ligands

Boxmi iron(II) complexes were also applied to the ketone hydrosilylation, yielding alcohols in high yields and with high enantioselectivities for aryl alkyl ketones (yields = 95–99%, ee = 94–99%, with ligand with R=Ph and R'=H).[56, 57] The same iron complexes were also efficient in the hydroboration of functionalized ketones to yield halohydrines, oxaheterocycles, or amino alcohols[58] and in the imine hydroboration to yield chiral amines with very high yields and with high enantioselectivities (ee in the 90–99% range for most substrates with ligands where R=Ph and R'=Me).[59] Boxmi manganese(II) complexes were also used in the ketone hydroboration, producing alcohols in quantitative yields and high enantioselectivities for both aryl aryl and alkyl alkyl ketones (ee = 18–99% with ligands where R=Ph and R'=H).[60,61] Cobalt(II) Boxmi complexes were also used in the asymmetric acetophenone hydrosilylation (ee up to 79 %, with ligands where R=Ph and R'=H).[62]

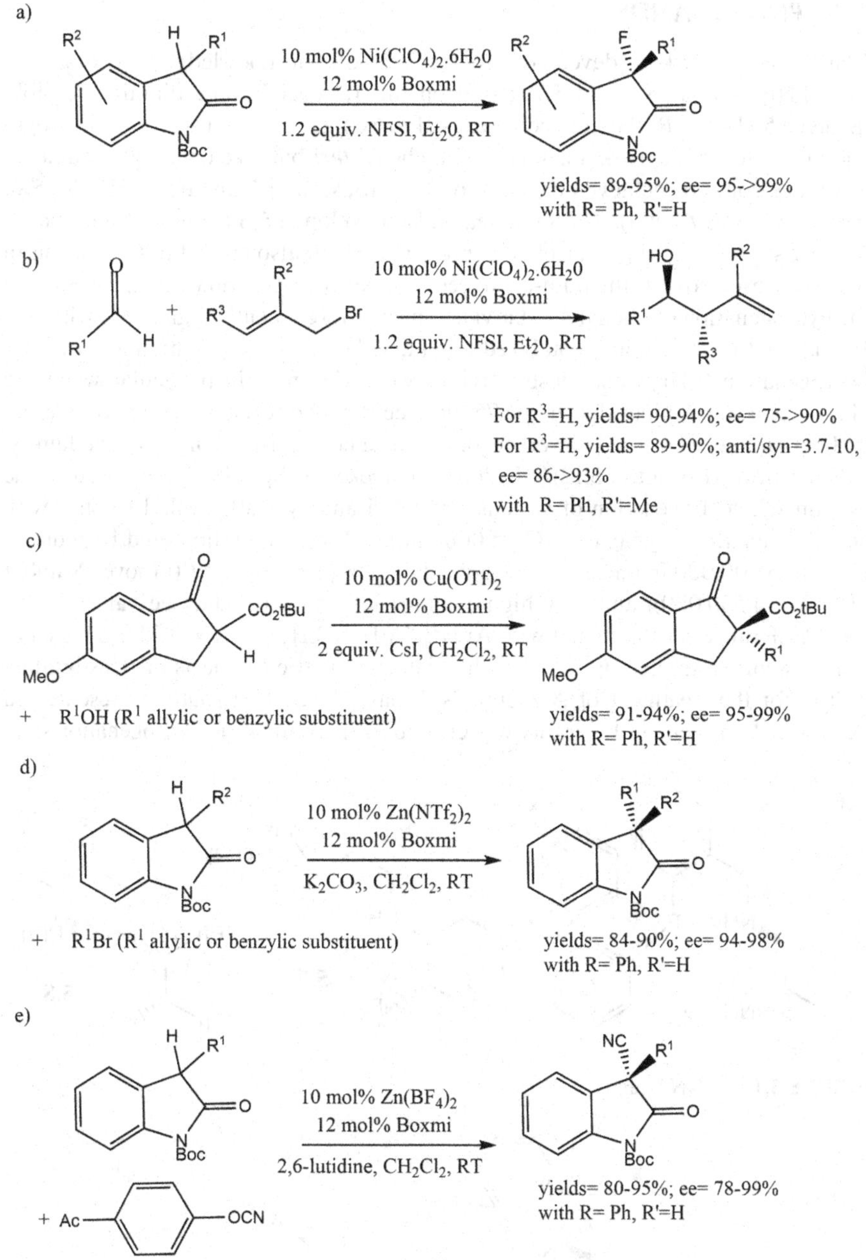

a)

yields= 89-95%; ee= 95->99%
with R= Ph, R'=H

b)

For R³=H, yields= 90-94%; ee= 75->90%
For R³=H, yields= 89-90%; anti/syn=3.7-10,
ee= 86->93%
with R= Ph, R'=Me

c)

+ R¹OH (R¹ allylic or benzylic substituent)

yields= 91-94%; ee= 95-99%
with R= Ph, R'=H

d)

+ R¹Br (R¹ allylic or benzylic substituent)

yields= 84-90%; ee= 94-98%
with R= Ph, R'=H

e)

+ Ac—⬡—OCN

yields= 80-95%; ee= 78-99%
with R= Ph, R'=H

SCHEME 5.3 Some asymmetric reactions with Boxmi ligands.

5.3 PNN LIGANDS

Clarke and coworkers developed, to the best of our knowledge, the first efficient PNN ligands **5.6** and **5.7,** based on the 1,2-cyclohexane diamine scaffold
(Scheme 5.11).[63–65] Ruthenium complexes of these ligands were used in the hydrogenation of a few phenyl alkyl ketones; with phenyl *tert*-butyl ketone, high enantioselectivities could be achieved (ee up to 67% for (*S,S,Sp*)-**5.7** and up to 85% for **5.6,**
with X = 3,5-di-*t*-butyl). The same team also developed **5.8** (analog of **5.6**), based
on the *t*-stilbenediamine scaffold (Figure 5.11), which also proved to be efficient in
the enantioselective ruthenium-catalyzed ketone hydrogenation (ee up to 80% in
the hydrogenation of phenyl *tert*-butyl ketone).[66] More recently, ligand **5.6** with X =
H was used in the iridium-catalyzed alkyl aryl ketone hydrogenation and transfer
hydrogenation.[67] High enantioselectivities were obtained, in particular when the
alkyl group was 4-piperidyl (in the 75–96% ee range) or cyclohexyl (up to 98%).

In 2011, Xie, Zhou, and coworkers described a new spiro tridentate ligand family,
called SpiroPAP (Figure 5.12).[68] Iridium complexes of SpiroPAP were used in the
asymmetric hydrogenation of various aryl alkyl and cycloalkyl alkyl ketones with
very low loadings (typically S/C= 5,000, although the S/C ratio could be reduced
down to 5,000,000 in one case with a yield of 91% after 15 days (Turnover Number
(TON) = 4,550,000)) and very high enantioselectivities (product ee values in the
88–99.9% range for the ligand with Ar = 3,5-(*t*Bu)$_2$C$_6$H$_3$ and X = 3-Me). This catalytic system was scaled up for the industrial asymmetric synthesis of rivastigmine,
a drug for the treatment of Alzheimer's disease.[69] The best route is presented in
Scheme 5.4: the center of chirality was created in an asymmetric hydrogenation step,

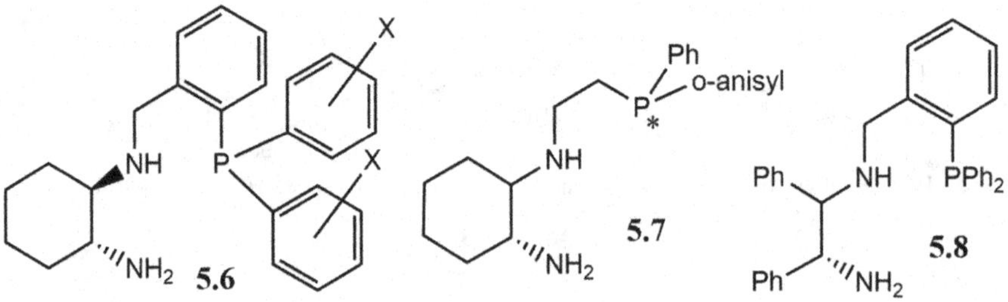

FIGURE 5.11 PNN ligands.

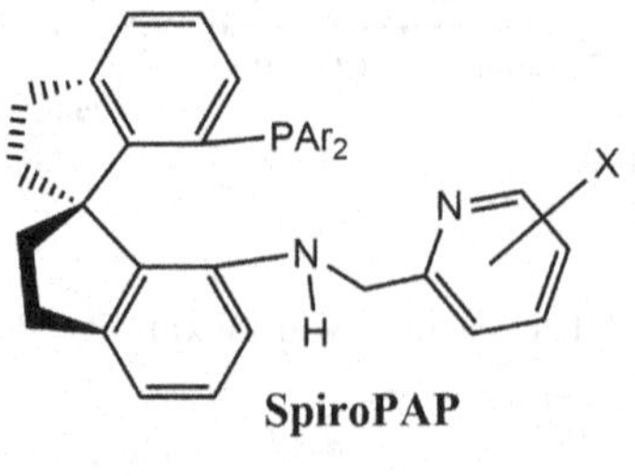

SpiroPAP

FIGURE 5.12 SpiroPAP ligands.

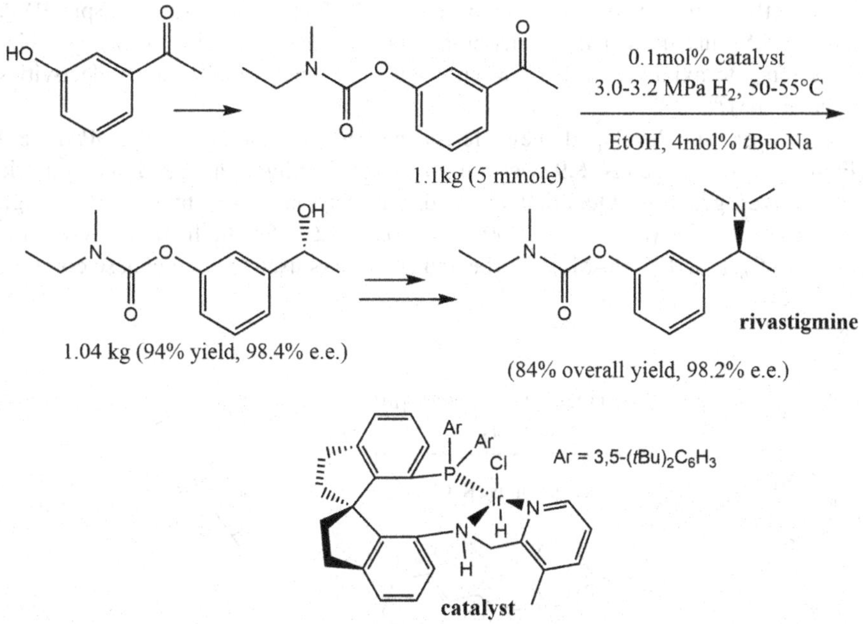

SCHEME 5.4 Asymmetric synthesis of rivastigmine.

using a SpiroPAP ligand with a high yield (isolated yield = 94% at the 5 mmole scale) and enantioselectivity (ee = 99.4%). The same catalytic system was used in the asymmetric synthesis of a chiral intermediate in the synthesis of montelukast, a leukotriene-receptor antagonist, used for the treatment of asthma and seasonal allergies at the 30 kg scale (yield = 94%, ee = 99.5%).[70]

SpiroPAP iridium complexes were also used in the β-ketoester asymmetric hydrogenation, once again with very high catalytic activities (typically substrate/catalyst ratio (S/C) = 1,000 but a TON of 1,230,000 could be obtained using a S/C ratio of 1,500,000) and high enantioselectivities (product ee in the 88–95.8% range; 95–99.8% for β-aryl β-ketoesters) for the ligand with Ar = 3,5-(tBu)$_2$C$_6$H$_3$ and X = 3-Me.[71] They were also used in the asymmetric hydrogenation of aryl and heteroaryl δ-ketoesters (for S/C = 1,000, yield in the 92–97% range and ee in the 96.7–99.9% range for the ligand with Ar = 3,5-(tBu)$_2$C$_6$H$_3$ and X = 3-Me)[72] and in the asymmetric hydrogenation of alkyl and aryl α-ketoacids (for S/C = 1,000, yield in the 92–98% range and ee in the 56–98% range for the ligand with Ar = 3,5-(tBu)$_2$C$_6$H$_3$ and X = 4-tBu).[73] SpiroPAP iridium complexes were also used in the asymmetric hydrogenation of racemic α-substituted lactones to yield chiral diols by dynamic kinetic resolution. The best results were again obtained with the ligand where Ar = 3,5-(tBu)$_2$C$_6$H$_3$ and X = 3-Me (yields = 82–95%, ee = 69–95%).[74] The same SpiroAP iridium complexes were also efficient in the asymmetric transfer hydrogenation of ketones (for S/C = 1,000, yields = 90–99% range and ee = 80–98% range for ligands with Ar = 3,5-(tBu)$_2$C$_6$H$_3$ and X = H).[75]

Recently, Yin, *et al.* developed a SpiroPAP analog (oxa (O)-SpiroPAP; Scheme 5.5) and used it in the iridium-catalyzed hydrogenation of bridged biaryl lactones to axially chiral compounds with high yields and enantioselectivities (Scheme 5.5).[76]

Chen, et al. developed new ferrocene-based analogs of SpiroPAP **5.9** (Figure 5.13).[77] Ligands **5.9** were used in the iridium-catalyzed asymmetric ketone hydrogenation. Quantitative yields of alcohols were obtained with high enantioselectivities (for S/C = 2,000, ee = 60.8–86.6% for the ligand, where Ar = 3,5-(tBu)$_2$C$_6$H$_3$ and X = 6-Me). Clarke and coworkers used the manganese complex

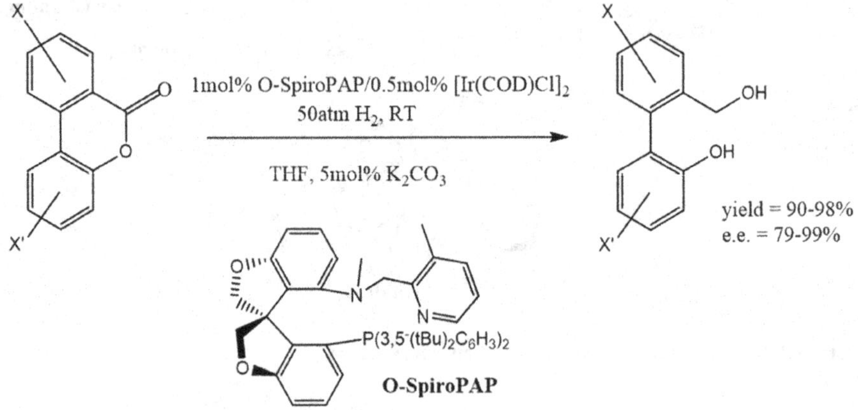

SCHEME 5.5 O-SpiroPAP.

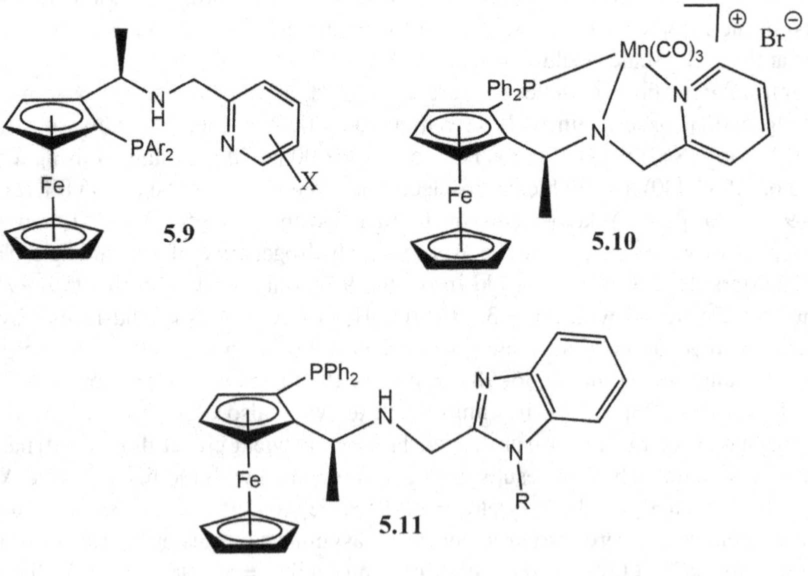

FIGURE 5.13 Ferrocene-based analogs of SpiroPAP ligands.

5.10 (Figure 5.14) in the hydrogenation of various aryl alkyl ketones to yield the corresponding alcohols in quantitative yields, with ee in the 20–91% range (> 80% for crowded substrates).[78] Recently, Zhong *et al.* synthesized ligands **5.11** (Figure 5.13), a new family of analogs of ligands **5.9**, where the pyridine moiety is replaced by a benzimidazole group.[79] These ligands were also used in the manganese-catalyzed asymmetric hydrogenation of ketones; for S/C = 1,000, yields were in the 95–99% range and ee in the 49.7–88.5% range for the ligand with R = Bz.

In 2012, Hu and coworkers used the chiral PNN ligands **5.12**, previously synthesized by the same research group,[80] and also the similar ligand **5.13** (with R = H), in the copper-catalyzed asymmetric [3+3] cycloaddition of racemic propargylic esters with cyclic enamines (Scheme 5.6). The highest selectivities were obtained with ligand **5.12**, yielding less than 2% of alkylation by-product **5.14** (yields = 58–88%, endo:exo > 98/2, ee = 67–98%; Scheme 5.6a).[81] The authors proposed that a copper allenylidene complex is a key reaction intermediate and tested their catalytic systems against many different substrates. The same complexes were also used for the asymmetric copper-catalyzed [3+3] cycloaddition of racemic propargylic esters with 5-pyrazolones, resulting in high yields and very high enantioselectivities (ee = 71–97% for **5.12**),[82] as well as for the decarboxylative alkylation of propargyl β-ketoesters (ee = 83–98% for **5.13** with R = Ph).[83] The same research group also exploited the ability of copper complexes of ligands **5.12** and **5.13** to activate propargylic esters to develop various asymmetric substitution reactions of racemic propargylic esters with different nucleophiles, namely nitrogen nucleophiles, such as primary and secondary anilines (the best ligand being **5.12** for aromatic substrates with ee = 87–97%, and **5.13** (R=H) for aliphatic ones with ee = 78–95%; Scheme 5.6b), [84] oxygen nucleophiles such as oximes (ee = 77–92% with **5.13** (R=Ph)),[85] but also various carbon nucleophiles, such as β-ketoesters (ee = 77–97% with **5.13** (R=Ph)),[86] 1,3-dicarbonyl compounds (ee = 94–99% with **5.13** (R=Ph)),[87] enamines (ee = 96–98%, with **5.12** for acyclic enamines[88] and 80–97% with **5.13** (R=Ph) for cyclic enamines),[89] oxindoles (de > 90%, ee = 95–99% with **5.12**),[90] and coumarins (ee = 77–97% with **5.13** (R=Ph)).[91] With phenols as substrates, a Friedel-Crafts reaction,[92] a [3+2] addition[93], or even a dearomatization reaction[94] could occur, each with high enantioselectivities (Scheme 5.6c).

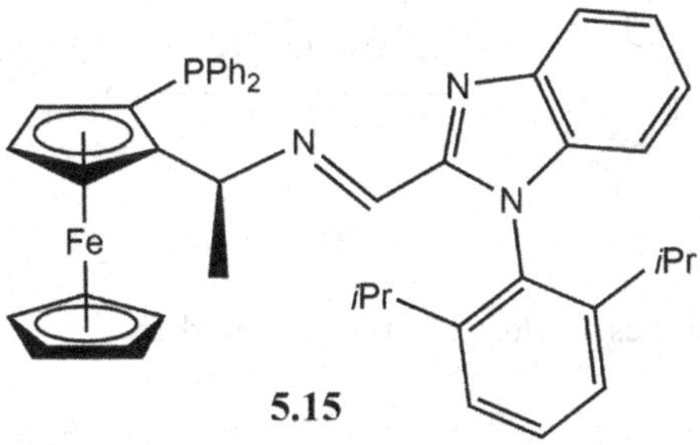

5.15

FIGURE 5.14   Ferrocene-based PNN ligands, developed by Zhong and coworkers.

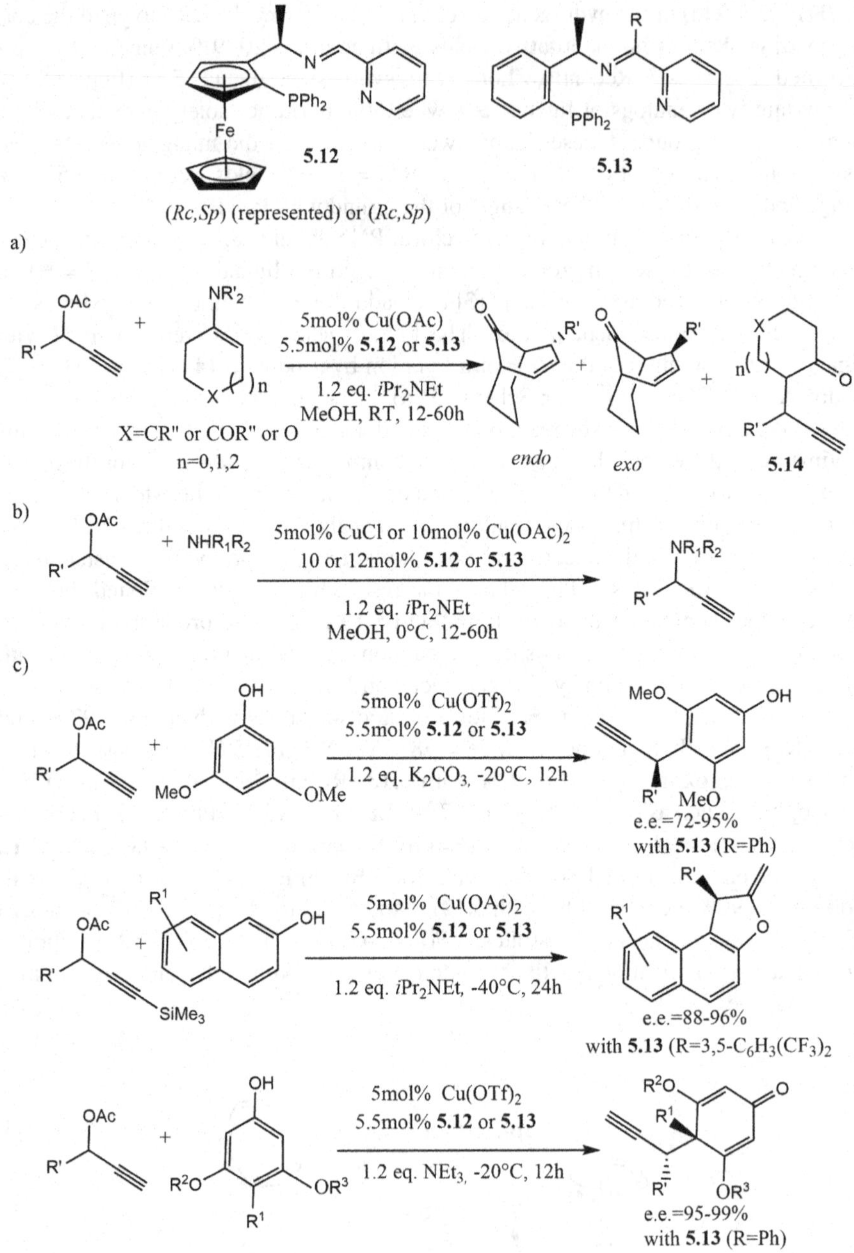

5.12

5.13

(*Rc,Sp*) (represented) or (*Rc,Sp*)

a)

b)

c)

SCHEME 5.6 New ligands developed by Hu and coworkers.

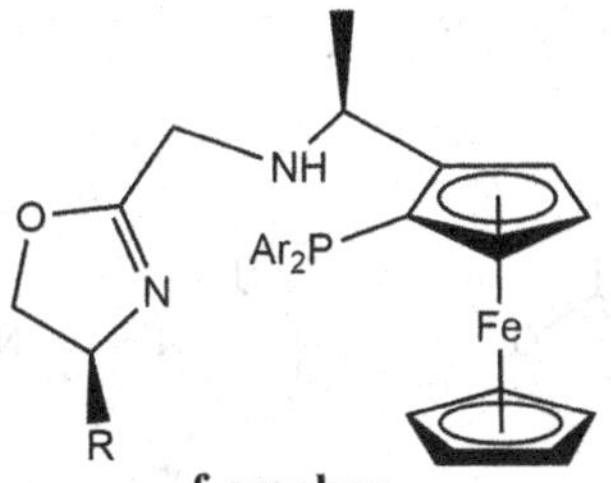

f-amphox

FIGURE 5.15 f-amphox ligands.

Recently, Zhong and coworkers synthesized **5.15**, the unsaturated analog of **5.11** and an analog of **5.12,** with a benzimidazole group in place of the pyridine moiety (Figure 5.14).[95] Manganese complexes of this ligands were used in the asymmetric hydrogenation of benzophenones, showing high catalytic activities (TON up to 13,000) and high enantioselectivities (ee up to 99%).

Lan, Dong, Zhang, and coworkers developed new ferrocene-based PNN ligands, bearing an oxazoline group (f-amphox; Figure 5.15).[96] These ligands were used in the iridium-catalyzed asymmetric hydrogenation of ketones: for S/C = 1,000, quantitative alcohol yields and ee in the 96–99.9% range were obtained for the ligand with Ar = Ph and R = tBu. The same catalytic system was applied with low catalytic loadings (TON up to 1,000,000) to the asymmetric hydrogenation of α-hydroxy ketones[97] and α-amino ketones[98] to yield 1,2-diols (ee = 98–99%) and 1,2-aminoalcohols (ee > 99%), respectively, in quantitative yields. β-Ketoesters could also be efficiently hydrogenated with high yields (80–98%) and with high enantioselectivities (71–95%), using the same catalytic system, although higher catalyst loadings were necessary (1 mol%).[99]

5.4 NPN LIGANDS

In 1997, Zhang and coworkers described the synthesis of three new chiral NPN ligands **5.16** (R^1 = Ph, R^2 = iPr, R^3 =H) and **5.17** (R^1 = Ph, R^2 = H, R^3 = Ph or R^1 = Ph, R^2 = iPr, R^3 = H) (Scheme 5.7) and used them in the ruthenium-catalyzed transfer hydrogenation of ketones.[100] The best results (ee in the 16–92% range) were obtained with **5.17** (R^1 = Ph, R^2 = H, R^3 = Ph).

Recently, Li and coworkers tested ligands **5.16** and **5.17** in the cobalt-catalyzed asymmetric allylic amination of racemic branched allylic carbonates (Scheme 5.7a).[101] Perfect regioselectivities and very high enantioselectivities were obtained, but the reaction was sluggish in certain cases. The best results (yield = 66–99%, ee = 84–99%) were obtained with **5.16** (R^1 = Ph, R^2 = Ph, R^3 = H). The same research team extended the use of ligands **5.16** to the cobalt-catalyzed asymmetric allylic alkylation of carbon nucleophiles, in particular malonitriles (Scheme 5.7b), where high yields and very high enantioselectivities were obtained (yield = 61–96%, ee =

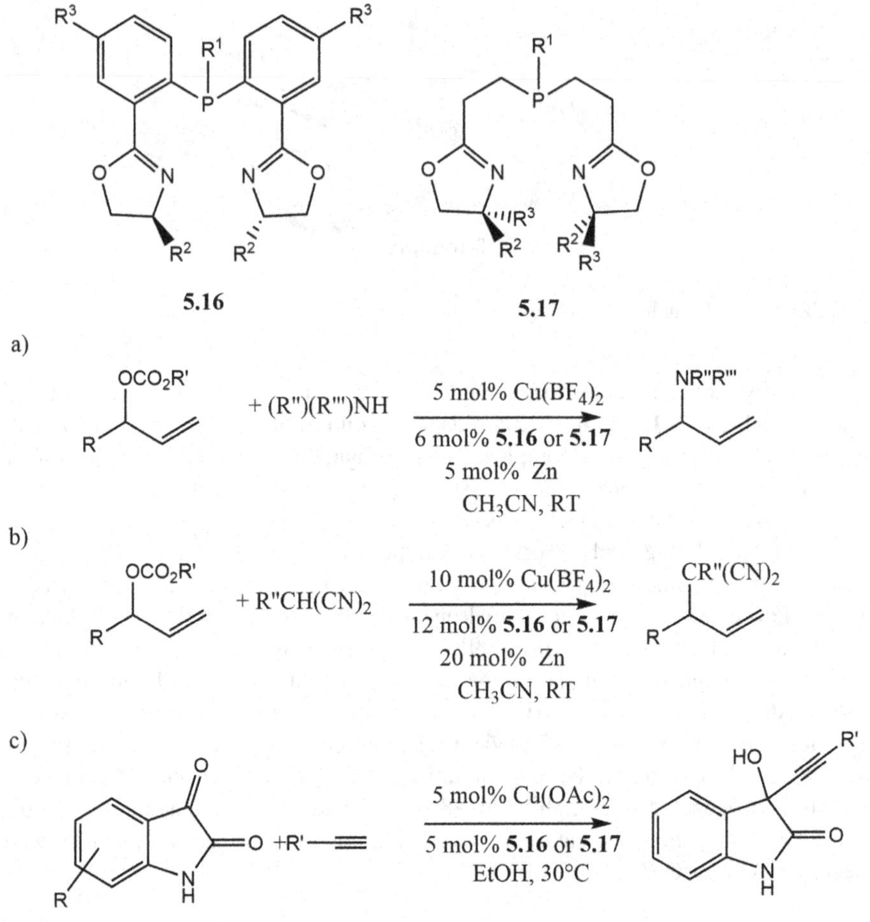

SCHEME 5.7 Phosphine-bisoxazoline ligands.

90–99% with **5.16** (R^1 = Me, R^2 = Me, R^3 = OMe).[102] These ligands were also successfully used in the cobalt-catalyzed asymmetric allylic alkynylation of isatins (Scheme 5.7c), with yield = 71–96% and ee = 92–99% with **5.16** (R^1 = Ph, R^2 = Me, R^3 = H).[103]

5.5 PNP LIGANDS

Morris and coworkers synthesized the chiral PNP ligands **5.18** (see Figure 5.16).[104, 105] The iron complexes [FeBr(CO)$_2$(**5.18**)]$^+$, BF$_4$$^-$, after treatment with LiAlH$_4$, followed by *tert*-amyl alcohol, yield the monohydride complexes [FeH(O-*tert*-amyl)(CO)(**5.19**)], which are pre-catalysts in the asymmetric hydrogenation of ketones. After activation with a base (KO-*tert*-Bu) under dihydrogen pressure, complexes [FeH(O-*tert*-amyl)(CO)(**5.19**)] yield the dihydride complexes [FeH$_2$(CO)(**5.19**)], which are

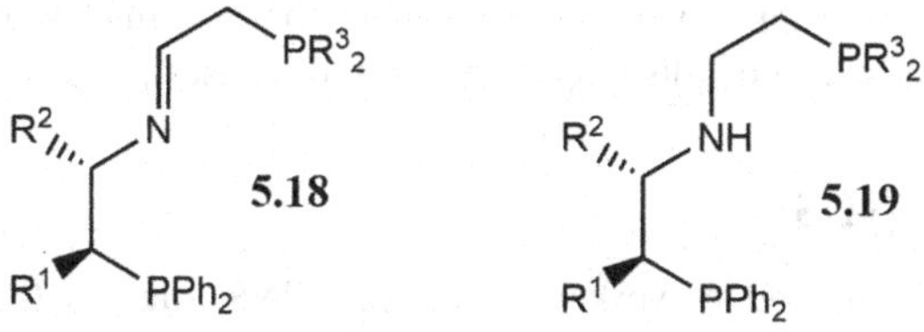

FIGURE 5.16   PNP ligands, developed by Morris and coworkers.

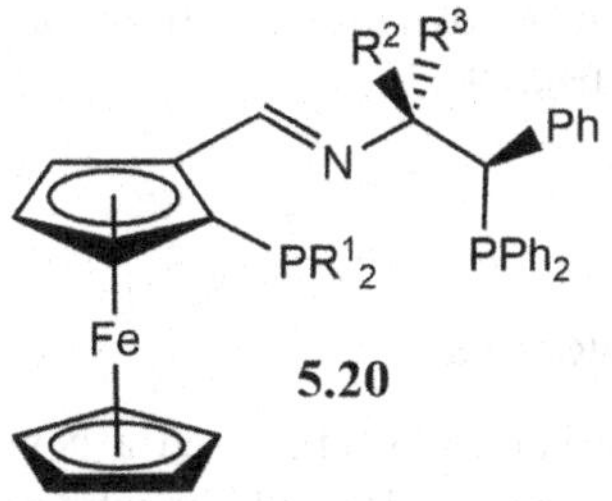

FIGURE 5.17   PNP ligands, developed by Zirakzadeh and coworkers.

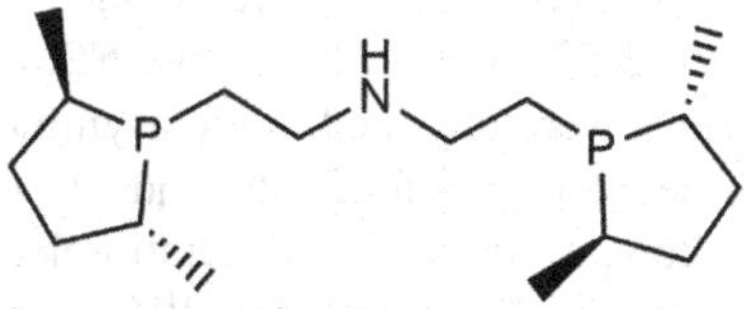

FIGURE 5.18   PNP ligands, developed by Beller and coworkers.

proposed to be the active hydrogenation catalysts.[106] Similar catalytic systems could also be obtained from [FeH(Cl)(CO)(**5.19**)] with a base (KO-*tert*-Bu) under dihydrogen pressure.[107] These catalytic systems proved to be very efficient (ee = 86–96%) in the asymmetric hydrogenation of aryl alkyl ketones with low catalyst loadings (typically 0.1 mol%), as well as in the asymmetric hydrogenation of activated imines (*N*-phosphinoyl and *N*-tosylimines) to yield the corresponding amines with high enantiomeric excesses (in the 90–98% range for most substrates).[108]

Zirakzadeh and coworkers developed ferrocenyl ligands **5.20** (Figure 5.17), which were used in the iron-catalyzed hydrogenation of ketones, and high to excellent alcohol yields could be obtained, with ee values up to 81%.[109] The same ligands were also tested in the manganese-catalyzed transfer hydrogenation of ketones, yielding ee values in the 20–85% range.[110]

Beller *et al.* synthesized a new chiral PNP **5.21** ligand bearing two stereogenic phospholane rings (Figure 5.18).[111] This ligand was tested in the asymmetric hydrogenation of ketones with various metals (Mn, Re, Fe, Ru), [112] with the manganese- and iron-based systems being especially efficient in the asymmetric hydrogenation of cyclic aliphatic ketones (ee values up to 99% in both cases).

5.6   PNS LIGANDS

Recently, Xie, Zhou, and coworkers developed PNS analogs (SpiroSAP) of the SpiroPAP ligands (Figure 5.19).[113] These ligands were used in the asymmetric iridium-catalyzed hydrogenation of β-alkyl-β-ketoesters (yield = 91–98%, ee = 95–99.9% with SpiroSAP(a)) and in the asymmetric iridium-catalyzed hydrogenation of racemic β-ketolactams by dynamic kinetic resolution (yield = 87–99%, TON up to 5,000, syn-/anti-ratio from 97/3 to 99/1, ee = 83–99.9% with SpiroSAP(a)).[114]

5.7   NNO LIGANDS

5.7.1   PSEUDODIPEPTIDE LIGANDS

In 2002, Adolfsson and coworkers synthesized a new highly modular family of NNO ligands **5.22**, which were named pseudodipeptides, by coupling of N-protected α-amino acids and various amino alcohols (Figure 5.20).[115–118] These ligands were used in the ruthenium-catalyzed transfer hydrogenation of ketones with isopropanol under basic conditions. The screening of the different ligands led to the identification of two privileged structures, **5.22a** and **5.22b** (Figure 5.20). With these two ligands, very high enantioselectivities were obtained in the asymmetric transfer hydrogenation of aryl alkyl ketones (ee = 84–96% for **5.22a**, and 94–99% for **5.22b**). The addition of lithium salts was shown to increase the enantioselectivities when *i*PrONa or *i*PrOK were used as a base, at which point a bimetallic outer-sphere-type mechanism was then proposed.[119, 120] The same ligands were also employed in the rhodium-catalyzed transfer hydrogenation of ketones with isopropanol and a base.[121] In this case, the best ligand was **5.22c** (Figure 5.20; ee = 61–98% for aryl or heteroaryl alkyl ketones).

Adolfsson, Dieguez, Pamies, and coworkers extended the pseudodipeptide NNO ligand family using carbohydrate-based amino alcohols (Figure 5.21).[122–124] In particular, ligands **5.23** were identified as being very efficient in the ruthenium-catalyzed

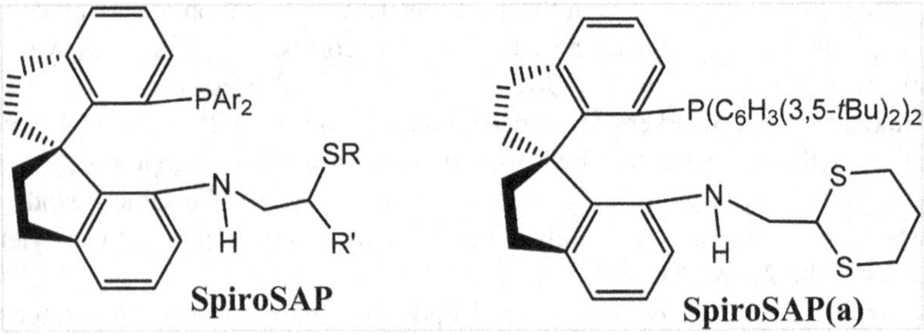

FIGURE 5.19   SpiroSAP ligands.

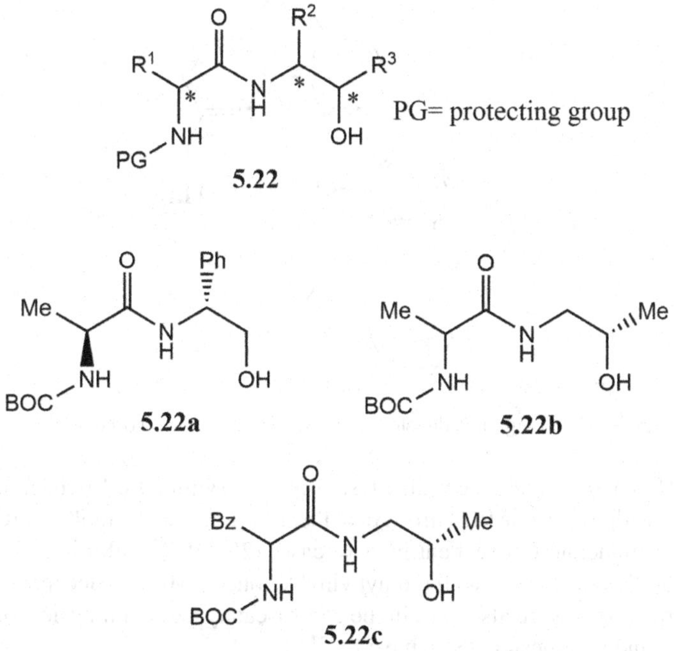

FIGURE 5.20 Pseudodipeptide NNO ligands, developed by Adolfsson and coworkers.

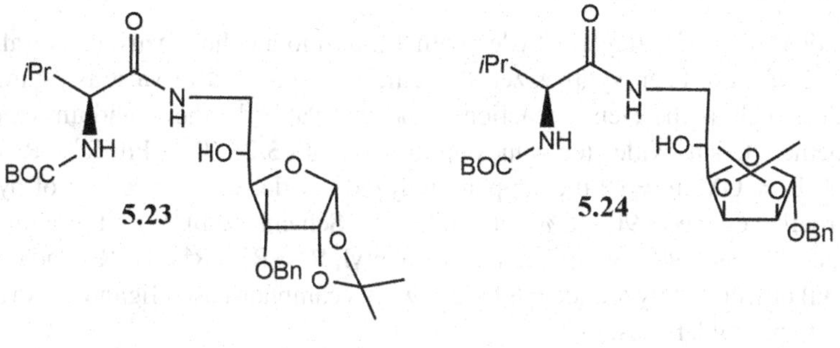

FIGURE 5.21 Carbohydrate-based pseudodipeptide NNO ligands, developed by Dieguez, Pamies, and Adolfsson.

transfer hydrogenation of ketones in isopropanol (ee > 99% for aryl alkyl ketones with **5.23**, 51–99% for aryl and heteroaryl alkyl ketones with **5.24**).

5.7.2 OTHER NNO LIGANDS

Nishiyama and coworkers developed chiral atropoisomeric ligands BinThro (for **Bin**ol-derived phenan**thro**line; Figure 5.22), bearing a phenanthroline and a phenol function and two stereogenic axes.[125] These ligands were used in the asymmetric addition of ZnEt$_2$ to aldehydes; yield = 32–94%, ee = 59–95%, when Ar = 3,5-(F$_3$C)$_2$C$_6$H$_3$

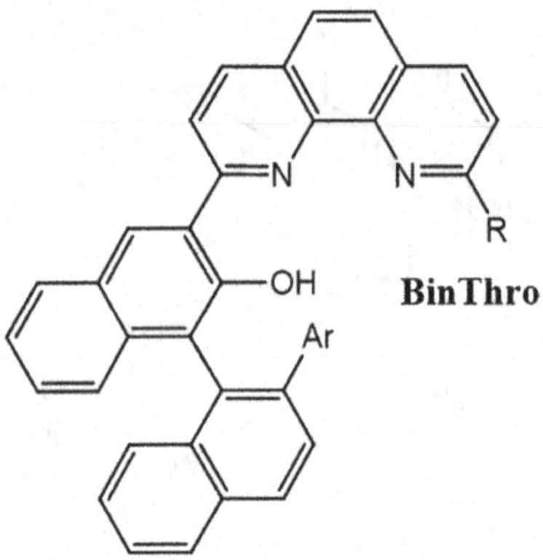

FIGURE 5.22 BinThro ligands, developed by Nishiyama and coworkers.

and R = H. These ligands were also tested in the asymmetric Michael addition to oxindoles; with 12 mol% BinThro (Ar = Ph, R = Ph) and 10 mol% Ni(OAc)$_2$, the yields were moderate to excellent in most cases (25–99%), with high enantioselectivities (65–87%), achieved with methyl vinyl ketone as Michael acceptor.[126]

BinThro ligands were also used in the copper-catalyzed asymmetric oxidations of oxindole[127] and β-ketoesters (Scheme 5.8).[128]

5.8 ONO LIGANDS

By condensation of salicylaldehydes with 1,2-amino alcohols, various chiral ONO ligands **5.25** have been synthesized (Scheme 5.9) and used in various asymmetric reactions, such as the Henry reaction (Scheme 5.9a),[129–131] the vanadium oxidation of thioethers to sulfoxides (ee values up to 94% with **5.25** (R^1 = Ph; R^2 = R^3 = R^4 = R^5 = H, R^6 = OMe)),[132–135] the copper-catalyzed Friedel-Crafts reaction of pyrroles on nitroalkenes (ee = 90–98% with **5.25** (see Scheme 5.9b),[136] or the aluminum-catalyzed (ee = 95–99% with **5.25** (R^1 = isobutyl; R^2 = R^3 = R^4 = H, R^5 = adamantyl, R^6 = tBu) or iron-catalyzed (ee = 83–95% with a camphor-based ligand)) hydrophosphonylation of aldehydes.[137, 138]

5.9 NCN LIGANDS

Designed as an anionic analog (after deprotonation) of Pybox, Phebox (3,5- dimethylphenyl-2,6-bis(oxazolinyl)) was originally developed independently by three different research teams (Scheme 5.10).[139–141] This family of ligands has been used successfully in various asymmetric reactions, such as alkene hydrosilylation, the aldol reaction, the reductive aldol reaction, or the β-boration of α,β-unsaturated esters.[142–144] Phebox continues to attract interest for the development of new catalytic systems, such as for ruthenium-catalyzed asymmetric three-component coupling reactions of alkyne, enone, and aldehyde (Scheme 5.10a),[145] or the rhodium-catalyzed asymmetric alkynylation of α-ketiminoesters (Scheme 5.10b).[146, 147]

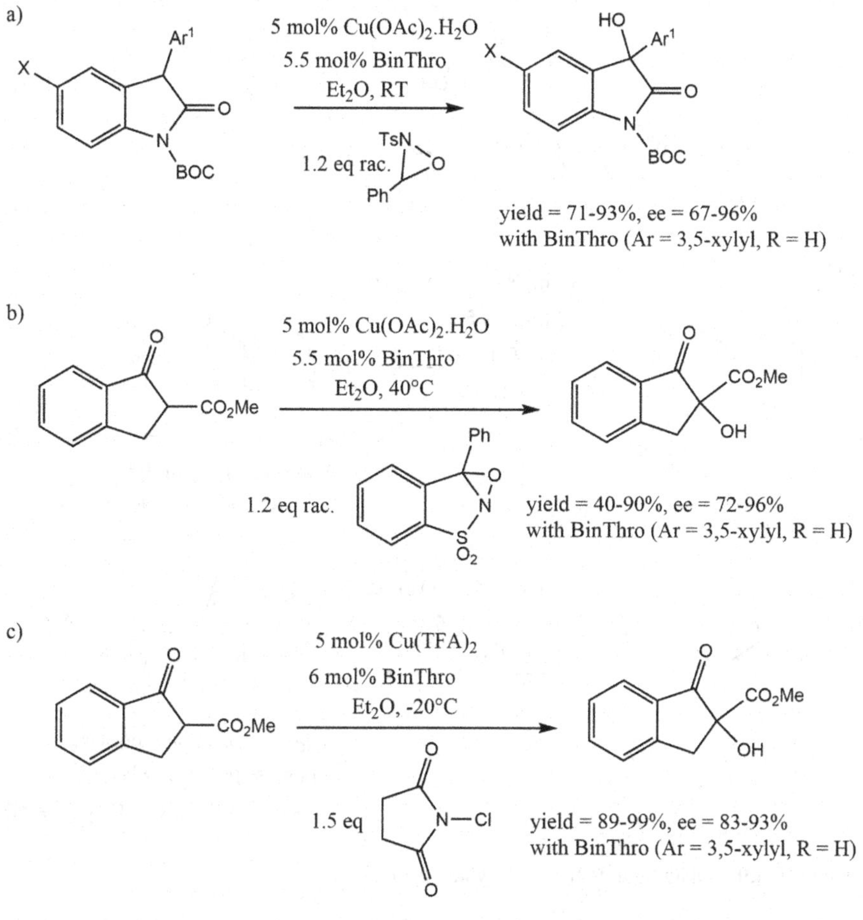

SCHEME 5.8 Oxidation reactions with BinThro ligands.

In 2006, Gong, Song, and coworkers described a new, highly modular family of analogs of the Phebox ligands called Phebim, where oxazoline rings have been replaced by imidazoline rings (Figure 5.23).[148] Phebim ligands were used in the platinum-catalyzed alkylation of indoles with nitroalkenes (yield = 61–99%, ee = 34–83% for the ligand with $R^1 = i$Pr, $R^2 = R^3 = R^4 = R^6 = H$, $R^5 = $Cy)[149, 150] in the palladium-catalyzed aza-Morita-Baylis-Hillman reaction of acrylonitrile with imines (yield = 75–98%, ee = 76–98% for the ligand with $R^1 = R^3 = $1-naphthyl, $R^2 = R^4 = R^6 = H$, $R^5 = $Ac);[151] in the rhodium-catalyzed allylation of aldehydes with allyltributyltin (yield = 78–99%, ee = 37–97% for the ligand with $R^1 = $Bn, $R^2 = R^3 = R^4 = R^6 = H$, $R^5 = p$-tolyl);[152] in the rhodium-catalyzed carbonyl-ene of ethyl trifluoropyruvates with 2-arylpropene (yield = 60–89%, ee = 65–94% for the ligand with $R^1 = t$Bu, $R^2 = R^3 = R^4 = R^6 = H$, $R^5 = p$-tolyl);[152] in the rhodium-catalyzed alkynylation of ethyl trifluoropyruvates (with aromatic terminal alkynes: yield = 80–99%, ee = 94–99%

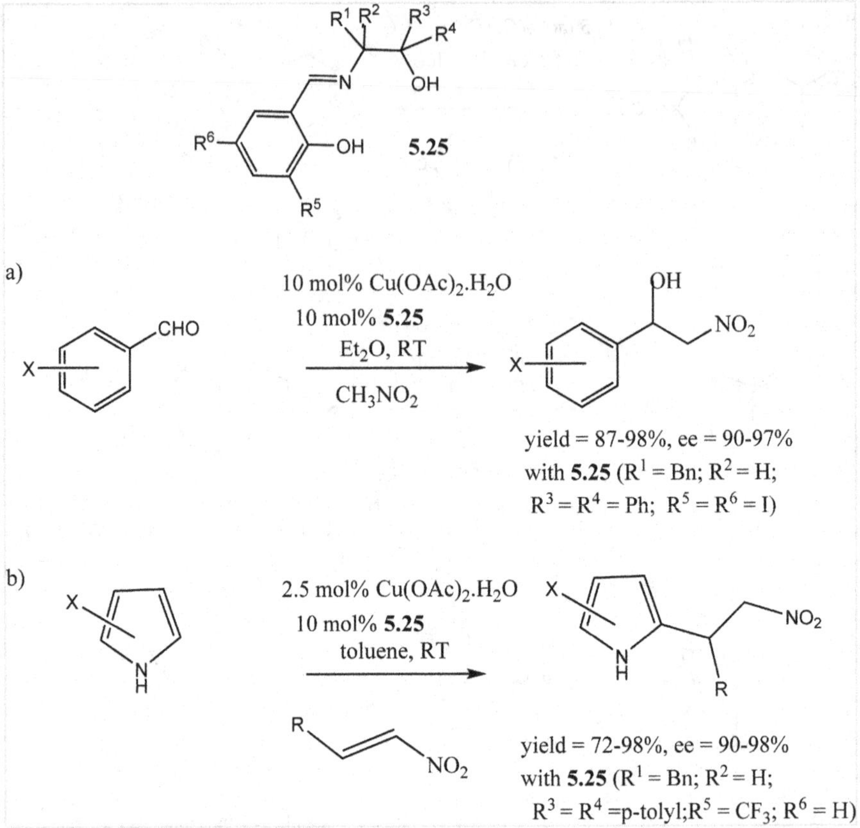

SCHEME 5.9   ONO ligands from salicylaldehyde.

for the ligand with $R^1 = R^3 = Ph$, $R^2 = R^4 = H$, $R^5 = p$-tolyl, $R^6 = NO_2$); with aliphatic terminal alkynes, yield = 68–86%, ee = 52–99% for the ligand with $R^1 = R^3 = Ph$, $R^2 = R^4 = H$, $R^5 = p$-tolyl, $R^6 = NO_2$)[153] in the palladium-catalyzed hydrophosphination of enones with diphenylphosphine (yield = 20–90%, ee = 79–94% for the ligand with $R^1 = R^3 = Ph$, $R^2 = R^4 = R^6 = H$, $R^5 = p$-tolyl)[154]; and, in the iridium-catalyzed C-H functionalization of indoles with α-aryl-α-diazoacetates (yield = 21–99%, ee = 6–86% for the ligand with $R^1 = R^3 = Ph$, $R^2 = R^4 = H$, $R^5 = p$-tolyl, $R^6 = tBu$).[155]

5.10   CONCLUSION

A very wide range of asymmetric catalytic systems for a very large panel of chemical reactions has already been developed, using numerous chiral tridentate ligands of various shapes and coordination patterns. Chiral tridentate ligands are now well-established tools for synthetic chemists and for catalyst developers and should continue to attract great interest in the following decades.

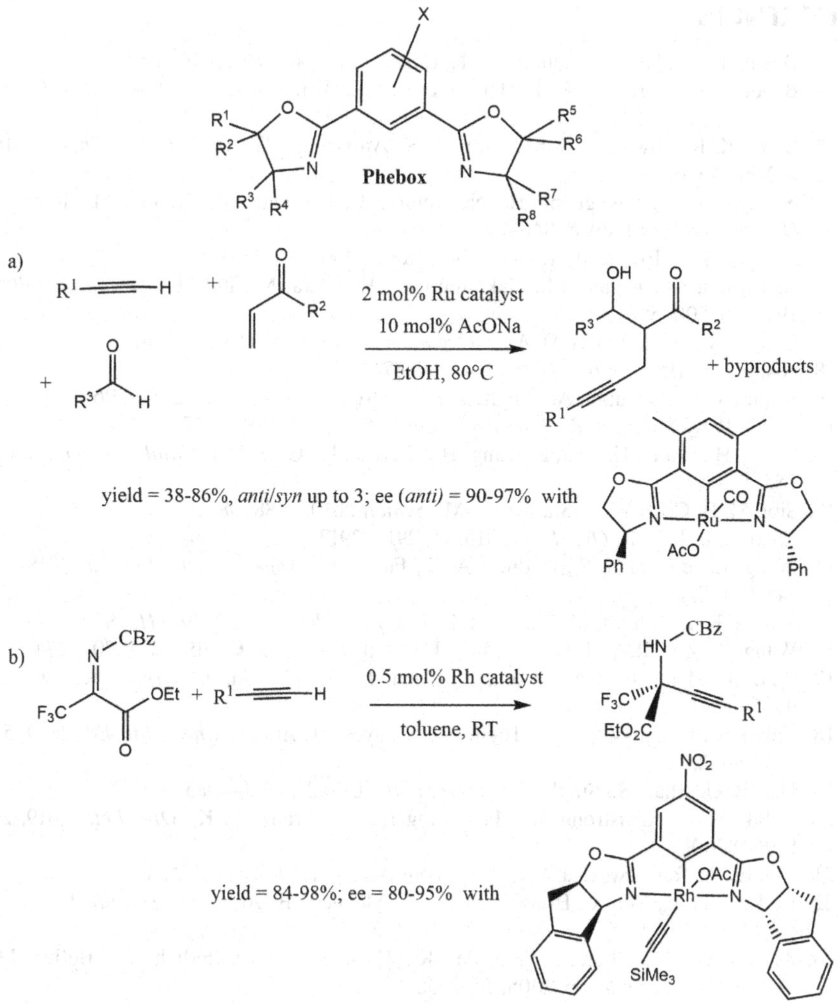

a)

yield = 38-86%, *anti/syn* up to 3; ee (*anti*) = 90-97% with

b)

yield = 84-98%; ee = 80-95% with

SCHEME 5.10 Phebox ligands.

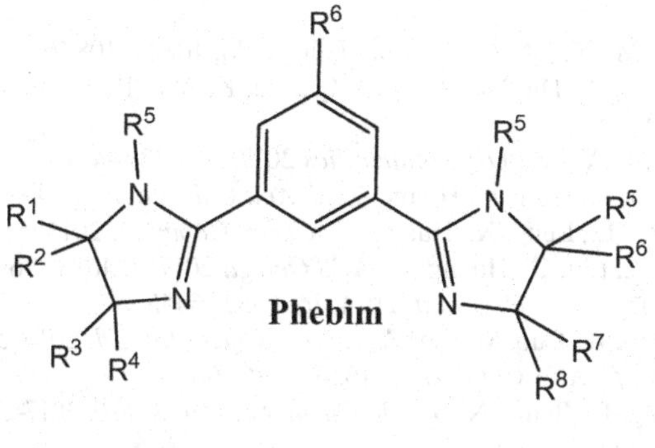

FIGURE 5.23 Phebim ligands.

REFERENCES

1. Desimoni, G., Faita, G., Quadrelli, P., *Chem. Rev.* 2003, *103*, 3119–3154.
2. Babu, S. A., Khrisnan, K. K., Ujwaldev, S. M., Anilkumar, G., *Asian J. Org. Chem.* 2018, *7*, 1033–1053.
3. Rohit, K. R., Ujwaldev, S. M., Saranya, S., Anilkumar, G., *Asian J. Org. Chem.* 2018, *7*, 2338–2356.
4. Nishiyama, H., Sakaguchi, H., Nakamura, T., Horihata, M., Kondo, M., Itoh, K., *Organometallics* 1989, *8*, 846–846.
5. Johnson, J. S., Evans, D. A., *Acc. Chem. Res.* 2000, *33*, 325–335.
6. Nishiyama, H., Kondo, M., Nakamura, T., Horihata, M., Itoh, K., *Organometallics* 1991, *10*, 500–508.
7. Uvarov, V. M., de Vekki, D. A., *J. Organomet. Chem.* 2020, *923*, 121415.
8. Nishiyama, H., *Enantiomer* 1999, *4*, 569–574.
9. Sekar, G., Datta Gupta, A., Singh, V. K., *J. Org. Chem.* 1998, *63*, 2961–2967.
10. Singh, P. K., Singh, V. K., *Pure Appl. Chem.* 2012, *84*, 1651–1657
11. Wang, H., Wang, H., Liu, P., Yang, H., Xiao, J., Li, C., *J. Mol. Catal. A Chem.* 2008, *285*, 128–131.
12. Sibi, M. P., Cruz, W. Jr, Stanley, L. M., *Synlett* 2010, *6*, 889–892.
13. Zhao, B, Loh, T.-P., *Org. Lett.* 2013, *15*, 2914–2917.
14. Wang, Z., Bachman, S., Dudnik, A. S., Fu, G. C., *Angew. Chem. Int. Ed.* 2018, *57*, 14529–14532.
15. Garcia, K. J., Gilbert, M. M., Weix, D. J., *J. Am. Chem. Soc.* 2019, *141*, 1822–1827.
16. Wang, C., Zhu, R.-Y., Liao, K., Zhou, F., Zhou, J., *Org. Lett.* 2019, *22*, 1270–1274.
17. Zhu, Q., Meng, B., Gu, C., Xu, Y., Chen, J., Lei, C., Wu, X., *Org. Lett.* 2019, *21*, 9985–9989.
18. Calvo, R., Comas-Vives, A., Togni, A., Katayev, D., *Angew. Chem. Int. Ed.* 2019, *58*, 1447–1452.
19. Das, B. G., Shah, S., Singh, V. K., *Org. Lett.* 2019, *21*, 4981–4985.
20. Cobo, A. A., Armstrong, B. M., Fettinger, J. C., Franz, A. K., *Org. Lett.* 2019, *21*, 8196–8200.
21. Liu, E.-C., Topczewski, J. J., *J. Am. Chem. Soc.* 2019, *141*, 5135–5138.
22. Li, L., Han, F., Nie, X., Hong, Y., Ivlev, S., Meggers, E., *Angew. Chem. Int. Ed.* 2020, *59*, 12392–12395.
23. Anilkumar, G., Bhor, S., Tse, M. K., Klawonn, M., Bitterlich, B., Beller, M., *Tetrahedron: Asymmetry* 2005, *16*, 3536–3561.
24. Junge, K., Möller, K., Wendt, B., Das, S., Gördes, D., Thurow, L., Beller, M., *Chem. Asian J.* 2012, *7*, 314–320.
25. Monge, D., Bermejo, A., Vasquez, J., Fernandez, R., Lassaletta, J. M., *Arkivoc* 2013, 33–45.
26. Ye, W., Zhao, M., Yu, Z., *Chem. Eur. J.* 2012, *18*, 10843–10846.
27. Ye, W. J., Zhao, M., Du, W. M., Jiang, Q., Wu, K., Wu, P., Yu, Z., *Chem. Eur. J.* 2011, *17*, 4737–4741.
28. Chai, H., Liu, T., Yu, Z., *Organometallics* 2017, *36*, 4136–4144.
29. Zhang, L., Zuo, Z., Wan, X., Huang, Z., *J. Am. Chem. Soc.* 2014, *136*, 15501–15504.
30. Zuo, Z., Zhang, L., Leng, X., Huang, Z., *Chem. Commun.* 2015, *51*, 5075–5076.
31. Ma, X., Zuo, Z., Liu, G., Huang, Z., *ACS Omega* 2017, *2*, 4688–4692.
32. Chen, J., Xi, T., Lu, Z., *Org. Lett.* 2014, *16*, 6452–6455.
33. Chen, J., Cheng, B., Cao, M., Lu, Z., *Angew. Chem. Int. Ed.* 2015, *54*, 4661–4664.
34. Zhang, H., Lu, Z., *ACS Catalysis* 2016, *6*, 6596–6600.
35. Jiang, Y., Jiang, J., Zhang, X., *J. Am. Chem. Soc.* 1998, *120*, 3817–3818.

36. Mc Manus, H. A., Guiry, P. J., *J. Org. Chem.* 2002, *67*, 8566–8573.
37. Lu, S.-F., Du, D.-M., Wang, S.-W., Xu, J., *Tetrahedron: Asymmetry* 2004, *15*, 3433–3441.
38. Du, D.-M., Lu, S.-F., Fang, T., Xu, J., *J. Org. Chem.* 2005, *70*, 3712–3715.
39. Nishiyama, H., Furuta, A., *Chem. Commun.* 2007, 760–762.
40. Inagaki, T., Phong, L. T., Furuta, A., Ito, J.-i, Nishiyama, H., *Chem. Eur. J.* 2010, *16*, 3090–3096.
41. Inagaki, T., Ito, A., Ito, J., Nishiyama, H., *Angew. Chem. Int. Ed.* 2010, *49*, 9384–9387.
42. Nixon, T. D., Ward, B. D., *Chem. Commun.* 2012, *48*, 11790–11792.
43. Bennett, S. D., Core, B. A., Blake, M. P., Pope, S. J. A., Mountford, P., Ward, B. D., *Dalton Trans.* 2014, *43*, 5871–5885.
44. Bauer, G., Cheung, C. W., Hu, X., *Synthesis* 2015, *47*, 1726–1732.
45. Li, C., Liu, F.-L., Zhou, Y.-Q., Chen, J.-R., Xiao, W.-J., *Synthesis* 2013, *45*, 601–608.
46. Chu, J. C. K., Dalton, D. M., Rovis, T., *J. Am. Chem. Soc.* 2015, *137*, 4445–4452.
47. Lei, G., Zhang, H., Chen, B., Xu, M., Zhang, G., *Chem. Sci.* 2020, *11*, 1623–1628.
48. Mo, X., Chen, B., Xu, M., Zhang, G., *Angew. Chem. Int. Ed.* 2020, *59*, 13998–14002.
49. Inoue, M., Suzuki, T., Nakada, M., *J. Am. Chem. Soc.* 2003, *125*, 1140–1141.
50. Inoue, M., Nakada, M., *J. Am. Chem. Soc.* 2007, *129*, 4164–4165.
51. Inoue, M., Nakada, M., *Org. Lett.* 2004, *6*, 2977–2980.
52. Deng, Q.-H., Wadepohl, H., Gade, L. H., *Chem. Eur. J.* 2011, *17*, 14922–14928.
53. Deng, Q.-H., Wadepohl, H., Gade, L. H., *J. Am. Chem. Soc.* 2012, *134*, 2946–2949.
54. Bleih, T., Deng, Q.-H., Wadepohl, H., Gade, L. H, *Angew. Chem. Int. Ed.* 2016, *55*, 7852–7856.
55. Qiu, J., Wu, D., Karmaker, P. G., Qi, G., Chen, P., Yin, H., Chen, F.-X., *Org. Lett.* 2017, *19*, 4018–4021.
56. Bleih, T., Wadepohl, H., Gade, L. H, *J. Am. Chem. Soc.* 2015, *137*, 2456–2459.
57. Bleih, T., Gade, L. H, *J. Am. Chem. Soc.* 2016, *138*, 4972–4983.
58. Blasius, C. K., Vasilenko, V., Gade, L. H, *Angew. Chem. Int. Ed.* 2018, *57*, 10231–10235.
59. Blasius, C. K., Heinrich, N. F., Vasilenko, V., Gade, L. H, *Angew. Chem. Int. Ed.* 2020, *59*, 15974–15977.
60. Vasilenko, V., Blasius, C. K., Wadepohl, H., Gade, L. H, *Angew. Chem. Int. Ed.* 2017, *56*, 8393–8397.
61. Vasilenko, V., Blasius, C. K., Gade, L. H, *J. Am. Chem. Soc.* 2018, *140*, 9244–9254.
62. Blasius, C. K., Wadepohl, H., Gade, L. H, *Eur. J. Inorg. Chem.* 2020, 2335–2342.
63. Clarke, M. L., Diaz-Valenzuela, M. B., Slawin, M. Z., *Organometallics* 2007, *26*, 16–19.
64. Diaz-Valenzuela, M. B., Phillips, S. D., France, M. B., Gunn, M. E., Clarke, M. L., *Chem. Eur. J.* 2009, *15*, 1227–1232.
65. Phillips, S. D., Andersson, K. H. O., Kann, N., Kuntz, M. T., France, M. B., Wawrzyniak, P., Clarke, M. L., *Catal. Sci. Technol.* 2011, *1*, 1336–1339.
66. Fuentes, J. A., Phillips, S. D., Clarke, M. L., *Chemistry Central J.* 2012, *6*, 151.
67. Fuentes, J. A., Carpenter, I., Kann, N., Clarke, M. L., *Chem. Commun.* 2013, *49*, 10245–10247.
68. Xie, J.-H., Liu, X.-Y., Xie, J.-B., Wang, L.-X., Zhou, Q.-L, *Angew. Chem. Int. Ed.* 2011, *50*, 7329–7332.
69. Yan, P.-C., Zhu, G.-L., Xie, J.-H., Zhang, X.-D., Zhou, Q.-L, Li, Y.-Q., Shen, W.-H., Che, D.-Q., *Org. Process Dev.* 2013, *17*, 307–312.
70. Zhu, G.-L., Zhang, X.-D., Yang, L.-J., Xie, J.-H., Che, D.-Q., Zhou, Q.-L, Yan, P.-C., Li, Y.-Q., *Org. Process Dev.* 2016, *20*, 81–85.
71. Xie, J.-H., Liu, X.-Y., Yang, X.-H., Xie, J.-B., Wang, L.-X., Zhou, Q.-L., *Angew. Chem. Int. Ed.* 2012, *51*, 201–203.

72. Yang, X.-Y., Xie, J.-H., Liu, W.-P, Zhou, Q.-L., *Angew. Chem. Int. Ed.* 2013, *52*, 7833–7836.

73. Yan, P.-C., Xie, J.-H., Zhang, X.-D., Chen, K., Li, Y.-Q., Zhou, Q.-L., Che, D.-Q., *Chem. Commun.* 2014, *50*, 15987–15990.

74. Yang, X.-Y., Yue, H.-T., Yu, N., Li, Y.-P., Xie, J.-H., Zhou, Q.-L., *Chem. Sci.* 2017, *8*, 1811–1814.

75. Liu, W.-P, Yuan, M.-L., Yang, X.-H., Li, K., Xie, J.-H., Zhou, Q.-L., *Chem. Commun.* 2015, *51*, 6123–6125.

76. Chen, G.-Q., Lin, B.-J., Huang, J.-M., Zhao, L.-Y., Chen, Q.-S., Jia, S.-P., Yin, Q., Zhang, X., *J. Am. Chem. Soc.* 2018, *140*, 8064–8068.

77. Nie, H., Zhou, G., Wang, Q., Chen, W., Zhang, S., *Tetrahedron: Asymmetry* 2013, *24*, 1567–1571.

78. Widegreen, M. H., Harkness, G. J., Slawin, A. M. Z., Cordes, D. B., Clarke, M. L., *Angew. Chem. Int. Ed.* 2017, *56*, 5825–5828.

79. Ling, F., Chen, J., Nian, S., Hou, H., Yi, X., Wu, F., Xu, M., Zhong, W., *Synlett* 2020, *31*, 285–289.

80. Hu, X., Dai, H., Bai, C., Chen, H., Zheng, Z., *Tetrahedron: Asymmetry* 2004, *1*, 1065–1068.

81. Zhang, C., Hu, X.-H., Wang, Y.-H., Zheng, Z., Xu, J., Hu, X.-P., *J. Am. Chem. Soc.* 2012, *134*, 9585–9588.

82. Li, L., Liu, Z.-T., Hu, X.-P., *Chem. Commun.* 2018, *54*, 12033–12036.

83. Zhu, F.-L., Wang, Y.-H., Zhang, D.-Y., Xu, J., Hu, X.-P., *Angew. Chem. Int. Ed.* 2014, *53*, 1410–1414.

84. Zhang, C., Wang, Y.-H., Hu, X.-H., Zheng, Z., Xu, J., Hu, X.-P., *Adv. Synth. Catal.* 2012, *354*, 2854–2858.

85. Wei, D.-Q., Liu, Z.-T., Wang, X.-M., Hou, C.-J., Hu, X.-P., *Tetrahedron Lett.* 2019, *60*, 151305.

86. Zhu, F.-L., Wang, Y.-H., Zhang, D.-Y., Xu, J., Hu, X.-P., *Angew. Chem. Int. Ed.* 2014, *53*, 10223–10227.

87. Han, F.-Z., Zhu, F.-L., Wang, Y.-H., Zou, Y., Hu, X.-H., Chen, S., Hu, X.-P., *Org. Lett.* 2014, *16*, 588–591.

88. Wang, B., Liu, C., Guo, H., *RSC Adv.* 2014, *4*, 53216–53219.

89. Zhang, C., Hui, Y.-Z., Zhang, D.-Y., Hu, X.-P., *RSC Adv.* 2016, *6*, 14763–14767.

90. Xia, J.-T., Hu, X.-P., *Org. Lett.* 2020, *22*, 1102–1107.

91. Xu, H., Laraia, L., Schneider, L., Louven, K., Strohmann, C., Antonchick, A. P., Waldmann, H., *Angew. Chem. Int. Ed.* 2017, *56*, 11232–11236.

92. Shao, L., Hu, X.-P., *Org. Biomol. Chem.* 2017, *15*, 9837–9844.

93. Shao, L., Wang, Y.-H., Zhang, D.-Y., Xu, J., Hu, X.-P., *Angew. Chem. Int. Ed.* 2016, *55*, 5014–5018.

94. Shao, L., Hu, X.-P., *Chem. Commun.* 2017, *53*, 8192–8195.

95. Ling, F., Hou, H., Chen, J., Nian, S., Yi, X., Wang, Z., Song, D., Zhong, W., *Org. Lett.* 2019, *21*, 3937–3941.

96. Wu, W., Liu, S., Duan, M., Tan, X., Chen, C, Xie, Y., Lan, Y., Dong, X.-Q., Zhang, X., *Org. Lett.* 2016, *18*, 2938–2941.

97. Wu, W., Xie, Y., Li, P., Li, X., Liu, Y., Dong, X.-Q., Zhang, X., *Org. Chem. Front.* 2017, *4*, 555–559.

98. Hu, Y., Wu, W., Dong, X.-Q., Zhang, X., *Org. Chem. Front.* 2017, *4*, 1499–1502.

99. Qin, C., Chen, X.-S., Hou, C.-J., Liu, H., Liu, Y.-J., Huang, D.-Z., Hu, X.-P., *Synth. Commun.* 2018, *48*, 672–676.

100. Jiang, Y., Jiang, Q., Zhu, G., Zhang, X., *Tetrahedron Lett.* 1997, *38*, 215–218.

101. Ghorai, S., Chirke, S. S., Xu, W.-B., Chen, J.-F., Li, C., *J. Am. Chem. Soc.* 2019, *141*, 11430–11434.
102. Ghorai, S., Chirke, S. S., Xu, W.-B., Chen, J.-F., Li, C., *Org. Lett.* 2020, *22*, 3519–3523.
103. Chen, J.-F., Li, C., *Org. Lett.* 2020, *22*, 4686–4691.
104. Lagiditis, P. O., Sues, P. E., Sonnenberg, J. F., Wan, K. Y., Lough, A. J., Morris, R. H., *J. Am. Chem. Soc.* 2014, *136*, 1367–1380.
105. Sonnenberg, J. F., Lough, A. J., Morris, R. H., *Organometallics*. 2014, *33*, 6452–6465.
106. Sonnenberg, J. F., Wan, K. Y., Sues, P. E., Morris, R. H., *ACS Catal.* 2017, *7*, 316–326.
107. Smith, S. A. M., Lagiditis, P. O., Lüpke, A., Lough, A. J., Morris, R. H., *Chem. Eur. J.* 2017, *23*, 7212–7216.
108. Seo, C. S. G., Tannoux, T., Smith, S. A. M., Lough, A. J., Morris, R. H., *J. Org. Chem.* 2019, *84*, 12040–12049.
109. Zirakzadeh, A., Kirchner, K., Roller, A., Stöger, B., Widhalm, M., Morris, R. H., *Organometallics* 2016, *35*, 3781–3787.
110. Zirakzadeh, A., de Aguiar, S. R. M. M., Stöger, B., Widhalm, M., Kirchner, K., *ChemCatChem* 2017, *9*, 1744–1748.
111. Garbe, M., Junge, K., Walker, S., Wei, Z., Jiao, H., Spannenberg, A., Bachmann, S., Scalone, M., Beller, M., *Angew. Chem. Int. Ed.* 2017, *56*, 11237–11241.
112. Garbe, M., Wei, Z., Tannert, B., Spannenberg, A., Jiao, H., Bachmann, S., Scalone, M., Junge, K., Beller, M., *Adv. Synth. Catal.* 2019, *361*, 1913–1920.
113. Bao, D.-H., Wu, H.-L., Liu, C.-L., Xie, J.-H., Zhou, Q.-L., *Angew. Chem. Int. Ed.* 2015, *54*, 8791–8794.
114. Bao, D.-H., Gu, X.-S., Xie, J.-H., Zhou, Q.-L., *Org. Lett.* 2017, *19*, 118–121.
115. Pastor, I. M., Västilä, P, Adolfsson, H., *Chem. Commun.* 2002, 2046–2047.
116. Pastor, I. M., Västilä, P, Adolfsson, H., *Chem. Eur. J.* 2003, *9*, 4031–4045.
117. Bogevig, A., Pastor, I. M., Adolfsson, H., *Chem. Eur. J.* 2004, *10*, 294–302.
118. Västilä, P., Wettergren, J., Adolfsson, H., *Chem. Commun.* 2005, 4039–4041.
119. Västilä, P., Zaytsev, A. B., Wettergren, J., Privalov, T., Adolfsson, H., *Chem. Eur. J.* 2006, *12*, 3218–3225.
120. Wettergren, J., Buitrago, E., Ryberg, P., Adolfsson, H., *Chem. Eur. J.* 2009, *15*, 5709–5718.
121. Wettergren, J., Zaytsev, A. B., Adolfsson, H., *Adv. Synth. Catal.* 2007, *349*, 2556–2562.
122. Coll, M., Pàmies, O., Adolfsson, H., Dieguez, M., *Chem. Commun.* 2011, *47*, 12188–12190.
123. Coll, M., Ahlford, K., Pàmies, O., Adolfsson, H., Dieguez, M., *Adv. Synth. Catal.* 2012, *354*, 415–427.
124. Margalef, J., Slagbrand, T., Tinnis, F., Adolfsson, H., Dieguez, M., Pàmies, O., *Adv. Synth. Catal.* 2016, *358*, 4006–4018.
125. Naganawa, Y., Namba, T., Aoyama, T., Shoji, K., Nishiyama, H., *Chem. Commun.* 2014, *50*, 13224–13227.
126. Naganawa, Y., Abe, H., Nishiyama, H., *Synlett* 2016, *27*, 1973–1978.
127. Naganawa, Y., Aoyama, T., Nishiyama, H., *Org. Biomol. Chem.* 2015, *13*, 11499–11406.
128. Naganawa, Y., Aoyama, T., Kato, K., Nishiyama, H., *ChemistrySelect* 2016, *1*, 1938–1942.
129. Korkmaz, N., Astley, D., Astley, S. T, *Turk. J. Chem.* 2011, *35*, 361–374.
130. Qiang, G. R., Shen, T. H., Zhou, X.-C., An, X.-X., Song, Q.-B., *Chirality* 2014, *26*, 780–783.
131. Song, Q., An, X., Xia, T., Zhou, X., Shen, T., *C. R. Chimie* 2015, *18*, 215–222.
132. Romanowski, G., Kira, J., *Polyhedron* 2013, *50*, 172–178.
133. Romanowski, G., Lis, T., *Inorganica Chimica Acta* 2013, *394*, 627–634.

134. Romanowski, G., Wera, M., *Polyhedron* 2013, *50*, 179–186.
135. Romanowski, G., Kira, J., Wera, M., *J. Mol. Cat. A: Chem.* 2014, *381*, 148–160.
136. Guo, F., Chang, D., Lai, G., Zhu, T., Xiong, S., Wang, S., Wang, Z., *Chem. Eur. J.* 2011, *17*, 11127–11130.
137. Wang, C., Xu, C., Tan, X., Peng, H., He, H. *Org. Biomol. Chem.* 2012, *10*, 1680–1685.
138. Boobalan, R., Chen, C., *Adv. Synth. Catal.* 2013, *355*, 3443–3450.
139. Motoyama, Y., Makihara, N., Mikami, Y., Aoki, K., Nishiyama, H., *Chem. Lett.* 1997, *26*, 951–952.
140. Denmark, S. E., Stavenger, R. A., Faucher, A.-M., Edwards, J. P., *J. Org. Chem.* 1997, *62*, 3375–3389.
141. Stark, M. A., Richards, C. J., *Tet. Lett.* 1997, *38*, 5881–5884.
142. Nishiyama, H., *Chem. Soc. Rev.* 2007, *36*, 1333–1141.
143. Ito, J.-i., Nishiyama, H., *Top. Organomet. Chem.* 2011, *37*, 185–205.
144. Ito, J.-i., Nishiyama, H., *Top. Organomet. Chem.* 2013, *40*, 243–270.
145. Ubukata, S., Ito, J.-i., Oguri, R., Nishiyama, H., *J. Org. Chem.* 2016, *81*, 3347–3355.
146. Morisaki, K., Sawa, M., Nomaguchi, J.-y, Morimoto, H., Takeuchi, Y., Mashima, K., Ohshima, T., *Chem. Eur. J.* 2013, *19*, 8417–8420.
147. Morisaki, K., Sawa, M., Yonesaki, R., Morimoto, H., Mashima, K., Ohshima, T., *J. Am. Chem. Soc.* 2016, *138*, 6194–6203.
148. Hao, X.-Q., Gong, J.-F., Du, C.-X, Wu, L.-Y., Wu, Y.-J., Song, M. P., *Tetrahedron Lett.* 2006, *47*, 5033–5036.
149. Wu, L.-Y., Hao, X.-Q., Xu, Y.-X., Jia, M.-Q., Wang, Y.-N., Gong, J.-F., Song, M. P., *Organometallics* 2009, *28*, 3369–3380.
150. Hao, X.-Q., Xu, Y.-X., Yang, M.-J., Wang, L., Niu, J.-L., Gong, J.-F., Song, M. P., *Organometallics* 2012, *31*, 835–846.
151. Hyodo, K, Nakamura, S., Shibata, N., *Angew. Chem. Int. Ed.* 2012, *51*, 10337–10341.
152. Wang, T., Hao, X.-Q., Huang, J;-J., Niu, J.-L., Gong, J.-F., Song, M. P., *J. Org. Chem.* 2013, *78*, 8712–8721.
153. Wang, T., Niu, J.-L, Liu, S.-L., Huang, J.-J., Gong, J.-F., Song, M. P., *Adv. Synth. Catal.* 2013, *355*, 927–937.
154. Hao, X.-Q., Zhao, Y.-W., Yang, J.-J., Niu, J.-L., Gong, J.-F., Song, M. P., *Organometallics* 2014, *33*, 1801–1811.
155. Li, N., Zhu, W.-J., Huang, J. J., Hao, X.-Q., Gong, J.-F., Song, M. P., *Organometallics* 2020, *39*, 2222–2234.

Chiral *N*-Heterocyclic
Carbene-Based Ligands

Vincent César, Christophe Fliedel,
and Agnès Labande

CONTENTS

6.1   INTRODUCTION

N-heterocyclic carbene (NHC) ligands have become key ligands in organometallic chemistry and catalysis since the isolation of the first stable free NHC by Arduengo in 1991,[1] and the pioneering work of Herrmann in homogeneous catalysis as early as 1995.[2,3] The implementation of NHCs as outstanding ancillary ligands in organometallic catalysis is often attributed to their very strong σ-electron-donating properties, supplemented by excellent steric protection, which allow for very strong metal–carbene bonds, bringing air and moisture stability to the complexes and preventing decomposition of the catalytic species, without the need for an excess of the ligand.[4] The great diversity and variability of the ligand structures, as well as the ease of preparation of the NHC complexes and their – mostly air-stable – salt precursors are additional beneficial features which help chemists in successful development of this research field.[5] Chemists rapidly started to investigate chiral versions, as exemplified by the first reports, in 1996, by Herrmann[6] and Enders.[7] However, due to their peculiar fan-like geometry, the development of efficient families of chiral NHCs for catalysis required the consideration of new ligand design. To this end, several synthetic methodologies have been proposed and optimized over the past 20 years. The first strategy aimed at designing bulky monodentate chiral NHCs, that could efficiently transfer chiral information; the second relied on the introduction of a second coordinating unit, often a heteroatom, to stabilize the complex and to prevent rotation around the metal–carbon (M–C) bond and thus to prevent potential loss of chiral information. An additional coordination unit also has the advantage of bringing a different stereo-electronic environment near the metal center.

Whereas many early reports dealt with polyfunctional chiral NHCs, the development of new, more efficient families of monodentate chiral NHCs has become more prominent in the past decade. This chapter aims to achieve an overview of the most successful chiral NHC families developed since 2011 and their applications in catalysis. Before 2011, we encourage the reader to refer to the numerous excellent reviews published in the area.[8,9] In the past decade, the structures of chiral NHC ligands increased in diversity, and new families of very efficient ligands emerged. On the other hand, chiral NHC ligands that had already proven their usefulness were used to remove bottlenecks in homogeneous catalysis. Both directions will be explored and illustrated in this chapter. New trends and challenges in ligand design will also be discussed.

6.2   MONODENTATE CHIRAL NHC LIGANDS

Being strongly σ–electron-donating, and forming strong, non-labile metal–carbene bonds, monodentate NHC ligands have found great significance in the synthesis of stable NHC complexes, as well as in outstanding organometallic catalytic systems, with the Ru-based metathesis catalysts and the Pd-based cross-coupling catalysts being the most famous examples. Moreover, unlike the cone angle of tertiary phosphines pointing away from the metal center, the exocyclic N-substituents of the NHCs are directed toward the inner metal's coordination sphere. These beneficial features have motivated intense research efforts toward the design and development of monodentate chiral NHC ligands and have led to the development of several classes of efficient stereo-directing monodentate NHC ligands. A notable drawback to overcome, caused by the one-point binding nature of monodentate NHCs, arises from the σ-bond character of the metal-carbon bond and the subsequent possible rotation around it. This may lead to an erosion of enantio-induction, since the active chiral environment at the metal center may be relatively ill-defined. Several efficient strategies have appeared over the past 15 years and are described herein, with an emphasis on the developments made since 2011.

6.2.1 MONODENTATE NHC LIGANDS WITH CHIRAL EXOCYCLIC NITROGEN SUBSTITUENTS

6.2.1.1 With an Asymmetric Carbon Atom in α-Position to the N-Atoms

This was the strategy employed by Enders and Herrmann in 1996 for the construction of the first chiral NHC ligands, as they can readily be prepared from enantiopure chiral primary amines.[6,7] The low enantioselectivities obtained at that time were ascribed to the possible rotation of the chiral *N*-substituents around the C–N bonds, again leading to the presence of multiple rotamers and an ill-defined chiral active environment.

A significant improvement was achieved by Kündig and co-workers in 2007 with the development of the very bulky, C_2–symmetric chiral NHCs **1** (Scheme 6.1). The key modification is here the installation of the *tert*-butyl group as a substituent of the stereogenic carbon center in place of a methyl group in Herrmann's previous version.[10] The potential of this ligand class was proved in the Pd-catalysed intramolecular α-arylation of amides, with excellent enantioselectivities (up to 96% enantiomeric excess [ee]) to form oxindoles (Scheme 6.1, Equation 1).[10,11] This ligand class was then used in the Pd-catalyzed intramolecular arylation of unactivated C(sp³)–H bonds leading to fused tricyclic indolines (Scheme 6.1, Equation 2).[12] The catalytic system, formed *in situ* by mixing the imidazolium precursors **1**·HI with [Pd-Cl(η³-cinnamyl)]₂, was found to be remarkably robust and gave indolines with very high enantioselectivities (up to 99% ee). Later, the same group reported that the racemic mixture of chiral *N*-aryl, *N*-branched alkyl carbamates reacts with a similar catalytic system in a regiodivergent reaction to yield enantioenriched indolines, resulting from either a methyl C–H activation or an asymmetric methylene C–H activation, the two products being highly enantioenriched (up to 99% ee) (Scheme 6.1, Eq. 3).[13]

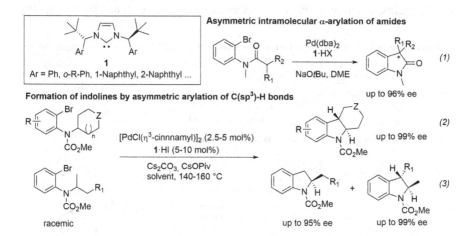

SCHEME 6.1 C_2-symmetric, bulky NHC with (*tert*-butyl)arylmethyl as *N*-substituents and their use as stereo-directing ligands in Pd-catalyzed cyclizations.

Combined theoretical and experimental studies showed that the Pd–NHC catalyzed C(sp³)–H arylation proceeds through a concerted metalation-deprotonation (CMD) mechanism and that the CMD step determines selectivity, giving a rationalization to the observed results.

More recently, Glorius and Toste [14,15] independently reported the successful use of chiral, C_2-symmetric imidazolinylidene (R,R)-SINpEt in highly efficient asymmetric transformations, by means of a precise definition of the chiral active species (Scheme 6.2). In 2017, Toste and co-workers reported the first highly enantioselective transformation catalyzed by a well-defined cationic gold(III) catalyst.[14] The latter is generated by abstraction of the chloride ligand from the stable pre-catalyst **2**, supported by the chiral NHC (R,R)-SINpEt and a chelating biphenylene moiety, the main role of which is to complete the square-planar coordination sphere and to prevent the undesired reduction of gold(III) into gold(I) species during catalysis. They developed the direct enantio-convergent cyclo-isomerization of 1,5-enynes into highly enantioenriched bicyclo[3.1.0]hexenes (Scheme 6.2, Eq. 1), and showed that a kinetic resolution of the starting 1,5-enynes takes place during the course of the reaction, with a selectivity or "s-factor" as high as 48 (the s-factor characterizes the relative reactivity rates between the chiral catalyst and the two enantiomers of the substrate). By mixing the ruthenium precursor [Ru(2-methylallyl)₂(COD)] with the imidazolinium precursor (R,R)-SINpEt·HBF₄ and an enantiopure (R,R)-diarylethylenediamine, Glorius and co-workers successfully isolated a series of air- and moisture-stable Ru-NHC-diamine complexes **3**, featuring an unusual coordination sphere.[15] Indeed, a cyclometalation occurs on the diamine during the synthesis, which, in turn, becomes a tridentate, facial L₂X-type ligand.[15] Moreover, to complete the unsaturated coordination sphere, the NHC ligand acts as a chelate ligand with an additional η^2-coordination of the naphthyl ring to ruthenium. Complexes **3** are versatile pre-catalysts for the enantioselective hydrogenation of isocoumarins, benzothiophene 1,1-dioxides (Scheme 6.2, Eq. 2) and ketones, giving

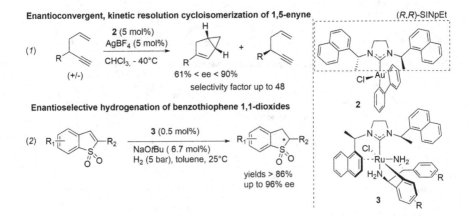

SCHEME 6.2 C_2-symmetric, chiral NHC ligand (R,R)-SINpEt in highly efficient asymmetric catalysis.

high levels of stereo-induction. Based on the identification of reaction intermediates, NMR studies and density-functional theory (DFT) calculations, the proposed catalytic cycle is meant to proceed *via* an outer-sphere bifunctional catalysis mechanism, similar to Noyori's system.[16] It is noteworthy that, in both cases, by judicious catalyst design, the active site is located in the *cis* position, relative to the NHC ligand, and the approach of the incoming substrates is directed orthogonally to the carbenic plane.

6.2.1.2 With C_2-Symmetric, Chiral Ortho-Disubstituted Aryl Groups

In 2011, Gawley and co-workers reported the C_2-symmetrical, chiral NHC ligand (*R,R,R,R*)-IPE, which constituted a real breakthrough in the field of chiral monodentate NHCs (Scheme 6.3).[17] Indeed, it can be considered to be the direct, effective chiral version of the "gold-standard" carbene IPr [IPr = 1,3-bis(2,6-diisopropylphenyl) -2H-imidazol-2-ylidene], which leads to the most effective catalytic systems in most of the NHC-catalyzed reactions. The introduction of chirality was done through the formal replacement of the *ortho*-isopropyl substituents by chiral, enantiopure (*R*)-1-phenylethyl substituents. The potential of this new family of chiral ligands was confirmed in the enantioselective copper-catalyzed hydrosilylation of aryl–alkyl ketones and even dialkylketones, with excellent enantioselectivities (up to 98% ee).[17] This class of bulky chiral IPE-type NHCs experienced a second and recent boom with the groups of Cramer and Shi, which independently developed structural analogs of IPE and successfully implemented them in challenging asymmetric, metal-catalyzed reactions.[18,19] This stereo-electronic tuning was carried out by varying the substituents at different locations on the NHCs. The substitution of the imidazolyl backbone, by either methyl or chloro groups, to give IPE^X (X = Cl or Me), or by annulation with an acenaphthene moiety, was shown to improve selectivity in most of the studied

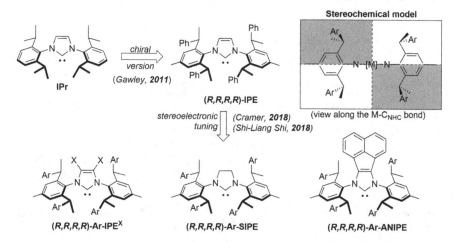

SCHEME 6.3 IPr ligand evolution into the C_2-symmetric, chiral, and bulky IPE versions and its subsequent stereo-electronic optimization. The stereochemical model is at the top right (gray quadrants are buried by ligand substituents).

catalytic reactions, as they slightly push the flanking groups toward the metal center ("buttressing effect" in NHC catalysis[20]). The same effect was observed by the formal saturation of the C=C double bond, leading to the SIPE-type ligand. Moreover, as two of the phenyl groups on the side arms in IPE are directed toward the coordination sphere, replacement of the phenyl groups by larger aryl groups, such as 3,5-xylyl or 3,5-di(*tert*-butyl) phenyl groups, turned out to be highly effective for increasing the enantioselectivity. The principle for the design of this class of ligands is based on a stereochemical model in which the C_2-symmetric NHC ligand occupies two diametrically opposed quadrants and leaves the two others accessible for the incoming substrate.

The C_2-symmetric, chiral anilines **5** are key for the synthesis of the imidazol(in)ium precursors of this class of ligands (Scheme 6.4). They are obtained in a two-step sequence. In the first step, the 2,6-divinyltoluidine **4** is obtained, either by Friedel–Craft condensation between phenylacetylene and *p*-toluidine, or by a Pd–NHC catalyzed Suzuki–Miyaura cross coupling between 2,6-dibromotoluidine and a vinylboronate ester (obtained from arylacetylene by nickel-catalyzed hydroalumination and Al/B *trans*-metalation). The latter option allows variation of the aryl group, leading to a high degree of modularity of the strategy. The enantiopure C_2-symmetric aniline (*R*,*R*)-**5** is obtained by enantioselective hydrogenation of the two alkene moieties in **4**, using the catalytic system [Rh(nbd)$_2$](BF$_4$)/(*1R,1'R,2S,2'S*)-DuanPhos, with an optical purity of more than 99% ee and in multi-gram quantities.[21] The other enantiomer, (*S*,*S*)-**5**, can also be obtained by using the opposite enantiomer of the DuanPhos ligand. From the enantiopure (*R*,*R*)-**5** aniline, the salt precursors are synthesized through classical condensation/cyclization methods.[22]

Over the past three years, this class of bulky chiral NHCs has gained considerable significance as being highly efficient, stereo-directing ligands in metal-catalyzed asymmetric transformations (Scheme 6.5). The flexible electronic and

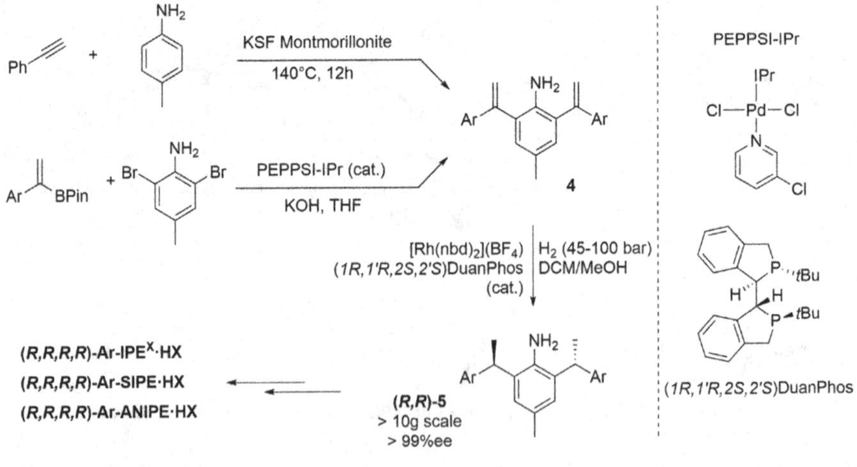

SCHEME 6.4 Synthetic access toward the imidazol(in)ium precursors of NHC ligands **Ar-IPE^X**, **Ar-SIPE**, and **Ar-ANIPE**.

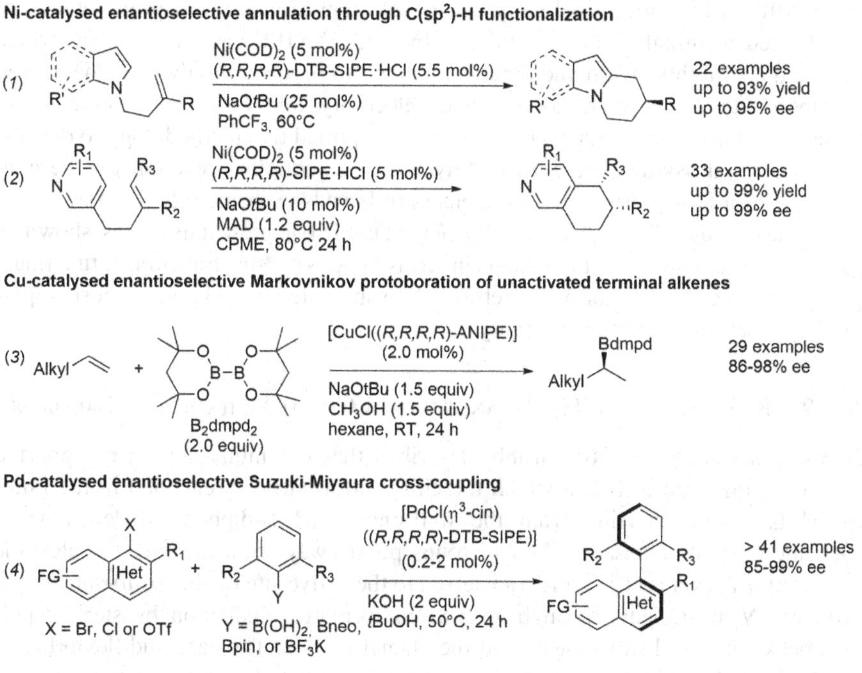

Ni-catalysed enantioselective annulation through C(sp²)-H functionalization

Cu-catalysed enantioselective Markovnikov protoboration of unactivated terminal alkenes

Pd-catalysed enantioselective Suzuki-Miyaura cross-coupling

SCHEME 6.5 Examples of enantioselective transformations catalyzed by **IPE**, **SIPE**, and **ANIPE**-based systems. DTB = 3,5-di-(*tert*-butyl)phenyl, MAD = methylaluminum bis(2,6-di(*tert*-butyl)-4-methylphenolate).

steric properties of these chiral NHC ligands appeared critical and beneficial to the development of highly enantioselective nickel-catalyzed annulation transformations through C(sp²)–H functionalization. The latter include the C–H functionalization of 2- and 4-pyridinones,[18] and the C–H cyclization of pyridines with olefins (Scheme 6.5, Eq. 2),[23] catalyzed by chiral Ni–NHC catalysts in the presence of the bulky Lewis acid MAD (methylaluminum bis(2,6-di(*tert*-butyl)-4-methylphenolate)), which promotes the C–H activation through coordination to the heterocyclic substrate. Similar monoligated Ni–NHC chiral systems were reported to catalyze the undirected C–H functionalization of indoles and pyrroles, providing access to valuable tetrahydropyridoindoles and tetrahydroindolizines in high yields and with high enantioselectivity under mild reaction conditions (Scheme 6.5, Eq. 1).[24] Here again, the peculiar stereo-electronic properties of this class of ligands were key to favoring the endo-cyclization, yielding the six-membered ring products, with the best results being obtained using the saturated DTB-SIPE. Analogously, an unprecedented enantioselective C–H alkylation of polyfluoroarenes with alkenes was reported using ANIPE- or SIPE-derived catalysts, enabling exclusive activation of C–H bonds over C–F bonds, with excellent enantioselectivity.[25] Finally, the enantioselective redox-neutral coupling of aldehydes and alkynes to produce chiral allylic alcohols was reported, using the Ni-SIPE catalyst.[26]

In 2018, Shi reported the highly enantioselective Markovnikov protoboration of unactivated terminal alkenes, catalyzed by a Cu-ANIPE catalytic system, which provided a straightforward and practical method to produce highly enantioenriched alkylboronate esters with good enantioselectivity and excellent enantiocontrol (Scheme 6.5, Eq. 3).[19] In 2019, the same group applied this methodology to develop a Pd-catalyzed asymmetric Suzuki–Miyaura cross-coupling reaction for the construction of a large variety of atropisomers of biaryls (Scheme 6.5, Eq. 4).[27] After optimization, the [PdCl(η^3-cin)((R,R,R,R)-DTB-SIPE)] pre-catalyst was shown to surpass all previous catalytic systems in terms of activity, substrate range, functional group tolerance, applicability to tetra-*ortho*-substituted biaryls, and, most importantly, induction of enantioselectivity.

6.2.2 MONODENTATE NHC LIGANDS WITH A CHIRAL HETEROCYCLIC BACKBONE

In a seminal study in 2001, Grubbs described the ruthenium catalyst **6** supported by a C_2-symmetric NHC, in which the chiral information is encoded in the imidazolyl backbone, starting from the well-known (R,R)-diphenylethylenediamine (Scheme 6.6). Whereas the phenyl groups point away from the metal center, the chiral information is efficiently transferred to the active site by the *ortho*-substituted aromatic *N*-substituents through an *anti*-conformation induction by steric repulsion between *o*-aryl substituents and the phenyl groups. The ease and flexibility of synthesis of the imidazolinium salt precursors, associated with the usually good to excellent chiral transfer in asymmetric catalysis, contributed to the popularity of this approach and to its evolution into what is arguably the most prevalent class of chiral NHCs in the decade 2000–2009. This chiral scaffold strategy matured and found numerous applications in various transition-metal-catalyzed enantioselective transformations before 2011, and the reader is invited to refer to the excellent pre-2011 reviews on chiral NHCs for further information.[8] Some more relevant advances in this field are described hereafter.

In a logical continuation of their research in *N*-(naphthalen-1-yl)-substituted NHCs, Dorta, and co-workers reported the C_2-symmetric chiral ligand **7**, in which the *N*-aryl groups were replaced by 2-cycloalkyl-naphthalen-1-yl groups, the main consequence of which is a greater dissymmetry and good chiral shaping around the metal center.[28] The [PdCl(η^3-cin)(**7**)] complex was a highly effective pre-catalyst in

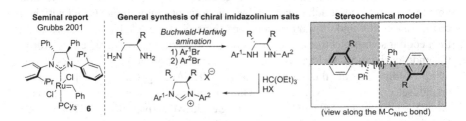

SCHEME 6.6 Overview of the NHCs bearing a chiral backbone, their synthesis, and the stereochemical model.

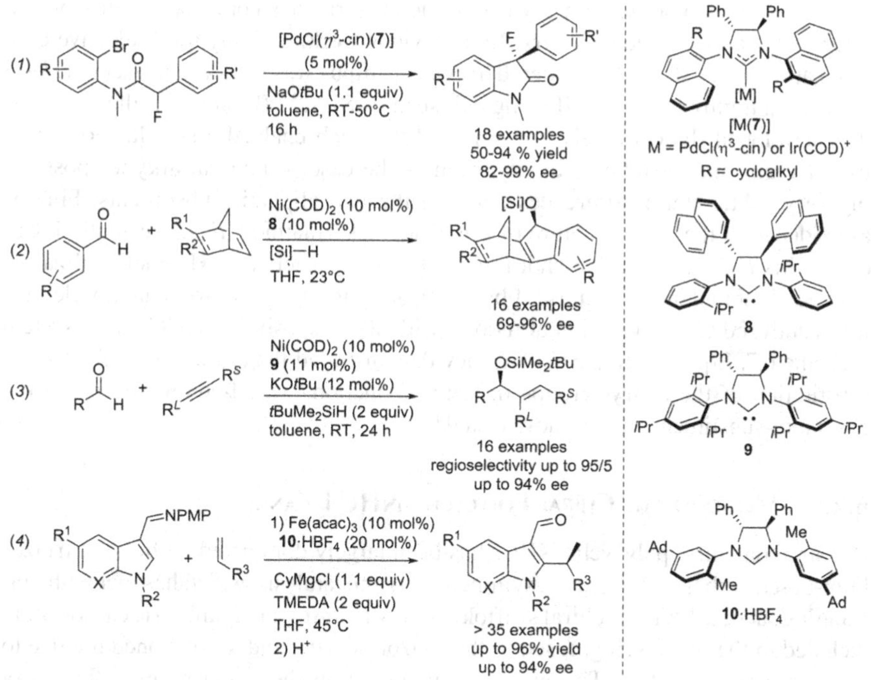

SCHEME 6.7 Selected examples of chiral-backbone NHC ligands and their use as efficient stereodirecting groups in asymmetric metal-catalyzed transformations.

the enantioselective, intramolecular α-arylation of α-fluoroamides to obtain 3-fluoro-substituted oxindoles in high yields and with excellent ee values (Scheme 6.7, Eq. 1).[29] Later on, this research group developed the asymmetric intramolecular hydroamination of unactivated amino-olefins with the catalyst [Ir(COD)(7)]⁺ and obtained enantiomeric excesses for the pyrrolidine products which were amongst the highest values reported for this reaction.[28] More recently, Cramer reported the nickel(0)-catalyzed reductive three-component coupling between an aromatic aldehyde and a norbornene derivative in the presence of a hydrosilane, and demonstrated that the key is the use of a 1,2-di(naphthalen-1-yl)ethylene diamine backbone in ligand **8**. Indeed, it strongly influenced the reaction outcome by providing the desired annulated silyl-protected indanol product as single diastereoisomer with high enantioselectivity (Scheme 6.7, Eq. 2).[30]

Interestingly, while rotationally unsymmetrical *N*-aryl groups are usually preferred for an efficient transfer of the chiral information, a recent report by Liu and Montgomery showed that even the symmetrical – yet highly sterically hindered – 2,4,6-tris(isopropyl)phenyl groups, such as in ligand **9**, are effective chiral transmitters in the nickel-catalyzed reductive coupling between aldehydes and internal alkynes (Scheme 6.7, Eq. 3).[31] Indeed, a benchmark challenge to the regiocontrol of the latter reaction is presented by internal alkynes that possess two different alkyl groups, as they do not feature any electronic or steric bias nor directing groups that effectively control regiochemistry. In a previous work, a highly regioselective coupling at the more sterically hindered alkyne terminus was achieved in racemic form by using a highly hindered NHC ligand, such as SIPr.[32] Building on these results, they developed the extremely bulky ligand 9, which enabled a simultaneous control of regio- and enantioselectivity, even in the case of internal alkynes possessing very subtle steric differences between the two aliphatic substituents. Finally, as evidence of the mature nature of this field, Ackermann *et al.* successfully introduced this NHC class in the "hot topic" of enantioselective C–H functionalization with earth-abundant 3d metals,[33] by describing, in 2017, the first enantioselective iron-catalyzed C–H alkylation of indoles with alkenes, using an NHC-based system (Scheme 6.7, Eq. 4).[34] In particular, they demonstrated the critical role of the *meta* substitution of the *N*-aryl groups in stereo-induction, with 1-adamantyl being the best *meta*-substituent among those tested.[34]

6.2.3 MONODENTATE CHIRAL POLYCYCLIC NHC LIGANDS

Research on chiral polycyclic NHCs has been largely dominated by bicyclic triazolium-based carbenes, for use in asymmetric organocatalysis, which has met with outstanding success.[35] Their chiral scaffold usually features an asymmetric carbon atom included in the second ring, fused to the triazolium ring and α-positioned relative to the shared nitrogen atom. This geometry greatly limits the conformational flexibility around the asymmetric center and leads to a more rigid and clearly described definition of the chiral space. However, as NHC-organocatalysis and metal-NHC catalysis require different stereo-electronic properties of the NHCs, the chiral triazolylidene-based ligands have seldom been used in asymmetric metal-catalyzed transformations. In that regard, Guo and Glorius described an important step in the use of such chiral bicyclic triazolylidenes as ligands, by showing that the triazol-3-ylidene 11 serves as both an organocatalyst and a stereodirecting ligand in the highly enantioselective Pd(PPh$_3$)$_4$/NHC-catalyzed annulation of vinyl benzoxazinanones and enals (Scheme 6.8, Eq.1).[36] More recently, Sigman and Toste took advantage of the great flexibility of this platform to optimize the gold(III)-catalyzed stereoselective γ,δ-selective Diels-Alder reaction of 2,4-dienals with cyclopentadiene, which represents a major challenge, because the site of the asymmetric induction is distant from that of the Lewis acid/base interaction (Scheme 6.8, Eq. 2).[37] Starting from their previous report, that the gold(III) pre-catalyst [AuCl(biphenylene)(IPr)] leads to an exquisite regioselectivity for this reaction,[38] an extensive screening of the chiral NHC ligands revealed that complex 12 is the lead catalyst, with enantioselectivities up to 98% ee. Computational studies showed the critical role of non-covalent interactions (NCIs) in the selectivity issue.[37]

In 2005, Lassaletta[39a] and Glorius[39b] independently reported the rigid, heterobicyclic imidazo[1,5-*a*]pyridin-3-ylidene (IPy), formed by the formal fusing of pyridinyl and imidazolyl rings,[39] and this class of aromatic *N*-fused heterobicyclic carbenes was completed with the report, by You, of the 1,2,4-triazolo[4,3-*a*]pyridin-3-ylidene (TriPy) in 2008.[40] Thanks to the advantages of IPy, linked to its peculiar

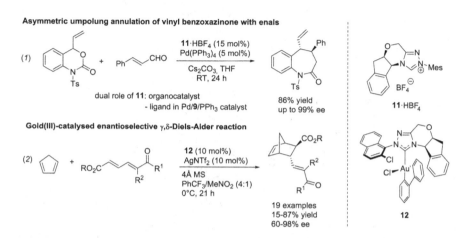

SCHEME 6.8   Chiral bicyclic triazolylidene ligands in asymmetric catalysis.

geometry and electronic structure, this bicyclic scaffold has become an emerging area over the past ten years, especially with respect to the design of original chiral NHC ligands. An excellent minireview on this topic appeared in 2016.[41] As a key milestone, in a collaborative work, the groups of Lassaletta and Mascarenas reported the highly enantioselective, gold(I)-catalyzed [4+2] Diels–Alder cycloaddition between allenamides and dienes, using [AuCl(**13**)] as pre-catalyst (Scheme 6.9, Eq. 1).[42] The challenge in asymmetric gold(I) catalysis resides in the di-coordinate, linear coordination geometry of the gold(I) center, which places the active site in the *trans* position to the supporting ligand, so that only a specific design of the ligand, bringing the chiral information close to the gold center, was shown to lead to efficient asymmetric gold(I) catalysis. Here, contrary to classical monocyclic NHCs, the bicyclic scaffold TriPy in **13** brings the lateral C5-naphthyl substituent into the direct environment of the gold center, favoring an efficient transfer of the axial chirality encoded in the ligand framework. The same catalytic system was further implemented in the enantioselective, intermolecular [2+2+2] cycloaddition between allenamides, alkenes, and aldehydes, resulting in highly substituted tetrahydropyrans.[43] In 2015, Scheidt reported the enantiopure, planar-chiral IPy-based ligand **14** (R = Me, Ar = Mes).[44] Starting from a racemic, pyridine-fused ferrocene derivative, the enantiomeric resolution was performed through a simple column chromatography separation of the diastereoisomers, obtained from a pseudoephedrine amide. The enantiopure azolium pro-ligand was then accessed after removal of the chiral auxiliary and subsequent cyclization. The ligand **14** was then successfully employed in the organocatalytic annulation of ketones with enals, as a ligand in the nickel-catalyzed reductive coupling of aldehydes and alkynes, and in the copper-catalyzed borylation of activated alkenes, demonstrating its potential as an effective stereodirecting NHC.[44] The same group subsequently developed the first highly enantioselective, catalytic hydroboration of *alkyl*-substituted aldimines, generated *in situ* from readily accessible *N*-benzoyl-protected α-tosylamines, to provide α-amidoboronates, which are important pharmacophores (Scheme 6.9, Eq. 2).[45] Optimization of the

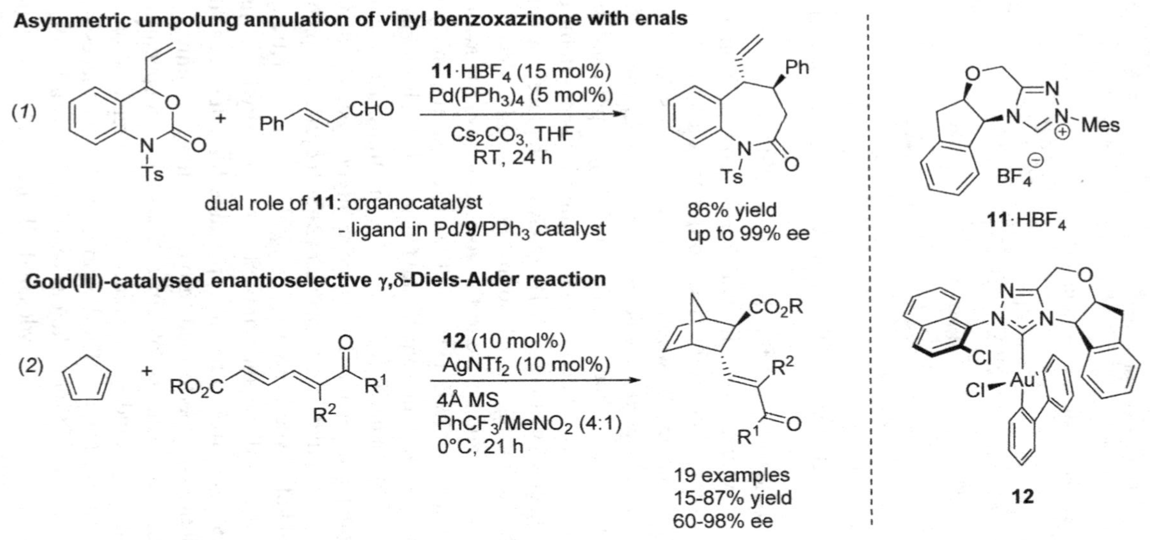

Asymmetric umpolung annulation of vinyl benzoxazinone with enals

11·HBF₄

(1)

11·HBF₄ (15 mol%)
Pd(PPh₃)₄ (5 mol%)
————————————————
Cs₂CO₃, THF
RT, 24 h

86% yield
up to 99% ee

dual role of **11**: organocatalyst
- ligand in Pd/**9**/PPh₃ catalyst

Gold(III)-catalysed enantioselective γ,δ-Diels–Alder reaction

(2)

12 (10 mol%)
AgNTf₂ (10 mol%)
————————————————
4Å MS
PhCF₃/MeNO₂ (4:1)
0°C, 21 h

19 examples
15–87% yield
60–98% ee

12

SCHEME 6.9 Chiral aromatic N-fused heterobicyclic carbenes in asymmetric catalysis.

ligand structure through modulation of the Cp base of the planar NHC scaffold and of the *N*-aryl substituents led to the selection of ligand **14**, with pentaphenylcyclopentadienyl and *N*-(2,6-dimethoxyphenyl) groups, as the best ligand, and the reaction proceeded with excellent enantio-induction (up to 98% ee).

6.2.4 PERSPECTIVES AND NEW TRENDS/DIRECTIONS

Whereas some earlier-developed strategies have already reached a certain maturity and led to chiral scaffolds used not only by the NHC community, but also in synthetic chemistry (such as the class of chiral backbone NHCs), some new trends have appeared in the past three years, since 2017, showing that the quest for the "ideal chiral NHC ligand" has not ended, and that new, exciting directions remain to be explored in this field. We can already foresee a glorious future for the class of monodentate chiral (S)IPE-type NHC ligands in synthetic methodologies. Moreover, this design principle, of rendering aryl groups chiral in ligands and the development of a reliable and scalable protocol to produce the chiral, C_2-symmetric anilines, are all new ways to achieve future developments in asymmetric organometallic catalysis, that are not restricted to either the monodentate mode or to NHC chemistry. Nonetheless, other new strategies and alternatives, which appeared since 2018, also deserve consideration, as they provide promising new trends in the design of monodentate chiral NHC ligands. Three such avenues are described below.

In 2020, Mauduit and Clavier reported the synthesis of well-defined enantiopure metal-NHC complexes [M(**15**)], bearing C_1- or C_2-symmetric NHC ligands, starting from prochiral imidazolium precursors of type **15·HX** (Scheme 6.10).[46] Indeed, while the rotation around the N–C$_{Ar}$ is free at room temperature in the precursors, the complexation induces the formation of stable atropoisomers, which can be separated by preparative chiral HPLC. This methodology presents the advantages of producing both enantiomers (along with the *meso* compounds for C_2-symmetric NHCs) at a very late stage of the synthesis, and of being highly modular, since the

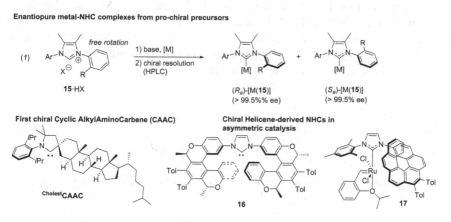

SCHEME 6.10 New trends in chiral monodentate NHC-metal complexes in asymmetric catalysis.

substitution pattern of imidazolium salts **15·HX** is virtually unlimited, with the existence of robust synthetic protocols. An obvious drawback is nevertheless the necessity for the complexes [M(**15**)] to be stable under chiral HPLC conditions, which *de facto* excludes many air-sensitive complexes. In our opinion, this strategy is similar to and should complement the approach of the chiral (S)IPE-type NHC ligands.

Although the search for chiral NHC ligands is almost exclusively focused on the standard imidazol(in)-2-ylidenes, many other types of NHCs have been reported over the past 15 years, featuring different and/or enhanced steric and electronic properties.[4a] In that regard, the recent report by Mauduit, Jazzar and Bertrand of the first chiral CyclicAlkylAminoCarbene (CAAC) CholestCAAC is of note (Scheme 6.10),[47] as it can open up a large field of new investigations with promising potential goals, in view of the outstanding performances of the CAAC ligands in organometallic chemistry and catalysis.[48]

Finally, the type of chirality introduced within the chiral monodentate NHC ligands could be diversified; whereas central chirality largely prevails in design strategies, planar and axial chiral elements have also found interesting uses in NHC chemistry. The past few years have witnessed an increasing interest in helicene-NHC-based metal complexes, which are the archetypes of helicoidal molecules. Although the field is still in its infancy, very promising results have already been obtained, concerning their chiroptical properties, such as in circularly polarized luminescence (CPL).[49] In asymmetric catalysis, only the two systems **16** and **17** have been reported up to now (Scheme 6.10),[50,51] with modest to good enantiomeric excesses. Undoubtedly, this research area is set to develop in the coming years.

6.3 POLYFUNCTIONAL CHIRAL NHC LIGANDS

As mentioned earlier, chiral *N*-heterocyclic carbenes, bearing one or several functional groups, were developed very early on for several reasons.[52] First, the chelating character of the ligand would avoid rotation about the C–M bond, that would potentially lead to loss of chiral information. Second, the introduction of a functional group, with a potentially very different stereo-electronic contribution to the metal, could have a strong influence on its reactivity; last but not least, the introduction of chirality can be achieved at various locations of the ligand (Figure 6.1).

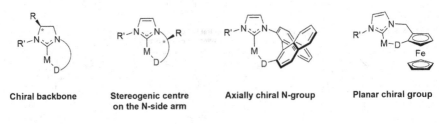

| Chiral backbone | Stereogenic centre on the N-side arm | Axially chiral N-group | Planar chiral group |

D = C, N ,O, P, S

FIGURE 6.1   Different types of polyfunctional chiral NHC ligands.

6.3.1 POLYFUNCTIONAL CHIRAL NHC-N, NHC-P, AND NHC-S LIGANDS

Mixed NHC-N, NHC-P, and NHC-S chiral ligands were investigated very early in NHC ligand research.[52] If they often proved as efficient or slightly more efficient than previously developed, analogous, chelating ligands, they recently suffered from comparison with chiral monodentate NHCs. The latter have often been easier to synthesize, more adaptable, and have generated extremely efficient catalysts in terms of reactivity and enantioselectivity (see Section 6.2).

Chiral NHC-oxazolines, analogs of the well-known phosphine-oxazolines, were among the first bifunctional chiral NHC ligands to be successfully used in asymmetric catalysis, mainly for rhodium-catalyzed hydrosilylation of carbonyl groups with **Rh-18** or iridium-catalyzed hydrogenation of olefins, using **Ir-19** (Figure 6.2).[53]

However, no real breakthrough in their design has been made since their initial development,[54] nor has any recent major development in catalytic applications been described either. One interesting, but isolated, example was described by Burgess *et al.* in 2013, with the synthesis of homo-Roche ester derivatives by Ir-catalyzed hydrogenation with complex **Ir-19**.[55] The same authors reported a comparative study of chiral NHC-oxazoline ligands in Ir-catalyzed hydrogenation of alkenes in a separate paper in 2013; they concluded that changing the imidazol-2-ylidene unit in **19,** to its saturated imidazolin-2-ylidene or annulated benzimidazol-2-ylidene derivatives, did not notably affect its activity nor its enantioselectivity.[56] More occasional examples of chiral NHC-N ligands, where the chelating group is a pyridine[57] or amine,[58] have been described in the past decade. Thus, Pfaltz *et al.* showed that chiral NHC-pyridine ligands were very efficient at catalyzing hydrogenation of various alkenes, and were better adapted to the hydrogenation of acid-sensitive substrates than was the chiral phosphinite-pyridine ligand (*S*)-**Phos** (Scheme 6.11).[59] Ligands **20a** and **20b**, bearing 5- or 6-membered carbocyclic rings, performed better than ligand **20c**. Only Burgess' catalyst **Ir-19** showed a performance equal to those of **20a** and **20b**.

Apart from the latter example, the efficiency in asymmetric catalysis of such NHC-N ligands remained somewhat lower than those observed with NHC-oxazolines. Chiral NHC-amide ligands have been described by the group of Sakaguchi,[60] but

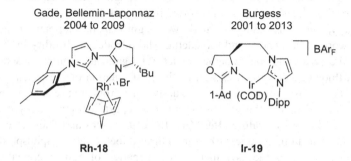

Gade, Bellemin-Laponnaz
2004 to 2009

Burgess
2001 to 2013

Rh-18   **Ir-19**

FIGURE 6.2 The most successful NHC-oxazoline ligands developed for asymmetric catalysis.

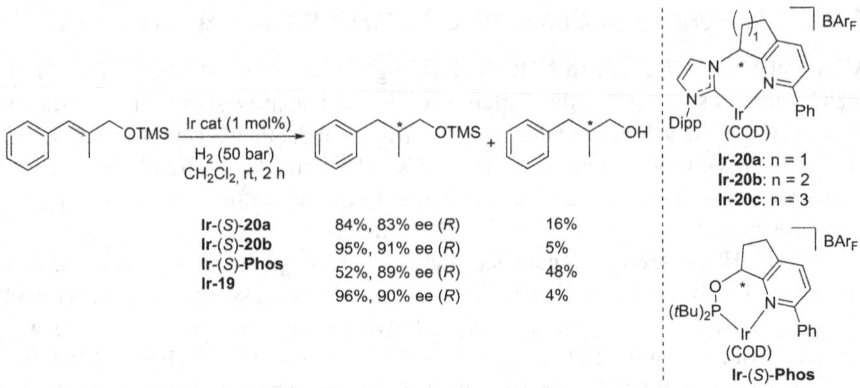

SCHEME 6.11 Chiral NHC-pyridine-based iridium complexes for enantioselective hydrogenation. Dipp = 2,6-diisopropylphenyl.

the coordination of the amide *via* the nitrogen atom has not been clearly demonstrated, according to the different metals used.

Given the popularity of diphosphines in asymmetric catalysis, mixed chiral NHC-P ligands were expected to be effective surrogates, with the replacement of a phosphine by a stronger σ-donor unit. In 2004, the group of Helmchen used a mixed chiral NHC-P ligand in the rhodium-catalyzed enantioselective hydrogenation of acrylates, with ee values up to 99%.[61] This imidazolin-2-ylidene is related to Grubbs' monodentate ligands of type **6**,[62] where one aryl group at nitrogen has been replaced by a naphthylphosphine. The same complex was used for enantioselective conjugate additions of boronic acids to enones and α,β-unsaturated esters, where it showed much better activity than previously described systems, and very high enantioselectivities (generally higher than 94% ee).[63] After this, many different ligands were designed on the basis of existing scaffolds: [2.2]paracyclophane,[64] binaphthyl,[65] ferrocene,[66] and also phosphine-NHC ligands, bearing the chirality on the alkyl spacer[67] or directly on the phosphorus atom.[68] Unfortunately, although interesting activities and enantioselectivities were obtained in some cases, compared with analogous non-NHC ligands, no real breakthrough in catalysis, nor real novelty in terms of ligand design, has been reported in the past decade.

Similarly, chiral NHC-S ligands underwent significant development in the 2000s, with the work of Fernandez and Lassaletta, who introduced chirality in the spacer between the NHC and the thioether moiety.[69] Again, the nature of substituents on NHC, thioether, or spacer, as well as the conformation of the chelate, strongly influenced enantioselectivities. However, despite the promising results obtained and the potential to diversify the structures, only one report involving catalytic applications appeared in the literature after 2011. In 2019, Pàmies and Diéguez compared the activity of iridium complexes bearing chiral thioether-NHC, -phosphinite, and -phosphite ligands for the asymmetric hydrogenation of non-functionalized olefins and cyclic β-enamides.[70] The NHC-thioether ligand design was inspired by the early work of Fernandez. Imidazolium precursors were synthesized in five steps

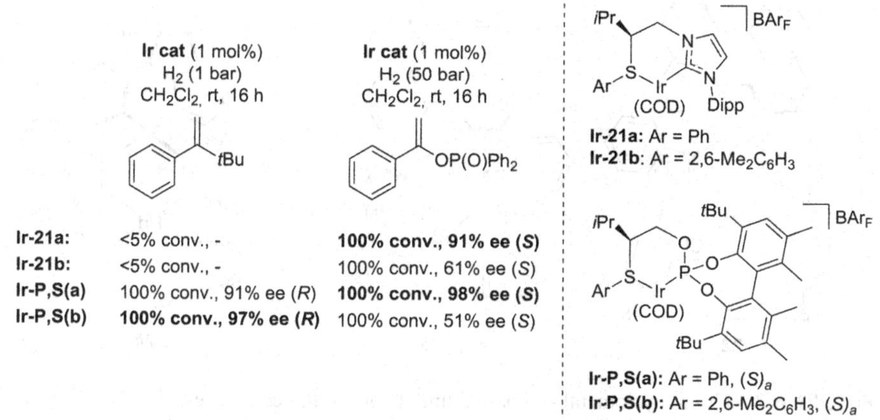

Ir cat (1 mol%) H₂ (1 bar) CH₂Cl₂, rt, 16 h	Ir cat (1 mol%) H₂ (50 bar) CH₂Cl₂, rt, 16 h

Ir-21a:	<5% conv., -	100% conv., 91% ee (S)
Ir-21b:	<5% conv., -	100% conv., 61% ee (S)
Ir-P,S(a)	100% conv., 91% ee (R)	100% conv., 98% ee (S)
Ir-P,S(b)	100% conv., 97% ee (R)	100% conv., 51% ee (S)

Ir-21a: Ar = Ph
Ir-21b: Ar = 2,6-Me₂C₆H₃

Ir-P,S(a): Ar = Ph, (S)ₐ
Ir-P,S(b): Ar = 2,6-Me₂C₆H₃, (S)ₐ

SCHEME 6.12 Chiral NHC-thioether-based iridium complexes for enantioselective hydrogenation.

from the readily available Evans' *N*-acyl carboxamide. NHC-thioether-based Ir complexes proved less efficient than other complexes, except for hydrogenation of some functionalized olefins (enol phosphonate derivatives), where the efficiency of **Ir-21a** was similar to that of **Ir-P,S(a–b)** (Scheme 6.12). These low performances were explained by high steric constraints imposed by the NHC-thioether design, which favored the formation of inactive trinuclear hydride species.

The lack of reports in recent years of the synthesis of functionalized, chelating NHC ligands could be partly explained by challenges with their synthesis, which can be made more difficult by the presence of several functional groups, and by the fact that it is more difficult to rationalize enantioselectivity levels with such NHC ligands than with monodentate NHCs.

6.3.2 POLYFUNCTIONAL CHIRAL NHC-C LIGANDS

Zhang and co-workers developed a series of palladium(II) complexes of chiral bis-NHC ligands, the chirality originating from the linker between the two NHC moieties (e.g., (1*R*,2*R*)-cyclohexene, binaphthylene…), for use in asymmetric Suzuki–Miyaura coupling. Within this series, cyclometalated complexes were isolated in the case of *N*,*N*'-bisaryl-bis(NHC)-Pd complexes with an axially chiral binaphthylene (**22, 23**) or spirobi[indene] (**24**) linker (Figure 6.3).[71] These compounds were evaluated in the asymmetric Suzuki-Miyaura coupling of various naphthyl halides and arylboronic acids, affording the desired axially chiral biaryl products with moderate enantioselectivities (up to 74% ee), the highest ee values being obtained with the axially chiral binaphthylene derivatives (*vs.* spiro).

From 2011, Glorius and co-workers developed a highly efficient and versatile chiral Ru-NHC catalyst for the enantioselective hydrogenation of (hetero)aromatic compounds (Scheme 6.13).[72] The catalytic system is generated *in situ* by mixing the ruthenium precursor [Ru(COD)(2-methallyl)₂] and the imidazolinium salt

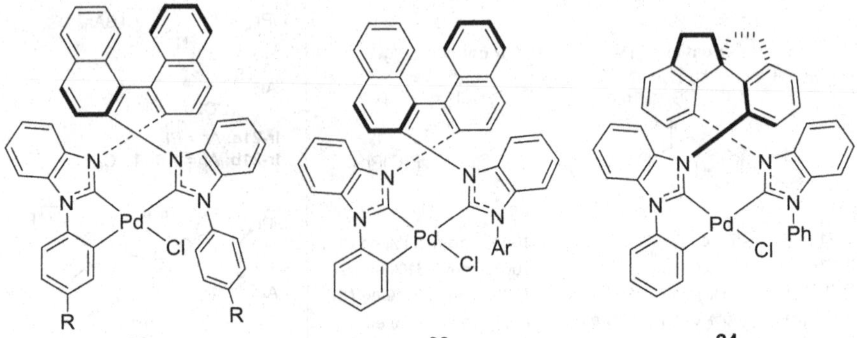

FIGURE 6.3 Chiral cyclometalated palladium(II)/bis(NHC) complexes.

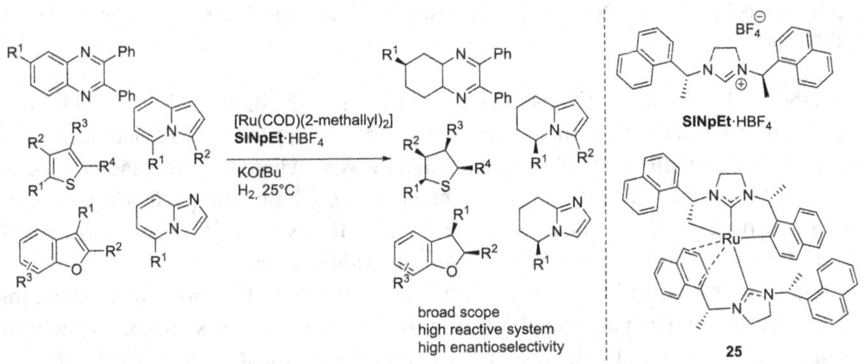

SCHEME 6.13 Ru-SINpEt enantioselective hydrogenation of heteroaromatic compounds.

SINpEt·HBF₄ in the presence of potassium *tert*-butoxide as the base. More impressively, under a hydrogen atmosphere and at 25°C, this catalytic system selectively reduces the heteroaromatic rings of a large number of heteroaromatics, including, for example, (benzo)furans, (benzo)thiophenes, indolizines, imidazo[1,2-*a*]pyridines, and the carbocyclic ring of quinoxalines (Scheme 6.13). As a result of a structural investigation aimed at achieving a better understanding of the actual active species, the ruthenium complex **25**, bearing two NHC ligands and featuring an unusual doubly cyclometalated NHC ligand, was isolated and identified as a pre-catalyst for the hydrogenation catalysis.[73] Although the nature of the active catalyst could not be precisely determined, the interplay between cyclometalated and monodentate ligands, as well as the partial hydrogenation of the naphthyl substituents of the NHC ligands, were shown to be key processes in this catalysis.

In 2016, the Glorius group designed a new class of chiral NHC precursors (**26a–g·HCl**), which underwent cyclometalation upon coordination to rhodium(I), suppressing the possible rotation around the C_{NHC}–Rh bond; it was therefore expected to exhibit high site- and enantioselectivities (Scheme 6.14).[74] Most of these

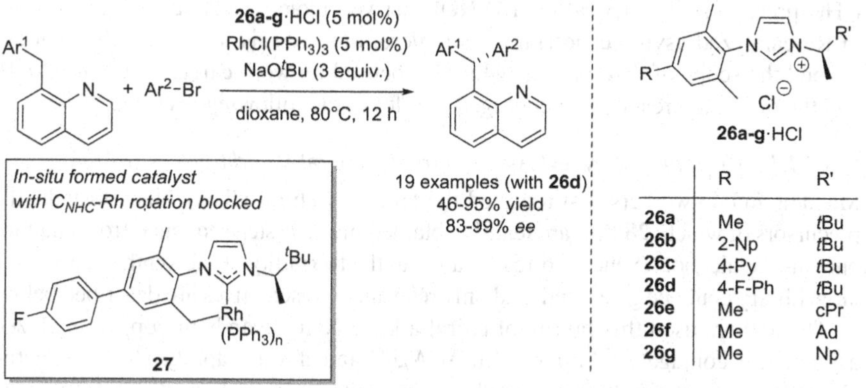

	R	R'
26a	Me	tBu
26b	2-Np	tBu
26c	4-Py	tBu
26d	4-F-Ph	tBu
26e	Me	cPr
26f	Me	Ad
26g	Me	Np

SCHEME 6.14 Chiral imidazolium salts designed to suppress the C_{NHC}-Rh rotation, *via* cyclometalation, and application to enantioselective arylation of C(sp^3)-H bonds.

pro-ligands generated efficient chiral Rh-based catalysts for the site- and enantioselective arylation of C(sp^3)–H bonds. The pro-ligand **26d**·HCl was found to achieve the best results and was successfully applied to a wide range of substrates and aryl bromides, giving access to enantioenriched triarylmethanes at 82–99% ee. Most importantly, the proposed mechanism is meant to proceed through the intermediary of the cyclometalated active species **27**, generated in an initial step by coordination of the free NHC **26d**, followed by intramolecular C–H bond activation of an aryl *ortho*-methyl group.

Although the number of reports dealing with the preparation and use in asymmetric catalysis of cyclometalated NHC complexes remains limited, some of these species have shown impressive performances. However, this approach appears to be limited to a combination of (the few) NHC scaffolds offering an easy-to-activate C–H bond and metal centers that readily undergo C–H activation.

6.3.3 Polyfunctional Chiral NHC-O Ligands

The development of chiral oxygen-functionalized NHC (NHC-O) ligands started in the early 2000s, and these compounds rapidly emerged as ligands of choice for several enantioselective metal-mediated transformations.[52,74] This class of ligand has already achieved several breakthroughs, and continues to attract much attention, especially in copper-catalyzed asymmetric reactions. The present section will highlight the most important advances accomplished by both aryl/alkoxide- and sulfonate-NHCs over the past decade.

6.3.3.1 Polyfunctional Chiral NHC-Aryl/alkoxide Ligands

All (pro-)ligands discussed in the present chapter exhibit a central chirality, that can be located: i) on the *N*-substituent(s), ii) on the NHC backbone (asymmetric *endo*-cyclic carbon(s)), or iii) at both positions. Nevertheless, the group of Hoveyda also exploited axial chirality in the past, *via* the synthesis of a series of chiral

NHC precursors, incorporating a BINOL-derived moiety as *N*-substituent, for use in Ru-catalyzed asymmetric ring-opening/cross-metathesis. However, this work is beyond the scope of this chapter (which is focused on work carried out after 2011), and the interested reader is encouraged to refer to the following reviews.[75b,75d]

6.3.3.1.1 Copper-Catalyzed Asymmetric Conjugate Additions (Cu-ACA)

Mauduit and co-workers first reported the family of chiral alkoxy-NHC ligands, the precursors of which, **28·PF₆**, are readily obtained in a five-step sequence from enantiopure amino alcohols (Scheme 6.15). Thanks to the flexibility of the synthetic access, a large library could be generated, and only relevant representatives are described below.

The authors used this family of chiral alkoxy-NHC ligands in copper-catalyzed asymmetric conjugate additions (Cu-ACA),[76] and it was rapidly shown that the Cu-ACA reaction may be performed on numerous cyclic and acyclic enones, using various nucleophiles, such as organozinc, organoaluminum, or Grignard reagents, with good yields and selectivities.[52, 75a,75c] Although the first generation of pro-ligands was based on *N*-aryl substituents, Alexakis and Mauduit showed, in 2012, that optimization could be carried out by varying this substituent, with the best results being obtained with pro-ligand **28b·PF₆**, bearing a *N*-CH₂Mes arm. The latter was applied to the Cu-ACA of Grignard reagents to 3-substituted cyclic enones at low catalyst loading (0.75 mol% [Cu(OTf)₂] and 1.00 mol% pro-ligand) and to large-scale reactions (60 mmol) with improved performances (1,4-regioselectivity, up to 90% yield and 93% ee).[77]

Pro-ligand **28c·PF₆**, in combination with [Cu(OTf)₂], efficiently mediated the addition of dimethylzinc to α,β- and α,β,γ,δ-unsaturated 2-acyl-*N*-methylimidazole derivatives with high yields and enantioselectivities, and excellent regio-selectivity for a polyenic substrate (1,4:1,6 ratio = 96/4) (Scheme 6.16, Eq. 1).[78]

Such 1,4-ACA products can be converted into the corresponding aldehydes, esters, or ketones, and this methodology was applied to the synthesis of natural products

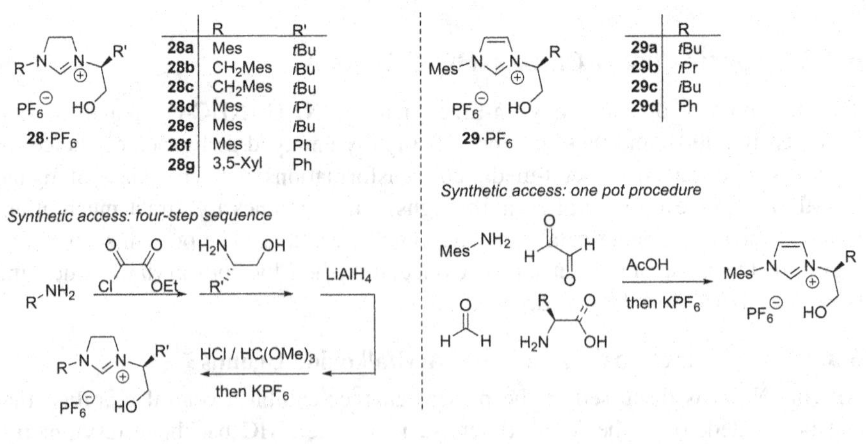

	R	R'
28a	Mes	*t*Bu
28b	CH₂Mes	*i*Bu
28c	CH₂Mes	*t*Bu
28d	Mes	*i*Pr
28e	Mes	*i*Bu
28f	Mes	Ph
28g	3,5-Xyl	Ph

	R
29a	*t*Bu
29b	*i*Pr
29c	*i*Bu
29d	Ph

Synthetic access: four-step sequence

Synthetic access: one pot procedure

SCHEME 6.15 Selected examples of the imidazol(in)ium salts derived from amino alcohols and precursors to the bidentate alkoxy-NHC ligands.

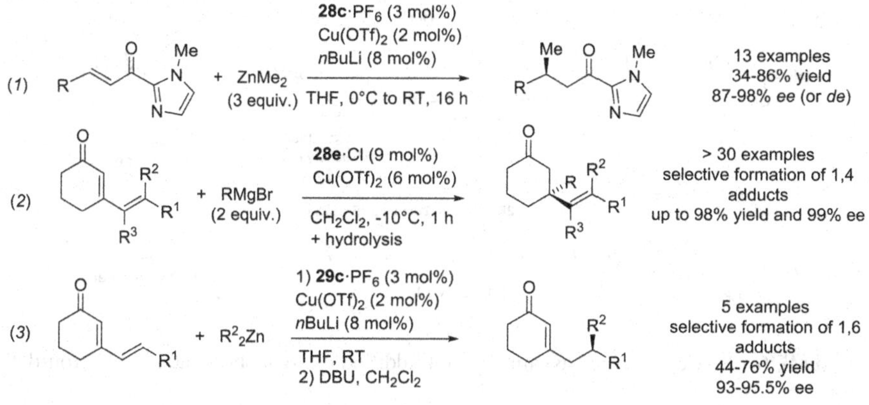

SCHEME 6.16   1,4-and 1,6-selective Cu-ACA to α,β- and $\alpha,\beta,\gamma,\delta$-unsaturated carbonyl compounds.

or, in an iterative process, leading to 1,3-desoxypropionate skeletons. The chiral pro-ligand **28e·Cl** exhibited excellent performances in the copper-catalyzed regio- and enantioselective 1,4-conjugate addition, i.e., on the most sterically hindered position, of Grignard reagents on polyconjugated cyclic dienones, trienones, and enynones, giving access to all-carbon quaternary stereogenic centers (Scheme 6.16, Eq. 2).[79] This unusual regioselectivity is of particular interest, because the remaining double bond may be further involved in several transformations, leading to useful building blocks.

In 2014, Mauduit's research group reported a one-pot multicomponent procedure, as an alternative to the initial four-step synthesis, to generate a large library of chiral hydroxyalkyl imidazolium salts **29·PF$_6$** from the corresponding enantiopure amino alcohols or amino acids in low to moderate yields (15–59% yield) (Scheme 6.15).[80] They observed that the nature of the carbenic heterocycle had little influence on the outcome of the Cu-ACA of ZnEt$_2$ to 3-methyl-2-cyclohexen-1-one, as the corresponding saturated and unsaturated pro-ligands gave similar results in terms of yields and enantioselectivity. Interestingly, the copper complexes generated from Cu(OTf)$_2$ and the leucine-derived pro-ligands **29c·PF$_6$** and **28e·PF$_6$** were found to catalyze the unprecedented 1,6-selective Cu-ACA of dialkylzinc reagents to $\alpha,\beta,\gamma,\delta$-unsaturated six-membered cyclic dienones in moderate to good yields and with excellent enantioselectivities (Scheme 6.16, Eq. 3).

Shintani and Hayashi reported in 2011 that organoboronates could be used as nucleophiles instead of highly reactive organometallic reagents (Grignard, organozinc, organoaluminum reagents), to mediate Cu-ACA reactions.[81] Among a selection of chiral hydroxyalkyl-imidazolinium precursors, the salt **30·PF$_6$**, in combination with CuBr (5 mol% cat), achieved the best results in the asymmetric 1,4-addition of organoboronic acids to various aryl-, heteroaryl-, and alkenyl-substituted methylidenecyanoacetate substrates in high yields and with good to excellent enantioselectivities (Scheme 6.17, Eq. 1). More recently, similar systems were applied

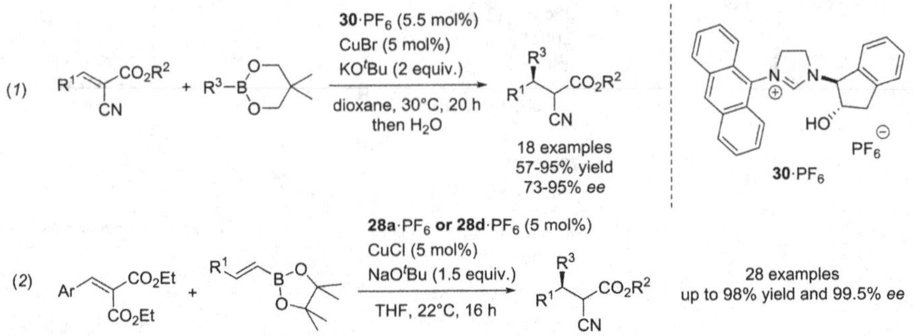

SCHEME 6.17   Cu-catalyzed asymmetric 1,4-addition of organoboronates to electron-deficient alkenes.

to the Cu-ACA of alkenyl-B(pin) reagents to α,β-unsaturated substrates, allowing the introduction of a wide range of functional groups onto the products that were obtained in high yields and with high enantioselectivities (Scheme 6.17, Eq. 2).[82]

Hoveyda and co-workers reported that chiral hydroxyalkyl-imidazolinium pro-ligands 28·PF₆ performed well in the Cu-catalyzed regio- and enantioselective 1,6-conjugate addition of propargyl and 2-boryl-substituted allyl groups to acyclic dienoates, whereas chiral phosphines gave unsatisfactory results (Scheme 6.18, Eq. 1).[83] With the optimal pro-ligand 28f·PF₆, the protocol was widely applicable, with excellent yields and enantio-inductions. Noteworthy, the crucial role of the hydroxyl group in the process was highlighted by the fact that the O-TBDMS analog of 28f·PF₆ led to a near-racemic mixture of the 1,6-adduct. The authors then successfully applied this NHC-O/Cu catalytic system to the multicomponent enantioselective 1,6-addition of 2-B(pin)-substituted allyl groups, using a mixture of a dienoate, a mono-substituted allene, and B₂(pin)₂ (Scheme 6.18, Eq. 2).[83] In this case, the less-hindered 3,5-xylyl pro-ligand 28g·PF₆ gave the best results. This multicomponent approach was also successfully applied by the group of Hoveyda to the Cu-catalyzed enantioselective conjugate addition of allyl moieties to enoate substrates, starting from butadienes, α,β-unsaturated carbonyl compounds, and B₂(pin)₂, with pro-ligand 28f·PF₆ performing best for this transformation.[84]

More recently, this copper-catalyzed 1,4-conjugate addition was extended to 1,1-diborylmethane (pinBCH₂Bpin) as the nucleophilic reagent, and was added to a series of α,β-unsaturated diesters, generating the corresponding alkylboronates (20 examples) in good yields (46–84%) and with high enantioselectivities (84–98% ee), with 28a·PF₆ as pro-ligand.[85] The interest of the approach resides in possible further post-functionalization of the product.

6.3.3.1.2   Copper-Catalyzed Asymmetric Allylic Alkylations (Cu-AAA) and Other Substitution or Addition Reactions

In-situ-generated copper catalysts, based on chiral alkoxide NHCs, were also shown to efficiently promote the reaction between allylphosphate substrates and dialkyl-zinc reagents, *via* Cu-AAA. This protocol allowed the formation of all-carbon

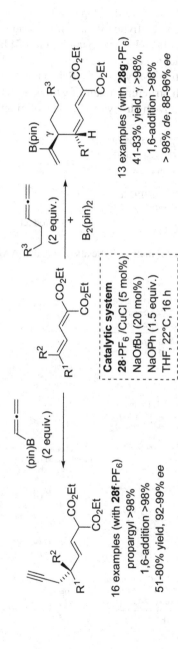

SCHEME 6.18 Cu-catalyzed enantioselective 1,6-conjugate additions of propargyl and 2-boryl-substituted allyl groups to acyclic dienoates.

quaternary centers in the resulting addition products, with a very high regioselectivity (S_N2' product *vs.* linear product) and, in most cases, high enantioselectivities (Scheme 6.19).[80, 86]

From 2014, Ohmiya and Sawamura developed efficient catalytic systems for the copper-catalyzed enantioselective allylic alkylation of various carbon nucleophiles through the use of the new chiral NHC pro-ligand family **31**·BF$_4$, based on a chiral (*S,S*)-diphenylenediamine backbone and bearing a phenolic hydroxy group in the *ortho* position of one of the two *N*-aryl groups (Scheme 6.20).[87] Most importantly, the authors showed that both structural ligand features are critical to the reaction outcome, as neither the chiral *N*-(hydroxyalkyl)imidazolinium salts of type **28**·PF$_6$,

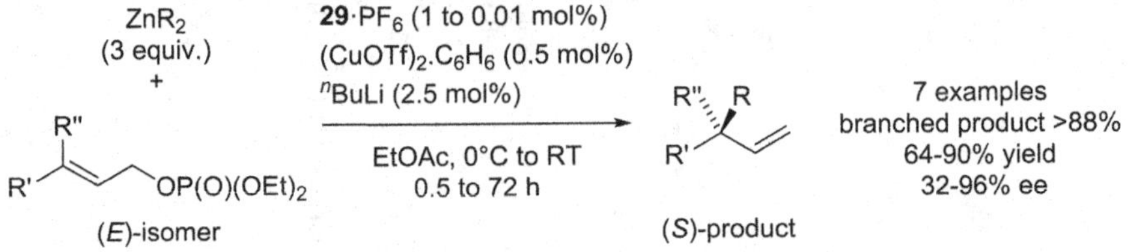

SCHEME 6.19 Cu-catalyzed asymmetric allylic alkylation (Cu-AAA) of allylphosphates to form all-carbon quaternary centers.

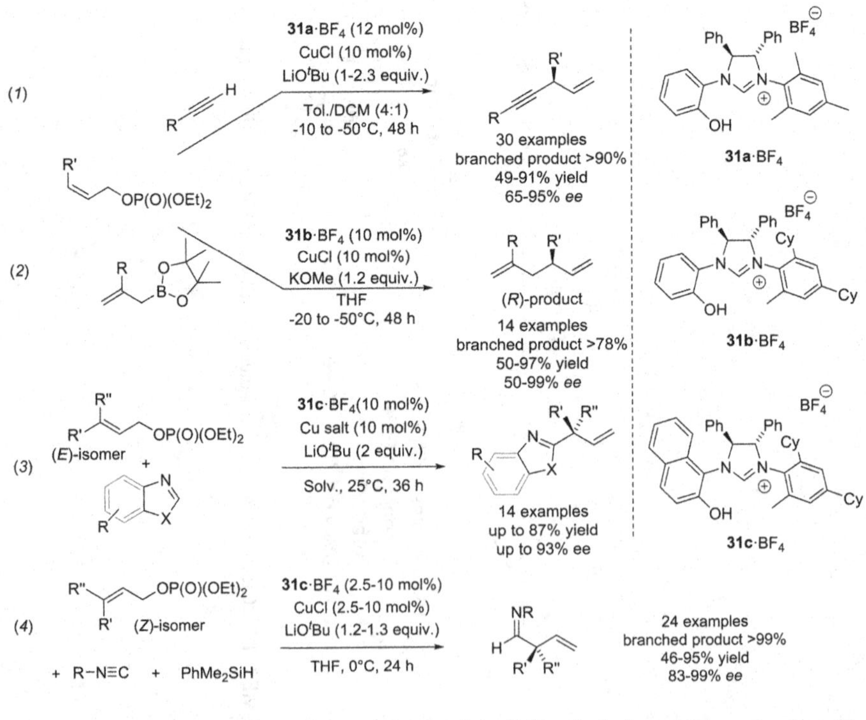

SCHEME 6.20 Copper-catalyzed enantioselective allylic alkylation reactions developed by Ohmiya and Sawamura.

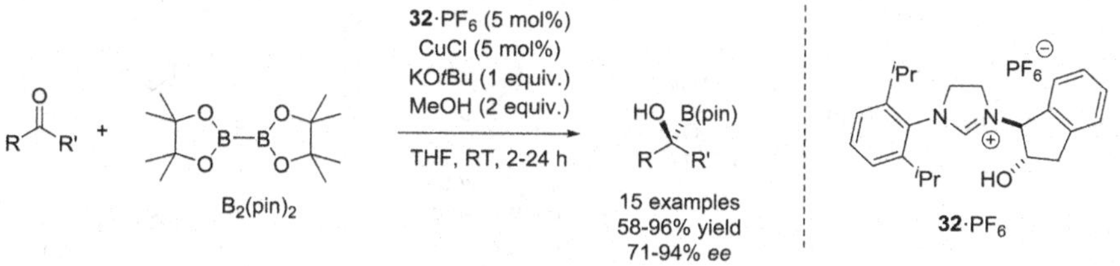

SCHEME 6.21   Cu-catalyzed enantioselective nucleophilic borylation of aliphatic ketones.

the monodentate chiral 'Grubbs-type' NHCs, nor the *O*-methylated analogs of **31·BF₄** led to good conversions or regio- and enantiocontrol of the reaction. In a first approach, terminal alkynes were used as the carbon nucleophiles and the reaction occurred with excellent γ-branched regioselectivity and high enantioselectivity, forming a stereogenic center at the propargylic/allylic position (Scheme 6.20, Eq. 1). Using the optimal imidazolinium salt **31b·BF₄**, the authors further developed an allyl-allyl coupling, using allylboronates as nucleophiles to provide chiral 1,5-dienes with a tertiary carbon atom at the allylic/homoallylic position (Scheme 6.20, Eq. 2).[88] Extension of the methodology to azole nucleophiles (benzothiazoles or (benz)oxazoles) also proved to be successful, generating a controlled all-carbon quaternary stereogenic center at the α-position of the heteroaromatic ring with excellent S_N2'-regioselectivity and high enantioselectivity (Scheme 6.20, Eq. 3).[89] In this case, the best pro-ligand was identified as **31c·BF₄**, having 2-naphthol as a *N*-substituent. Eventually, capitalizing on the previously known, metal-catalyzed 1,1-insertion of isocyanides into metal-H bonds, forming formimidoylmetal complexes, the authors developed the highly regio- and stereoselective copper-catalyzed three-component coupling between isocyanides, hydrosilanes, and γ,γ-disubstituted allylic phosphates to form chiral α-quaternary formimines, which are useful and highly functional synthetic intermediates (Scheme 6.20, Eq. 4).[90]

The copper(I) complex of the chiral NHC-O ligand derived from **32·PF₆** was found to efficiently catalyze the enantioselective nucleophilic borylation of various aliphatic ketones, therefore providing ready access to α-hydroxyboronates with good enantioselectivities (71–94% ee) (Scheme 6.21).[91] Such compounds may be further involved in a stereospecific C-C bond-forming reaction, such as the Aggarwal cross-coupling, paving the way to readily produce chiral tertiary alcohols.

6.3.3.1.3 Copper-Free Asymmetric Allylic Alkylation (AAA)

Following the pioneering work of Lee and Hoveyda on copper-free asymmetric allylic alkylation (AAA),[92] Alexakis and co-workers applied a large library of NHC-O pro-ligands (more than 50), with one or two endo-cyclic asymmetric carbons, to the AAA of allyl bromide substrates (e.g. cinnamyl bromide), with various Grignard reagents.[93] The formation of stereogenic quaternary centers was found to be highly regioselective, for both aliphatic and aromatic derivatives, and the products were obtained with good to excellent enantioselectivities (up to 96% ee). Within the series of chiral NHC-O precursors tested, compounds **33a-c·Cl** (Scheme 6.22) exhibited

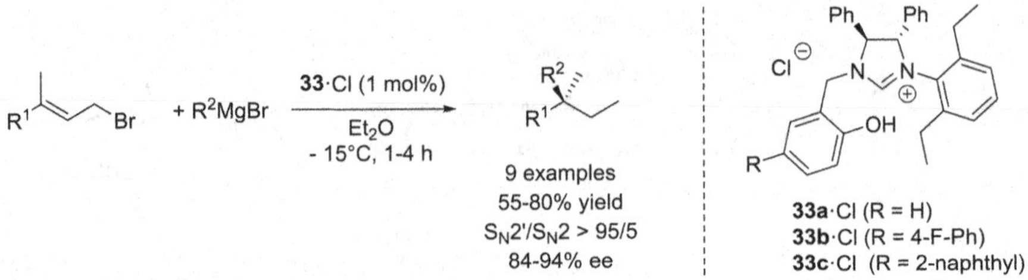

SCHEME 6.22   Enantioselective copper-free allylic alkylation of allylic bromides.

the best performance/accessibility ratio. NMR investigations allowed the identification of a magnesium complex, supported by a chelating monoanionic NHC-O ligand, as a catalytically active species.

Over the past decade, chiral NHC-O-supported metal complexes were used in several other processes, such as intramolecular hydroamination and arylation/cyclization reactions,[94] asymmetric hydrogenation reactions,[95] and asymmetric reductive Heck reactions.[96] However, their performances remained very modest and will not be detailed in the present chapter.

6.3.3.2 Polyfunctional Chiral NHC-Sulfonate Ligands

Until 2007, the field of asymmetric catalysis, using chiral NHC-O ligands, had been devoted entirely to alkoxide-functionalized NHC ligands (see Section 6.3.3.1). The group of Hoveyda had largely contributed to it but they noted that many challenges remained, especially in the field of copper-catalyzed asymmetric reactions. In 2007, the Hoveyda group described the synthesis of a chiral sulfonate-NHC precursor (**34a**·H, Scheme 6.23, Eq. 1), that allowed them to achieve significantly greater efficiency and asymmetric induction than was achieved by the previously reported systems of copper-catalyzed asymmetric conjugate additions (ACA),[97] or shortly after, in asymmetric allylic alkylations (AAA).[98] For the latter, their initial assumptions were based on the fact that a sulfonate moiety should facilitate oxidative addition of substrates on copper, compared with alkoxide-based ligands, while reductive elimination should remain as efficient as before. Although their initial hypotheses later proved inaccurate,[99] these ligands allowed achievement of unprecedented efficiencies and selectivities, particularly in the areas of ACA and AAA reactions. Most ligand variations (such as aryl substituents on nitrogen or phenyl substituents on the backbone) were investigated in the years following this first report.[97] Although most of the structural variations remained minor, from a synthetic point of view, the authors showed that a subtle change in the substituents present on the NHC led to dramatic changes in activity and selectivity, depending on the substrates involved. Thus, many synthetic bottlenecks were overcome during the past decade, mainly by the Hoveyda group, but also by other groups. The first sulfonate-imidazolinium salt **34a**·H, the precursor of the sulfonate-NHC ligand, could be generated on a multi-gram scale *via* a four-step procedure (Scheme 6.23, Eq. 1).

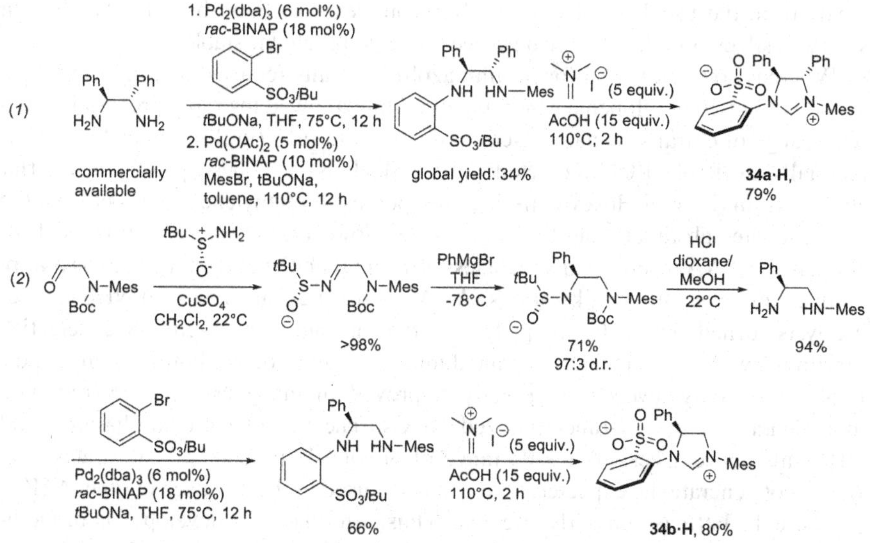

SCHEME 6.23 Synthesis of chiral sulfonate-imidazolinium salts, precursors of NHC ligands.

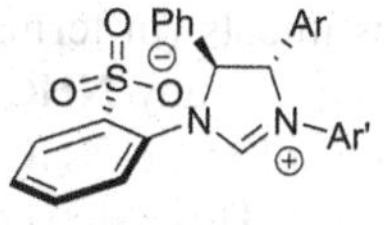

NHC	Ar	Ar'
34a	Ph	Mes
34b	H	Mes
34c	H	Dipp
34d	Ph	3,5-(2,4,6-(*i*Pr)$_3$C$_6$H$_2$)$_2$C$_6$H$_3$
34e	Ph	3,5-(*t*Bu)$_2$C$_6$H$_3$
34f	Ph	Dipp
34g	H	2,6-Et$_2$C$_6$H$_3$
34h	Ph	2-Ph-6-MeC$_6$H$_3$
34i	Ph	2,4,6-(*i*Pr)$_3$C$_6$H$_2$

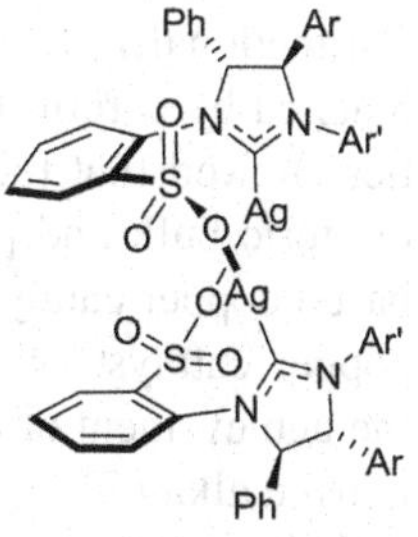

FIGURE 6.4 Privileged chiral sulfonate-imidazolinium salts and sulfonate-NHC silver(I) complexes.

This involved two successive Pd-catalyzed C–N coupling reactions between the commercially available chiral diamine and the aryl bromides. Cyclization, to give the imidazolinium salt, was carried out with formaldehyde in acetic acid, which also promoted the cleavage of the *i*BuO group on the sulfonate group.[97] This step was later performed with Eschenmoser's salt. The synthesis of imidazolinium salts bearing only one phenyl group on the imidazolinium backbone, such as **34b·H**, was less straightforward, because the starting chiral diamine is non-symmetrical and had to be prepared in six steps from commercial mesitylamine (Scheme 6.23, Eq. 2).[100] Silver(I) complexes, most often used as chiral ligand-to-copper transfer agents, were easily prepared by reaction of the imidazolinium salt with silver(I) oxide (Figure 6.4).

Apart from the usual sensitivity of silver complexes to light, both imidazolinium salts and silver complexes were described as indefinitely air stable.

Whereas some of the sulfonate-imidazolinium salts required lengthy syntheses, Woodward *et al.* developed a general method that allowed the rapid preparation of a wide range of chiral sulfonate-based imidazolinium salts (in three steps from commercially available BOC-protected amino alcohols).[101] These pro-ligands differ slightly from those of Hoveyda in that they possess a methylene linker between the arylsulfonate substituent and the heterocycle. However, they were mainly used for copper-free AAA reactions and were less efficient than the ubiquitous ligands **34a,b**.

In asymmetric allylic alkylations (AAA), sulfonate-functionalized NHC-copper catalysts turned out to be uniquely effective at catalyzing highly S_N2'-selective reactions with organozinc, organoaluminum, and organoboron compounds. Enantioselectivity levels were greatly improved in many cases, compared with monodentate or alkoxy-functionalized NHCs. The use of sulfonate-based chiral NHCs also gave access to a wide range of previously problematic substrates, that would not generate the expected products when used in the presence of other NHCs, or where the latter were poorly effective. Thus, after the early development of methods for asymmetric allylic substitutions, with alkyl aluminum,[98] and aryl–Zn,[102] as well as (silyl-substituted)alkenyl–Al,[103] alkynyl–Al compounds could be successfully used.[104] With allenyl boronates,[105] the reactions were S_N2'-selective with sulfonate-based NHCs, in contrast to what had been observed with monodentate or O-functionalized NHCs. However, high levels of enantioselectivity could only be reached by careful ligand screening: a thorough examination of the ligand properties showed that there was a strong influence of the *N*-aryl substituents (in terms of steric bulk and position) on enantioselectivity. On the other hand, many NHC-based copper catalysts gave S_N2"-selective reactions, but with only sulfonate-NHC-copper catalysts were high levels of enantioselectivity achieved.[106] This allowed the achievement of an advanced intermediate in the total synthesis of fasicularin, a marine alkaloid. Again, these catalysts were compatible with alkenyl boronates [107] and silyl-protected propargyl boronates,[106, 108] and many intermediates of natural products were efficiently prepared by this method. With bis(boryl)methane, sulfonate-NHCs were unique at giving S_N2'-selectivity (Scheme 6.24). In other cases, only S_N2 products were observed.[99, 109] Again, this allowed the synthesis of an intermediate of rhopaloic acid A.

Extensive mechanistic studies showed that the sulfonate moiety was not coordinated to copper, contrary to what had initially been assumed, the Lewis acidity of the copper(I) center being lower than that of zinc(II) or aluminum(III).[99] However, the unique influence of the sulfonate moiety was explained by the possible involvement of a metal bridge between sulfonate, Lewis acid cation (coming from the metal alkoxide), and Lewis base leaving group of the substrate. This was particularly true to explain high levels of S_N2' selectivity. DFT, coupled to experimental studies, showed that C–O bond cleavage and subsequent generation of a π-allyl intermediate complex was probably stereo-determining. Finally, the differences in levels and sense of enantioselectivity observed with the different sulfonate-NHC-based copper complexes were explained by a subtle interplay between electronic interactions

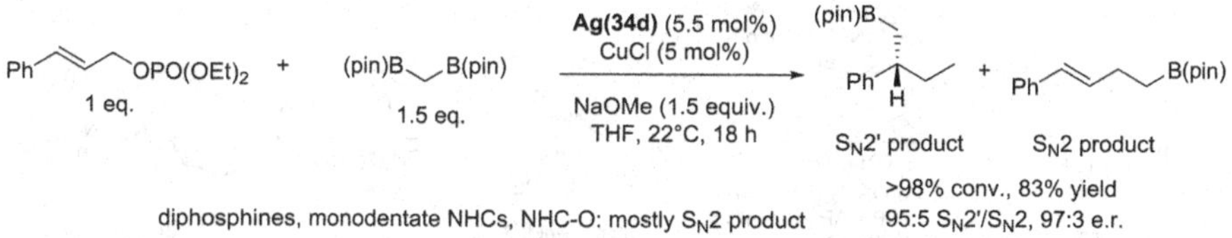

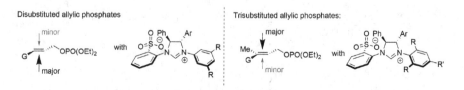

SCHEME 6.24 S_N2'- and enantioselective allylic substitution with diborylmethane.

Disubstituted allylic phosphates

Trisubstituted allylic phosphates:

FIGURE 6.5 Role of the sulfonate NHC ligands in the stereochemical outcome of the reaction of silyl-containing propargyl-B(pin) onto disubstituted or trisubstituted allylic phosphates.

(sulfonate/metal/phosphonate bridge) and steric repulsions between the ligand and the substrate. As an example, Hoveyda carried out a thorough mechanistic study to understand the origin of enantioselectivity in the reaction of silyl-containing propargyl-B(pin) onto disubstituted or trisubstituted allylic phosphates.[108] The role of the sulfonate ligand substituents was highlighted. The stereochemical outcome differed, depending on whether the N-aryl substituents were situated in a *meta* position, leading to preferential addition to the disubstituted phosphates from the *si* face, or in the *ortho/para* positions, leading to preferential *re* face addition to the trisubstituted phosphates (Figure 6.5).

As demonstrated earlier, copper-catalyzed asymmetric conjugate additions (ACA) similarly benefited from the use of NHC-based ligands, as exemplified by the work of Mauduit (see Section 6.3.3.1). At the time of Hoveyda's report of the first chiral sulfonate-NHC precursor, **34a·H**,[97] these reactions suffered mainly from a limited substrate range, despite the great advances made by several research teams using Rh-based[110] or Cu-based systems.[110b,111] For instance, activated electrophiles were often required, with cyclohexenones being mostly used.

Notably difficult substrates, such as β-ester-substituted cyclohexenones, cyclopentenones, or even acyclic enones, benefited from the use of sulfonate-NHC-based copper catalysts. Thus, after the initial report of successful ACA of alkyl-Zn compounds onto β-ester-substituted cyclohexenones,[97] alkyl-Al compounds reacted with acyclic trisubstituted enones,[112] or cyclopentenones to give the expected products in high yields and enantioselectivities. Snyder *et al.* used this strategy with the cyclopentenones to carry out total syntheses of the natural products, conidiogenones, and waihoensene (Scheme 6.25).[113]

Similarly, sulfonate-NHC ligands were effective for the copper-catalyzed ACA of alkenyl-Al compounds with disubstituted acyclic enones,[114] trisubstituted cyclopentenones, or cyclohexenones[99] (total synthesis of *trans*-clerodane by Overman

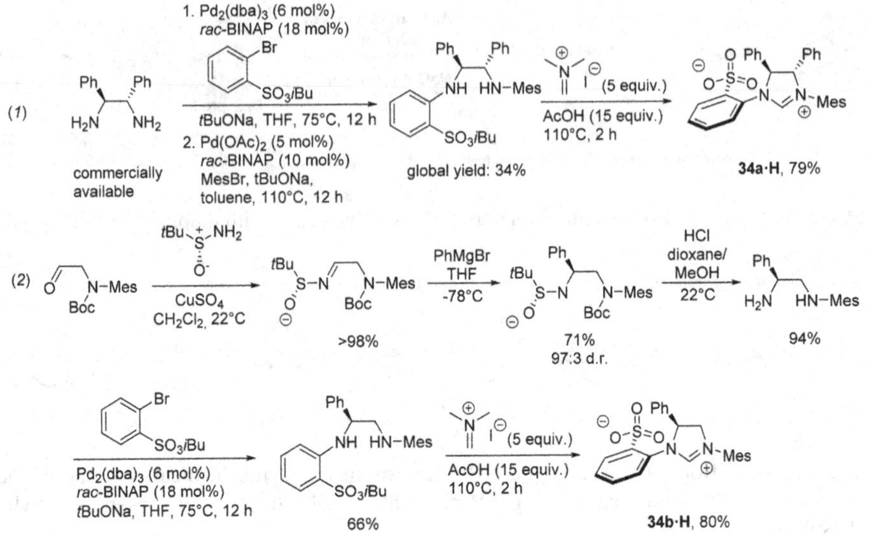

SCHEME 6.25 Sulfonate NHC-copper catalyzed ACA of AlMe₃ on a cyclopentenone in a total synthesis of waihoensene.

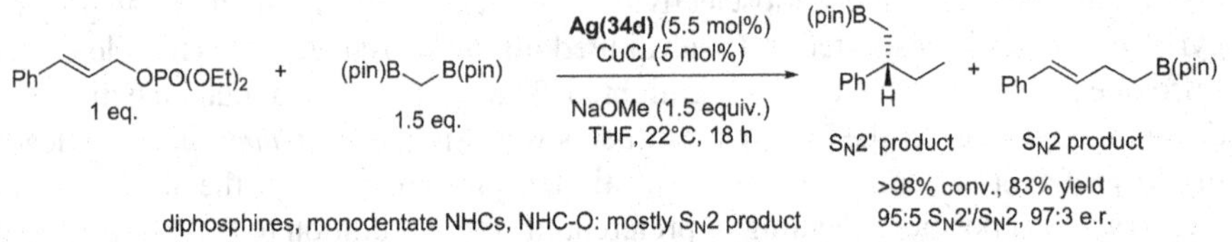

diphosphines, monodentate NHCs, NHC-O: mostly S_N2 product

SCHEME 6.26 ACA reactions of alkenyl-Al compounds with trisubstituted cyclohexenones.

et al.[115]), and trisubstituted acyclic enones (synthesis of enokipodin B).[116] However, the nature of the alkenyl-Al compound proved crucial, as the phenoxy-NHC-based silver precursor [Ag(**35**)] performed better than those with sulfonate-NHCs [Ag(**34g**)] in the presence of a phenyl substituent (Scheme 6.26, Eq. 1).[99] Changing from a phenyl to a silyl group reversed the efficiency of the catalysts since sulfonate-NHC-based catalysts became more efficient. This was used in the synthesis of an isomer of riccardiphenol B with **Ag(34g)** (Scheme 6.26, Eq. 2).[117]

This shows that, however efficient, the sulfonate-NHC-based systems are not the "alpha and omega" of chiral NHC ligands, and careful catalyst screening is always necessary. With aryl- and heteroaryl–aluminum compounds, ACA of a heteroaryl moiety to acyclic trisubstituted enones was reported for the first time in 2013, through the use of sulfonate-NHC-based systems and allowed the synthesis of a serotonin receptor inhibitor.[112]

Copper-catalyzed enantioselective B–H additions across alkenes have also gained from the use of chiral sulfonate-NHC ligands. The development of efficient methods for enantioselective hydroboration reactions is synthetically very useful, as organoboranes can readily be further functionalized.[118] Classical methods involve the use of rhodium or iridium catalysts associated with phosphine ligands, to furnish intermediates with a benzylic C–B bond and a homobenzylic C–H bond from aryl alkenes. With copper, depending on the mechanism involved, additions to olefins allow the formation of either a C–B(pin) bond (proto-boryl addition) or a C–H bond (boron-hydride addition) in the homobenzylic position. These are complementary methods, as they can grant access to different regioisomers. In 2009, Hoveyda *et al.* reported the first example of enantioselective proto-boryl addition, catalyzed by the sulfonate NHC-copper complex **Cu(34i)**.[119] Following this report, copper-catalyzed hydroboration of challenging substrates became possible, including acyclic and exocyclic 1,1-disubstituted aryl alkenes[120] and allenes.[121] Mechanistic studies have shown that, contrary to AAA and ACA, there is probably no sulfonate/substrate interaction,[99] but the influence of the sulfonate is still crucial for enantioselectivity.

In the case of aryl-substituted vinylsilanes, the use of sulfonate-NHC Ag(I) complexes, such as **Ag(34h)**, selectively conferred geminal borosilyl products with very high enantioselectivities, while other NHC ligands were less selective.[122] However, in the case of alkyl-substituted vinylsilanes, the selectivity was reversed and vicinal borosilyl products were obtained, with enantioselectivity being highest with monodentate NHCs (here, pro-ligand **36 H**) (Scheme 6.27).

Sulfonate-NHCs are also key to obtaining chemo-, regio-, and enantioselective sequential proto-boryl additions to alkynes; this method has been applied to a synthesis of an antioxidant plasmalogen phospholipid, C18 (plasm)-16:0 (PC).[123] Finally, in 2020, the group of Yun showed that the sulfonate NHC-copper complex **Cu(34i)** was effective for the regio- and enantioselective Cu–B(pin) addition to an alkenyl-B(dan) substrate, followed by stereospecific Pd-catalyzed cross coupling of the organocopper nucleophile with an aryl bromide.[124]

Chiral sulfonate NHC ligands have also been used in other copper-catalyzed reactions, such as enantioselective boryl or silyl substitutions. Whereas most findings concerning boryl substitutions were reported before 2011,[125] enantioselective silyl substitutions appeared later, with the first examples of NHC-based Cu-catalyzed reactions in 2013.[126] Although previous reports showed high S_N2'- and

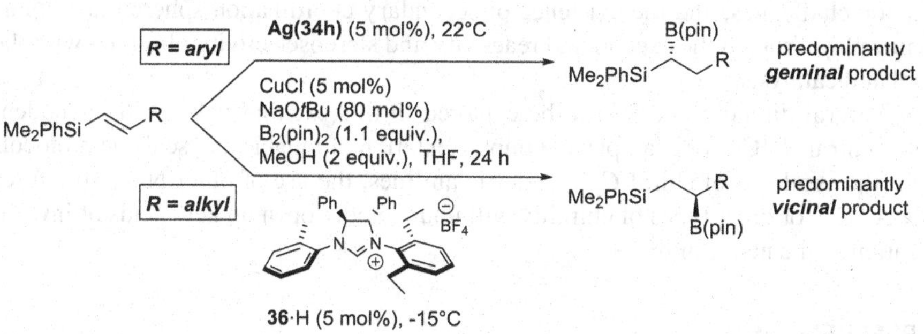

SCHEME 6.27 Catalyst-dependent regio- and enantioselective proto-boryl addition to alkenylsilanes.

enantioselectivities, the use of sulfonate-NHC ligands achieved the lowering of the catalytic charge to 1 mol%, and alkyl-, as well as aryl-substituted allylic phosphates, could be successfully used.[99]

6.3.3.3 Summary and Future Trends

In summary, a plethora of chiral alkoxy-NHC (pro-)ligands have been reported to the present day in the literature, and, for most of them, the chirality originates from the presence of one (or more) asymmetric carbon atoms on the *N*-side arm or within the imidazol(in)ium ring (C4, C5 positions). Whereas a few studies are still oriented to the development of new chiral NHC-O precursors, or to the improvement of the reported synthetic procedures, the majority of the recent contributions are mostly dedicated to the development of new reactions, targeting specific valuable skeletons, but using previously reported (pro-)ligands/complexes.

On the other hand, it has been shown that in copper-catalyzed reactions using chiral sulfonate-NHC ligands, the sulfonate acted rather as a secondary coordination sphere group, and that its role was essential in achieving high enantioselectivities. Thus, secondary coordination sphere interactions, which can help to orient substrates and to act in bond-breaking events, might prove more useful for asymmetric catalysis in the future than having an additional functional group that forms a chelate with an NHC. As seen in Section 6.2, the future will also tell whether chiral monodentate NHCs are only complementary ligands to sulfonates or whether they can compete for the remaining synthetic challenges.

6.4 CONCLUSION

While polyfunctional chiral ligands were very successful in the 2000s, they have faced strong competition from chiral monodentate ligands in the past decade. Indeed, the structures of the latter have diversified enormously and have made it possible to identify privileged ligands, that are extremely efficient in many asymmetric catalysis reactions. On the other hand, functional NHC ligands, such as alkoxy/aryloxy-NHCs or sulfonate-NHCs, have allowed the tackling of several synthetic challenges, and the existence of secondary coordination sphere interactions may play a role in the exceptional reactivity and stereoselectivity observed with the related catalysts.

Several directions remain to be explored in this exciting field. With monodentate chiral NHCs, for example, the implementation of reliable and scalable protocols for the synthesis of chiral C_2-symmetric anilines, the use of other NHC structures (CAACs), or other types of chirality will undoubtedly open up new fields of investigation in the near future.

REFERENCES

1. A. J. Arduengo, R. L. Harlow, M. Kline, *J. Am. Chem. Soc.* **1991**, *113*, 361–363.
2. W. A. Herrmann, M. Elison, J. Fischer, C. Köcher, G. R. J. Artus, *Angew. Chem. Int. Ed. Engl.* **1995**, *34*, 2371–2374.

3. a) D. Bourissou, O. Guerret, F. P. Gabbaï, G. Bertrand, *Chem. Rev.* **2000**, *100*, 39–92; b) C. S. J. Cazin, *N-Heterocyclic Carbenes in Transition Metal Catalysis and Organocatalysis*, 1st ed., Springer, **2011**; c) S. Díez-González, N. Marion, S. P. Nolan, *Chem. Rev.* **2009**, *109*, 3612–3676; d) F. E. Hahn, M. C. Jahnke, *Angew. Chem. Int. Ed.* **2008**, *47*, 3122–3172; e) E. Peris, *Chem. Rev.* **2018**, *118*, 9988–10031.

4. a) H. V. Huynh, *Chem. Rev.* **2018**, *118*, 9457–9492; b) A. Gomez-Suarez, D. J. Nelson, S. P. Nolan, *Chem. Commun.* **2017**, *53*, 2650–2660.

5. a) L. Benhamou, E. Chardon, G. Lavigne, S. Bellemin-Laponnaz, V. César, *Chem. Rev.* **2011**, *111*, 2705–2733; b) D. J. Nelson, *Eur. J. Inorg. Chem.* **2015**, *2015*, 2012–2027.

6. W. A. Herrmann, L. J. Goossen, C. Köcher, G. R. J. Artus, *Angew. Chem. Int. Ed. Engl.* **1996**, *35*, 2805–2807.

7. D. Enders, H. Gielen, G. Raabe, J. Runsink, J. H. Teles, *Chem. Ber.* **1996**, *129*, 1483–1488.

8. a) F. Wang, L.-j. Liu, W. Wang, S. Li, M. Shi, *Coord. Chem. Rev.* **2012**, *256*, 804–853; b) V. Cesar, S. Bellemin-Laponnaz, L. H. Gade, *Chem. Soc. Rev.* **2004**, *33*, 619–636.

9. D. Janssen-Muller, C. Schlepphorst, F. Glorius, *Chem. Soc. Rev.* **2017**, *46*, 4845–4854.

10. E. P. Kündig, Thomas M. Seidel, Y.-X. Jia, G. Bernardinelli, *Angew. Chem. Int. Ed.* **2007**, *46*, 8484–8487.

11. D. Katayev, Y.-X. Jia, A. K. Sharma, D. Banerjee, C. Besnard, R. B. Sunoj, E. P. Kündig, *Chem. Eur. J.* **2013**, *19*, 11916–11927.

12. M. Nakanishi, D. Katayev, C. Besnard, E. P. Kündig, *Angew. Chem. Int. Ed.* **2011**, *50*, 7438–7441.

13. a) D. Katayev, M. Nakanishi, T. Burgi, E. P. Kundig, *Chem. Sci.* **2012**, *3*, 1422–1425; b) E. Larionov, M. Nakanishi, D. Katayev, C. Besnard, E. P. Kundig, *Chem. Sci.* **2013**, *4*, 1995–2005; c) D. Katayev, E. Larionov, M. Nakanishi, C. Besnard, E. P. Kündig, *Chem. Eur. J.* **2014**, *20*, 15021–15030.

14. P. T. Bohan, F. D. Toste, *J. Am. Chem. Soc.* **2017**, *139*, 11016–11019.

15. a) W. Li, T. Wagener, L. Hellmann, C. G. Daniliuc, C. Mück-Lichtenfeld, J. Neugebauer, F. Glorius, *J. Am. Chem. Soc.* **2020**, *142*, 7100–7107; b) W. Li, M. P. Wiesenfeldt, F. Glorius, *J. Am. Chem. Soc.* **2017**, *139*, 2585–2588.

16. R. Noyori, T. Ohkuma, *Angew. Chem. Int. Ed.* **2001**, *40*, 40.

17. A. Albright, R. E. Gawley, *J. Am. Chem. Soc.* **2011**, *133*, 19680–19683.

18. J. Diesel, A. M. Finogenova, N. Cramer, *J. Am. Chem. Soc.* **2018**, *140*, 4489–4493.

19. Y. Cai, X.-T. Yang, S.-Q. Zhang, F. Li, Y.-Q. Li, L.-X. Ruan, X. Hong, S.-L. Shi, *Angew. Chem. Int. Ed.* **2018**, *57*, 1376–1380.

20. Y. Zhang, G. Lavigne, N. Lugan, V. César, *Chem. Eur. J.* **2017**, *23*, 13792–13801.

21. E. Spahn, A. Albright, M. Shevlin, L. Pauli, A. Pfaltz, R. E. Gawley, *J. Org. Chem.* **2013**, *78*, 2731–2735.

22. L. Benhamou, E. Chardon, G. Lavigne, S. Bellemin-Laponnaz, V. César, *Chem. Rev.* **2011**, *111*, 2705–2733.

23. W.-B. Zhang, X.-T. Yang, J.-B. Ma, Z.-M. Su, S.-L. Shi, *J. Am. Chem. Soc.* **2019**, *141*, 5628–5634.

24. J. Diesel, D. Grosheva, S. Kodama, N. Cramer, *Angew. Chem. Int. Ed.* **2019**, *58*, 11044–11048.

25. Y. Cai, X. Ye, S. Liu, S.-L. Shi, *Angew. Chem. Int. Ed.* **2019**, *58*, 13433–13437.

26. Y. Cai, J.-W. Zhang, F. Li, J.-M. Liu, S.-L. Shi, *ACS Catal.* **2019**, *9*, 1–6.

27. D. Shen, Y. Xu, S.-L. Shi, *J. Am. Chem. Soc.* **2019**, *141*, 14938–14945.

28. G. Sipos, A. Ou, B. W. Skelton, L. Falivene, L. Cavallo, R. Dorta, *Chem. Eur. J.* **2016**, *22*, 6939–6946.

29. L. Wu, L. Falivene, E. Drinkel, S. Grant, A. Linden, L. Cavallo, R. Dorta, *Angew. Chem. Int. Ed.* **2012**, *51*, 2870–2873.

30. J. S. E. Ahlin, N. Cramer, *Org. Lett.* **2016**, *18*, 3242–3245.
31. H. Wang, G. Lu, G. J. Sormunen, H. A. Malik, P. Liu, J. Montgomery, *J. Am. Chem. Soc.* **2017**, *139*, 9317–9324.
32. E. P. Jackson, J. Montgomery, *J. Am. Chem. Soc.* **2015**, *137*, 958–963.
33. J. Loup, U. Dhawa, F. Pesciaioli, J. Wencel-Delord, L. Ackermann, *Angew. Chem. Int. Ed.* **2019**, *58*, 12803–12818.
34. J. Loup, D. Zell, J. C. A. Oliveira, H. Keil, D. Stalke, L. Ackermann, *Angew. Chem. Int. Ed.* **2017**, *56*, 14197–14201.
35. D. M. Flanigan, F. Romanov-Michailidis, N. A. White, T. Rovis, *Chem. Rev.* **2015**, *115*, 9307–9387.
36. C. Guo, D. Janssen-Müller, M. Fleige, A. Lerchen, C. G. Daniliuc, F. Glorius, *J. Am. Chem. Soc.* **2017**, *139*, 4443–4451.
37. J. P. Reid, M. Hu, S. Ito, B. Huang, C. M. Hong, H. Xiang, M. S. Sigman, F. D. Toste, *Chem. Sci.* **2020**.
38. C.-Y. Wu, T. Horibe, C. B. Jacobsen, F. D. Toste, *Nature* **2015**, *517*, 449.
39. a) M. Alcarazo, S. J. Roseblade, A. R. Cowley, R. Fernández, J. M. Brown, J. M. Lassaletta, *J. Am. Chem. Soc.* **2005**, *127*, 3290–3291; b) C. Burstein, C. W. Lehmann, F. Glorius, *Tetrahedron* **2005**, *61*, 6207–6217.
40. Y. Ma, S. Wei, J. Lan, J. Wang, R. Xie, J. You, *J. Org. Chem.* **2008**, *73*, 8256–8264.
41. J. Iglesias-Siguenza, C. Izquierdo, E. Diez, R. Fernandez, J. M. Lassaletta, *Dalton Trans.* **2016**, *45*, 10113–10117.
42. J. Francos, F. Grande-Carmona, H. Faustino, J. Iglesias-Sigüenza, E. Díez, I. Alonso, R. Fernández, J. M. Lassaletta, F. López, J. L. Mascareñas, *J. Am. Chem. Soc.* **2012**, *134*, 14322–14325.
43. I. Varela, H. Faustino, E. Díez, J. Iglesias-Sigüenza, F. Grande-Carmona, R. Fernández, J. M. Lassaletta, J. L. Mascareñas, F. López, *ACS Catal.* **2017**, *7*, 2397–2402.
44. C. T. Check, K. P. Jang, C. B. Schwamb, A. S. Wong, M. H. Wang, K. A. Scheidt, *Angew. Chem. Int. Ed.* **2015**, *54*, 4264–4268.
45. C. B. Schwamb, K. P. Fitzpatrick, A. C. Brueckner, H. C. Richardson, P. H. Y. Cheong, K. A. Scheidt, *J. Am. Chem. Soc.* **2018**, *140*, 10644–10648.
46. L. Kong, J. Morvan, D. Pichon, M. Jean, M. Albalat, T. Vives, S. Colombel-Rouen, M. Giorgi, V. Dorcet, T. Roisnel, C. Crévisy, D. Nuel, P. Nava, S. Humbel, N. Vanthuyne, M. Mauduit, H. Clavier, *J. Am. Chem. Soc.* **2020**, *142*, 93–98.
47. D. Pichon, M. Soleilhavoup, J. Morvan, G. P. Junor, T. Vives, C. Crévisy, V. Lavallo, J.-M. Campagne, M. Mauduit, R. Jazzar, G. Bertrand, *Chem. Sci.* **2019**, *10*, 7807–7811.
48. M. Melaimi, R. Jazzar, M. Soleilhavoup, G. Bertrand, *Angew. Chem. Int. Ed.* **2017**, *56*, 10046–10068.
49. a) E. S. Gauthier, L. Abella, N. Hellou, B. Darquié, E. Caytan, T. Roisnel, N. Vanthuyne, L. Favereau, M. Srebro-Hooper, J. A. G. Williams, J. Autschbach, J. Crassous, *Angew. Chem. Int. Ed.* **2020**, *59*, 8394–8400; b) N. Hellou, M. Srebro-Hooper, L. Favereau, F. Zinna, E. Caytan, L. Toupet, V. Dorcet, M. Jean, N. Vanthuyne, J. A. G. Williams, L. Di Bari, J. Autschbach, J. Crassous, *Angew. Chem. Int. Ed.* **2017**, *56*, 8236–8239.
50. I. G. Sanchez, M. Samal, J. Nejedly, M. Karras, J. Klivar, J. Rybacek, M. Budesinsky, L. Bednarova, B. Seidlerova, I. G. Stara, I. Stary, *Chem. Commun.* **2017**, *53*, 4370–4373.
51. M. Karras, M. Dąbrowski, R. Pohl, J. Rybáček, J. Vacek, L. Bednárová, K. Grela, I. Starý, I. G. Stará, B. Schmidt, *Chem. Eur. J.* **2018**, *24*, 10994–10998.
52. C. Fliedel, A. Labande, E. Manoury, R. Poli, *Coord. Chem. Rev.* **2019**, *394*, 65–103.
53. a) L. H. Gade, V. César, S. Bellemin-Laponnaz, *Angew. Chem. Int. Ed.* **2004**, *43*, 1014–1017; b) M. T. Powell, D.-R. Hou, M. C. Perry, X. Cui, K. Burgess, *J. Am. Chem. Soc.* **2001**, *123*, 8878–8879.
54. C. A. Swamy, P. A. Varenikov, G. de Ruiter, *Organometallics* **2020**, *39*, 247–257.

55. S. Khumsubdee, H. Zhou, K. Burgess, *J. Org. Chem.* **2013**, *78*, 11948–11955.
56. S. Khumsubdee, Y. Fan, K. Burgess, *J. Org. Chem.* **2013**, *78*, 9969–9974.
57. a) T. Soeta, Y. Hatanaka, T. Ishizaka, Y. Ukaji, *Tetrahedron* **2018**, *74*, 4601–4605; b) L. Yang, Y. Li, P. Mao, J. Yuan, Y. Xiao, *Phosphorus, Sulfur Silicon Relat. Elem.* **2019**, *194*, 780–788.
58. a) K. Y. Wan, A. J. Lough, R. H. Morris, *Organometallics* **2016**, *35*, 1604–1612; b) K. Y. Wan, M. M. H. Sung, A. J. Lough, R. H. Morris, *ACS Catal.* **2017**, *7*, 6827–6842; c) K. Y. Wan, F. Roelfes, A. J. Lough, F. E. Hahn, R. H. Morris, *Organometallics* **2018**, *37*, 491–504.
59. A. Schumacher, M. Bernasconi, A. Pfaltz, *Angew. Chem. Int. Ed.* **2013**, *52*, 7422–7425.
60. a) H. Chiyojima, S. Sakaguchi, *Tetrahedron Lett.* **2011**, *52*, 6788–6791; b) J. Kondo, A. Harano, K. Dohi, S. Sakaguchi, *J. Mol. Catal. A: Chem.* **2014**, *395*, 66–71; c) K. Shinohara, S. Kawabata, H. Nakamura, Y. Manabe, S. Sakaguchi, *Eur. J. Org. Chem.* **2014**, *2014*, 5532–5539; d) Y. Nakano, S. Sakaguchi, *J. Organomet. Chem.* **2017**, *846*, 407–416.
61. E. Bappert, G. Helmchen, *Synlett* **2004**, *2004*, 1789–1793.
62. T. J. Seiders, D. W. Ward, R. H. Grubbs, *Org. Lett.* **2001**, *3*, 3225–3228.
63. J.-M. Becht, E. Bappert, G. Helmchen, *Adv. Synth. Catal.* **2005**, *347*, 1495–1498.
64. T. Focken, G. Raabe, C. Bolm, *Tetrahedron: Asymmetry* **2004**, *15*, 1693–1706.
65. a) P. Gu, J. Zhang, Q. Xu, M. Shi, *Dalton Trans.* **2013**, *42*, 13599–13606; b) P. Gu, Q. Xu, M. Shi, *Tetrahedron* **2014**, *70*, 7886–7892.
66. a) S. Gischig, A. Togni, *Eur. J. Inorg. Chem.* **2005**, *2005*, 4745–4754; b) F. Visentin, A. Togni, *Organometallics* **2007**, *26*, 3746–3754; c) N. Debono, A. Labande, E. Manoury, J.-C. Daran, R. Poli, *Organometallics* **2010**, *29*, 1879–1882.
67. a) R. Hodgson, R. E. Douthwaite, *J. Organomet. Chem.* **2005**, *690*, 5822–5831; b) S. Nanchen, A. Pfaltz, *Helv. Chim. Acta* **2006**, *89*, 1559–1573.
68. N. Toselli, D. Martin, G. Buono, *Org. Lett.* **2008**, *10*, 1453–1456.
69. a) A. Ros, D. Monge, M. Alcarazo, E. Álvarez, J. M. Lassaletta, R. Fernández, *Organometallics* **2006**, *25*, 6039–6046; b) S. J. Roseblade, A. Ros, D. Monge, M. Alcarazo, E. Álvarez, J. M. Lassaletta, R. Fernández, *Organometallics* **2007**, *26*, 2570–2578; c) A. Ros, M. Alcarazo, J. Iglesias-Sigüenza, E. Díez, E. Álvarez, R. Fernández, J. M. Lassaletta, *Organometallics* **2008**, *27*, 4555–4564; d) J. Iglesias-Sigüenza, A. Ros, E. Díez, A. Magriz, A. Vázquez, E. Álvarez, R. Fernández, J. M. Lassaletta, *Dalton Trans.* **2009**, 8485–8488.
70. P. de la Cruz-Sánchez, J. Faiges, Z. Mazloomi, C. Borràs, M. Biosca, O. Pàmies, M. Diéguez, *Organometallics* **2019**, *38*, 4193–4205.
71. a) D. Zhang, J. Yu, *Organometallics* **2020**, *39*, 1269–1280; b) J. Tang, Y. He, J. Yu, D. Zhang, *Organometallics* **2017**, *36*, 1372–1382.
72. a) S. Urban, N. Ortega, F. Glorius, *Angew. Chem. Int. Ed.* **2011**, *50*, 3803–3806; b) N. Ortega, S. Urban, B. Beiring, F. Glorius, *Angew. Chem. Int. Ed.* **2012**, *51*, 1710–1713; c) S. Urban, B. Beiring, N. Ortega, D. Paul, F. Glorius, *J. Am. Chem. Soc.* **2012**, *134*, 15241–15244; d) C. Schlepphorst, M. P. Wiesenfeldt, F. Glorius, *Chem. Eur. J.* **2018**, *24*, 356–359.
73. D. Paul, B. Beiring, M. Plois, N. Ortega, S. Kock, D. Schlüns, J. Neugebauer, R. Wolf, F. Glorius, *Organometallics* **2016**, *35*, 3641–3646.
74. J. H. Kim, S. Gressies, M. Boultadakis-Arapinis, C. Daniliuc, F. Glorius, *ACS Catal.* **2016**, *6*, 7652–7656.
75. a) T. E. Schmid, S. Drissi-Amraoui, C. Crévisy, O. Baslé, M. Mauduit, *Beilstein J. Org. Chem.* **2015**, *11*, 2418–2434; b) A. H. Hoveyda, *J. Org. Chem.* **2014**, *79*, 4763–4792; c) J. Wencel, H. Hénon, S. Kehrli, M. Mauduit, A. Alexakis, *Aldrichimica Acta* **2009**, *42*, 43–50; d) A. H. Hoveyda, D. G. Gillingham, J. J. Van Veldhuizen, O. Kataoka, S. B. Garber, J. S. Kingsbury, J. P. A. Harrity, *Org. Biomol. Chem.* **2004**, *2*, 8–23.

76. a) D. Rix, S. Labat, L. Toupet, C. Crévisy, M. Mauduit, *Eur. J. Inorg. Chem.* **2009**, *2009*, 1989–1999; b) H. Clavier, L. Coutable, L. Toupet, J.-C. Guillemin, M. Mauduit, *J. Organomet. Chem.* **2005**, *690*, 5237–5254; c) H. Clavier, L. Coutable, J.-C. Guillemin, M. Mauduit, *Tetrahedron: Asymmetry* **2005**, *16*, 921–924.
77. N. Germain, M. Magrez, S. Kehrli, M. Mauduit, A. Alexakis, *Eur. J. Org. Chem.* **2012**, *2012*, 5301–5306.
78. S. Drissi-Amraoui, M. S. T. Morin, C. Crévisy, O. Baslé, R. Marcia de Figueiredo, M. Mauduit, *Angew. Chem. Int. Ed.* **2015**, *54*, 11830–11834.
79. M. Tissot, D. Poggiali, H. Hénon, D. Müller, L. Guénée, M. Mauduit, A. Alexakis, *Chem. Eur. J.* **2012**, *18*, 8731–8747.
80. C. Jahier-Diallo, M. S. T. Morin, P. Queval, M. Rouen, I. Artur, P. Querard, L. Toupet, C. Crévisy, O. Baslé, M. Mauduit, *Chem. Eur. J.* **2015**, *21*, 993–997.
81. K. Takatsu, R. Shintani, T. Hayashi, *Angew. Chem. Int. Ed.* **2011**, *50*, 5548–5552.
82. Q. Chong, Z. Yue, S. Zhang, C. Ji, F. Cheng, H. Zhang, X. Hong, F. Meng, *ACS Catal.* **2017**, *7*, 5693–5698.
83. F. Meng, X. Li, S. Torker, Y. Shi, X. Shen, A. H. Hoveyda, *Nature* **2016**, *537*, 387–393.
84. X. Li, F. Meng, S. Torker, Y. Shi, A. H. Hoveyda, *Angew. Chem. Int. Ed.* **2016**, *55*, 9997–10002.
85. W. J. Jang, J. Yun, *Angew. Chem. Int. Ed.* **2019**, *58*, 18131–18135.
86. R. Tarrieu, A. Dumas, J. Thongpaen, T. Vives, T. Roisnel, V. Dorcet, C. Crévisy, O. Baslé, M. Mauduit, *J. Org. Chem.* **2017**, *82*, 1880–1887.
87. A. Harada, Y. Makida, T. Sato, H. Ohmiya, M. Sawamura, *J. Am. Chem. Soc.* **2014**, *136*, 13932–13939.
88. Y. Yasuda, H. Ohmiya, M. Sawamura, *Angew. Chem. Int. Ed.* **2016**, *55*, 10816–10820.
89. H. Ohmiya, H. Zhang, S. Shibata, A. Harada, M. Sawamura, *Angew. Chem. Int. Ed.* **2016**, *55*, 4777–4780.
90. K. Hojoh, H. Ohmiya, M. Sawamura, *J. Am. Chem. Soc.* **2017**, *139*, 2184–2187.
91. K. Kubota, S. Osaki, M. Jin, H. Ito, *Angew. Chem. Int. Ed.* **2017**, *56*, 6646–6650.
92. Y. Lee, A. H. Hoveyda, *J. Am. Chem. Soc.* **2006**, *128*, 15604–15605.
93. a) D. Grassi, A. Alexakis, *Adv. Synth. Catal.* **2015**, *357*, 3171–3186; b) D. Grassi, C. Dolka, O. Jackowski, A. Alexakis, *Chem. Eur. J.* **2013**, *19*, 1466–1475; c) O. Jackowski, A. Alexakis, *Angew. Chem. Int. Ed.* **2010**, *49*, 3346–3350.
94. a) C. Michon, F. Medina, M.-A. Abadie, A. Agbossou-Niedercorn, *Organometallics* **2013**, *32*, 5589–5600; b) M. A. Gilani, E. Rais, R. Wilhelm, *Synlett* **2015**, *26*, 1638–1641.
95. a) M. Yoshimura, R. Kamisue, S. Sakaguchi, *J. Organomet. Chem.* **2013**, *740*, 26–32; b) S. Urban, N. Ortega, F. Glorius, *Angew. Chem. Int. Ed.* **2011**, *50*, 3803–3806.
96. S. Raoufmoghaddam, S. Mannathan, A. J. Minnaard, J. G. de Vries, B. de Bruin, J. N. H. Reek, *ChemCatChem* **2018**, *10*, 266–272.
97. M. K. Brown, T. L. May, C. A. Baxter, A. H. Hoveyda, *Angew. Chem. Int. Ed.* **2007**, *46*, 1097–1100.
98. D. G. Gillingham, A. H. Hoveyda, *Angew. Chem. Int. Ed.* **2007**, *46*, 3860–3864.
99. A. H. Hoveyda, Y. Zhou, Y. Shi, K. Brown, H. Wu, S. Torker, *Angew. Chem. Int. Ed.* DOI: 10.1002/anie.202003755.
100. Y. Lee, K. Akiyama, D. G. Gillingham, M. K. Brown, A. H. Hoveyda, *J. Am. Chem. Soc.* **2008**, *130*, 446–447.
101. C. M. Latham, A. J. Blake, W. Lewis, M. Lawrence, S. Woodward, *Eur. J. Org. Chem.* **2012**, *2012*, 699–707.
102. M. A. Kacprzynski, T. L. May, S. A. Kazane, A. H. Hoveyda, *Angew. Chem. Int. Ed.* **2007**, *46*, 4554–4558.

103. a) K. Akiyama, F. Gao, A. H. Hoveyda, *Angew. Chem. Int. Ed.* **2010**, *49*, 419–423; b) F. Gao, A. H. Hoveyda, *J. Am. Chem. Soc.* **2010**, *132*, 10961–10963; c) F. Gao, K. P. McGrath, Y. Lee, A. H. Hoveyda, *J. Am. Chem. Soc.* **2010**, *132*, 14315–14320.
104. a) J. A. Dabrowski, F. Gao, A. H. Hoveyda, *J. Am. Chem. Soc.* **2011**, *133*, 4778–4781; b) J. A. Dabrowski, F. Haeffner, A. H. Hoveyda, *Angew. Chem. Int. Ed.* **2013**, *52*, 7694–7699.
105. B. Jung, A. H. Hoveyda, *J. Am. Chem. Soc.* **2012**, *134*, 1490–1493.
106. Y. Zhou, Y. Shi, S. Torker, A. H. Hoveyda, *J. Am. Chem. Soc.* **2018**, *140*, 16842–16854.
107. a) F. Gao, J. L. Carr, A. H. Hoveyda, *Angew. Chem. Int. Ed.* **2012**, *51*, 6613–6617; b) F. Gao, J. L. Carr, A. H. Hoveyda, *J. Am. Chem. Soc.* **2014**, *136*, 2149–2161.
108. Y. Shi, B. Jung, S. Torker, A. H. Hoveyda, *J. Am. Chem. Soc.* **2015**, *137*, 8948–8964.
109. Y. Shi, A. H. Hoveyda, *Angew. Chem. Int. Ed.* **2016**, *55*, 3455–3458.
110. a) T. Hayashi, K. Yamasaki, *Chem. Rev.* **2003**, *103*, 2829–2844; b) D. Müller, A. Alexakis, *Chem. Commun.* **2012**, *48*, 12037–12049; c) M. M. Heravi, M. Dehghani, V. Zadsirjan, *Tetrahedron: Asymmetry* **2016**, *27*, 513–588.
111. a) J. Wencel, H. Hénon, S. Kehrli, M. Mauduit, A. Alexakis, *Aldrichimica Acta* **2009**, *42*, 43–50; b) D. Pichon, J. Morvan, C. Crévisy, M. Mauduit, *Beilstein J. Org. Chem.* **2020**, *16*, 212–232.
112. J. A. Dabrowski, M. T. Villaume, A. H. Hoveyda, *Angew. Chem. Int. Ed.* **2013**, *52*, 8156–8159.
113. a) P. Hu, H. M. Chi, K. C. DeBacker, X. Gong, J. H. Keim, I. T. Hsu, S. A. Snyder, *Nature* **2019**, *569*, 703–707; b) C. Peng, P. Arya, Z. Zhou, S. A. Snyder, *Angew. Chem. Int. Ed.* **2020**, *59*, 13521–13525.
114. K. P. McGrath, A. K. Hubbell, Y. Zhou, D. P. Santos, S. Torker, F. Romiti, A. H. Hoveyda, *Adv. Synth. Catal.* **2020**, *362*, 370–375.
115. D. S. Müller, N. L. Untiedt, A. P. Dieskau, G. L. Lackner, L. E. Overman, *J. Am. Chem. Soc.* **2015**, *137*, 660–663.
116. K. P. McGrath, A. H. Hoveyda, *Angew. Chem. Int. Ed.* **2014**, *53*, 1910–1914.
117. a) T. L. May, J. A. Dabrowski, A. H. Hoveyda, *J. Am. Chem. Soc.* **2011**, *133*, 736–739; b) T. L. May, J. A. Dabrowski, A. H. Hoveyda, *J. Am. Chem. Soc.* **2014**, *136*, 10544–10544.
118. B. S. L. Collins, C. M. Wilson, E. L. Myers, V. K. Aggarwal, *Angew. Chem. Int. Ed.* **2017**, *56*, 11700–11733.
119. Y. Lee, A. H. Hoveyda, *J. Am. Chem. Soc.* **2009**, *131*, 3160–3161.
120. R. Corberán, N. W. Mszar, A. H. Hoveyda, *Angew. Chem. Int. Ed.* **2011**, *50*, 7079–7082.
121. H. Jang, B. Jung, A. H. Hoveyda, *Org. Lett.* **2014**, *16*, 4658–4661.
122. F. Meng, H. Jang, A. H. Hoveyda, *Chem. – Eur. J.* **2013**, *19*, 3204–3214.
123. S. J. Meek, R. V. O'Brien, J. Llaveria, R. R. Schrock, A. H. Hoveyda, *Nature* **2011**, *471*, 461–466.
124. H. Lee, S. Lee, J. Yun, *ACS Catal.* **2020**, *10*, 2069–2073.
125. A. Guzman-Martinez, A. H. Hoveyda, *J. Am. Chem. Soc.* **2010**, *132*, 10634–10637.
126. a) L. B. Delvos, D. J. Vyas, M. Oestreich, *Angew. Chem. Int. Ed.* **2013**, *52*, 4650–4653; b) M. Takeda, R. Shintani, T. Hayashi, *J. Org. Chem.* **2013**, *78*, 5007–5017.

7 Chiral Monophosphorus Ligands

Wenjun Tang, Feng Wan, and He Yang

CONTENTS

7.1   INTRODUCTION

The development of efficient methods for producing enantiomerically enriched compounds from readily available achiral or racemic starting materials has been one of the most important tasks in modern organic synthesis.[1] Among various synthetic methods, transition-metal-catalyzed asymmetric transformations have attracted considerable attention and have become powerful tools for the synthesis of valuable chiral products. Chiral ligands have played a crucial role in asymmetric catalysis, because the chirality and enantiomeric purity of the products are governed by the catalyst, bearing chiral ligands. The development of effective chiral ligands has thus become one of most vigorous and dynamic research areas in search for high reactivity, enantioselectivity, and efficiency. Chiral phosphorus ligands were among the earliest to be explored and continue to be among the most effective in asymmetric catalysis. In comparison with chiral bisphosphorus ligands, which are among the earliest and most frequently used ligands in asymmetric catalytic reactions,[2] the development of chiral monophosphorus ligands is relatively underdeveloped, and their catalytic properties and synthetic applications have rarely been reviewed.[3]

Initial research on chiral monophosphorus ligands was documented half a century ago, when the *P*-chiral monophosphorus ligands CAMP and NMDPP were first introduced in the early 1970s by Knowles[4] and Morrison,[5] respectively. Nevertheless, the development of chiral monophosphorus ligands was slow for quite a long time and it was not until the late 1990s that the development and applications of chiral monophosphorus ligands started to flourish. Based on the origin of the chirality, chiral phosphorus ligands can be classified into two categories: ligands with *P*-chirality, and those with backbone chirality. Recently, a number of electron-rich, conformationally rigid, and air-stable *P*-chiral monophosphorus ligands have been developed, despite difficulties with their synthesis (**Scheme 7.1a**).[6] Most well-known monophosphorus ligands exhibit backbone chirality, equipped with privileged frameworks, including chiral 1,1'-binaphthyl, BINOL, TADDOL, and

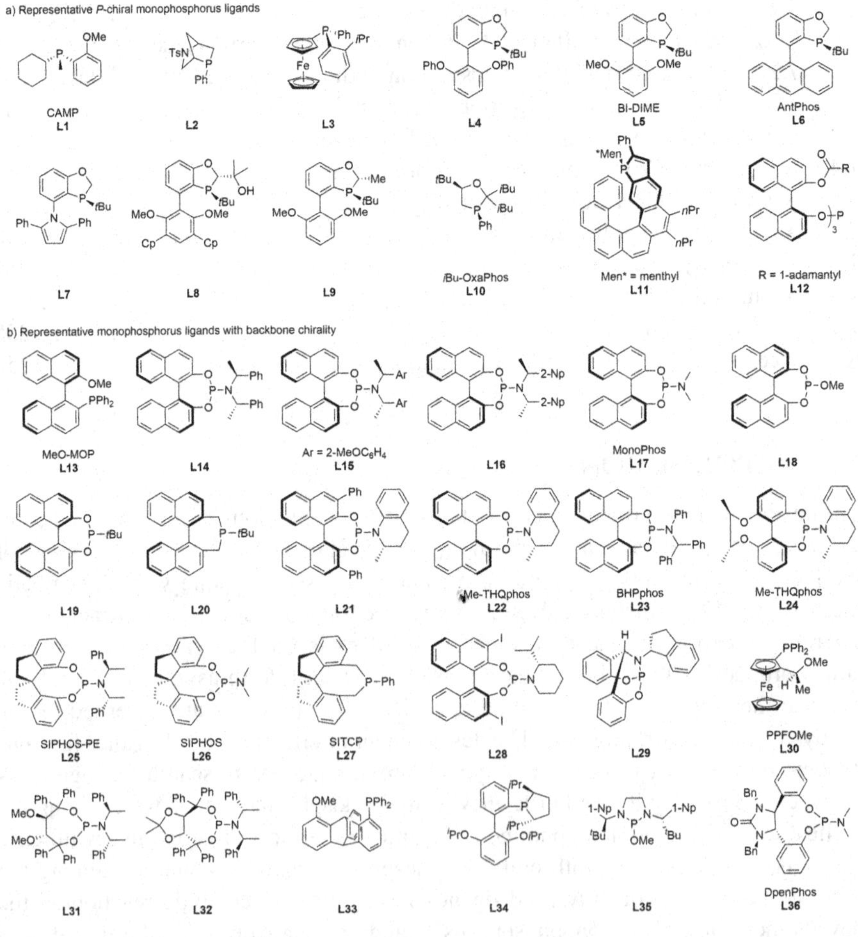

SCHEME 7.1 Representative monophosphorus ligands with *P*-chirality and backbone chirality.

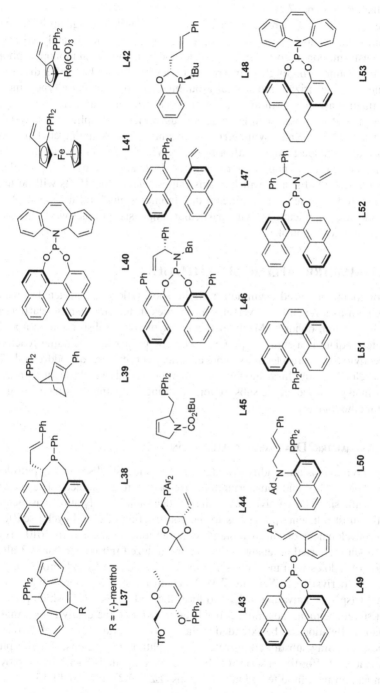

SCHEME 7.2 Representative chiral monophosphorus ligands, possessing an olefin moiety.

spirobiindane, etc. (**Scheme 7.1b**).[7] It should be noted that chiral monophosphorus ligands possessing an olefin moiety[8] as a labile coordination site are also included in this chapter (**Scheme 7.2**).

There have been a few reviews on chiral monophosphorus ligands.[9] To avoid redundancy, this chapter covers the progress of transition-metal-catalyzed asymmetric transformation, reported in the past decade by using chiral monophosphorus ligands, but examples highlighted in a recent review[10] will not be further discussed. The content is categorized into several groupings, based on reaction types, including asymmetric allylic substitution, asymmetric dearomative and Heck-type cyclization, asymmetric cross-coupling reaction, asymmetric coupling of π-systems, asymmetric addition, and asymmetric hydrogenation. It is anticipated that this review should provide readers with a brief overview of recent developments and applications of chiral monophosphorus ligands. Instead of listing all reported chiral monophosphorus ligands, which seems impossible to do, efforts will be made to describe in detail the most significant developments on chiral monophosphorus ligands and their applications, with particular emphasis on reactivity, selectivity, and stereochemical control.

7.2 ASYMMETRIC ALLYLIC SUBSTITUTION

Transition-metal-catalyzed asymmetric allylic substitution has attracted considerable attention as a result of its versatile properties in forming carbon–carbon and carbon–heteroatom bonds.[11] Mechanistically, asymmetric substitution on a π-allyl metal intermediate is a complex process, whereby properties, including reactivity, regioselectivity, diasteroselectivity, and enantioselectvity, are all addressed. The recent development in chiral monophosphoramide ligands has brought significant progress in asymmetric allylic substitution, resulting in a number of interesting and efficient methodologies.

7.2.1 ASYMMETRIC DEAROMATIVE ALLYLATION

Aromatic compounds are readily available chemical feedstocks from the petroleum industry. Their utility and transformation have been a major subject of research in organic synthesis. The catalytic asymmetric dearomative allylation has become one of the most efficient approaches to the construction of chiral cyclic scaffolds, with the development of chiral monophosphoramidite ligands.[12] In 2010, You[13] and coworkers reported an enantioselective Ir-catalyzed intramolecular C-3 allylic alkylation of indoles with monophosphoramidite **L22** as the ligand, producing spiroindolenine derivative **2** (**Scheme 7.3**). Mechanistic studies offered evidence on the initial C(sp²)–H activation of **L22** to form the real catalyst **C1**, which provided efficient stereochemical induction. It has been shown that the catalytic asymmetric dearomative allylation can be extended to the dearomatization of other electron-rich aromatics, including substituted pyrroles,[14] β-naphthols,[15] phenols,[16] and 2-phenylpropanoates,[17] affording a series of chiral spiro compounds **3–6**[18] by employing chiral monophosphoramidite ligands THQphos (**L22**), **L21**, **L14**, and **L15**.

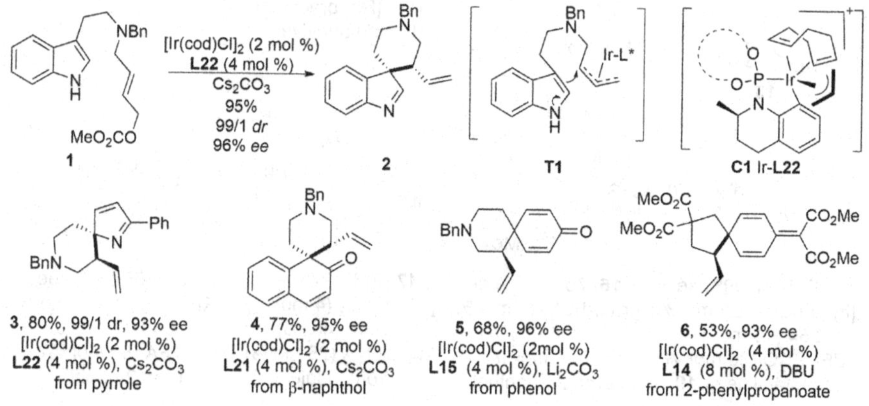
NBn
MeO2CO
1
[Ir(cod)Cl]2 (2 mol %)
L22 (4 mol %)
Cs2CO3
95%
99/1 dr
96% ee
Bn
N
2
Bn
N
Ir-L*
T1
O
O-P-Ir
N
C1 Ir-L22
Ph
BnN
3, 80%, 99/1 dr, 93% ee
[Ir(cod)Cl]2 (2 mol %)
L22 (4 mol %), Cs2CO3
from pyrrole
Bn
N
O
4, 77%, 95% ee
[Ir(cod)Cl]2 (2 mol %)
L21 (4 mol %), Cs2CO3
from β-naphthol
BnN
O
5, 68%, 96% ee
[Ir(cod)Cl]2 (2mol %)
L15 (4 mol %), Li2CO3
from phenol
MeO2C
MeO2C
CO2Me
CO2Me
6, 53%, 93% ee
[Ir(cod)Cl]2 (4 mol %)
L14 (8 mol %), DBU
from 2-phenylpropanoate

SCHEME 7.3 The Ir-catalyzed intramolecular asymmetric allylic dearomatization reaction.

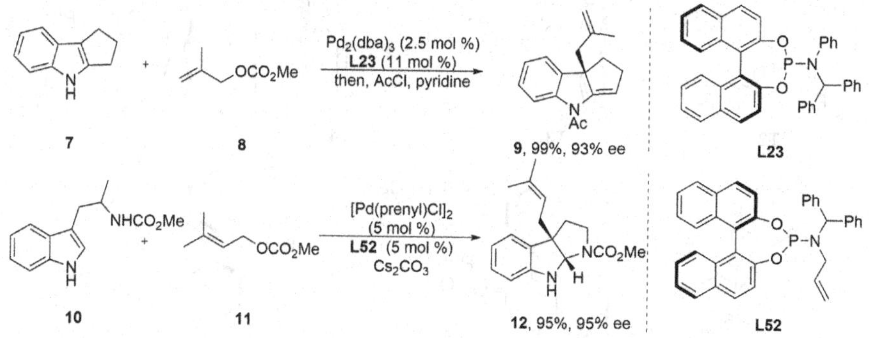
N
H
7
OCO2Me
8
Pd2(dba)3 (2.5 mol %)
L23 (11 mol %)
then, AcCl, pyridine
N
Ac
9, 99%, 93% ee
O
P-N
O
Ph
Ph
Ph
L23
NHCO2Me
10
OCO2Me
11
[Pd(prenyl)Cl]2
(5 mol %)
L52 (5 mol %)
Cs2CO3
NCO2Me
12, 95%, 95% ee
Ph
Ph
P-N
L52

SCHEME 7.4 The Pd-catalyzed intermolecular asymmetric allylic dearomatization reaction.

The Pd-catalyzed intermolecular dearomative allylation between substituted indoles and allylic carbonates was realized by You[19] and coworkers with the chiral monophosphoramidite ligand **L23** (BHQphos), forming a series of cyclic imines, bearing an all-carbon quaternary stereocenter, with excellent yields and enantioselectivities (**Scheme 7.4**). By using a monophosphoramidite ligand bearing an olefin moiety, allylphos (**L52**), a palladium-catalyzed asymmetric dearomative prenylation was developed to furnish chiral prenylated indoline alkaloid skeletons with excellent enantioselectivities.

Branched allylic alcohols are challenging electrophiles in transition-metal-catalyzed allylic reactions. These substrates could be activated by Lewis or Brønsted acids for oxidative addition by transition metals to generate the corresponding metal π-allyl species. As shown in **Scheme 7.5**, the Ir π-allyl species, coordinated with a chiral monophosphoramidite ligand containing an olefin moiety, was attacked by various nucleophiles, such as anilines, β-naphthols, and indoles, generating the corresponding products **15–17**, respectively, with excellent enantioselectivities.[21–23]

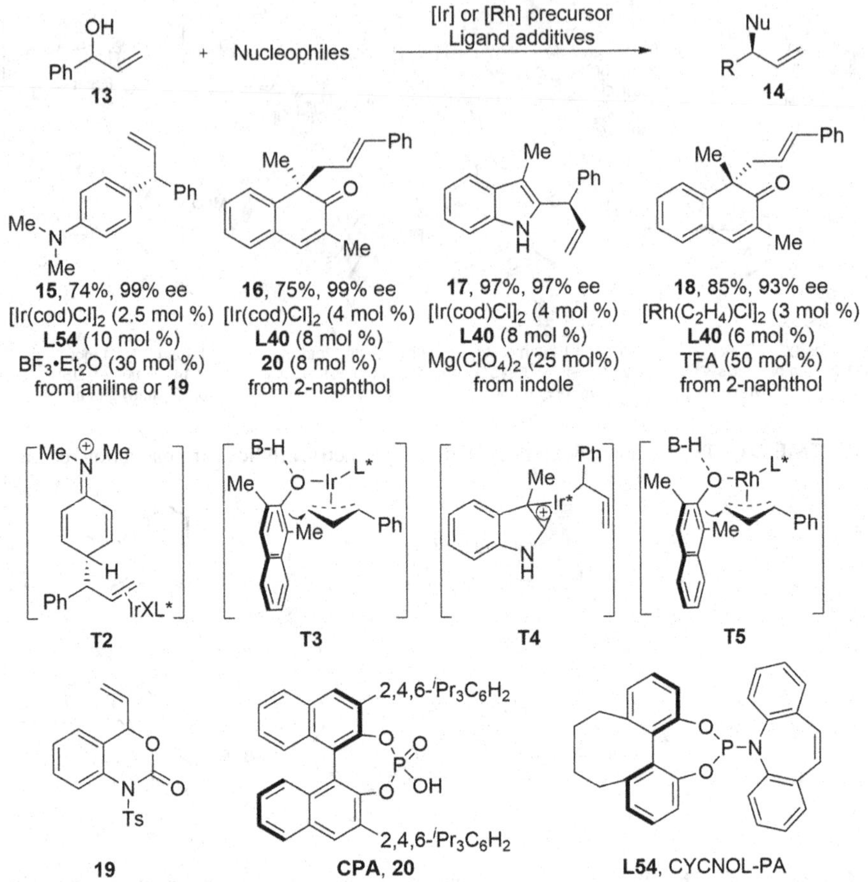

SCHEME 7.5 The asymmetric allylic dearomatization reaction, catalyzed by a chiral monophosphoramidite ligand containing an olefin moiety.

A Rh π-allyl species-mediated asymmetric allylic dearomatization reaction of β-naphthol was also reported.[24]

7.2.2 ASYMMETRIC ALLYLIC SUBSTITUTION WITH OTHER CARBON NUCLEOPHILES

Various stabilized carbon nucleophiles are applicable for Ir-, Pd-, or Rh-catalyzed asymmetric allylic substitution. By employing chiral monophosphorus ligands, the iridium-catalyzed allylic alkylation, between linear or branched allylic carbonates and various carbon nucleophiles, allowed the synthesis of olefins bearing chiral tertiary or quaternary carbon centers with excellent yields and enantioselectivities (**Scheme 7.6**).[25–36]

Ligand **L40** has proved to be an efficient ligand for a number of asymmetric allylic substitutions. When a branched allylic alcohol or allylic carbonate is used as

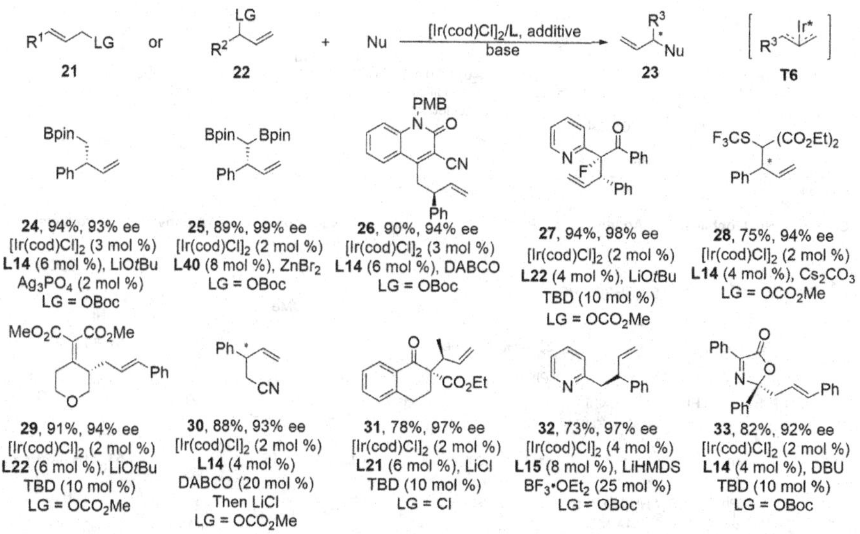

SCHEME 7.6 The Ir-catalyzed asymmetric allylic substitution with various carbon nucleophiles.

the substrate, a Lewis or Brønsted acid ($Sc(OTf)_3$, $Zn(OTf)_2$, $ZnBr_2$, etc.) as additive is usually required to facilitate the oxidative addition process. The combination of Ir-**L40** has turned out to be a versatile catalyst for intermolecular allylic substitutions with various carbon nucleophiles, including organozinc reagents, boron reagents, enolates, enamides, hydrazones, and allyl silanes (**Scheme 7.7**).[37–45] Work by Carreira[37,38a,39,41,43], You[42], Yang[38b,40,42,44,45], and their respective coworkers showed that various allylic alkylation products were formed in high yields and with ee (enantiomeric excess) values exceeding 99%. Moreover, the Ir-catalyzed allylic alkylation generated acyclic products bearing two vicinal all-carbon quaternary centers. Stoltz [36, 46, 47] described the effectiveness of Ir-**L40** catalyst in asymmetric allylic alkylation of trisubstituted allylic carbonate (up to 95% ee).

In 2012, Carreira and coworkers described an enantioselective cation–π polyene cyclization enabled by Ir-**L40** with $Zn(OTf)_2$ as the Lewis-acid promoter. Under optimal conditions, the polyene cyclization proceeded to form the corresponding decaline derivatives, with excellent ee values and yields (**Scheme 7.8**). This cyclization is applicable to total synthesis of a number of natural products.[48, 49]

The Ir-catalyzed asymmetric umpolung allylation/2-aza-Cope rearrangements of ketimines, using **L14** as the ligand, were realized by Wang[50, 51] and Niu[52], respectively. This cascade provided 1,4-disubstituted homoallylic amines at high yields and with excellent enantioselectivities (**Scheme 7.9**). Mechanistic studies revealed **T11** as the key intermediate in the initial umpolung allylation, which was followed by a stereospecific 2-aza-Cope rearrangement.

Besides imines, aldehydes were also suitable substrates for transition-metal-catalyzed allylation. Carreira[53, 54] and coworkers described an Ir-catalyzed branched

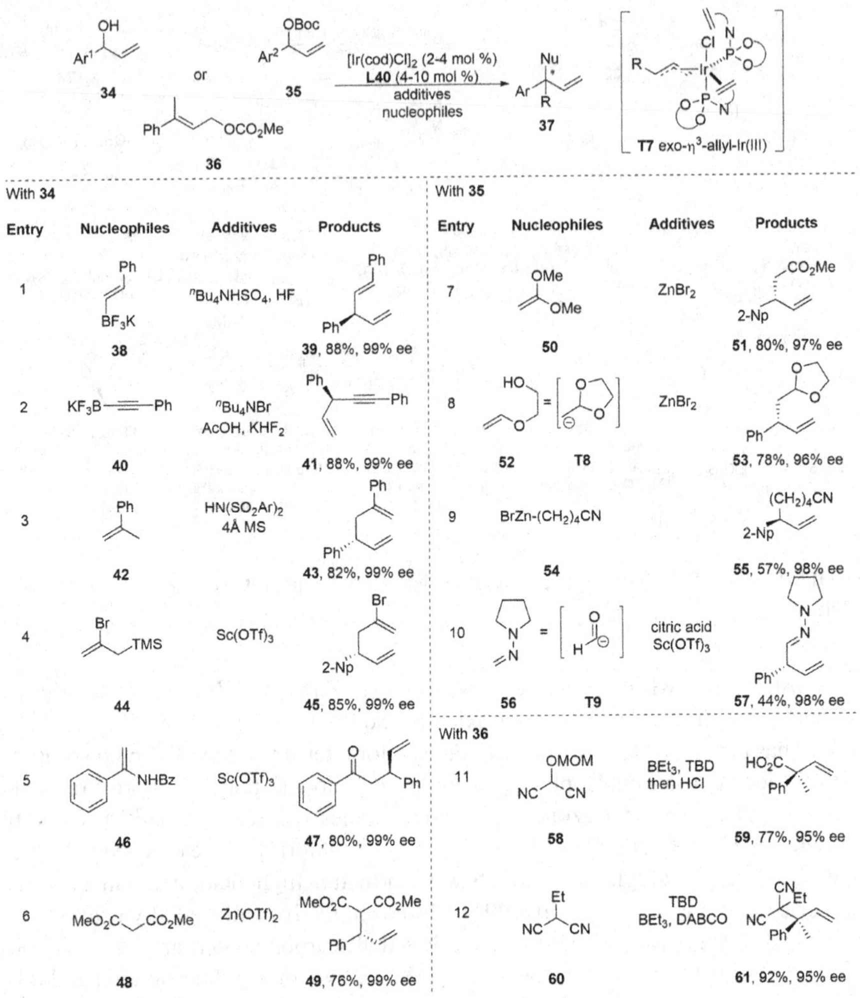

Scheme reagents: **34** (Ar¹ allylic alcohol, OH) or **35** (Ar² allylic carbonate, OBoc) or **36** (Ph–CH=CH–CH₂OCO₂Me) → [Ir(cod)Cl]₂ (2–4 mol %), **L40** (4–10 mol %), additives, nucleophiles → **37** (Ar–C(R)(Nu)–CH=CH₂). **T7** exo-η³-allyl-Ir(III).

With 34

Entry	Nucleophiles	Additives	Products
1	**38** (Ph–CH=CH–BF₃K)	ⁿBu₄NHSO₄, HF	**39**, 88%, 99% ee
2	**40** (KF₃B–C≡C–Ph)	ⁿBu₄NBr, AcOH, KHF₂	**41**, 88%, 99% ee
3	**42** (isopropenyl/Ph)	HN(SO₂Ar)₂, 4Å MS	**43**, 82%, 99% ee
4	**44** (Br, TMS allyl)	Sc(OTf)₃	**45**, 85%, 99% ee
5	**46** (Ph–C(NHBz)=CH₂)	Sc(OTf)₃	**47**, 80%, 99% ee
6	**48** (MeO₂C–CH₂–CO₂Me)	Zn(OTf)₂	**49**, 76%, 99% ee

With 35

Entry	Nucleophiles	Additives	Products
7	**50** (ketene dimethyl acetal, OMe/OMe)	ZnBr₂	**51**, 80%, 97% ee
8	**52** = T8 (HO–CH=CH–O, 1,3-dioxolane)	ZnBr₂	**53**, 78%, 96% ee
9	**54** (BrZn–(CH₂)₄CN)		**55**, 57%, 98% ee
10	**56** = T9 (pyrrolidine enamine)	citric acid, Sc(OTf)₃	**57**, 44%, 98% ee

With 36

Entry	Nucleophiles	Additives	Products
11	**58** (NC–C(OMOM)–CN)	BEt₃, TBD then HCl	**59**, 77%, 95% ee
12	**60** (NC–C(Et)–CN)	TBD, BEt₃, DABCO	**61**, 92%, 95% ee

SCHEME 7.7 The asymmetric allylic substitution powered by the Ir-**L40** catalyst.

α-allylation of linear aldehydes with racemic allylic alcohols, in which an enamine intermediate **T12** was generated *in situ* from the reaction of aldehyde and chiral diarylsilyl prolinol ether co-catalyst (**Scheme 7.10a**). Sunoj[55] and coworkers offered insights on the stereodivergence, while density functional theory (DFT) calculations showed the "recognition/interaction" between the two chiral catalysts in diastereocontrolling the C–C bond formation through a series of weak noncovalent interactions, playing a pivotal role in influencing the stereochemistry. For the strong competitiveness of palladium in forming linear products, Jørgensen[18] revealed the roles of both Pd-**L55** catalyst and diphenylprolinol silyl ether in dual catalytic γ-allylation of unsaturated aldehyde **77** and allylic ester **78** (**Scheme 7.10b**). Gong[56–58] and coworkers also described a Pd-catalyzed asymmetric linear allylation

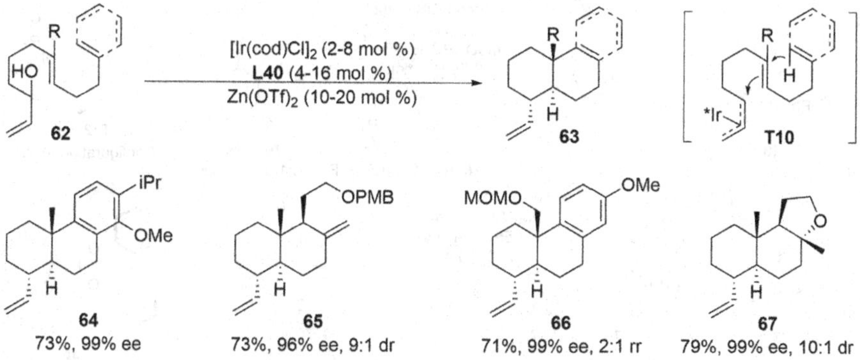

SCHEME 7.8 Ir-catalyzed asymmetric polyene cyclization powered with the Ir-**L40** catalyst.

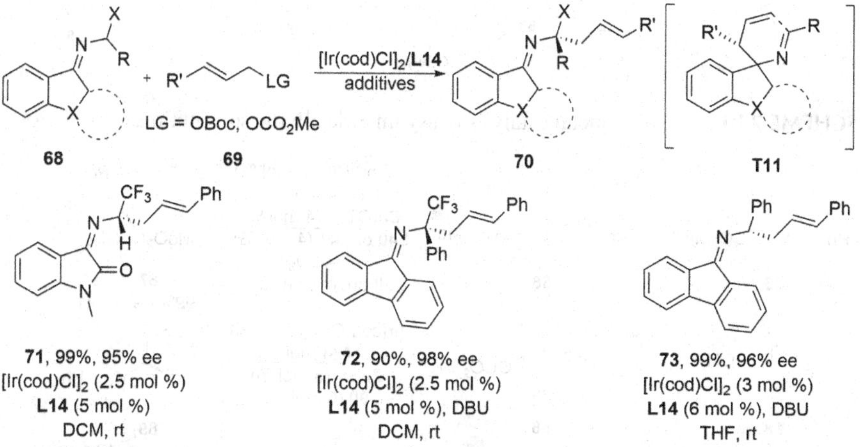

SCHEME 7.9 Ir-catalyzed asymmetric umpolung allylation of imines.

reaction at the benzylic position of furfurals enabled by the synergistic stereochemical control of a chiral TADDOL-based phosphoramidite ligand and a chiral diphenylprolinol silyl ether **83** (**Scheme 7.10c**).

Zhang[59, 60] and Wang[61] later established a dual-metal catalysis mode for the stereodivergent α-allylation of aldimine esters by employing Ir/**L14** or Cu/*P, N* ligand **L57** (Wang) or **L58** (Zhang), respectively, as the catalysts, accomplishing the preparation of a series of nonproteinogenic α-amino acids (α-AAs), bearing two contiguous stereogenic centers with high stereoselectivity (**Scheme 7.11**). Zhang[62] and coworkers also developed a dual Ir/Zn catalyst for the α-allylation of unprotected α-hydroxylketones under mild conditions. According to the proposed catalysis mode, the contributions of both Cu catalyst **T13** and Zn catalyst **T14** were similar to the role of the diphenylprolinol silyl ether in Carreira's research (**Scheme 7.10a**),

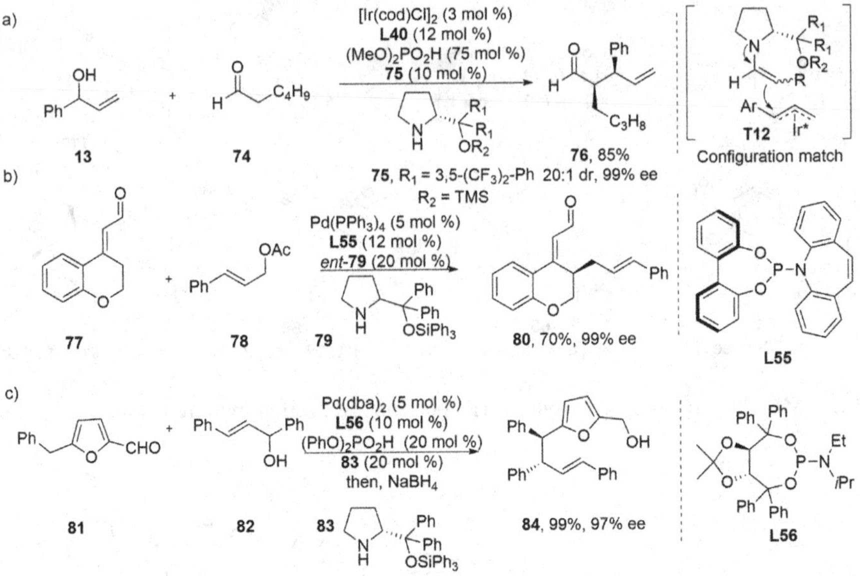

SCHEME 7.10 The dual-metal catalysis in asymmetric allyllic substitution.

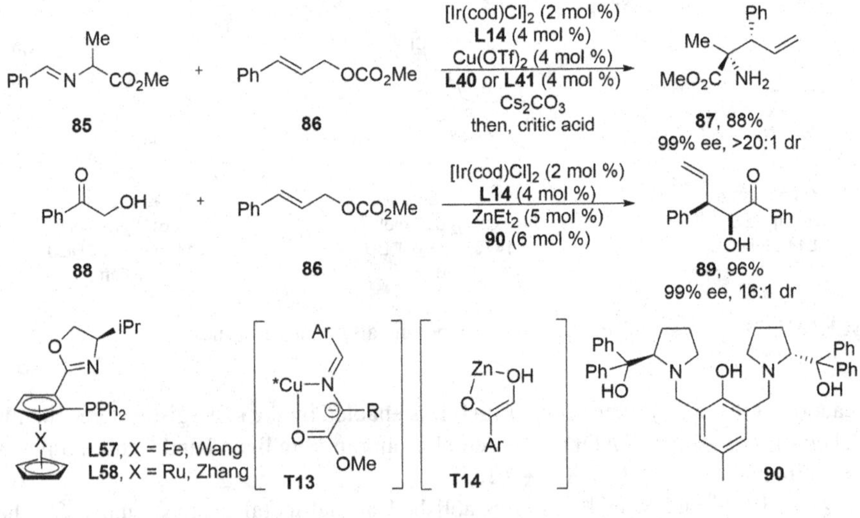

SCHEME 7.11 The dual-metal catalysis of asymmetric allyllic substitution.

leading to high values of ee when the absolute configuration of chiral Ir-catalyst and Cu/Zn catalyst matched.

Copper-catalyzed asymmetric allylic alkylation of organometallic reagents with di- and tri-substituted allylic halides was accomplished by Feringa[10] and coworkers, as exemplified by the highly enantioselective desymmetrization of **91** (**Scheme 7.12**). The reaction generated enantio-enriched bromocycloalkenes **92**, which were

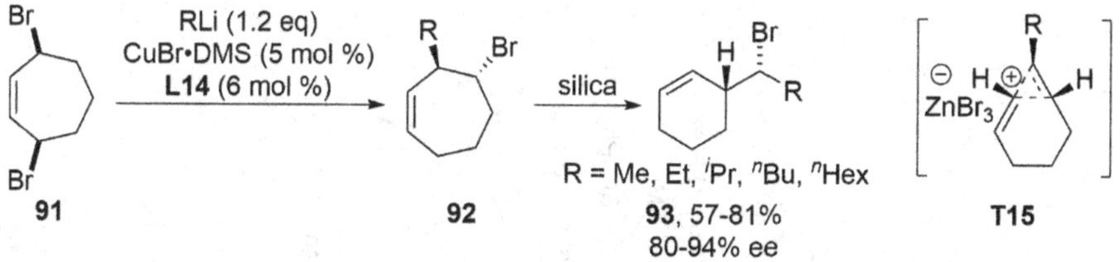

SCHEME 7.12 The Cu-catalyzed S_N2' type asymmetric allyllic substitution.

converted to cyclohexene derivatives after an unusual type I dyotropic rearrangement through **T15**.[63, 64]

7.2.3 ASYMMETRIC ALLYLIC SUBSTITUTION WITH HETEROATOM NUCLEOPHILES

Ir-catalyzed inter- and intra-molecular asymmetric allylic aminations, etherifications, and sulfonations have been developed by Hartwig and Feringa,[10] powered by chiral monophosphoramidite ligands **L14** or **L15**. Asymmetric allylic substitution with heteroatom nucleophiles, employing an oxidative activation energy and Ir-catalyzed allylic substitution of branched alcohols with heteroatom nucleophiles, has been reviewed elsewhere[65] and will not be discussed further here.

Instead of employing a chiral iridium catalyst, Kleij[66, 67] and coworkers applied a chiral Pd-**L14** catalyst in asymmetric allylic substitution with heteroatom nucleophiles, forming chiral olefins bearing a quaternary stereocenter with excellent yields and enantioselectivities (**Scheme 7.13a**). Key to this success was the formation of postulated zwitterionic π-allyl palladium intermediates **T16–T17**, resulting from a Pd-catalyzed decarboxylation of vinyl cyclic carbonate **94**. In addition, Zhang[68] and coworkers reported an example of boron-assisted allylic etherification (**Scheme 7.13b**). Kleij[69] and coworkers reported a Pd-catalyzed allylic amination and sulfonation, forming products bearing a quaternary stereocenter with high ee values from allylic carbonate **98**.

7.2.4 PD-CATALYZED CYCLIZATION BASED ON ALLYLIC CHEMISTRY

A zwitterionic π-allyl palladium species could be attacked by an amphiphile to construct a cyclic product (**Scheme 7.14**). Employing chiral monophosphoramidite SIPHOS-PE (**L25**) as the ligand, Shi[75] and coworkers developed an intermolecular Pd-catalyzed dearomative [3+2] cycloaddition of vinyl cyclopropanes (**101**). The reaction underwent a dearomative process, via **T19**, and generated **105** at a high ee. Trost[70–73] and Deng[74] reported that several trimethylenemethane (TMM) variants **102** were activated by the Pd-**L14/L61** catalyst and transformed to **T20**. A subsequent step-wise stereoselective cyclization, with nitro olefin, ketone, and unsaturated imines, led to the formation of **106–108**, respectively.

Shi[123] and coworkers employed Pd-SITCP(**L27**) as the catalyst for the enantioselective decarboxylative [4+2] cyclization between vinyl benzoxazinanones and

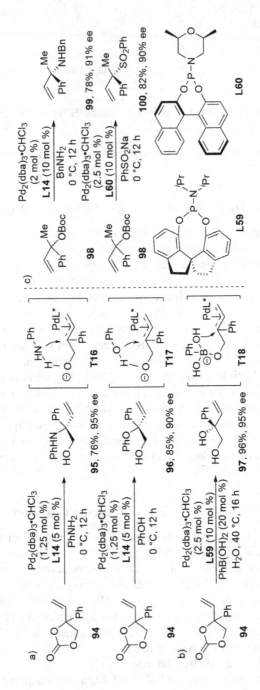

SCHEME 7.13 The Pd-catalyzed asymmetric allylic substitutions with heteroatom nucleophiles.

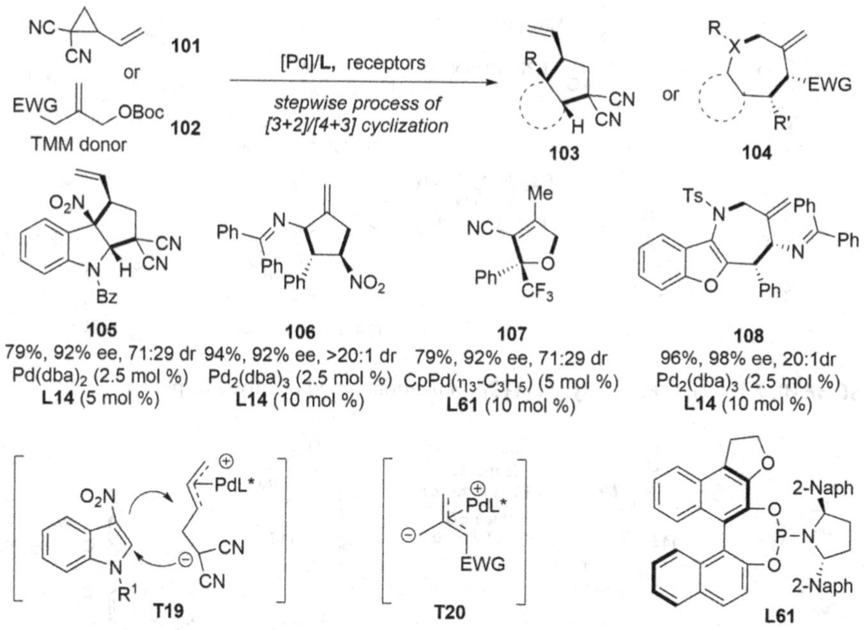

SCHEME 7.14 The Pd-catalyzed [3+2]/[4+3] cyclization, based on allyl Pd species.

isatins (**107**), and chiral tryptanthrin **111** was synthesized at 95% yield and 95% ee through the intermediate **T21** (**Scheme 7.15**). This strategy was also used by Song[76] and coworkers for the stereoselective synthesis of **114**.

Xiao[77, 78] and coworkers achieved a Pd-catalyzed asymmetric [5+2] cycloaddition between vinyl ethylene carbonate and α-diazoketone **115** (**Scheme 7.16**). The reaction was promoted by blue light-emitting diodes (LEDs) and proceeded *via* ketene **117**. Preliminary mechanistic studies suggested that the palladium species involving coordination of two phosphoramidite ligands was responsible for the high reactivity and enantioselectivity.

7.3 ASYMMETRIC DEAROMATIVE AND HECK-TYPE CYCLIZATION

Transition-metal-catalyzed asymmetric intramolecular dearomative arylation is one of the most efficient methods by which to construct polycyclic compounds bearing quaternary stereocenters. Although a few examples of intramolecular dearomative arylation have been reported, the asymmetric versions of these reactions have been less studied, due mainly to the lack of efficient chiral monophosphorus ligands.

Excellent enantioselectivities were achieved by Tang and coworkers on asymmetric dearomative arylation of phenols. By employing a *P*-chiral biaryl monophosphorus ligand **L7**, an enantioselective Pd-catalyzed dearomative cyclization of substituted phenols was established.[79] The reaction generated a series of chiral tricyclic phenanthrenone derivatives bearing all-carbon quaternary centers, with

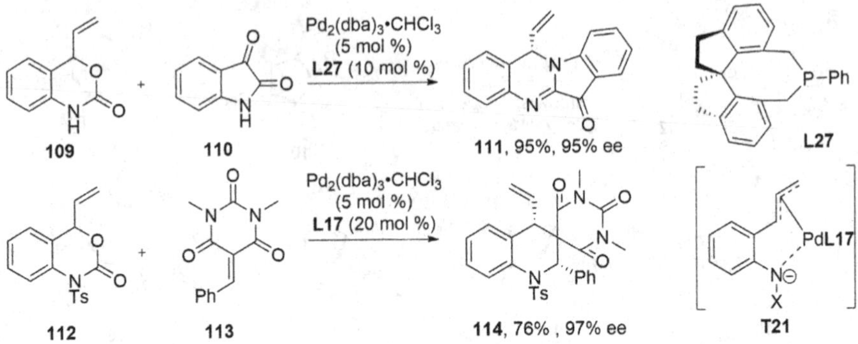

SCHEME 7.15 The Pd-catalyzed [4+2] cyclization, based on allyl-Pd species.

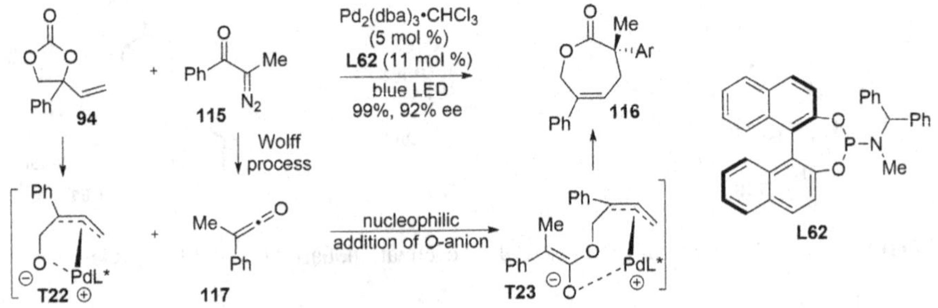

SCHEME 7.16 The Pd-catalyzed [5+2] cyclization with α-diazoketones.

excellent enantioselectivities (**Scheme 7.17a**). A stereochemical model was proposed and the stereochemistry of the product was determined during the reductive elimination step. The 2,5-diphenylpyrrole moiety of **L7** effectively blocked the backside of the complex, while the bulky *tert*-butyl group dictated the orientation of the substrate. Tang[80] and coworkers also developed a facile production of bridged tetracyclic skeletons, with up to 99% ee, by employing a chiral Pd-**L5** catalyst (**Scheme 7.17b**). The dearomative cyclization was successfully applied to enantioselective synthesis of several natural products.

The asymmetric Heck reaction has become one of the most important carbon–carbon bond-forming reactions and has been successfully used to construct chiral tertiary or quaternary carbon centers with excellent enantioselectivity. Recently, You[81] and Jia[82] developed Pd-catalyzed enantioselective intramolecular dearomative Heck reactions of indoles, benzofurans, pyrroles, and furans, offering the straightforward production of a range of chiral spiro- and benzo-fused heterocycles, bearing nitrogen-/oxygen-substituted quaternary stereocenters (**Scheme 7.18**). New chiral 1,1'-bi-2,2'-naphthol (BINOL) or *H*8-BINOL-based phosphoramidite ligands **L63–L65** were responsible for the excellent enantioselectivities.

Bower[83] and coworkers demonstrated that a Pd catalyst, supported by a SPINOL-derived phosphoramidate ligand, could promote a highly enantioselective aza-Heck

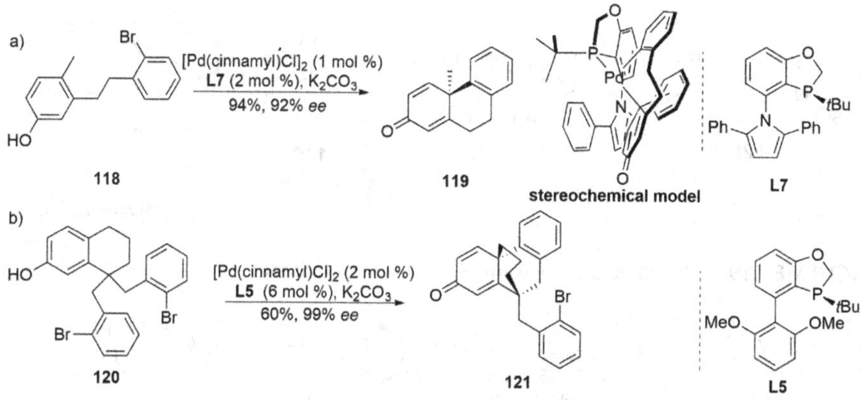

SCHEME 7.17 The Pd-catalyzed asymmetric dearomative cyclization of substituted phenols.

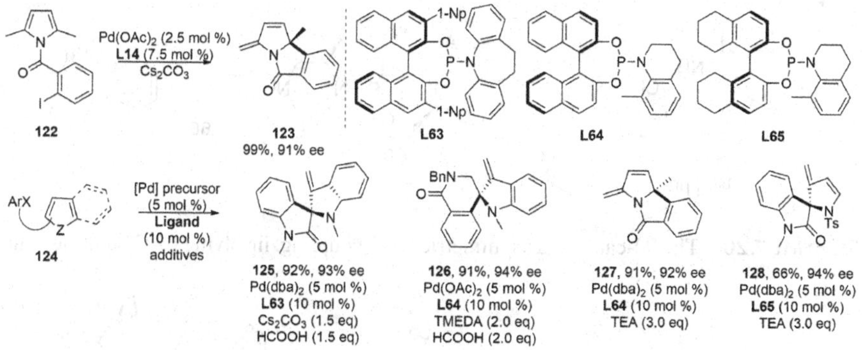

SCHEME 7.18 The Pd-catalyzed asymmetric dearomative Heck cyclizations.

cyclization of alkenyl N-(tosyloxy)carbamates (**Scheme 7.19**). This aza-Heck method is complementary to related oxidative aza-Wacker cyclization.

7.4 ASYMMETRIC CROSS-COUPLING REACTION

Transition-metal-catalyzed asymmetric cross-coupling reactions have played a major role in constructing chiral molecules in organic synthesis. Reactivity and selectivity remain the key issues in the field of cross coupling. The most effective strategy to address these problems is the rational design of efficient chiral ligands and catalysts. The past decade has witnessed the development of several prominent chiral mono-phosphorus ligands for asymmetric cross-coupling reactions.

Transition-metal-catalyzed asymmetric C(sp²)–C(sp²) cross coupling is one of the most efficient methods for constructing biaryl skeletons with axial chirality, which exist in the structures of numerous natural products and drugs. In 2019, Gu[84] and coworkers described a Pd-catalyzed cross coupling between aryl bromides and biaryl

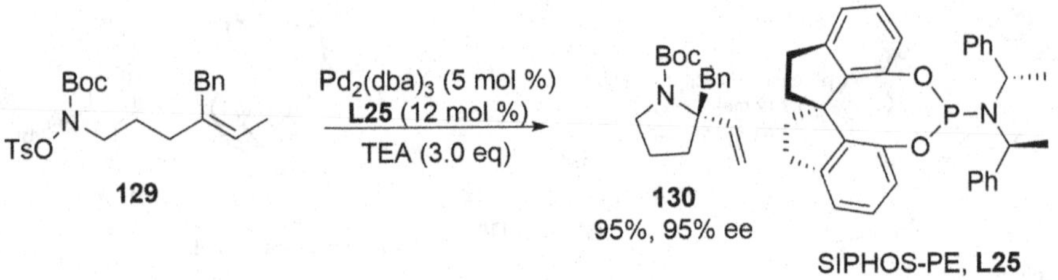

SCHEME 7.19   The Pd-catalyzed asymmetric Heck cyclization.

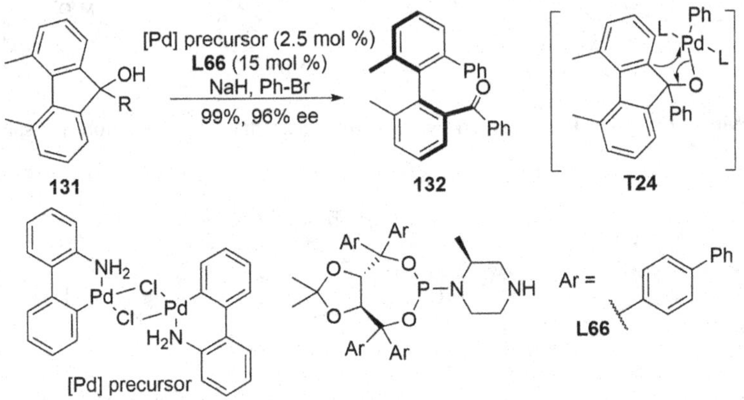

SCHEME 7.20   The Pd-catalyzed asymmetric cross coupling, involving C–C bond cleavage.

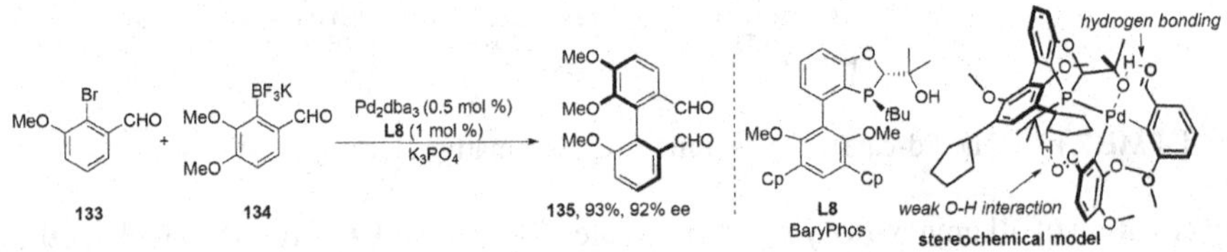

SCHEME 7.21   The Pd-catalyzed asymmetric Suzuki-Miyaura cross coupling, achieving tetra-*ortho*-substituted biaryls.

alcohols. The use of the TADDOL-based ligand **L66** enabled the C(sp²)–C(sp³) bond cleavage (**T24**) and a subsequent enantioselective aryl-aryl coupling (**Scheme 7.20**).

The direct synthesis of chiral tetra-*ortho*-substituted biaryl structures *via* asymmetric Suzuki-Miyaura coupling remains a challenging and unsolved problem. In 2020, Tang[85] and coworkers described a powerful Suzuki–Miyaura coupling, facilitated by a *P*-chiral monophosphorus ligand BaryPhos (**L8**). In addition to the increased reactivity for sterically hindered cross coupling, the rational design of BaryPhos also allowed a new catalysis mode of asymmetric cross coupling, involving noncovalent interactions between the ligand and two coupling partners, to effect efficient stereoinduction (**Scheme 7.21**).

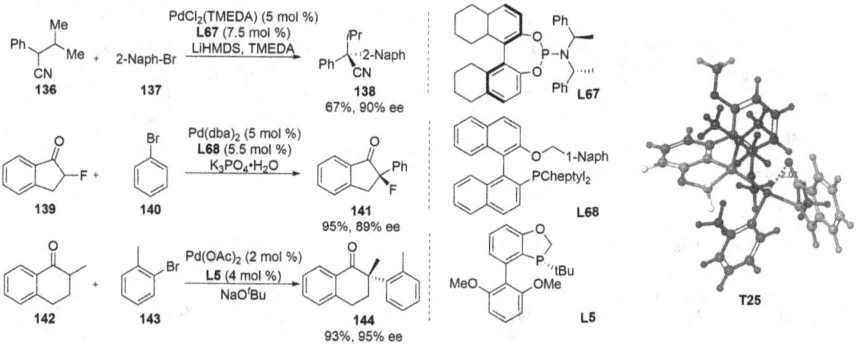

SCHEME 7.22 The Pd-catalyzed enantioselective α-arylation.

Effective asymmetric cross coupling, involving sp³ carbon, remains in high demand in organic synthesis. The enantioselective α-arylation of carbonyl compounds has been facilitated by the development of new chiral monophosphorus ligands. In 2016, Zhou[86] and Hartwig[87] established the asymmetric α-arylation of ketones with electron-withdrawing groups (-CN or -F) by using L67 or L68, respectively. Later, in 2018, the enantioselective α-arylation of sterically hindered substrates was achieved by Tang[88] and coworkers by employing a P-chiral monophosphorus ligand L5. DFT calculation of the palladium complex T25 during the reductive elimination stage revealed that the well-defined structure of BI-DIME (L5) effectively dictated the orientations of two coupling partners, leading to the arylation product 144 with the S configuration (Scheme 7.22).

7.5 ASYMMETRIC C–H BOND FUNCTIONALIZATION

C–H bond functionalization, as one of most atom- and step-economical processes, has attracted much attention. However, the transition-metal-catalyzed asymmetric C–H bond functionalization remains a very challenging area. Thanks to the development of several novel chiral monophosphorus ligands, the transition-metal-catalyzed asymmetric C–H functionalization was realized for both the C(sp²)–H and C(sp³)–H bonds (Scheme 7.23). Duan[89] and coworkers reported a Pd-catalyzed enantioselective C(sp³)–H arylation of N-(o-Br-aryl) anilides, using the chiral monophosphoramide ligand L69, and the resulting α-nitro amide was formed with up to 98% ee. Chen[90] and coworkers reported a Pd-catalyzed enantioselective benzylic C–H arylation of 3-arylpropanamides, using a Pd-L17 catalyst. By employing 8-aminoquinoline (AQ) as the directing group, the protocol provided a series of β,β-diaryl amides with high enantioselectivities. Gong[91] and coworkers employed an unusual hybrid catalyst, composed of both Pd-L70 and an anionic Co(III)-salen salt, which proved to be effective in asymmetric thioamide-directed C(sp³)–H arylation.

Recent examples of Pd-catalyzed C(sp2)–H bond functionalization by monophosphor (di)amide ligand were reported by Cramer,[92] Gu,[93] Zhu,[94, 95] and You[96, 97] (Scheme 7.24). Facilitated by the SPINOL-based ligand L73, the insertion of the

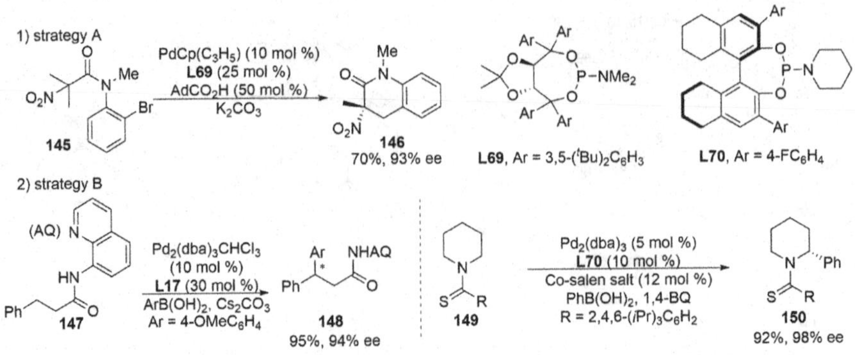

SCHEME 7.23 The Pd-catalyzed enantioselective C(sp³)–H bond functionalization.

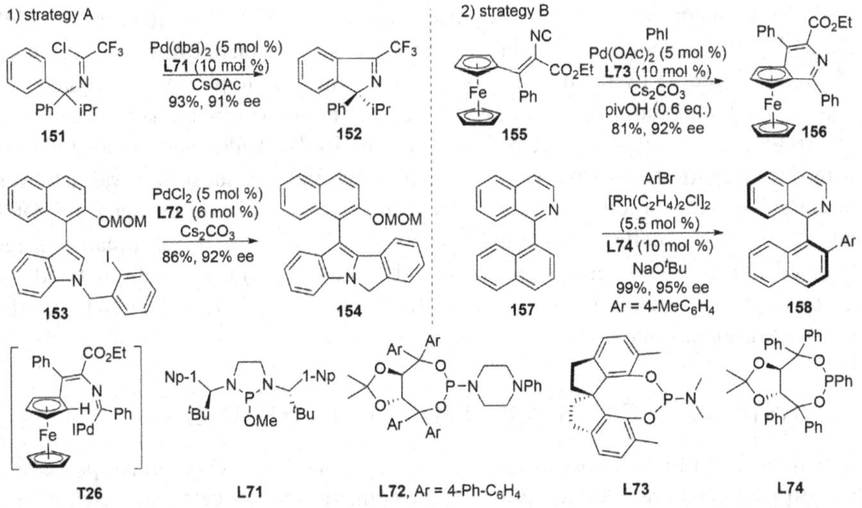

SCHEME 7.24 The Pd-catalyzed enantioselective C(sp²)–H bond functionalization.

isocyano group into aryl-Pd(II) iodide led to the formation of the intermediate **T26,** which underwent C(sp²)–H functionalization to yield the cyclic product **156** with high ee.

7.6 ASYMMETRIC COUPLING OF Π-SYSTEMS CATALYZED BY TRANSITION METALS

The transition-metal-catalyzed coupling of two or more π systems (alkenes, alkynes, allenes, aldehydes, aldimines, ketones, dienes, etc.) provides a convenient way to form multiple bonds in one step with great atom economy and efficiency. Among them, the coupling of two π systems is the most-studied example, providing chiral products of considerable complexity.

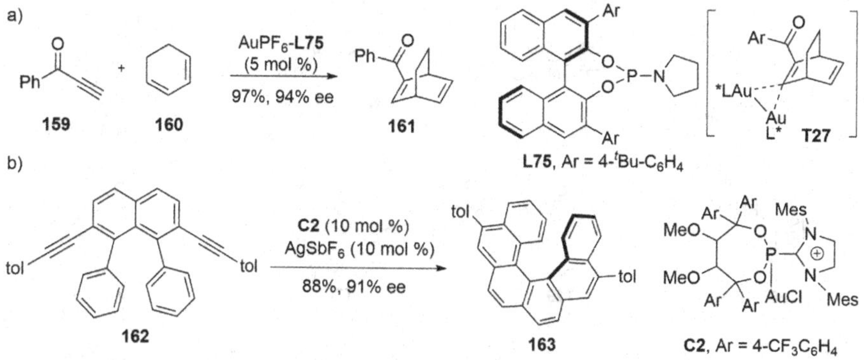

SCHEME 7.25   The Au-catalyzed enantioselective cycloaddition.

Gold(I)-catalyzed enantioselective intermolecular [4+2] cycloaddition between ynones and cyclohexadiene was developed by Mikami[98] and coworkers (**Scheme 7.25a**). Key to the success of this reaction was the generation of a geminal digold terminal alkyne capable of the ensuing [4+2] reaction, leading to the cycloaddition product **161** *via* the digold intermediate **T27**. Alcarazo[99, 100] and coworkers reported the use of α-cationic phosphonite derived from TADDOL as the ancillary chiral ligand, allowing the enantioselective synthesis of substituted [6]-carbohelicene **160** by sequential Au-catalyzed intramolecular hydroarylation of diyne **162** (**Scheme 7.25b**). Enhanced reactivity was observed when the gold complex **C2** was directly employed as the catalyst precursor.

Cycloaddition with a rhodium catalyst was among the most-studied reactions. Previous work by Hayashi, Nozaki, and Rovis[10] demonstrated that a monophosphorus ligand, such as MOP, was efficient for various [4+2], [4+3], and [2+2+2] cycloadditions. Several representative examples of Rh-catalyzed enantioselective cycloadditions are shown in **Scheme 7.26**. Dong[101] and coworkers described a rhodium-catalyzed [4+1] cyclization between cyclobutanones and allenes, affording a distinct [4.2.1]-bicyclic skeleton, containing two quaternary carbon centers. The reaction involved the carbon–carbon bond activation of cyclobutanones enabled by a TADDOL-derived phosphoramidite ligand. Anderson[102] and coworkers designed a highly reactive rhodium catalyst, Rh-**L14**, for the stereoselective cycloisomerization of ynamide-vinylcyclopropanes to form [5.3.0]-azabicycles. Theoretical investigations[102] disclosed an unexpected reaction pathway (through **T32** or **T33**), in which the electronic structure of the phosphoramidite ligand influenced the reaction rate and enantioselectivity dramatically.

7.7   ASYMMETRIC ADDITION

Transition-metal-catalyzed conjugate addition of a carbon-nucleophile to an activated double bond remains one of most important carbon–carbon bond-forming reactions. Pioneering work from the groups of Feringa, Fillion, and Alexakis[10] showed that

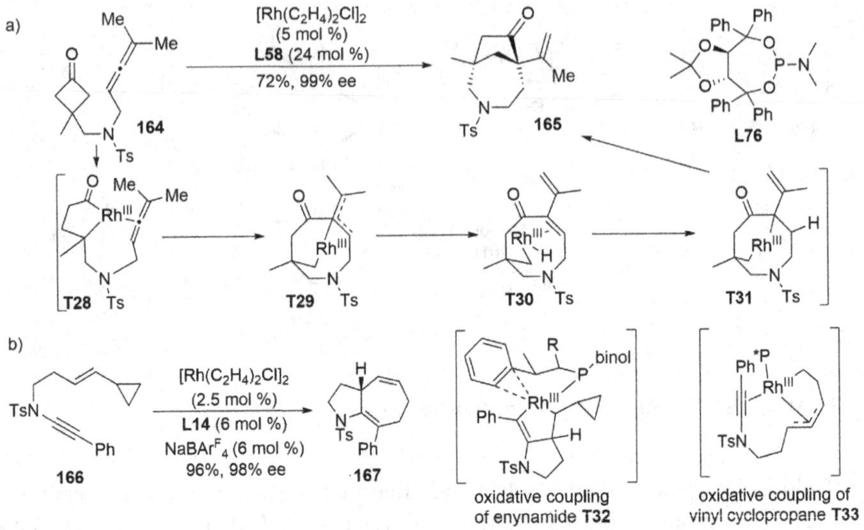

SCHEME 7.26   The Rh-catalyzed enantioselective cycloaddition.

chiral chelating ligands played a major role in the development of asymmetric conjugate addition. This section of the chapter focuses on recent developments in the field.

The use of organometallic reagents as nucleophiles in Cu-catalyzed asymmetric addition reactions has attracted significant attention. With a chiral BINOL-derived monophosphoramidite ligand **L77–L81**, Fletcher[103–108] and coworkers developed the Cu-catalyzed asymmetric conjugate addition of alkylzirconium reagents to α, β-unsaturated lactones, acyclic ketones, and cyclic ketones in moderate to good yields, with high enantioselectivities (**Scheme 7.27**). The alkylzirconium reagents in their studies were generated *in situ* from alkenes and the Schwartz reagent Cp_2ZrHCl. All-carbon quaternary stereocenters were constructed *via* 1,4-addition to tri-substituted enones. Mechanistic studies showed that the high efficiency was mainly attributed to the fine-tuning of the phosphoramidite ligands, which exerted a Cu–P, η^2 olefin coordination in **T34**.

Tan[109] and coworkers reported a copper-catalyzed atroposelective Michael-type addition between azonaphthalenes and arylboronic acids to prepare optically active biaryls (**Scheme 7.28**). The use of BINOL-derived phosphoramidite chiral ligand **L28** enabled the efficient construction of biaryl atropisomers. In particular, the routine 1,2-addition was effectively inhibited, and the formation of an aryl–aryl chiral axis was promoted. In addition, this strategy bypassed the use of an oxidant and the harsh conditions normally required for transition-metal-mediated arene C–H coupling with arylboronic acids.

The transition-metal-catalyzed asymmetric 1,2-addition of organometallic reagents to imines, ketones, aldehydes, and alkenes represents an attractive method in organic synthesis. Lin[110] and coworkers reported a Cu-catalyzed asymmetric allylation of chiral N-*tert*-butanesulfinyl imines achieved by **L40**. The combination

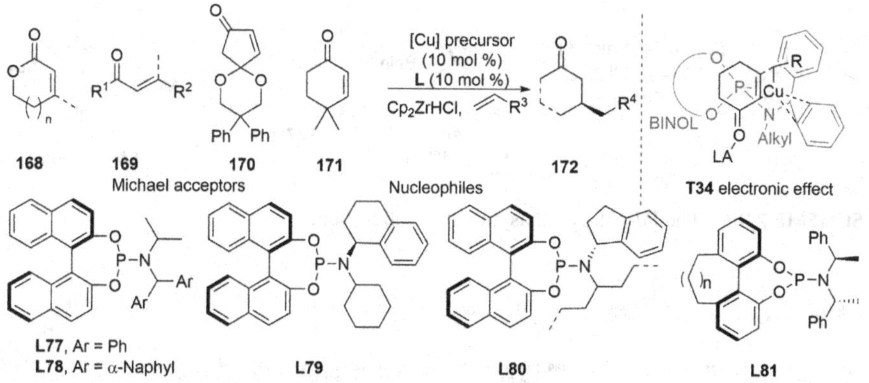

SCHEME 7.27 The Cu-catalyzed asymmetric conjugate addition of alkylzirconium reagents.

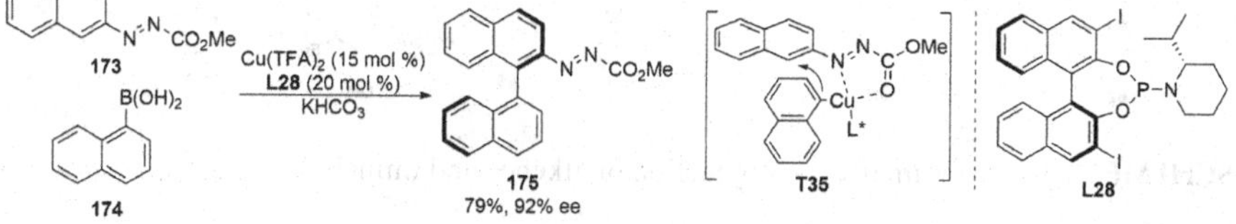

SCHEME 7.28 The Cu-catalyzed asymmetric conjugate addition of azonaphthalenes.

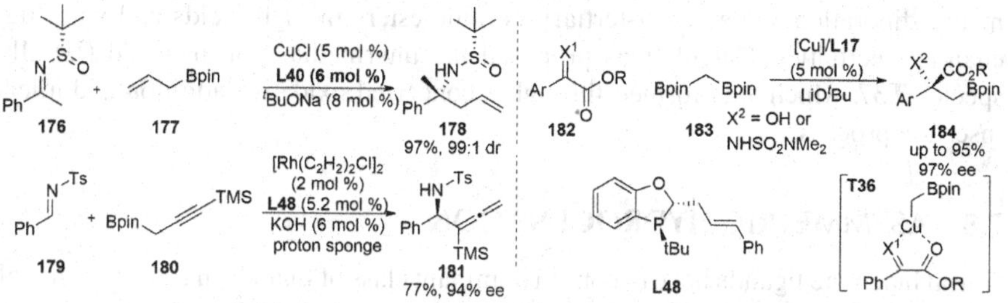

SCHEME 7.29 The Cu- or Rh-catalyzed 1,2-addition of C(sp^3)-Bpin chemicals.

of a chiral auxiliary and a chiral copper catalyst Cu-**L40** was proposed to be essential to achieve high diastereoselectivities (**Scheme 7.29**). The monophosphorus ligand JoshPhos (**L48**)[111] proved to be effective in Rh-catalyzed 1,2-addition of aldimines to form various chiral α-secondary amines. In addition, Zhou[112] and coworkers reported a Cu-SIPHOS-PE catalyzed asymmetric 1,2-addition, generating diarylamines with high ee values. Meek reported the use of *gem*-diborates as nucleophiles, offering opportunities to synthesize chiral β-boron-substituted alcohols or amines through a Cu/**L17**-catalyzed asymmetric addition[113–115], with excellent enantioselectivities.

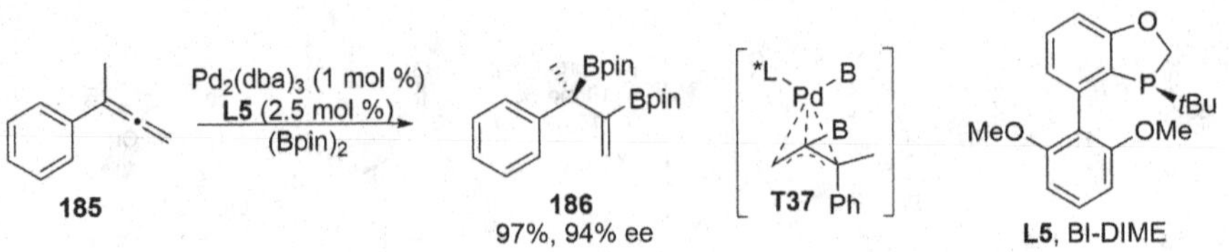
Pd2(dba)3 (1 mol %)
L5 (2.5 mol %)
(Bpin)2
185
186
97%, 94% ee
Bpin
Bpin
T37 Ph
L5, BI-DIME
MeO
OMe
tBu

SCHEME 7.30 The Pd-catalyzed asymmetric diboration.

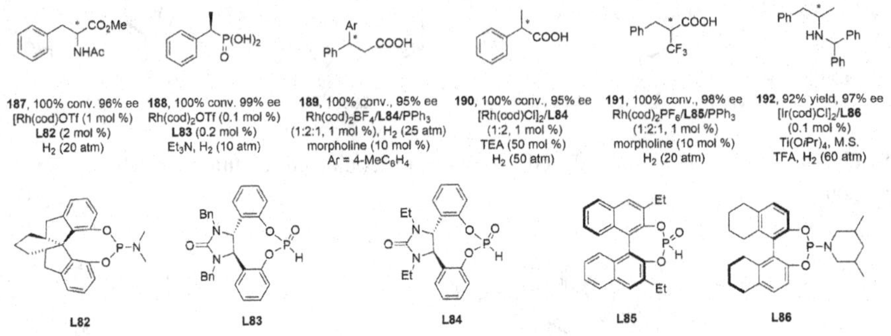
187, 100% conv. 96% ee
[Rh(cod)OTf (1 mol %)
L82 (2 mol %)
H2 (20 atm)
188, 100% conv. 99% ee
Rh(cod)2OTf (0.1 mol %)
L83 (0.2 mol %)
Et3N, H2 (10 atm)
189, 100% conv., 95% ee
Rh(cod)2BF4/L84/PPh3
(1:2:1, 1 mol %), H2 (25 atm)
morpholine (10 mol %)
Ar = 4-MeC6H4
190, 100% conv., 95% ee
[Rh(cod)Cl]2/L84
(1:2, 1 mol %)
TEA (50 mol %)
H2 (50 atm)
191, 100% conv., 98% ee
Rh(cod)2PF6/L85/PPh3
(1:2:1, 1 mol %)
morpholine (10 mol %)
H2 (20 atm)
192, 92% yield, 97% ee
[Ir(cod)Cl]2/L86
(0.1 mol %)
Ti(OiPr)4, M.S.
TFA, H2 (60 atm)
L82
L83
L84
L85
L86

SCHEME 7.31 Asymmetric hydrogenation of alkenes and imines.

P-Chiral monophosphorus ligands have also been successfully used in asymmetric 1,2-addition[116] (**Scheme 7.30**). By using (*S*)-BI-DIME as the ligand, the asymmetric diboration generated α-tertiary boronic esters in high yields and with high enantioselectivities. Calculations proposed the intermediary of an η^3-Pd (II)-allyl species **T37**, which was formed through a concerted oxidative addition and allene insertion process.

7.8 ASYMMETRIC HYDROGENATION

Chiral bidentate ligands have become a dominant class of ligands in transition-metal-catalyzed asymmetric hydrogenation. The past two decades have witnessed a surge in the use of chiral monophosphorus ligands for use in asymmetric hydrogenation, and significant progress has been achieved. The representative chiral monophosphorus ligands for Rh- and Ir-catalyzed asymmetric hydrogenation of functionalized alkenes and imines in recent years are shown in **Scheme 7.31**.[120]

Cationic iminium species could participate in asymmetric hydrogenation to provide chiral amine products. An efficient Ir-catalyzed asymmetric hydrogenation of racemic γ-lactams was reported[121], using chiral phosphoramidite ligand **L87** (**Scheme 7.32**). Mechanistic studies indicated that the reduced products **194** were obtained *via* the hydrogenation of the *N*-acyliminium cation **T38**, which was generated *in situ* by the elimination of β, γ-unsaturated γ-lactams.

The utility of axially chiral phosphoramidite ligands was shown by Takacs[122] and coworkers in Rh-catalyzed asymmetric hydrogenation of alkenes (**Scheme 7.33**).

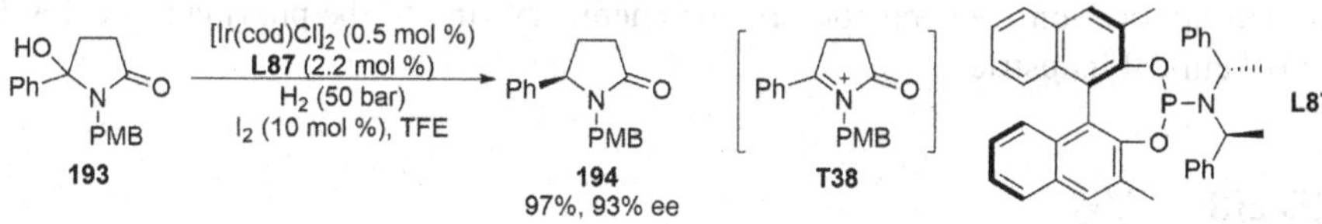

SCHEME 7.32 Elimination/iridium-catalyzed asymmetric hydrogenation of γ-hydroxy γ-lactams.

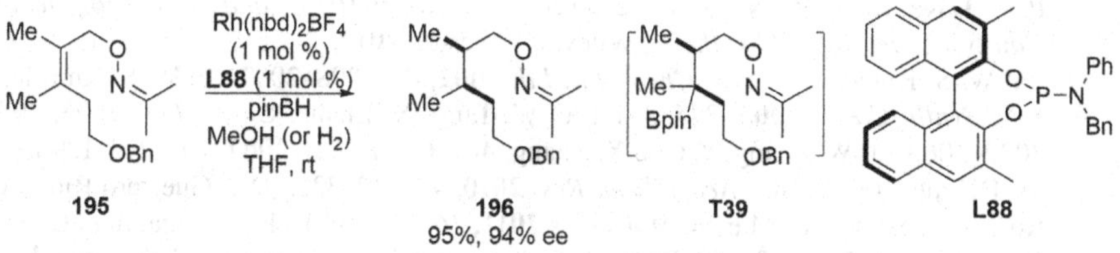

SCHEME 7.33 Rh-catalyzed hydrogenation *via* a hydroboration process.

The oxime functionality served as a directing group and the reaction proceeded through a catalytic asymmetric hydroboration, followed by deboronation upon the addition of a proton source, such as methanol, or by running the reaction under a hydrogen atmosphere.

7.9 CONCLUSIONS AND OUTLOOK

It is without doubt that chiral monophosphorus ligands have played significant roles in the development of various asymmetric catalytic transformations, such as asymmetric allylic substitution, asymmetric dearomative arylation, asymmetric Heck reactions, asymmetric cross coupling, asymmetric C–H bond functionalization, asymmetric coupling of π systems, asymmetric addition, asymmetric hydrogenation, and many others. By taking advantage of recent progress in both synthetic and computational chemistry, chiral monophosphorus ligands with new structural features have been designed and synthesized. In particular, *P*-chiral monophosphorus ligands have become an indispensable type of ligand, with increasing numbers of applications in transition-metal-catalyzed asymmetric catalysis.

The design and exploration of efficient chiral phosphorus ligands will continue to be at the heart of research into asymmetric transition-metal-catalyzed transformations. Further efforts are needed to expand the range of substrates, to reduce catalyst loadings, and to improve the selectivity of these available synthetic methods with the aid of novel chiral ligands. Systematic and in-depth investigations of reaction mechanisms will help to understand the role of chiral monophosphorus ligands in asymmetric catalysis and their structure/activity/selectivity relationship. Continuous efforts in design and applications of new chiral monophosphorus ligands will greatly enhance the impact of monophosphorus ligands in asymmetric catalysis, as well as in their applications to synthetic organic chemistry, and to the pharmaceutical and agrochemical industries.

REFERENCES

1. (a) R. Noyori, *Asymmetric Catalysis in Organic Synthesis*, Wiley: New York, **1993**. (b) A. N. Collins, S. Gary, J. Crosby, *Chirality in Industry: II: Developments in the Commercial Manufacture and Applications of Optically Active Compounds*, Wiley: Chichester, **1997**.
2. P. C. J. Kamer, P. W. N. M. V. Leeuwen, *Phosphorus(III) Ligands in Homogeneous Catalysis: Design and Synthesis*, Wiley: Chichester, **2012**.
3. (a) W. S. Knowles, *Angew. Chem. Int. Ed.* **2002**, *41*, 1998–2007. (b) W. S. Knowles, *Adv. Synth. Catal.* **2003**, *345*, 3–13. (c) W. Tang, X. Zhang, *Chem. Rev.* **2003**, *103*, 3029–3069. (d) W. Zhang, Y. Chi, X. Zhang, *Acc. Chem. Res.* **2007**, *40*, 1278–1290. (e) M. Diéguez, O. Pàmies, *Acc. Chem. Res.* **2010**, *43*, 312–322. (f) I. Guerrero Rios, A. Rosas-Hernandez, E. Martin, *Molecules* **2011**, *16*, 970. (g) B. L. Feringa, *Acc. Chem. Res.* **2000**, *33*, 346–353. h) J. Pedroni, N. Cramer, *Chem. Commun. (Cambridge, UK)* **2015**, *51*, 17647–17657. (i) G. Liu, G. Xu, R. Luo, W. Tang, *Synlett* **2013**, *24*, 2465–2471. (j) I. V. Komarov, A. Borner, *Angew. Chem. Int. Ed.* **2001**, *40*, 1197–1200. (k) T. Jerphagnon, J.-L. Renaud, C. Bruneau, *Tetrahedron: Asymmetry* **2004**, *15*, 2101–2111. (l) J.-H. Xie, S.- F. Zhu, Y. Fu, A.-G. Hu, Q.-L. Zhou, *Pure Appl. Chem.* **2005**, *77*, 2121–2132. (m) A. J. Minnaard, B. L. Feringa, L. Lefort, J. G. de Vries, *Acc. Chem. Res.* **2007**, *40*, 1267–1277. (n) L. Eberhardt, D. Armspach, J. Harrowfield, D. Matt, *Chem. Soc. Rev.* **2008**, *37*, 839–864. (o) X.-P. Hu, D.-S. Wang, C.-B. Yu, Y.-G. Zhou, Z. Zheng, *Top. Organomet. Chem.* **2011**, *36*, 313–354. (p) J. F. Teichert, B. L. Feringa, *Angew. Chem. Int. Ed.* **2010**, *49*, 2486–2528.
4. W. S. Knowles, M. J. Sabacky, B. D. Vineyard, *J. Chem. Soc., Chem. Commun.* **1972**, 10–11.
5. J. D. Morrison, R. E. Burnett, A. M. Aguiar, C. J. Morrow, C. Phillips, *J. Am. Chem. Soc.* **1971**, *93*, 1301–1303.
6. (a) F. Lagasse, H. B. Kagan, *Chem. Pharm. Bull.* **2000**, *48*, 315–324. (b) T. Hayashi, *Yuki Gosei Kagaku Kyokaishi* **1994**, *52*, 900–911.
7. Q.-L. Zhou, *Privileged Chiral Ligands and Catalysts*, Wiley: Weinheim, **2011**.
8. Y.-N. Yu, M.-H. Xu, *Acta Chimica Sinica* **2017**, *75*, 655.
9. (a) S. Gladiali, E. Alberico, K. Junge, M. Beller, *Chem. Soc. Rev.* **2011**, *40*, 3744–3763. (b) P. W. N. M. V. Leeuwen, P. C. J. Kamer, C. Claver, O. Pàmies, M. Diéguez, *Chem. Rev.* **2011**, 111, 2077–2118. (c) E. Alberico, K. Junge, M. Beller, *Chem. Soc. Rev.* **2011**, *40*, 3744–3763. (d) T. Hayashi, *Acc. Chem. Res.* **2000**, *33*, 354–362.
10. W. Fu, W. Tang, *ACS Catal.* **2016**, *6*, 4814–4858.
11. (a) J. F. Hartwig, L. M. Stanley, *Acc. Chem. Res.* **2010**, *43*, 1461–1475. (b) P. Tosatti, A. Nelson, S. P. Marsden, *Org. Biomol. Chem.* **2012**, *10*, 3147–3163. (c) C.-X. Zhuo, C. Zheng, S.-L. You, *Acc. Chem. Res.* **2014**, *47*, 2558–2573.
12. C. X. Zhuo, C. Zheng, S.-L. You, *Acc. Chem. Res.* **2014**, *47*, 2558–2573.
13. Q.-F. Wu, H. He, W.-B. Liu, S.-L. You, *J. Am. Chem. Soc.* **2010**, *132*, 11418–11419.
14. C.-X. Zhuo, W.-B. Liu, Q.-F. Wu, S.-L. You, *Chem. Sci.* **2012**, *3*, 205–208.
15. Q. Cheng, Y. Wang, S. L. You, *Angew. Chem. Int. Ed.* **2016**, *55*, 3496–3499.
16. Q. F. Wu, W. B. Liu, C. X. Zhuo, Z. Q. Rong, K. Y. Ye, S. L. You, *Angew. Chem. Int. Ed.* **2011**, *50*, 4455–4458.
17. Z. P. Yang, R. Jiang, Q. F. Wu, L. Huang, C. Zheng, S. L. You, *Angew. Chem. Int. Ed.* **2018**, *57*, 16190–16193.

18. L. Naesborg, K. S. Halskov, F. Tur, S. M. Monsted, K. A. Jørgensen, *Angew. Chem. Int. Ed.* **2015**, *54*, 10193–10197.
19. R. D. Gao, L. Ding, C. Zheng, L. X. Dai, S. L. You, *Org. Lett.* **2018**, *20*, 748–751.
20. H.-F. Tu, X. Zhang, C. Zheng, M. Zhu, S.-L. You, *Nat. Catal.* **2018**, *1*, 601–608.
21. J. A. Rossi-Ashton, A. K. Clarke, J. R. Donald, C. Zheng, R. J. K. Taylor, W. P. Unsworth, S. L. You, *Angew. Chem. Int. Ed.* **2020**, *59*, 7598–7604.
22. D. Shen, Q. Chen, P. Yan, X. Zeng, G. Zhong, *Angew. Chem. Int. Ed.* **2017**, *56*, 3242–3246.
23. H. Tian, P. Zhang, F. Peng, H. Yang, H. Fu, *Org. Lett.* **2017**, *19*, 3775–3778.
24. S. B. Tang, H. F. Tu, X. Zhang, S. L. You, *Org. Lett.* **2019**, *21*, 6130–6134.
25. Z. T. He, J. F. Hartwig, *Nat. Chem.* **2019**, *11*, 177–183.
26. A. Matsunami, K. Takizawa, S. Sugano, Y. Yano, H. Sato, R. Takeuchi, *J. Org. Chem.* **2018**, *83*, 12239–12246.
27. M. Zhan, R.-Z. Li, Z.-D. Mou, C.-G. Cao, J. Liu, Y.-W. Chen, D. Niu, *ACS Catal.* **2016**, *6*, 3381–3386.
28. R. Sarkar, S. Mukherjee, *Org. Lett.* **2019**, *21*, 5315–5320.
29. W. B. Liu, N. Okamoto, E. J. Alexy, A. Y. Hong, K. Tran, B. M. Stoltz, *J. Am. Chem. Soc.* **2016**, *138*, 5234–5237.
30. J. C. Hethcox, S. E. Shockley, B. M. Stoltz, *Angew. Chem. Int. Ed.* **2016**, *55*, 16092–16095.
31. X. J. Liu, S. Jin, W. Y. Zhang, Q. Q. Liu, C. Zheng, S. L. You, *Angew. Chem. Int. Ed.* **2020**, *59*, 2039–2043.
32. Y. Lee, J. Park, S. H. Cho, *Angew. Chem. Int. Ed.* **2018**, *57*, 12930–12934.
33. X. J. Liu, S. L. You, *Angew. Chem. Int. Ed.* **2017**, *56*, 4002–4005.
34. X. D. Bai, Q. F. Zhang, Y. He, *Chem. Commun. (Camb)* **2019**, *55*, 5547–5550.
35. T. Song, X. Zhao, X. Wang, *J. Org. Chem.* **2019**, *84*, 15648–15654.
36. J. C. Hethcox, S. E. Shockley, B. M. Stoltz, *Angew. Chem. Int. Ed.* **2018**, *57*, 8664–8667.
37. J. Y. Hamilton, N. Hauser, D. Sarlah, E. M. Carreira, *Angew. Chem. Int. Ed.* **2014**, *53*, 10759–10762.
38. (a). Breitler, E. M. Carreira, *J. Am. Chem. Soc.* **2015**, *137*, 5296–5299. (b). Liang, K. Wei, Y. R. Yang, *Chem. Commun.* **2015**, *51*, 17471–17474.
39. Y. Sempere, E. M. Carreira, *Angew. Chem. Int. Ed.* **2018**, *57*, 7654–7658.
40. C. Y. Meng, X. Liang, K. Wei, Y. R. Yang, *Org. Lett.* **2019**, *21*, 840–843.
41. Y. Sempere, J. L. Alfke, S. L. Rossler, E. M. Carreira, *Angew. Chem. Int. Ed.* **2019**, *58*, 9537–9541.
42. S. B. Tang, X. Zhang, H. F. Tu, S. L. You, *J. Am. Chem. Soc.* **2018**, *140*, 7737–7742.
43. J. Y. Hamilton, D. Sarlah, E. M. Carreira, *J. Am. Chem. Soc.* **2013**, *135*, 994–997.
44. Y. Zheng, B. B. Yue, K. Wei, Y. R. Yang, *Org. Lett.* **2018**, *20*, 8035–8038.
45. B. B. Yue, Y. Deng, Y. Zheng, K. Wei, Y. R. Yang, *Org. Lett.* **2019**, *21*, 2449–2452.
46. J. C. Hethcox, S. E. Shockley, B. M. Stoltz, *Org. Lett.* **2017**, *19*, 1527–1529.
47. S. E. Shockley, J. C. Hethcox, B. M. Stoltz, *Angew. Chem. Int. Ed.* **2017**, *56*, 11545–11548.
48. Y. Lai, N. Zhang, Y. Zhang, J. H. Chen, Z. Yang, *Org. Lett.* **2018**, *20*, 4298–4301.
49. S. Zhou, H. Chen, Y. Luo, W. Zhang, A. Li, *Angew. Chem. Int. Ed.* **2015**, *54*, 6878–6882.
50. C. Shen, R. Q. Wang, L. Wei, Z. F. Wang, H. Y. Tao, C. J. Wang, *Org. Lett.* **2019**, *21*, 6940–6945.
51. L. M. Shi, X. S. Sun, C. Shen, Z. F. Wang, H. Y. Tao, C. J. Wang, *Org. Lett.* **2019**, *21*, 4842–4848.
52. C.-G. Cao, B. He, Z. Fu, D. Niu, *Org. Process Res. Dev.* **2019**, *23*, 1758–1761.
53. S. Krautwald, M. A. Schafroth, D. Sarlah, E. M. Carreira, *J. Am. Chem. Soc.* **2014**, *136*, 3020–3023.

54. M. A. Schafroth, G. Zuccarello, S. Krautwald, D. Sarlah, E. M. Carreira, *Angew. Chem. Int. Ed.* **2014**, *53*, 13898–13901.
55. B. Bhaskararao, R. Sunoj, *J. Am. Chem. Soc.* **2017**, *139*, 6675–6685.
56. Y.-L. Su, Z.-Y. Han, Y.-H. Li, L.-Z. Gong, *ACS Catal.* **2017**, *7*, 7917–7922.
57. L. F. Fan, P. S. Wang, L. Z. Gong, *Org. Lett.* **2019**, *21*, 6720–6725.
58. L.-F. Fan, T.-C. Wang, P.-S. Wang, L.-Z. Gong, *Organometallics* **2019**, *38*, 4014–4021.
59. X. Huo, R. He, X. Zhang, W. Zhang, *J. Am. Chem. Soc.* **2016**, *138*, 11093–11096.
60. J. Zhang, X. Huo, B. Li, Z. Chen, Y. Zou, Z. Sun, W. Zhang, *Adv. Synth. Catal.* **2019**, *361*, 1130–1139.
61. L. Wei, Q. Zhu, S. M. Xu, X. Chang, C. J. Wang, *J. Am. Chem. Soc.* **2018**, *140*, 1508–1513.
62. R. He, P. Liu, X. Huo, W. Zhang, *Org. Lett.* **2017**, *19*, 5513–5516.
63. S. S. Goh, P. A. Champagne, S. Guduguntla, T. Kikuchi, M. Fujita, K. N. Houk, B. L. Feringa, *J. Am. Chem. Soc.* **2018**, *140*, 4986–4990.
64. S. S. Goh, S. Guduguntla, T. Kikuchi, M. Lutz, E. Otten, M. Fujita, B. L. Feringa, *J. Am. Chem. Soc.* **2018**, *140*, 7052–7055.
65. S. L. Rössler, D. A. Petrone, E. M. Carreira, *Acc. Chem. Res.* **2019**, *52*, 2657–2672.
66. A. Cai, W. Guo, L. Martinez-Rodriguez, A. W. Kleij, *J. Am. Chem. Soc.* **2016**, *138*, 14194–14197.
67. W. Guo, A. Cai, J. Xie, A. W. Kleij, *Angew. Chem. Int. Ed.* **2017**, *56*, 11797–11801.
68. A. Khan, S. Khan, I. Khan, C. Zhao, Y. Mao, Y. Chen, Y. J. Zhang, *J. Am. Chem. Soc.* **2017**, *139*, 10733–10741.
69. A. Cai., A. W. Kleij, *Angew. Chem. Int. Ed.* **2019**, *58*, 14944–14949.
70. B. M. Trost, Y. Wang, *Angew. Chem. Int. Ed.* **2018**, *57*, 11025–11029.
71. B. M. Trost, Z. Jiao, C. Hung, *Angew. Chem. Int. Ed.* **2019**, *58*, 15154–15158.
72. B. M. Trost, G. Mata, *Angew. Chem. Int. Ed.* **2018**, *57*, 12333–12337.
73. B. M. Trost, Z. Zuo, *Angew. Chem. Int. Ed.* **2020**, *59*, 1243–1247.
74. Y. Z. Liu, Z. Wang, Z. Huang, X. Zheng, W. L. Yang, W. P. Deng, *Angew. Chem. Int. Ed.* **2020**, *59*, 1238–1242.
75. M. Sun, Z. Q. Zhu, L. Gu, X. Wan, G. J. Mei, F. Shi, *J. Org. Chem.* **2018**, *83*, 2341–2348.
76. H. W. Zhao, N. N. Feng, J. M. Guo, J. Du, W. Q. Ding, L. R. Wang, X. Q. Song, *J. Org. Chem.* **2018**, *83*, 9291–9299.
77. M. M. Li, Y. Wei, J. Liu, H. W. Chen, L. Q. Lu, W. J. Xiao, *J. Am. Chem. Soc.* **2017**, *139*, 14707–14713.
78. Y. Wei, S. Liu, M. M. Li, Y. Li, Y. Lan, L. Q. Lu, W. J. Xiao, *J. Am. Chem. Soc.* **2019**, *141*, 133–137.
79. K. Du, H. Yang, P. Guo, L. Feng, G. Xu, Q. Zhou, L. W. Chung, W. Tang, *Chem. Sci.* **2017**, *8*, 6247–6256.
80. X. Mu, H. Yu, H. Peng, W. Xiong, T. Wu, W. Tang, *Angew. Chem. Int. Ed.* **2020**, *59*, 8143–8147.
81. P. Yang, S.-L. You, *Org. Lett.* **2018**, *20*, 7684–7688.
82. X. Li, B. Zhou, R. Z. Yang, F. M. Yang, R. X. Liang, R. R. Liu, Y. X. Jia, *J. Am. Chem. Soc.* **2018**, *140*, 13945–13951.
83. X. Ma, I. R. Hazelden, T. Langer, R. H. Munday, J. F. Bower, *J. Am. Chem. Soc.* **2019**, *141*, 3356–3360.
84. R. Deng, J. Xi, Q. Li, Z. Gu, *Chem* **2019**, *5*, 1834–1846.
85. H. Yang, J. Sun, W. Gu, W. Tang, *J. Am. Chem. Soc.* **2020**, *142*, 8036–8043.
86. Z. Jiao, K. W. Chee, J. S. Zhou, *J. Am. Chem. Soc.* **2016**, *138*, 16240–16243.
87. Z. Jiao, J. J. Beiger, Y. Jin, S. Ge, J. S. Zhou, J. F. Hartwig, *J. Am. Chem. Soc.* **2016**, *138*, 15980–15986.
88. X. Rao, N. Li, H. Bai, C. Dai, Z. Wang, W. Tang, *Angew. Chem. Int. Ed.* **2018**, *57*, 12328–12332.

89. W. X. Kong, S. J. Xie, C. Y. Cao, C. W. Zhang, C. Wang, W. L. Duan, *Chem. Commun.* **2020**, *56*, 2292–2295.
90. H.-R. Tong, S. Zheng, X. Li, Z. Deng, H. Wang, G. He, Q. Peng, G. Chen, *ACS Catal.* **2018**, *8*, 11502–11512.
91. H. J. Jiang, X. M. Zhong, J. Yu, Y. Zhang, X. Zhang, Y. D. Wu, L. Z. Gong, *Angew. Chem. Int. Ed.* **2019**, *58*, 1803–1807.
92. D. Grosheva, N. Cramer, *Angew. Chem. Int. Ed.* **2018**, *57*, 13644–13647.
93. C. He, M. Hou, Z. Zhu, Z. Gu, *ACS Catal.* **2017**, *7*, 5316–5320.
94. S. Luo, Z. Xiong, Y. Lu, Q. Zhu, *Org. Lett.* **2018**, *20*, 1837–1840.
95. J. Wang, D.-W. Gao, J. Huang, S. Tang, Z. Xiong, H. Hu, S.-L. You, Q. Zhu, *ACS Catal.* **2017**, *7*, 3832–3836.
96. Q. Wang, Z. J. Cai, C. X. Liu, Q. Gu, S. L. You, *J. Am. Chem. Soc.* **2019**, *141*, 9504–9510.
97. Z. J. Cai, C. X. Liu, Q. Wang, Q. Gu, S. L. You, *Nat. Commun.* **2019**, *10*, 4168.
98. M. Nanko, S. Shibuya, Y. Inaba, S. Ono, S. Ito, K. Mikami, *Org. Lett.* **2018**, *20*, 7353–7357.
99. E. Gonzalez-Fernandez, L. D. Nicholls, L. D. Schaaf, C. Fares, C. W. Lehmann, M. Alcarazo, *J. Am. Chem. Soc.* **2017**, *139*, 1428–1431.
100. L. D. M. Nicholls, M. Marx, T. Hartung, E. González-Fernández, C. Golz, M. Alcarazo, *ACS Catal.* **2018**, *8*, 6079–6085.
101. X. Zhou, G. Dong, *J. Am. Chem. Soc.* **2015**, *137*, 13715–13721.
102. R. N. Straker, Q. Peng, A. Mekareeya, R. S. Paton, E. A. Anderson, *Nat. Commun.* **2016**, *7*, 10109.
103. Z. Gao, S. P. Fletcher, *Chem. Sci.* **2017**, *8*, 641–646.
104. K. Garrec, S. P. Fletcher, *Org. Lett.* **2016**, *18*, 3814–3817.
105. M. Sidera, S. P. Fletcher, *Chem. Commun.* **2015**, *51*, 5044–5047.
106. R. Ardkhean, P. M. C. Roth, R. M. Maksymowicz, A. Curran, Q. Peng, R. S. Paton, S. P. Fletcher, *ACS Catal.* **2017**, *7*, 6729–6737.
107. R. Ardkhean, M. Mortimore, R. S. Paton, S. P. Fletcher, *Chem. Sci.* **2018**, *9*, 2628–2632.
108. A. V. Brethome, R. S. Paton, S. P. Fletcher, *ACS Catal.* **2019**, *9*, 7179–7187.
109. S. Yan, W. Xia, S. Li, Q. Song, S. H. Xiang, B. Tan, *J. Am. Chem. Soc.* **2020**, *142*, 7322–7327.
110. Y. S. Zhao, Q. Liu, P. Tian, J. C. Tao, G. Q. Lin, *Org. Biomol. Chem.* **2015**, *13*, 4174–4178.
111. J. D. Sieber, V. V. Angeles-Dunham, D. Chennamadhavuni, D. R. Fandrick, N. Haddad, N. Grinberg, H. Kurouski, H. Lee, J. J. Song, N. K. Yee, A. E. Mattson, C. H. Senanayake, *Adv. Synth. Catal.* **2016**, *358*, 3062–3068.
112. C. Wu, X. Qin, A. M. P. Moeljadi, H. Hirao, J. S. Zhou, *Angew. Chem. Int. Ed.* **2019**, *58*, 2705–2709.
113. J. Kim, K. Ko, S. H. Cho, *Angew. Chem. Int. Ed.* **2017**, *56*, 11584–11588.
114. J. Kim, C. Hwang, Y. Kim, S. H. Cho, *Org. Process Res. Dev.* **2019**, *23*, 1663–1668.
115. J. Kim, M. Shin, S. H. Cho, *ACS Catal.* **2019**, *9*, 8503–8508.
116. J. Liu, M. Nie, Q. Zhou, S. Gao, W. Jiang, L. W. Chung, W. Tang, K. Ding, *Chem. Sci.* **2017**, *8*, 5161–5165.
117. Z. Zheng, Y. Cao, D. Zhu, Z. Wang, K. Ding, *Chem. Eur. J.* **2019**, *25*, 9491–9497.
118. Z. Zheng, Y. Cao, Q. Chong, Z. Han, J. Ding, C. Luo, Z. Wang, D. Zhu, Q. L. Zhou, K. Ding, *J. Am. Chem. Soc.* **2018**, *140*, 10374–10381.
119. G. Gao, S. Du, Y. Yang, X. Lei, H. Huang, M. Chang, *Molecules* **2018**, *23*, 2207–2214.
120. H. Huang, X. Liu, L. Zhou, M. Chang, X. Zhang, *Angew. Chem. Int. Ed.* **2016**, *55*, 5309–5312.
121. Q. Yuan, D. Liu, W. Zhang, *Org. Lett.* **2017**, *19*, 1144–1147.
122. V. M. Shoba, J. M. Takacs, *J. Am. Chem. Soc.* **2017**, *139*, 5740–5743.
123. G. Meo, C. Bian, G. Li, S. Xu, W. Zheng, F. Shi, *Org. Lett.* **2017**, *19*, 3219–3222.

8 Solvent-Oriented Ligand and Catalyst Design in Asymmetric Catalysis – Principles and Limits

Armin Börner

CONTENTS

8.1 INTRODUCTION

The design of an asymmetric catalytic reaction is based on a very complex approach, taking into account the interplay between metal, chiral ligand, substrate, and reagent in a strictly determined environment and under singular conditions (Figure 8.1). This particular set-up is finally summarized under the term "reaction conditions".

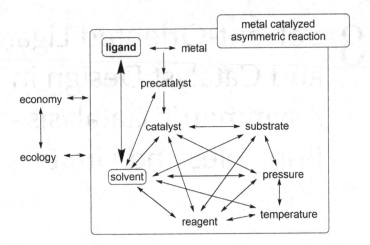

FIGURE 8.1 Network of parameters affecting the course of a metal-catalyzed asymmetric catalysis.

Due to various interdependencies, in principle, no simple relations exist among this complex network and can be therefore deduced. Nevertheless, in a search for optimization, chemists always tend to reduce this complexity by assuming isolated structure–effect relationships. It is astonishing that, in spite of these oversimplifications, this approach has led in the past decades to tremendous success in asymmetric catalysis, although our mechanistic understanding is far from complete.

Therefore, an evaluation of solvent effects on the role of ligand structures in asymmetric catalysis, as envisaged in this review (Figure 8.1), is associated with numerous uncertainties. Of course, the medium where the catalytic reaction takes place is decisive for the success of an asymmetric transformation. This concerns the physical and chemical properties of solvents, but also economic and ecologic reasons play a crucial role. Finally, for an industrial application, a compromise must be found, meeting all these demands.

In some cases, the reaction can be simply carried out without any solvent ("neat") and the reaction medium is the substrate or the product or a mixture of both. In this case, the question of the role of the solvent becomes more complicated to answer, because the properties of the "solvent" are changing with the progress of the reaction. However, in terms of work-up, this approach may be advantageous, provided a high conversion is achieved. Under these circumstances, the catalyst can be easily separated from the almost pure product. In most instances, it is true that a properly chosen solvent facilitates the separation protocol.

Perhaps one of the most important preconditions for every successful homogeneous catalysis is that the solvent must solubilize the catalyst, substrate, and reagent. This is expressed in the rule "similia similibus solvunture", or "like dissolves like".[1]

The solvent is not only responsible for the solubilization of substrate, catalyst, and reagent, but also interferes actively with the catalyst.[2] These effects are dependent on the temperature. Particularly polar solvents may temporarily occupy vacant coordination sites at the electronically unsaturated metal complex. In this manner, a solvent behaves as an excess ligand, and the homogeneous metal catalyst is stabilized and not subjected to clustering or other degradation reactions.

In asymmetric catalysis, the solvent may dramatically influence stereo-discriminating interactions in the diastereomeric catalyst–substrate complexes, due to different polarities. From the Eyring equation, it can be concluded that only small energy differences between diastereomeric transition states of stereoselective reactions exist ($\Delta\Delta G^{\#}$ = 1.5–3.4 kcal/mol for 85–99 % enantiomeric excess (abbreviated to % ee)), which may be influenced by the solvent in a positive or negative sense.

Due to a change in the reaction mechanism, a change of solvents may even induce opposite configurations in the chiral product. This phenomenon has been mentioned in the literature under a variety of names, such as "unexpected inversion of enantioselectivity", "change in the sense/direction of enantioselectivity", "chirality inversion", "switch of the expected chiral sense", or "unexpected reversal of the configuration".[3] Up to now, such effects cannot be anticipated. A rare example in asymmetric hydrogenation was reported, e.g., by Selke using a rhodium complex of an α-glucosidic bisphosphinite ligand (Me-β-glup). In the hydrogenation of a dehydroamino methyl ester in ethanol in the chiral product, 73% ee (S) was induced, but only 6% ee (R) in benzene (Scheme 8.1).[4] This effect could not be rationalized.

Another example was discovered also serendipitously by Feringa, de Vries and Minnaard, with a rhodium complex based on a monodentate phosphoramidite (PipPhos) in hand, which gave 90% ee (R) in dichloromlethane but 26% ee (S) in methanol during the hydrogenation of an enol acetate (Scheme 8.2).[5] Neither the reversal of chirality nor the degree of enantioselectivity could be explained.

Screening of solvents for use in asymmetric catalysis may follow varying motivations. One of the most important approaches is to identify solvents which have a positive impact on reactivity and enantioselectivity. In this regard, a change of the solvent would be expected to improve catalytic properties of ordinary chiral catalysts. This follows a simple trial-and-error approach. In an alternative approach, ligands and catalysts have been specifically "designed" for operation only in a particular solvent, with the same expectation. Another very important aim for a solvent change is to facilitate the final work-up procedure. The targeted solubility and insolubility,

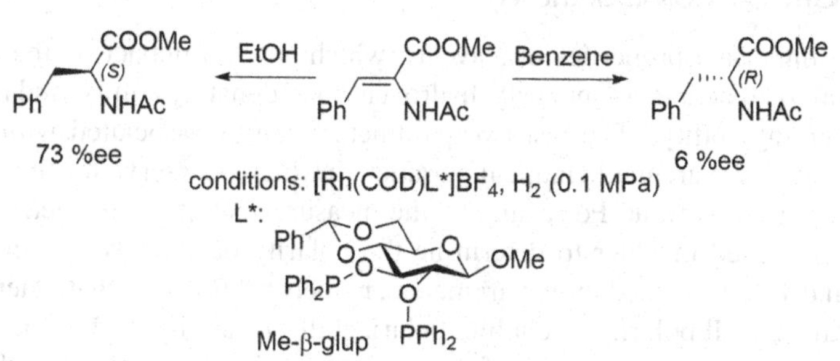

SCHEME 8.1 Reversal of chirality in the asymmetric hydrogenation of an amino acid precursor only by change of the solvent.

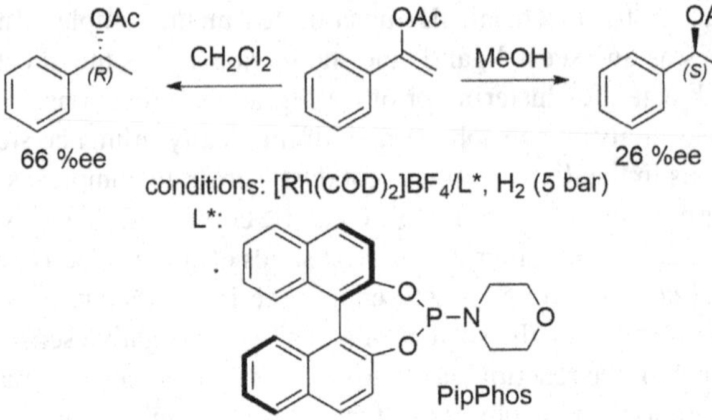

SCHEME 8.2 Reversal of chirality in the asymmetric hydrogenation of an enol ester only by change of the solvent.

respectively, can be achieved by attaching appropriate solubilizing groups to the ligand or to the pre-catalyst. Up to now, such a "rational" design approach, leading to better results, is in the minority. In most instances, mechanistic conclusions about success or failure have been drawn in a retrospective manner and forecasts for other set-ups are not really possible. The third case, which becomes more and more important with respect to economic requirements and ecological constraints (keyword: "green chemistry"[6]), directs the search for cheaper or more sustainable media.

Herein, all three cases will be considered. Sometimes, no clear differentiation between these three approaches can be made. Due to the chemical specialization of the author, mainly asymmetric hydrogenations will be considered. Another reason for this choice is that, of all chemical reactions, hydrogenations are among the most important from the industrial point of view.[7] Only when appropriate will other stereoselective transformations be included.

8.2 PROPERTIES OF SOLVENTS AND SOLVENT-ORIENTED LIGAND DESIGN

8.2.1 General Considerations

The most important properties of solvents, which have an impact on the course of every catalytic reaction, are polarity, hydrogen-bond donating ability, and hydrogen-bond accepting ability.[8] The last two parameters can be associated with proticity and basicity. There are mathematical parameters which may serve for the characterization of a given solvent. For example, the measurement of the dielectric constant is commonly used in order to determine the polarity of the solvent.[9] In addition, Kamlet and Taft developed three parameters, namely π^*, β, and α, that independently indicate the overall polarity (including polarizability), basicity, and proticity.[10] Such solvent parameters can provide useful correlations with the catalytic performance. Moreover, some conclusions about the catalytic mechanism *via* polar or nonpolar intermediates may be possible.

The following table (Table 8.1) is derived from a review by Dyson and Jessop, who collected the solubility in commonly used solvents of important reagent gases from different references.[11]

In recent decades, uncommon solvents, such supercritical CO_2 ($scCO_2$) or ionic liquids (ILs), started a "career" in asymmetric catalysis, with an example being given in Table 8.1 with [BMIM][BF$_4$] (BMIM = 1-butyl-3-methylimidazolium hexafluorophosphate). The use of such "alternative" media is driven by the intention to improve the sustainability of the catalytic reaction.[12]

For some years, reactions in fluorous solvents (perfluorinated alkanes), especially in combination with other solvents (liquid–liquid biphase processes) became attractive for chemists working in homogeneous catalysis.[13] However, the cost of the relevant tailor-made ligands containing fluorinated alkyl chains hampered their use on an industrial scale. Moreover, since solvents and ligands represent inert fluorinated materials, the question was raised about their correct residue-free disposal, in order to prevent the accumulation of even the smallest traces in the atmosphere.[14] This problem is probably the reason that the interest in this special approach declined. An exclusion from this general tendency concerns some recent academic examples in the field of metal-catalyzed asymmetric reactions with common, not specially designed, ligands and catalysts.[15] Reactions, e.g., copper(II)-catalyzed Friedel–Crafts alkylation of indoles with nitroalkanes,[16] cobalt(II)-catalyzed [2+1] cycloadditions of alkynes[17] or asymmetric palladium-catalyzed allylic alkylations[18] were conducted in benzotrifluoride (BTF), fluorobenzene, or hexafluorobenzene as solvents or cosolvents. These aromatic solvents are polar, and have no hydrogen-bonding acceptors atoms, like N or O. In comparison with commonly used solvents, such as dichloromethane, toluene, benzene, or acetonitrile, sometimes significant improvements or at

TABLE 8.1

The Solubility (mM) of Selected Reagent Gases in Common Solvents at 25°C and 0.101 MPa Partial Pressure of the Gas, Presented in Decreasing Order of Ability of the Solvent to Dissolve Reagent Gases

Solvent	H_2	CO	O_2	CO_2
$scCO_2$	117	117	117	18 500
Perfluoroheptane	6.2	17.2	24.8	94
Hexane	5.0	13.1	15.1	92
Acetone	4.1	10.5	11.4	259
Ethyl acetate	3.5	10.4	-	235
Methanol	4.0	9.2	10.2	160
Ethanol	3.5	8.4	10.0	125
Toluene	3.0	7.5	8.7	96
DMSO	1.1	-	1.9	129
[BMIM][BF$_4$]	0.86	2.0	-	98
Water	0.78	0.98	1.3	34

least similar percentages of conversion and ee values were reported. Irrespective of the decline of interest, several phosphorus ligands created for use in fluorous phases have seen a new application in reactions with supercritical carbon dioxide ($scCO_2$) as the solvent.

In general, for the choice of chiral ligands or catalysts, two approaches can be differentiated:

- The application of ordinary, sometimes commercially available ("privileged") chiral ligands and catalysts, in alternative reaction media. For this approach, either no changes or only marginal modifications in the structure of the ligand or the catalysts have been made.
- Ligands and catalysts were specifically modified for the reaction in the new medium, attempting to achieve greater solubility and therefore a better catalytic result. Ironically, such designed catalytic systems sometimes gave better results in common solvents than in the targeted alternative solvent, illustrating the cognitive problem associated with the concept of "rational design".[19]

In several cases, the development commences with the synthesis of an achiral version of the ligand followed by a chiral successor, according to the most common rule in asymmetric catalysis: "Give me first reactivity, I will give you enantioselectivity later."

A nice example to illustrate the heuristic approach connected with the design of appropriate ligands represents the decoration of the well-known Xantphos with long alkyl groups (Scheme 8.3).[20] The parent ligand, Xantphos itself, is one of the most prominent ligands in metal-catalyzed homogeneous catalysis.[21] In order to overcome the problem of low solubility in typical aromatic solvents used for rhodium-catalyzed hydroformylation or in the neat olefinic substrate, van Leeuwen and co-workers

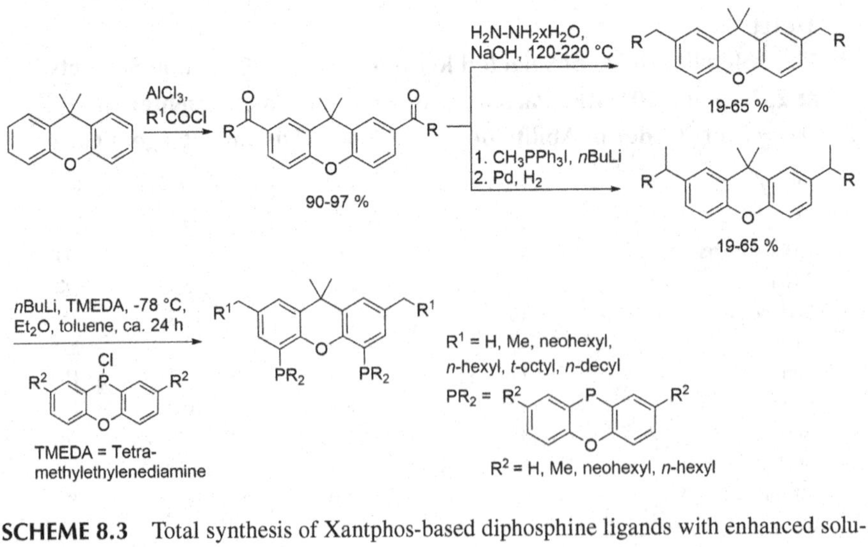

SCHEME 8.3 Total synthesis of Xantphos-based diphosphine ligands with enhanced solubility in aromatic solvents.

linked long-chain alkyl groups to the parent ligand. Alkyl groups of different length were introduced by a Friedel–Crafts acylation of 9,9-dimethylxanthene, followed by reduction of the keto group with hydrazine. The last step suffered from low yields. Finally, by the reaction with 10-chlorophenoxyphosphines, two bulky phosphine groups were incorporated, also bearing long alkyl chains.

As a result of this modification, the solubility of the parent ligand Xantphos in toluene was significantly enhanced. Differences from 2–500 mmol/L in toluene were observed in relation to the length of the alkyl chain. The decoration with long alkyl groups principally modified the molecular interactions found in the parent diphosphine, Xantphos. The longer carbon chain decreases the overall aromatic character and leads to more soluble systems, since π–π stacking interactions are hampered. Due to the extension of the carbon chain length, the number of possible conformers increases and consequently the configurational entropy of the system becomes greater. The latter effect holds especially for the neohexyl and t-octyl modified systems. In the series n-hexyl, neohexyl to t-octyl, the rigidity of the alkyl chain increases dramatically; this reduces the number of possible conformations and gives rise to a decrease in solubility. Ligands with a solubility <13 mmol/L showed precipitation upon cooling the reaction mixture to room temperature.

Meanwhile, because P^*-chirogenic Xantphos ligands have also been described,[22] this fact could stimulate the development of chiral versions of these particularly decorated ligands.

R = H, tBu

general structure of P^*-chirogenic Xantphos

However, as impressively shown by the preceding example, the tuning of ordinary ligands for selected solvents requires enhanced efforts in the synthesis. Moreover, due the change of the molecular structure in several cases, no or only marginal correlations to the original catalytic properties of the parent ligand can be identified.

8.2.2 MIXTURES OF MISCIBLE SOLVENTS

Mixtures of miscible solvents can support the efficiency of a catalytic reaction, even though pure solvents dominate the literature. A remarkable solvent effect was e.g., observed with Wilkinson's catalyst in the (non-asymmetric) chemoselective hydrogenation of alkenes linked to oxidatively sensitive aryl nitro groups (Scheme 8.4).[23] In order to prevent the reduction of the nitro group or dehalogenation from taking place, very mild reaction conditions are required. In benzene, under 1 atmosphere of H_2 and at room temperature, no reaction was observed, whereas, in methanol, the

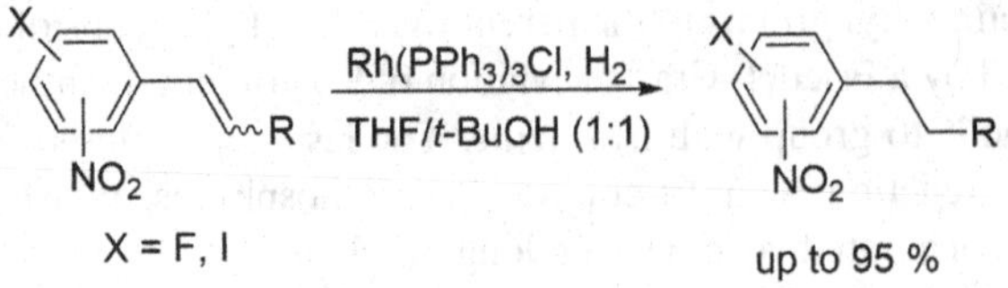

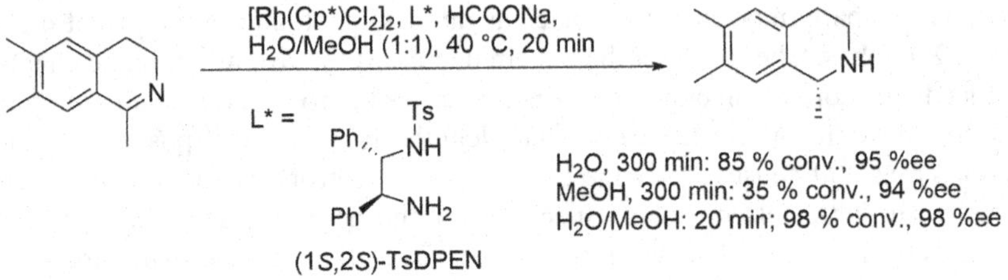

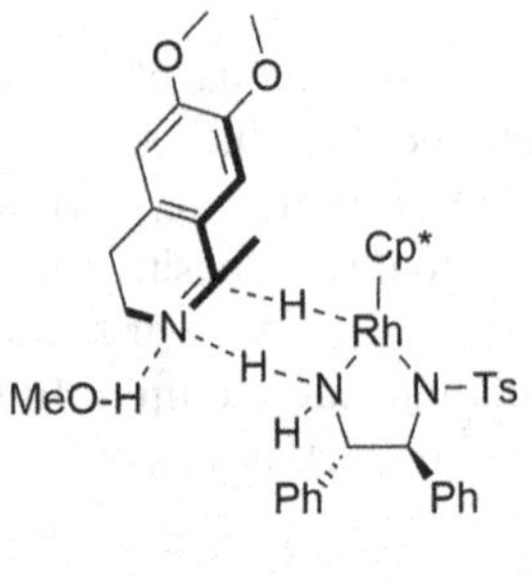

268   Chiral Ligands
```

**SCHEME 8.4** Chemoselectivity governed by the solvent.

**SCHEME 8.5** Different rates and ee values in the asymmetric hydrogenation, due to use of a solvent mixture.

alkene bond was reduced to afford the saturated product at 80% yield. In THF, the yield increased further to 91%. Finally, highest yields were obtained in a THF/*t*BuOH (1 : 1) mixture.

A related effect was observed in the asymmetric transfer hydrogenation of imines (Scheme 8.5).[24] The hydrogenation of various cyclic imines proceeded efficiently with a water/methanol co-solvent medium in 20 min, with excellent yields and enantioselectivities, by employing a rhodium complex based on the commercially available TsDPEN ligand, with sodium formate as a hydrogen donor. In contrast, the reaction in pure water or methanol proceeded much more slowly, and the ee-values were slightly less.

A similar beneficial effect was noted by employing mixtures of water with other alcohols, such as ethanol, propanol, butanol, or ethylene glycol. Polar aprotic solvents, like DMF, dimethyl sulfoxide (DMSO) or NMP gave conversions up to 95% and similarly high enantioselectivities within 30 min reaction time. However, with THF, just 25% of the product was formed in the same time. DFT calculations gave some evidence that, in the outer-sphere mechanism, methanol as a co-solvent reduces the activation barrier in transfer hydrogenation of imine by coordination to the imine nitrogen *via* an H-bond, and thus an increased reaction rate is macroscopically noted.
```

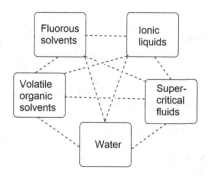

FIGURE 8.2 Important combinations of fluid phases for biphasic chemistry.

8.2.3 Mixtures of Non-Miscible Solvents

Immiscible solvents form multiphasic reaction media if their polarities are sufficiently different. The most common fluid multiphase systems are biphasic, but triphasic and even higher numbers of phases are known. Multiphasic approaches aim to overcome the major problems of homogeneous catalysis, that is catalyst recovery and product separation. In a biphasic fluid system, the catalyst is immobilized in one solvent, which is not miscible with the second one. Substrates and products are dissolved in the latter.

Of course, when gaseous reagents, e.g. hydrogen gas or carbon monoxide, are used, then solubility problems may come into play. Such liquid or liquid/liquid reactions with gases should be principally categorized as biphasic or triphasic, respectively, but, in the daily academic lab life, this aspect is usually addressed only in special cases, where special kinetic issues are in the focus. The situation may change in those cases where industrial application is intended.

In multiphase fluid catalysis, immiscible solvents hosting catalyst, substrate, and reagent are suspended by vigorous mixing. When the reaction is stopped, the catalyst is retained in one phase and the product is localized in the second. In the ideal case, the catalyst can be reused and the product in the second phase is not contaminated with the catalyst.

There are various combinations that give rise to multiphasic systems. Most important combinations of fluid phases for biphasic chemistry are given in Figure 8.2, taking into consideration ordinary volatile organic solvents, such as alcohols, ethers, or aromatic compounds, and more sophisticated solvents such as fluorous solvents, ionic liquids, supercritical fluids, and water (Figure 8.2).[25]

8.2.4 Temperature-Dependent Multicomponent Solvent Systems (TMS)

Miscibility of solvents is dependent on the temperature. In most cases, an increase in the temperature leads to greater miscibility of solvents. Therefore, undesired separation of catalyst and substrate can be avoided principally when the separation of two non-miscible phases is counterbalanced by a change in the temperature, leading to a single phase during the reaction. Such so-called temperature-dependent

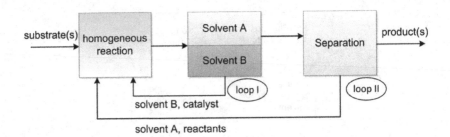

FIGURE 8.3 Concept for total integration of thermomorphic solvent systems in a continuous process (taken from Ref. 132).

multicomponent solvent systems (TMS) play an important role in catalysis and technology research.[26]

In general, a TMS consists of at least two solvents with different polarity. The solvents are immiscible at low temperature and form a single homogeneous phase at elevated temperatures. The reaction is conducted in a single phase by repression of the miscibility gap at reaction temperature. After completion of the catalytic reaction, a decrease in the temperature to room temperature results in separation of the two phases and an easy catalyst/product separation is possible. In general, the system consists of a nonpolar and a polar solvent (Figure 8.3). The catalyst is predominantly soluble in one solvent (solvent B) and the product is located in the other solvent (solvent A). Solvents A and B can be principally recycled by technological solutions (loop I and loop II). A typical composition of a technically approved solvent mixture are 1-dodecene/n-decane/N,N-dimethylformamide.

This methodology has been already used in pilot plants. The application scope in asymmetric catalysis, however, seems to be limited due to the elevated temperature during the reaction. Since, in most cases, energy differences in the transition state between diastereomeric catalyst–substrate complexes are very small (see above), they can be equalized at higher temperatures and the stereo-differentiating effect is lost. Therefore, only catalytic systems which operate and are also highly stereoselective at higher temperatures can be used successfully in this approach.

Quite recently, an example was given where a thermomorphic ionic-liquid-based microemulsion system was used for the ruthenium catalyzed asymmetric transfer hydrogenation of ketones.[27] As solvent, a ternary mixture of ionic liquid/n-heptane/water was used. A special feature of the ionic liquid was its property to form a microemulsion. The use as solvent of sulfonated ligands originally developed for catalytic reactions in water improved the immobilization of the catalysts and significantly decreased the catalyst leaching into the organic layer upon extraction of the product.

8.2.5 SOLVENT-DEPENDENT *A POSTERIORI* SEPARATION OF CATALYST AND PRODUCT

By formation of binary or ternary mixtures of solvents only after the reaction, by addition of a further solvent, separation of catalyst and product is possible. A typical

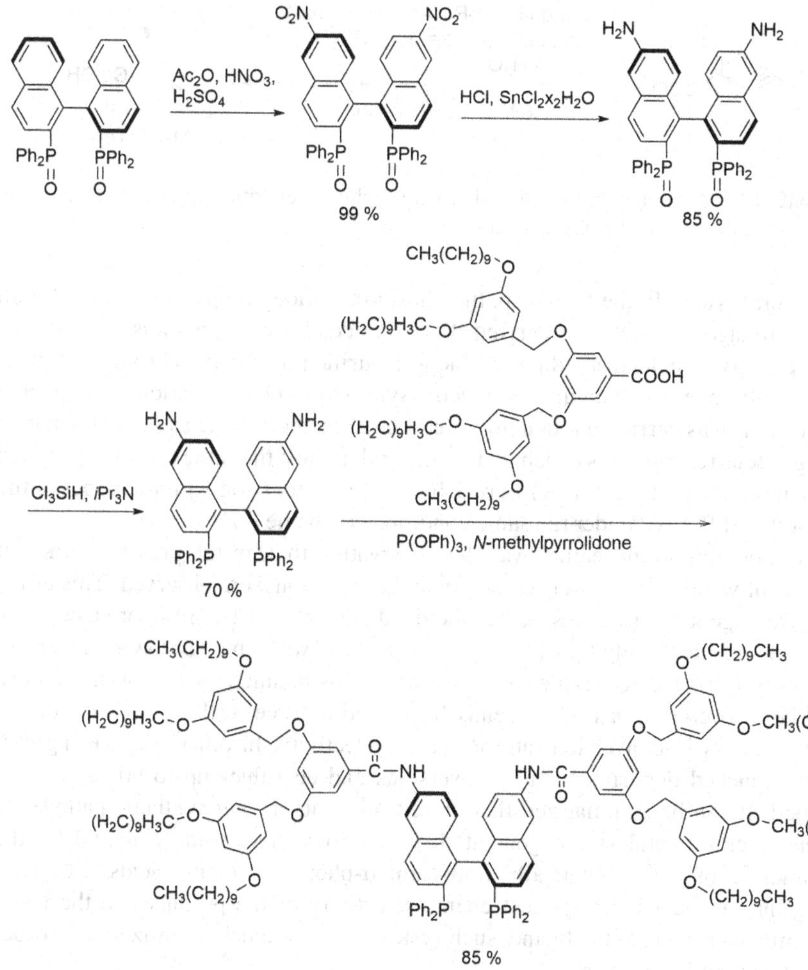

SCHEME 8.6 Synthesis of a "heterogenized" BINAP ligand.

example was illustrated in the asymmetric hydrogenation of prochiral acryl acid derivatives as model substrates, using a Fréchet-type dendrimer, namely a peripherally alkyl-functionalized dendritic Ru-BINAP catalyst (Scheme 8.6).[28] Subsequently, not only will the considered asymmetric reaction be detailed, but also the entire pathway of the synthesis of the ligand. The description of the synthetic route gives an idea of the considerable additional synthetic efforts which are frequently associated with the concept of "heterogenization" of homogeneous chiral catalysts. BINAP itself cannot be attached to a dendrimer. Therefore, firstly (R)-5,5'-diamino-BINAP (R-1) had to be synthesized.[29] It can be derived from enantiopure BINAP(O), which is part of the Takasago route to BINAP and therefore commercially available.[30] By nitration and subsequent reduction of the nitro groups, the relevant diamine is obtained in

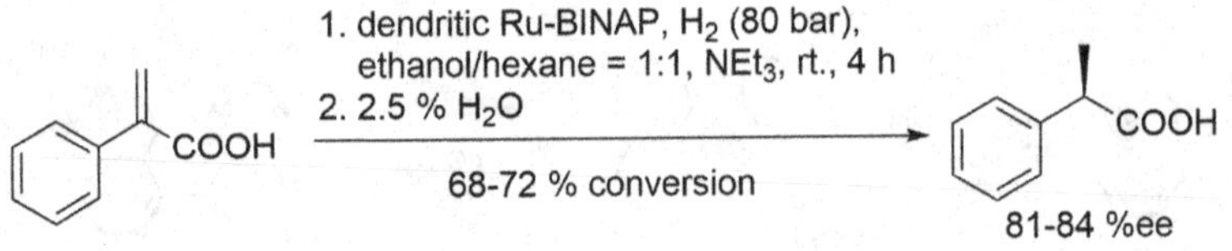

1. dendritic Ru-BINAP, H$_2$ (80 bar),
   ethanol/hexane = 1:1, NEt$_3$, rt., 4 h
2. 2.5 % H$_2$O

68-72 % conversion

81-84 %ee

SCHEME 8.7 Asymmetric hydrogenation with a chiral dendritic ruthenium catalyst and separation of the product by phase separation.

good overall yield. In the final step, the phosphine oxide groups were reduced with silane. Linkage of this amino-derived BINAP to dendritic wedges was achieved *via* amide groups. The dendritic diphosphine was purified by column chromatography.

The subsequent ruthenium-catalyzed asymmetric hydrogenation of α-phenyl acrylic acids was carried out in a mixture of ethanol and hexane as solvent, where a homogeneous reaction was concluded. In general, almost the same rate of conversion and enantioselectivity (89 % ee) were achieved in comparison with the effect of the non-modified BINAP under the same conditions (Scheme 8.7).

The most important feature was the observation that, in the presence of small amounts of water, much lower selectivity and conversion were observed. This effect was advantageously used to separate the dendritic catalyst from the product at the end of the reaction. Only by addition of 2.5 % water was a biphasic system formed. The catalyst was located in the hexane layer. In this manner, 99.3 % of the catalyst could be recovered. It has subsequently been used in three further runs without any significant loss in conversion rate or enantioselectivity. In other trials, using still higher branched dendrimers, full conversions and ee values up to 90 % ee were obtained. It can be summarized that, based on a rather long synthetic pathway, a recyclable chiral catalyst was generated, which could have some potential for the manufacture of pharmaceutically important α-phenyl propionic acids. However, taking into account the imperfect enantioselectivity of the product and the time-consuming synthesis of the ligand, such systems require much optimization in order to meet economic demands.

8.2.6   SPECIAL SOLVENT EFFECTS IN HOMOGENEOUS CATALYSIS

Polar solvents compete with prochiral substrates for vacant coordination sites at the catalytically active metal center. In this regard, bidentate coordinating substrates have a better chance of displacing solvent molecules than monodentate substrates, due to the chelate effect. This is one reason why, in the history of asymmetric hydrogenation (the most important reaction in asymmetric catalysis), mainly bidentate substrates were used as standard for such a long period. Typical examples are α-N-acyl dehydroamino acid derivatives, and itaconic acid and its esters. Preferred solvents were short alkyl alcohols, such as methanol or ethanol, with strong coordinating properties. Only later could monodentate coordinating substrates, like unfunctionalized olefins, ketones, or imines, be converted with high activity and high enantioselectivities.

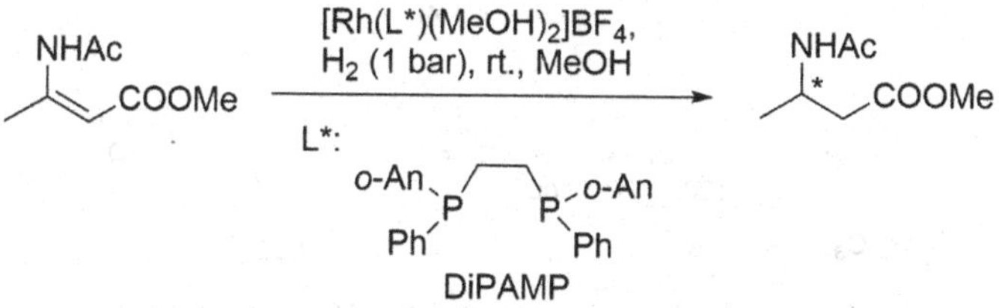

SCHEME 8.8 Asymmetric hydrogenation used for demonstrating the inhibitory effect of aromatic solvents.

Seemingly innocent nonpolar solvents may also have an influence on the course of the reaction. A striking example was investigated by Heller *et al.* in the asymmetric hydrogenation of an unsaturated β-dehydroamino acid precursor with chiral rhodium complexes, such as [Rh(DiPAMP)(MeOH)$_2$]BF$_4$ (Scheme 8.8).[31]

Only by addition of small amounts of *p*-xylene to the reaction in methanol was a dramatic decreasing effect noted. The decelerating effect was attributed to the formation of catalytically inactive rhodium–xylene complex of DiPAMP, which could be detected spectroscopically and by means of X-ray structural analysis.

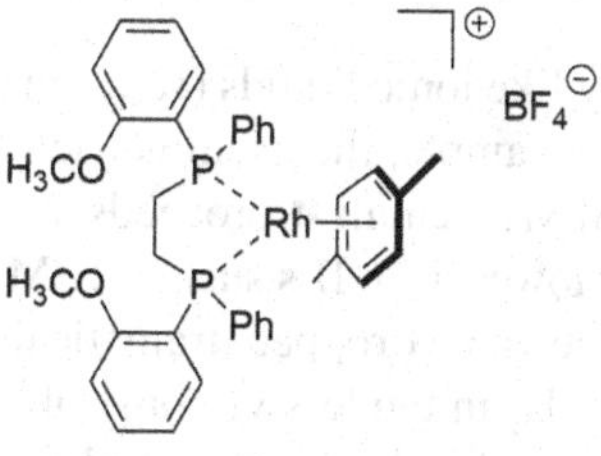

DiPAMP-Rh complex

The authors emphasized that in several cases, where very low reaction rates in aromatic solvents have been reclaimed, this phenomenon can possibly be traced back to this effect. Nevertheless, in numerous asymmetric reactions, the use of aromatic solvents, mainly toluene, is still usual, which gives evidence that this effect may be overridden by other effects. It is noteworthy, in the precedingly described example, that the deceleration did not affect the ee values of the asymmetric reaction.

When reagent mixtures, like synthesis gas, are used the different solubilities of H$_2$ and CO must be taken into consideration, especially when the reaction is conducted under low or ambient pressure. It was found, for example, that by using formaldehyde as surrogate for synthesis gas (CO/H$_2$) in the hydroformylation of olefin conversion, the yield of the desired aldehyde could be more than doubled in the presence of 10 bar additional hydrogen pressure (Scheme 8.9).[32] This effect can be clearly rationalized by the different solubilities of CO and H$_2$ in toluene (see also Table 8.1).

In special solvents, such as supercritical carbon dioxide (*sc*CO$_2$), hydrogen and nitrogen behave as "anti-solvents". This means that they decrease the solubility of other solutes. This feature must be taken into consideration for the design of a proper reaction set-up.[33]

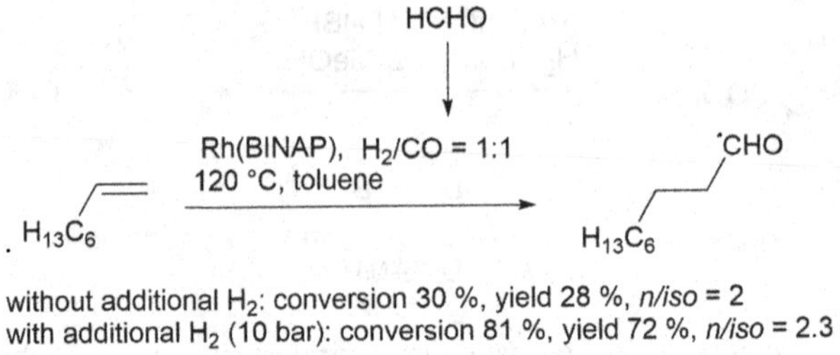

HCHO

Rh(BINAP), H_2/CO = 1:1
120 °C, toluene

$H_{13}C_6$ ⟶ $H_{13}C_6$ CHO

without additional H_2: conversion 30 %, yield 28 %, n/iso = 2
with additional H_2 (10 bar): conversion 81 %, yield 72 %, n/iso = 2.3

SCHEME 8.9   Influence of different solubilities of gas mixtures on the conversion.

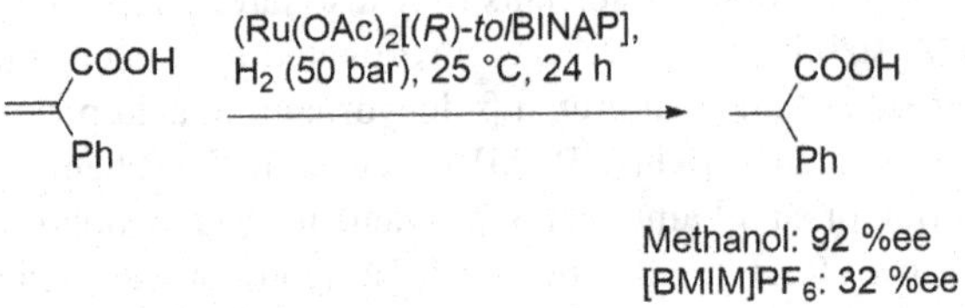

COOH   (Ru(OAc)$_2$[(R)-tolBINAP],
         H_2 (50 bar), 25 °C, 24 h   ⟶   COOH

Ph   Ph

Methanol: 92 %ee
[BMIM]PF$_6$: 32 %ee

SCHEME 8.10   Drop of the ee value by a change of the solvent from methanol to an IL.

Solvents of high viscosity like ionic liquids (ILs) can affect the ee values of asymmetric transformations. For example, the asymmetric hydrogenation of atropic acid by a chiral ruthenium tolBINAP catalyst proceeds with high enantioselectivity in methanol (Scheme 8.10).[34] However, in ILs such as [BMIM][PF$_6$] (BMIM = 1-butyl-3-methylimidazolium), the ee value dropped dramatically. This effect was explained by the greater availability of H_2 in the less viscous solvent methanol. In contrast, by submitting tiglic acid as the prochiral substrate, the best result in methanol (88 % ee) could be exceeded (93 % ee) by running the reaction in the ionic liquid.

Obviously, the concentration of gaseous reagents, e.g., hydrogen, in solution may have a strong impact on the mechanism and therefore on the enantioselectivity of an asymmetric hydrogenation, as shown by Blackmond and co-workers.[35] It is noteworthy that the decisive kinetic parameter affecting the ee value is the concentration of hydrogen in the liquid phase (here, the IL), rather than the hydrogen pressure in the gas phase. In general, effects can be observed whenever, under typical reaction conditions, the concentration of H_2 differs widely from its equilibrium saturation value, and gas–liquid diffusion is the rate-limiting step. In these cases, the intrinsic stereo-differentiating ability of the chiral catalyst may be masked by the pressure effect.

As already mentioned, solvents interact more or less with the catalyst or the catalyst–substrate complex. Polar solvents can coordinate to the metal center and thereby activate or deactivate the catalyst. Therefore, achiral counterions of charged chiral metal complexes may have a pronounced effect on the rate or enantioselectivity in dependence on the solvent. Likewise, chiral counterions can be used for this purpose to influence the stereo-discriminating ability of the catalyst.

In general, strong and weakly coordinating ions can be differentiated. Ion pairing is dependent on the polarity of the solvent. In solvents of low polarity, the cation and anion form a strong ion-pair. If the counter-ion blocks the active site of the catalyst, it may inhibit its activity. In contrast, in polar solvents, the two ions are solvated, which means that the cation and the anion are usually completely separated from each other, allowing catalysis to proceed without the hindrance of the counter-ion.

Ion pairing is dependent on the charge and the size of the ion. Extremely large anions are able to exist entirely isolated in solution. A typical specially designed example is tetrakis[3,5-bis(trifluoromethyl)phenyl]borate (BARF) with the chemical formula $[\{3,5\text{-}(CF_3)_2C_6H_3\}_4B]^-$.[36] Therefore, BARF has been successfully used in various asymmetric hydrogenations in supercritical CO_2.[37]

BARF$^{\ominus}$

8.3 UNUSUAL SOLVENTS – APPLICATIONS

Some asymmetric reactions will now be considered in detail, where the effect of less common solvents has been investigated or where a special solvent effect has been targeted. Due to the vast literature, only typical examples of alternative solvents and ligands/catalysts will be presented, without the intention to cover the entire literature.

8.3.1 REACTIONS IN IONIC LIQUIDS (ILs)

Ionic liquids (ILs) consist entirely of ions. They have melting points below 100°C and the vapor pressure is extremely low at near-ambient conditions.[38] Due to these properties, most ions forming ILs display low charge densities, resulting in low inter-ionic interaction. An important feature is the possibility of altering the solubility and coordination properties by varying the nature of the anions and cations systematically (Figure 8.4).

ILs with weakly coordinating or even inert anions, e.g., $[(CF_3SO_2)_2N]^-$, $[BF_4]^-$, or $[PF_6]^-$, and inert cations, which do not coordinate to the catalyst themselves, can be considered as "innocent" solvents in transition-metal catalysis. In this connectivity, the role of the ILs is solely to provide a more or less polar or a more or less weakly coordinating medium for the transition-metal catalyst. However, in contrast to most conventional solvents, many ILs combine high solvating power for polar catalyst complexes with weak coordination (nucleophilicity).[39] Therefore, in the biphasic reaction mode, catalysts can operate which are usually deactivated by water or polar organic solvents. Even more interesting, the biphasic approach allows, in principle,

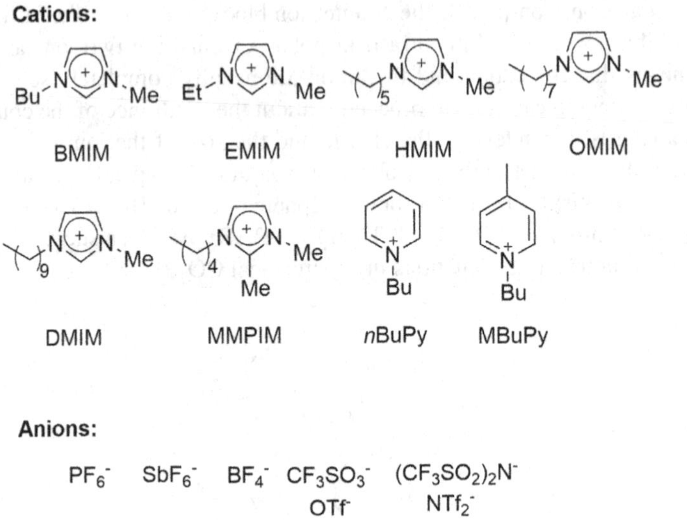

Cations:

| BMIM | EMIM | HMIM | OMIM |

| DMIM | MMPIM | nBuPy | MBuPy |

Anions:

PF_6^- SbF_6^- BF_4^- $CF_3SO_3^-$ $(CF_3SO_2)_2N^-$
OTf^- NTf_2^-

FIGURE 8.4 Important ionic liquids and their abbreviations.

the recycling of the ionic catalyst solution. Most investigations are predominantly focused on the immobilization of phosphorus ligands, bearing transition-metal catalysts in a biphasic system. The most common reactions are asymmetric hydrogenation and some C–C coupling reactions. Proper immobilization can be achieved either by ionic transition-metal complexes or by ionic ligands. Only a few examples are known where chiral phosphine ligands have been modified by ionic moieties.

For hydrogenation reactions, it is noteworthy that hydrogen solubility in ionic liquids is reduced compared with most traditional solvents, so that the actual hydrogen concentration may be low at the catalytic center. Mass transfer of hydrogen into the catalyst layer is affected by the viscosity of the ionic liquid (with low viscosities enhancing mass transfer) but has been found to be fast enough in most cases to reach acceptable reaction rates comparable with those obtained in organic media. For C–C coupling reactions, it is important to keep into account the polarity of reactants and products. Polar reactants and nonpolar products are preferred to ensure sufficient solubility of the reactants in the IL reaction phase, and good separation of the products due to high solubility in the polar phase. Good results were achieved with $scCO_2$ as the second, product-extracting phase.

Chauvin et al. pioneered the field describing the hydrogenation of the model substrate (Z)-α-acetamido cinnamic acid, using [Rh(COD)(DIOP)][PF_6] as the catalyst in a [BMIM][SbF_6] melt, generating the chiral product with 64% ee.[40] The protected amino acid was quantitatively separated and the ionic liquid could be reused. The loss of rhodium was less than 0.02% per run.

Other authors achieved 80 % ee in the hydrogenation of 2-arylacrylic acids with a chiral ruthenium catalyst in [BMIM][BF_4] melts (Scheme 8.11).[41] In both reactions, an organic solvent, e.g., iPrOH was added.

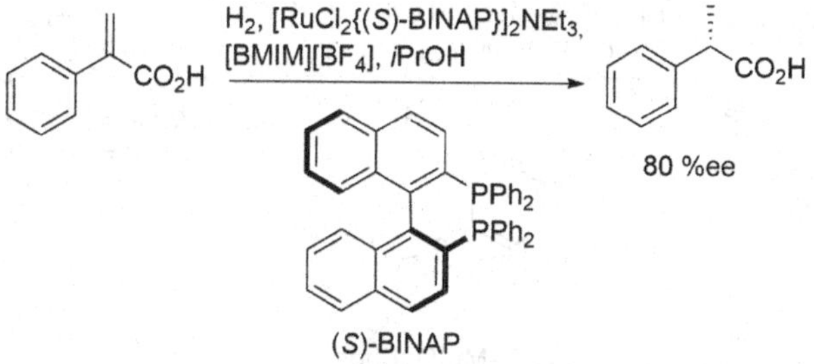

SCHEME 8.11 Hydrogenation of 2-phenylacrylic acid to (*S*)-2-phenylpropionic acid with a chiral ruthenium catalyst in [BMIM][BF$_4$].

These promising results stimulated the use of numerous other prominent diphosphine ligands, such as substituted BINAP and its congeners, as well as ligands of the DuPhos or JosiPhos series.[42] These ligands, mostly commercially available, were used without any additional modification. Rhodium, as well as ruthenium, complexes were investigated in most cases for typical unsaturated standard substrates.

Geresh *et al.* applied a chiral rhodium DuPhos catalyst dissolved in [BMIM] [PF$_6$] for the asymmetric hydrogenation of two enamides commonly used as standard substrates (Scheme 8.12).[43] The catalyst was fully soluble in the IL and the product was recovered in the isopropanol phase. A significant stabilization of the air-sensitive catalyst was observed, due to entrapment in the IL. Due to this property, recycling of the ionic catalyst solution was possible. Remarkably, with the small substrate (R = H) the ee values eroded much faster than with the corresponding phenyl derived enamide (R = Ph). Therefore, with only the latter were five efficient recyclings possible.

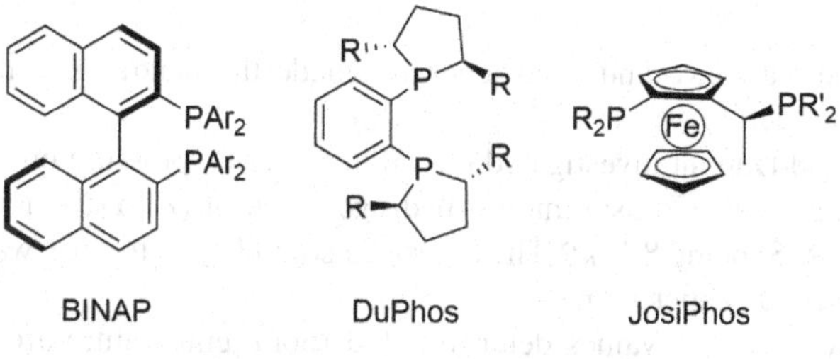

BINAP   DuPhos   JosiPhos

Giernoth *et al.* screened several ILs and compared them with toluene as the solvent in the hydrogenation of trimethylindolenine with an Ir-XyliPhos catalyst system (Scheme 8.13).[44] The ee values obtained were similar to those achieved in toluene as the solvent, but the reaction time could be almost cut in half. Compared with the reaction in conventional organic media, slightly higher temperatures (50°C) were required, which was rationalized by the higher viscosities of the ionic media. Probably the increase in temperature counterbalances the decelerating rate effect, which is caused by the limited solubility of hydrogen in the IL.

Furthermore, a stabilization of the ionic catalyst solution against atmospheric oxygen was noted. This stabilization effect facilitated the transfer of freshly prepared

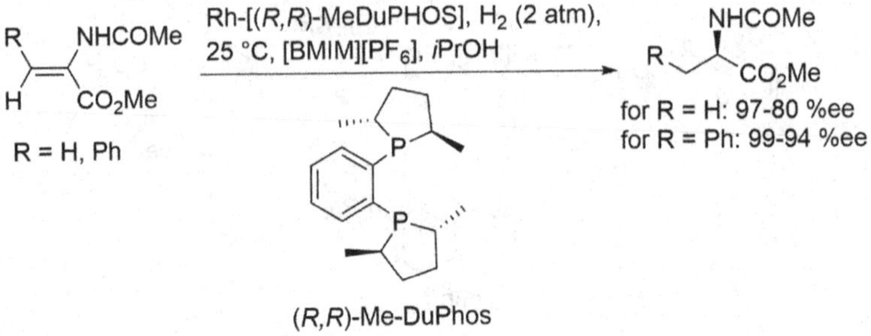

(R,R)-Me-DuPhos

SCHEME 8.12 Asymmetric hydrogenation of two standard substrates catalyzed by Rh(Me-DuPhos) catalyst immobilized in [BMIM][PF$_6$].

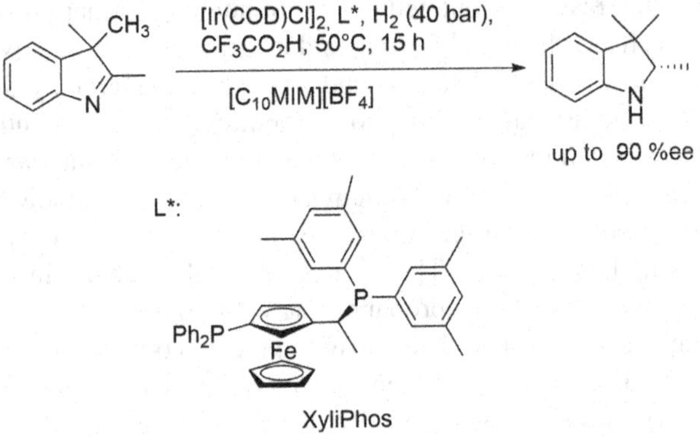

XyliPhos

SCHEME 8.13 Enantioselective hydrogenation of trimethylindolenine, using Ir(XyliPhos) as catalyst.

catalyst to the autoclave and – in general – made the handling of the IL/catalyst system much easier.

de Souza and Dupont investigated the influence of H$_2$ pressure on the conversion in the rhodium-catalyzed asymmetric hydrogenation of (Z)-α-acetamido cinnamic acid in two ILs (Scheme 8.14).[45] The hydrogen solubility in the ILs was determined using pressure drop experiments.

The Henry solubility values determined at room temperature are $K = 3.0 \times 10^{-3}$ mol L^{-1} atm^{-1} for H$_2$ in [BMIM][BF$_4$] and 8.8×10^{-4} mol L^{-1} atm^{-1} for H$_2$ in [BMIM][PF$_6$]. These values clearly illustrate that molecular hydrogen is almost four times more soluble in [BMIM][BF$_4$] than in [BMIM][PF$_6$] under the same pressure. This difference in solubility leads to different degrees of conversion and enantioselectivity: Thus, under 50 bar hydrogen pressure in both experiments, a conversion of 73% (93% ee) was found for the reaction in [BMIM][BF$_4$], whereas only 26% conversion (81% ee) was realized when using [BMIM][PF$_6$] as the reaction medium.

Jessop investigated the influence of ILs in comparison with methanol in the asymmetric hydrogenation of a range of prochiral acrylic acids, such as atropic acid and tiglic acid, with ruthenium BINAP-type catalysts.[35] In general a strong dependency

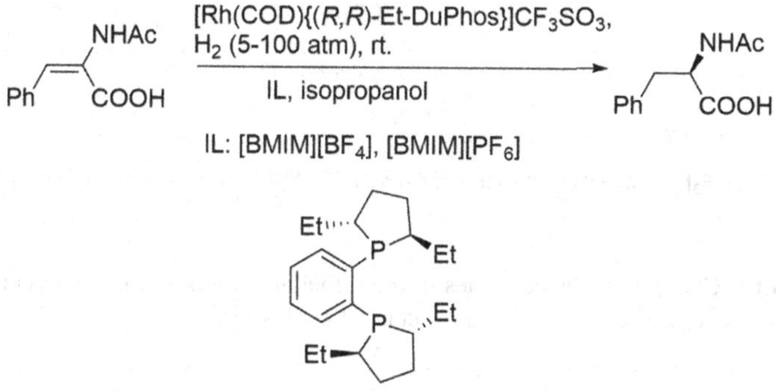

SCHEME 8.14 Hydrogenation of a prochiral model substrate in two ILs with a common diphosphine ligand.

$$[BMIM][BF_4] < [EMIM][O_3SCF_3] < [BMIM][PF_6] < [EMIM][N(O_3SCF_3)_2] < [DMPIM][N(O_3SCF_3)_2]$$

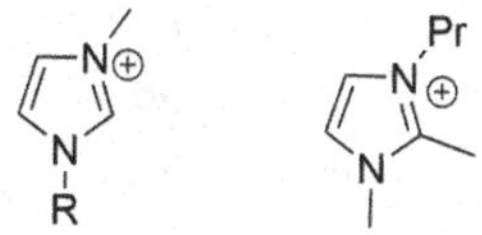

R = Bu (BMIM)   DMPIM
R = Et (EMIM)

FIGURE 8.5 Change of the ee-values in the ruthenium catalyzed hydrogenation of atropic acid in dependency on the IL used as solvent at 50 bar H_2-pressure.

of the ee values on the used solvent was found. High enantioselectivities (79–92 %) were consistently observed in methanol as solvent. With atropic acid as substrate, the ee values varied in the range of 15 % to 39 % and increased in the order shown in Figure 8.5.

Addition of methanol, isopropanol, or toluene improved the enantioselectivity to some extent. This effect is presumably due to the reduction in viscosity by methanol addition (i.e., enhanced mass transfer) and to increased hydrogen solubility compared with pure ILs.

In strong contrast, enantioselectivities were, in general, much higher with tiglic acid in ILs, showing once more the strong dependency of this parameter on slight structural modifications of the prochiral substrate (Figure 8.6).[35] For the reactions in neat ILs, the selectivity was in the range between 88 % and 95 % and increased in the order shown in Figure 8.6.

Interestingly, with this substrate, the addition of common organic solvents (alcohols or toluene) to the ILs had a largely negative effect. Results with [EMIN]+ as cation, using different anions e.g. [O_3SCF_3]− versus [N(O_2SCF_3)_2]− also showed the strong influence of the counter-ion.

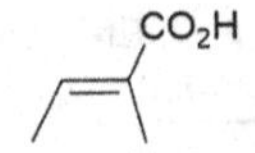

Increase of ee-values:

$$[EMIM][O_3SCF_3] < [mbpy][BF_4] = [BMIM][BF_4] < [DMPIM][N(CF_3SO_2)_2] < [BMIM][PF_6]$$
$$< [EMIM][N(O_2SCF_3)_2]$$

FIGURE 8.6 Changes in the ee values in the ruthenium-catalyzed hydrogenation of tiglic acid in dependency on the IL used as solvent at 5 bar H_2-pressure.

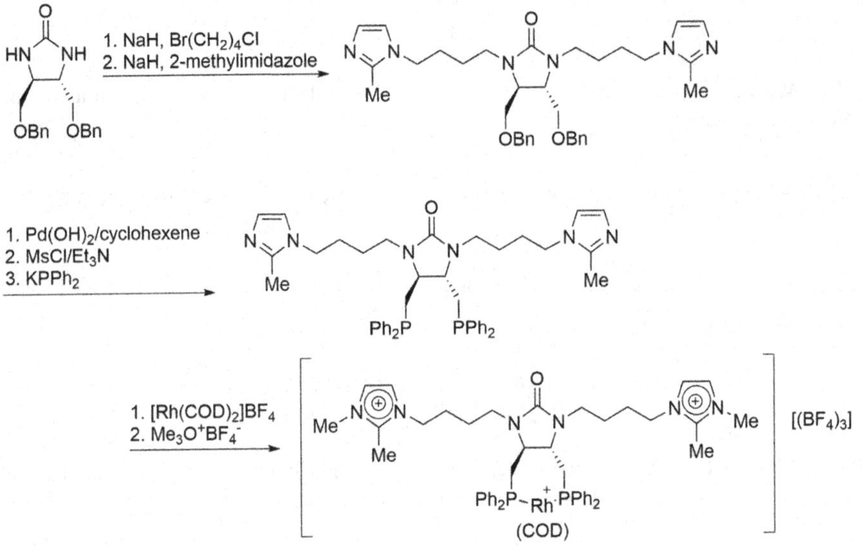

SCHEME 8.15 Preparation of a chiral rhodium catalyst designed for use in ILs.

These results clearly demonstrate that the effectiveness of the considered asymmetric hydrogenations in ILs is not only a function of H_2 availability. Many solvent parameters, including polarity, coordinating ability, and hydrophobicity, have to be taken into account and must be added to achieve a complex picture of the reaction that is still far away from being fully understood.

Attempts to improve the solubility and immobilization of chiral hydrogenation catalysts by application of specially designed ligands in ILs were published by Lee and co-workers, who synthesized a chiral rhodium complex with a diphosphine, containing a cation as ligand (Scheme 8.15).[46] The neutral diphosphine ligand was prepared from **1** by N,N′-dialkylation with 1-bromo-4-chlorobutane, followed by reaction with 2-methylimidazole. After deprotection of the O-benzyl groups, the resulting diol was mesylated and then reacted with potassium diphenylphosphide to generate the diphosphine. Alkylation of the neutral diphosphine, affording the tricationic complex was only possible after protection of the phosphorus atoms with the transition metal. Without complexation, no consistent product was obtained, possibly due to alkylation of the phosphorus atom. The ^{31}P NMR signal of the diphosphine was shifted after complexation and alkylation, from $\delta = 10.1$ ppm to $\delta = 32.6$ ppm with $J_{Rh-P} = 139.2$ Hz.

In the hydrogenation of N-acetyl phenylethenamine, immobilization of the tricationic complex in the biphasic system [BMIM][SbF$_6$]/iPrOH showed almost the same results compared with the non-modified methyl group-bearing complex (Scheme 8.16). Both catalysts could be reused three times. In the fourth run, conversion and ee values, particularly with the N-methyl-substituted ligand, decreased more markedly.

Sometimes, the performance of a chiral rhodium hydrogenation catalyst can be improved by employing an IL/water mixture.[47] So called "wet ionic liquids" can give superior results in comparison with ILs containing no additional co-solvent. This effect may, however, be less pronounced at higher hydrogen pressure.

Ionic derivatives of BINAP were suggested by Lemaire *et al.* in room temperature ILs (Scheme 8.17).[48] The requisite ammonium salts were assembled from

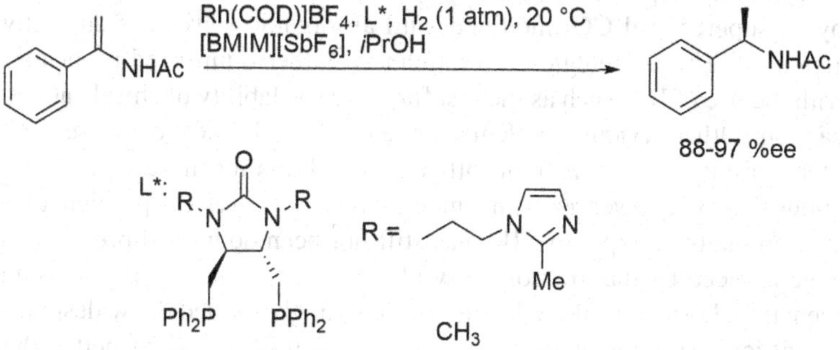

SCHEME 8.16 Rh-catalyzed asymmetric hydrogenation of N-acetylphenylethenamine in an IL, with a specially designed catalyst.

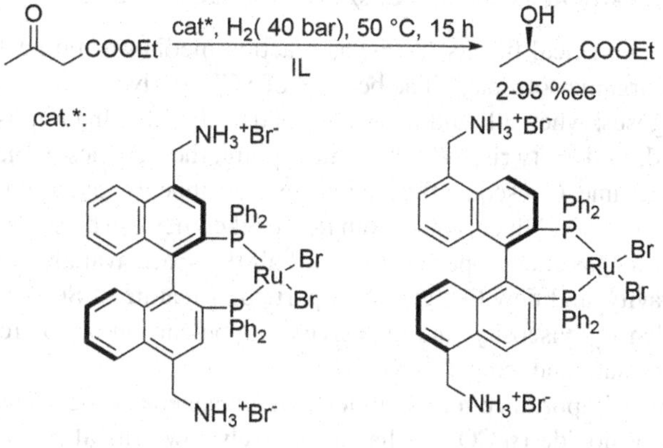

SCHEME 8.17 Hydrogenation of ethyl acetoacetate in ILs, with a specially designed ionic ligand.

enantiopure BINAP in five steps. Catalysts were prepared *in situ* from the respective bromohydrates and $[Ru(\eta^3\text{-}2\text{-methylallyl})_2(\eta^2\text{-COD})]$. Comparative studies of the hydrogenation of ethyl acetoacetate revealed the best results for imidazolium- and pyridinium-containing ILs. In contrast, no significant stereo-induction was observed with the phosphonium salt $[PCy_3(C_{14}H_{29})]^+$. This observation was attributed to problems of solubility and to the ability of complexation with the phosphonium ion. With respect to the selection of anions, $[BF_4]^-$ appeared to be superior to $[PF_6]^-$ and $[N(CF_3SO_2)_2]^-$.

It should be noted that impurities in IL may strongly affect the efficiency of the catalytic reaction. For example, in the asymmetric ruthenium-catalyzed enantioselective hydrogenation of methyl-3-oxobutanoate in an ethanol/[BMIM][PF$_6$] biphasic system, only traces of impurity in the IL decreased the ee values from 97 to 25 %, depending on the nature of the contamination of a particular supply.[49]

More recent developments in this field concerns the use of nanofiltration for the recovery of the catalyst or the immobilization of catalysts in the so-called ionic-liquid-tagged strategy.[50] Alternatively the immobilization of rhodium catalysts, based on common chiral diphosphine ligands in a supported ionic liquid phase (SILP) and employing supercritical CO$_2$ modified with an organic solvent, is currently under investigation.[51] These strategies aim to manage hitherto unsolved problems associated with the use of ILs, such as the insufficient recyclability of chiral catalysts after the reaction. Other serious problems are, e.g., the solidification of several ILs at lower temperatures, which is frequently a precondition for those asymmetric transformations that need lower reaction temperatures. Moreover, the problem of employing large amounts of expensive ILs has still not been addressed properly. Another challenge is faced by the low solubility of some gases, such as H$_2$, in asymmetric hydrogenation. Up to now, these features have strongly limited the widespread use of the present ionic systems in asymmetric catalysis. It is not clear whether the development of new chiral ligands can contribute to the management of these problems.

8.3.2 SUPERCRITICAL FLUENTS (SCFs) IN THE scCO$_2$ EXAMPLE

The use of supercritical fluents (SCFs) as reaction media was targeted at replacing conventional organic solvents.[52] The benefits of SCFs derive from a special effect of compressed gases: when a liquid is heated, its density falls. In contrast, when a gas is compressed, its density rises. At the critical point, the densities of the liquid and of the gas are the same. Consequently, the interface disappears, and a supercritical fluid is formed. Supercritical fluents have both liquid-like (they can dissolve many organic compounds) and gas-like properties (they fill all the space available to them, do not fall under gravity, and flow like gases). A particular feature of SCFs is that changes in polarity, density, viscosity, and diffusivity can be carried out by relatively small changes in pressure and temperature.

Of particular importance for asymmetric homogeneous metal catalysis is supercritical carbon dioxide (scCO$_2$). It has a relatively low critical point (critical temperature (Tc) = 31.0°C, critical pressure (Pc) = 73.75 atm), which can be adjusted by means of rather simple technological equipment. The miscibility of $scCO_2$ with numerous gases and the absence of a liquid/gas-phase boundary in the supercritical state allows for a maximum availability of gaseous reactants. Therefore, potential problems of mass-transfer limitations can be avoided.[53] Carbon dioxide is benign in nature and inert under most reaction conditions.

In several cases of asymmetric catalysis in $scCO_2$, ordinary ligands and pre-catalysts have been tested without additional modification. Sometimes, the rate of the reaction could be considerably improved, in comparison with common organic solvents, by a change of the medium. As found by Leitner and Pfaltz, the hydrogenation of prochiral imines with iridium complexes, based on chiral *P,N*-ligands, showed a remarkable enhancement of catalyst efficiency when the reaction was conducted in $scCO_2$ (Scheme 8.18).[54] The time required for full conversion was ~20 times shorter in $scCO_2$, as compared with dichloromethane. This allowed the reaction to be run in the latter solvent at much lower pressure. The ee value was only slightly affected by the change of the solvent.

The apparent zero-order kinetics in $scCO_2$ indicates that substrate binding is quantitative and much faster than the reaction with hydrogen. By proper choice of the counter-anion, the catalyst could be separated from the product after the reaction. Thus, by using the corresponding BARF-catalyst, it precipitated almost quantitatively from the reaction mixture and could be reused four times for the same transformation without loss of reactivity or enantioselectivity. The employment of BARF is crucial for the success of the reaction in $scCO_2$. In contrast, in the same reaction in dichloromethane, no difference in terms of enantioselectivity was noted when PF_6, BPh_4, or BARF was employed.

In the asymmetric hydrogenation of tiglic acid with ruthenium catalysts, based on half-saturated BINAP (H_8-BINAP), ee values were similar in $scCO_2$ to those observed in methanol or hexane (ca. 80 %) (Scheme 8.19).[55] However, at lower pressure, an erosion of the enantioselectivity in carbon dioxide was noted. In strong contrast, the enantioselectivity in conventional organic solvents increased to ca. 95 % ee at lower pressure.

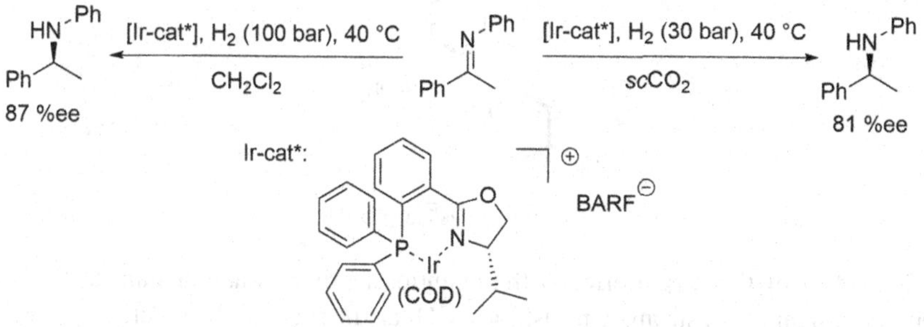

SCHEME 8.18 Comparison of the asymmetric hydrogenation with an iridium catalyst in dichloromethane and $scCO_2$.

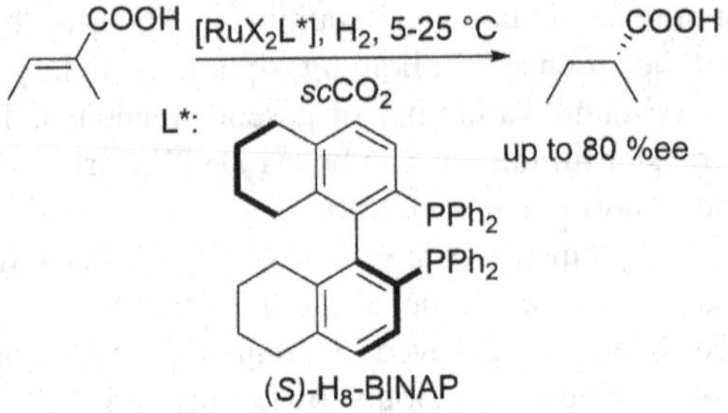

SCHEME 8.19   Asymmetric hydrogenation in *sc*CO₂.

Rate of conversion and enantioselectivity in this reaction decreased in the following order, because of the decreasing solubilities of the diphosphine ligand in $scCO_2$:

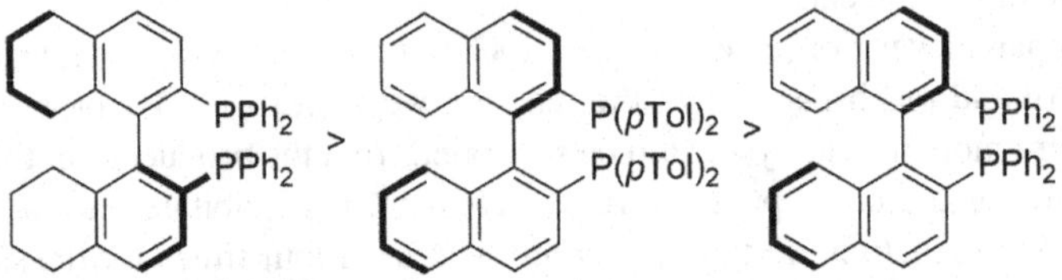

Addition of $CF_3(CH_2)_6CH_2OH$ was beneficial for both parameters. The authors speculated that the fluorinated alcohol may improve the solubility of aromatic compounds in $scCO_2$ by forming micelles.

When C_6F_{13} substituents were incorporated in the 6,6'-position of BINAP, the reaction rate of the ruthenium asymmetric hydrogenation of methyl itaconate in methanol as the solvent was slowed down.[56] The deceleration was not observed when the fluorine-containing unit was separated by an ethanediyl spacer. NMR measurements provided evidence that substitution with ethylene-spaced perfluoroalkyl groups in remote positions of aryl rings bearing phosphine groups keeps structural and electronic changes at the metal center to a minimum.[57]

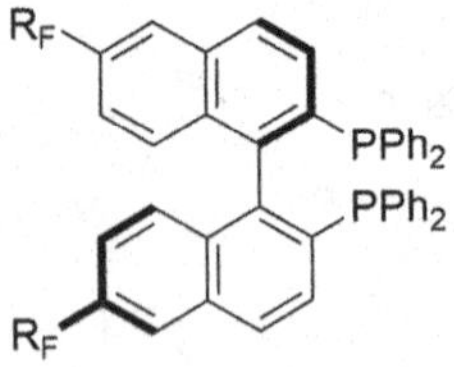

$R_F = C_6F_{13}, C_6F_{13}CH_2CH_2-$

Independent of these rate effects, the enantioselectivity was constant at 95 % ee with both ligands. In strong contrast, in $scCO_2$ as the solvent, the reaction proceeded much slower (24 h rather than 0.25 h for complete conversion) and the ee values eroded significantly, which was explained by the different polarities of the catalyst and $scCO_2$. By addition of small amounts of methanol, again high conversion and superior ee values could be realized. However, the authors concluded self-critically that this result could be due to fast hydrogenation (in methanol) before the conditions for the formation of the supercritical status of CO_2 had been reached.

Not only the polarity of the ligand, but also the lipophilicity of the whole catalyst is a crucial factor determining its solubility in $scCO_2$. Generally, a high lipophilicity causes a greater solubility. By using the R_f values derived from thin layer chromatography, the following tendency for different pre-catalysts was concluded, which indicates that the anion (BPh$_4$ *versus* BARF) has a larger influence on the lipophilicity of these cationic complexes than does the ligand substitution pattern, R = H *versus* $CH_2CH_2C_6F_{13}$ (Figure 8.7). This means that the more sophisticated and expensive synthesis of the fluorinated ligand can be avoided in this case.

As described above, ordinary chiral ligands, originally designed for asymmetric catalysis in common organic solvents, can also display excellent catalytic performance in $scCO_2$. A convincing example of this is the use of the bisphospholane ligand Et-DuPhos in the rhodium-catalyzed asymmetric hydrogenation of ethyl *N*-acetylamino acrylate (Scheme 8.20).[58] In comparison to the reaction in methanol or hexane, similarly high ee values were achieved in $scCO_2$.

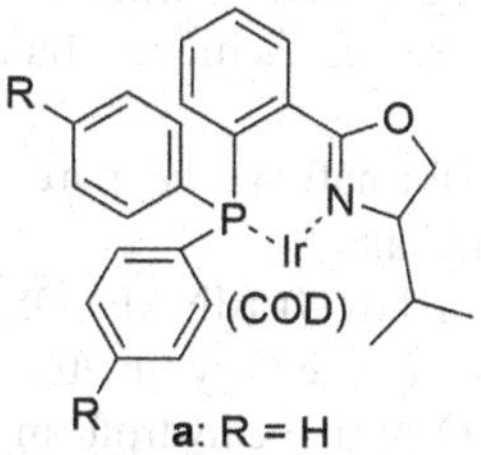

Liphophilicity:

$$[Ir(COD)(a)]Cl \leq [Ir(COD)(b)]Cl << [Ir(COD)(a)]BPh_4$$
$$< [Ir(COD)(b)]BPh_4 << [Ir(COD)(a)]BARF < [Ir(COD)(b)]BARF$$

a: R = H
b: R = $CH_2CH_2C_6F_{13}$

FIGURE 8.7 Lipophilicity of chiral iridium pre-catalysts in $scCO_2$ in terms of dependency on ligand structure and counter-ion.

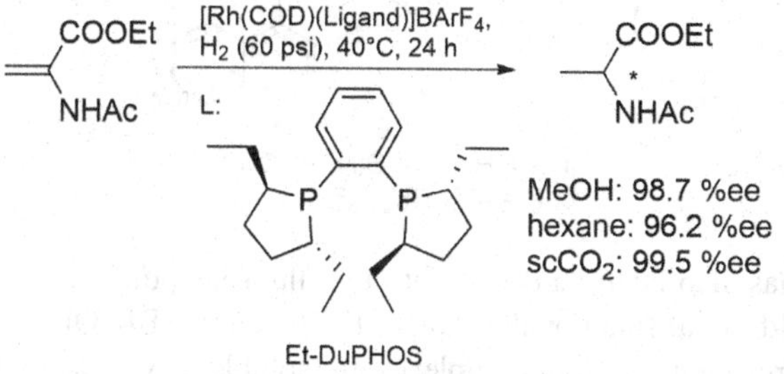

MeOH: 98.7 %ee
hexane: 96.2 %ee
scCO$_2$: 99.5 %ee

Et-DuPHOS

SCHEME 8.20 Comparison of results from the asymmetric hydrogenation with a common rhodium catalyst in dependency on the solvent.

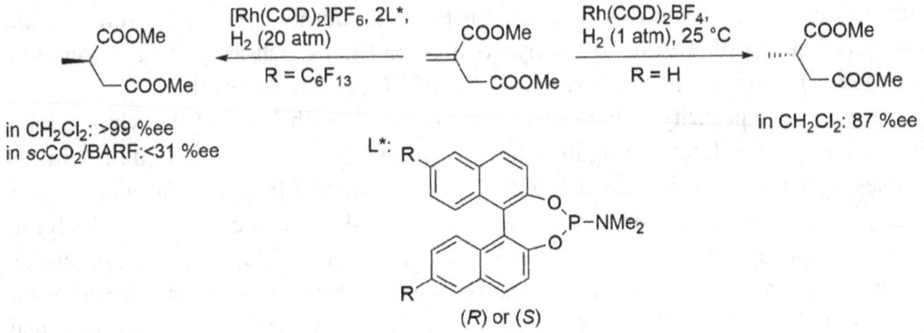

SCHEME 8.21 Asymmetric hydrogenation of dimethyl itaconate, with a specially designed ligand. Comparison of solvents.

However, there are also numerous examples in the literature which show that the expensive, special design of ligands, combined with the reaction in $scCO_2$, does not fulfill the high expectations. For example, Hope and co-workers tried to achieve benefits from a monodentate phosphoramidite decorated with fluorine-modified ponytails (Scheme 8.21).[59] In the hydrogenation of dimethyl itaconate in methylene chloride, almost perfect enantioselectivity was achieved, but, in $scCO_2$, the ee value dropped dramatically.

In comparison, in the same transformation, the parent (undecorated) ligand MonoPhos gave 87 % ee in dichloromethane.[60]

The chiral phosphine-phosphite ligand (R,S)-BINAPHOS (**a**), which is reputed to be highly efficient in several rhodium-catalyzed asymmetric hydroformylation reactions, was also tested in compressed CO_2.[61] In the hydroformylation of styrene at low CO_2-pressure, similarly high efficiencies were noted as in organic solvents. However, beyond the critical pressure of CO_2 the ee values dropped dramatically. Apparently, under these conditions, the chiral rhodium catalyst is not formed, and the reaction is catalyzed by the "naked" metal.[62]

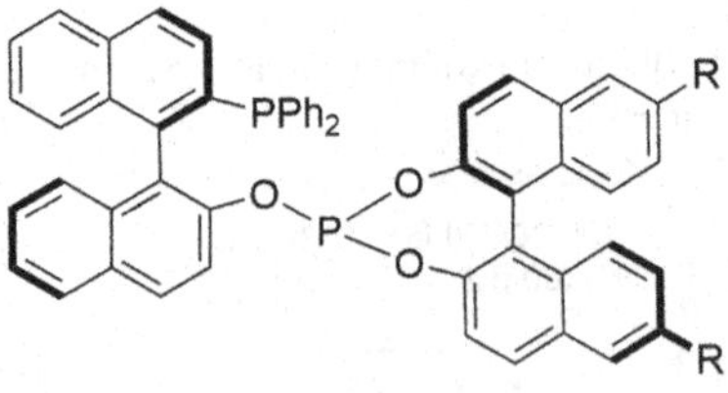

a, R = H; (R,S)-BINAPHOS
b, R = -$CH_2CH_2C_6F_{13}$

BINAPHOS has also been modified with two fluorous ponytails (**b**). It is noteworthy that the additional functionalization of the precursor BINOL required six steps. Although the relevant rhodium complexes are soluble in $scCO_2$, they gave lower ee values than those described with ligand **a** for styrene hydroformylation. Even superior results were obtained with the fluorinated ligand **b** in benzene as solvent.

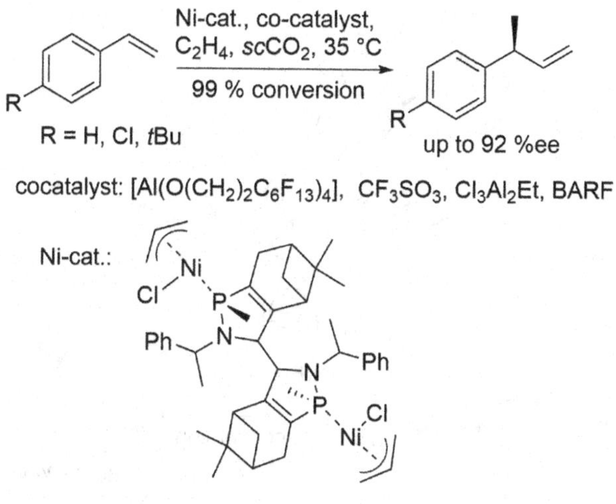

SCHEME 8.22   Asymmetric hydrovinylation in $scCO_2$.

By mixing ILs with supercritical carbon dioxide, or by reaction in CO_2-expanded solvents, new and mostly unexpected effects result (Figure 8.8).[35,] As a general trend for the range of the enantioselectivity in the rhodium-catalyzed hydrogenation of prochiral acryl acids, the order shown in Figure 8.8 was concluded.

Asymmetric hydrovinylation of styrenes was investigated by Wegner and Leitner in $scCO_2$ (Scheme 8.22).[63] The outcome of this reaction is strongly dependent on the nature of the anion, which acts as a co-catalyst. Triflate and $[Al(O(CH_2)_2C_6F_{13})_4]^-$ almost inhibited the reaction, whereas Cl_3Al_2Et allowed the reaction to take place, even at 1°C. The best results in terms of conversion, selectivity, and enantioselectivity were observed with BARF.

This observation inspired the same authors to run the reaction in $scCO_2$ in combination with ionic liquids (Scheme 8.23).[64] In this case the IL [EMIM][N(CF$_3$SO$_3$)$_2$] (EMIM) 1-ethyl-3-methylimidazolium) acts simultaneously as solvent and chloride abstracting agents to "switch on" the catalysis in the presence of $scCO_2$.

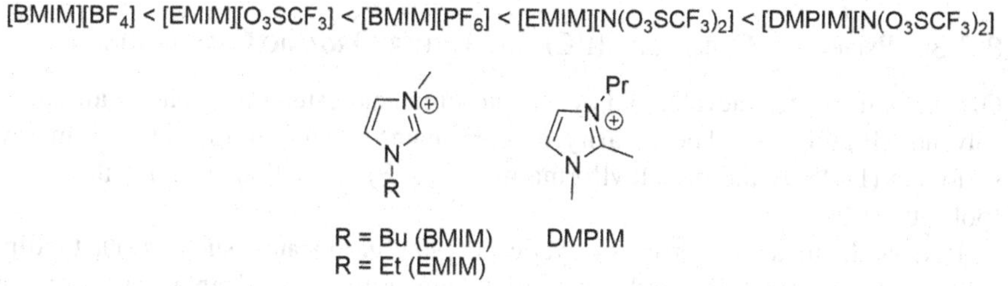

$[BMIM][BF_4] < [EMIM][O_3SCF_3] < [BMIM][PF_6] < [EMIM][N(O_3SCF_3)_2] < [DMPIM][N(O_3SCF_3)_2]$

R = Bu (BMIM)   DMPIM
R = Et (EMIM)

FIGURE 8.8   Increase in ee values of the asymmetric hydrogenation of prochiral acryl acids in a mixture of $scCO_2$/IL, in dependency on the nature of the IL.

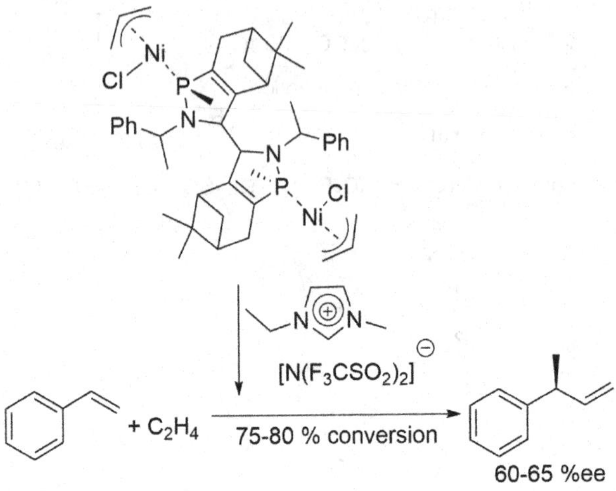

SCHEME 8.23 Asymmetric hydrovinylation in a mixture of $scCO_2$ and an IL.

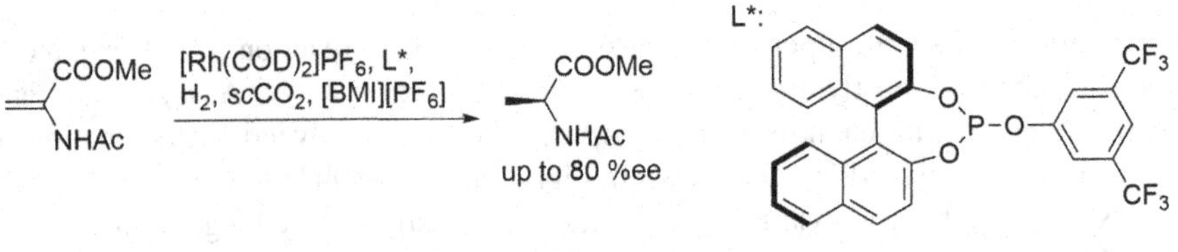

SCHEME 8.24 Asymmetric hydrogenation in a mixture of $scCO_2$ and an IL.

The catalyst was immobilized in this active and selective form by simply dissolving the stable precursor in the IL. This system remained stable over a 60-h period in a continuous flow system, with compressed CO_2 as the mobile phase.

In a similar manner, asymmetric hydrogenation with monodentate ligands can benefit from the use of $scCO_2$ and [BMI][PF$_6$] (Scheme 8.24).[65] The best catalytic system was achieved in an IL medium and could be recycled up to ten times without activity loss. It should be noted that the ligand was decorated with two CF_3 groups.

8.3.3 PROPYLENE CARBONATE (PC) AND OTHER ORGANIC CARBONATES

Organic carbonates (acyclic and cyclic carbonic acid esters) are volatile and polar solvents (Figure 8.9). The polarity of carbonates is similar to that of dimethyl sulfoxide (DMSO) and dimethylformamide (DMF), as well as acetonitrile (ACN) (polarity: 0.46).

Physicochemical properties of cyclic alkylene carbonates, such as outstanding solvency, biodegradability, and high boiling temperature, are advantageous for many industrial applications. Thus, they are widely used as solvents in the agricultural and textile industries. Meanwhile, large collections of physicochemical data, such as vapor pressure and vaporization enthalpies at different temperatures, are available

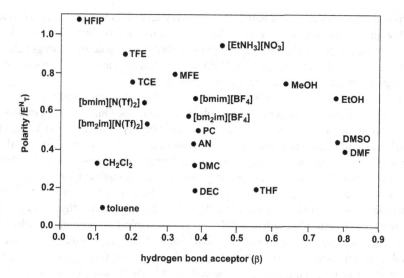

FIGURE 8.9 Polarity and hydrogen-bond acceptor values of selected solvents (HFIP = hexafluoroisopropanol, TFE = trifluoroethanol, MFE = monofluoroethanol, TCE = trichloroethanol, DMSO = dimethyl sulfoxide, DMF = dimethyl formamide, AN = acetonitrile, DMC = dimethyl carbonate, DEC = diethyl carbonate; for ionic liquids: bmim = 1-butyl-3-methyl-1,3-imidazolium ion, bm_2im = 1-butyl-2,3-dimethyl-1,3-imidazolium-ion, Tf = triflate).

TABLE 8.2
Classification of Carbonates

	Open Chain Carbonates	Cyclic	Carbonates
HO⌒OH (O) Carbonic acid	H_3CO⌒OCH_3 (O) Dimethyl carbonate	(O) Ethylene carbonate	(O)* Propylene carbonate
Melting point [°C]	2	34	−49
Boiling point [°C]	90	247	243
Density [g/ml]	1.07	1.32	1.2

in the literature.[66] The physical properties of carbonates can be greatly influenced by an appropriate choice of the alcohols for the ester synthesis. In particular, propylene carbonate (PC) has shown excellent solvent properties (Table 8.2).[67] It fulfills important requirements for a so-called "green solvent", like low flammability and volatility, as well as low toxicity. Ca. 40 % of its molecular weight represents one equivalent of carbon dioxide, so that it represents an interesting tool for fixation of carbon dioxide. Moreover, it is liquid over a very large temperature interval. Due to its low price, it is especially suitable for technical use. PC is chiral. Meanwhile, the relevant enantiopure precursor (S)-propan-1,2-diol is available by hydrogenation of lactic acid or polylactic acids on a large scale.[68] Racemization or inversion of the configuration gives access to the opposite enantiomer.[69]

Due to these advantages, it is rather surprising that, up to now, organic carbonates did not play a much stronger role as a solvent in catalysis. Only a few examples can be found in the literature, which mainly concern non-asymmetric transformations.

The research group of Behr showed the potential of PC in the platinum-catalyzed hydrosilylation of unsaturated fatty acid esters[70] and in rhodium-catalyzed hydroformylation reactions with temperature-dependent multi-component solvent systems.[71] Reetz et al. used PC for the stabilization of Pd clusters in Heck reactions.[72] With heterogeneous Pt/Al$_2$O$_3$ catalysts, Bradley et al. hydrogenated ethyl pyruvate and compared PC with other solvents.[73] PC is also suitable for continuous hydrogenation processes. Heller and co-workers demonstrated, in kinetic studies with various catalyst precursors, that conventional solvents, such as dichloromethane, methanol, or tetrahydrofuran, can readily be replaced by PC without any loss in the reactivity of the chiral catalyst.[74]

An innovative approach of product separation from propylene carbonate is the direct distillation from the reaction solution, as described by Börner et al.[75] The aim of the overall synthesis was located in the framework of the manufacture of enantiomerically pure lactic acid (Scheme 8.25). The rhodium-catalyzed asymmetric hydrogenation of a substituted acrylic acid ester was the decisive step in the whole sequence. In contrast to related DuPHOS-type ligands, a rhodium catalyst based on Et-catASium M led to full conversion of the substrate in PC and induced excellent enantioselectivities for both the ethyl ester and methyl ester substrates (>98%). Moreover, the undesired opening of the maleic anhydride moiety of catASium M, which is sometimes observed in the alcohols usually used as solvent, could be prevented under these conditions. At the end of this process, enantiomerically pure O-acylated lactic acid esters were distilled out of the solvent using a

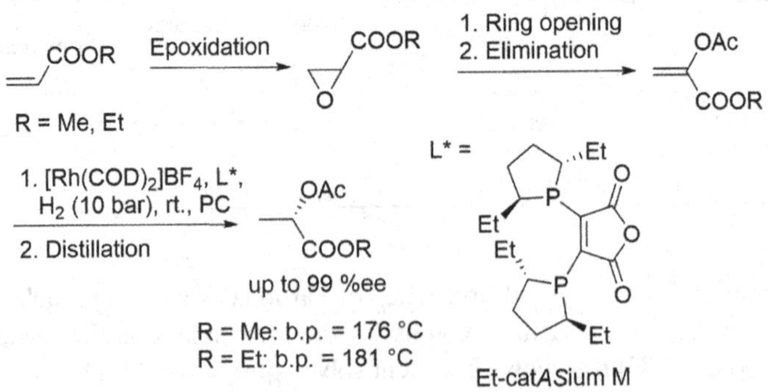

SCHEME 8.25 Synthesis of almost enantiopure lactic acid by beneficial use of propylene carbonate as solvent.

"Spaltrohrkolonne" at 7.5 mbar vacuum. Due to the large difference in boiling points of propylene carbonate and the chiral esters, clean separation was possible.

Because of the polar characters of PC, separation of a non-polar product can be carried out in two-phase systems, using a non-polar second solvent. Such a case was discovered coincidentally in the asymmetric hydrogenation of benzo-annulated methylidene cyclohexane, with an iridium catalyst based on a common chiral 1,3-oxazolidine phosphinite as ligand (Scheme 8.26).[76]

The reaction was conducted in either methylene chloride or in propylene carbonate (PC). During the reaction, remarkable differences were noted between the two solvents. Whereas fast isomerization of the external double bond of the substrate into the ring occurred in dichloromethane, in PC this migration was almost prevented. Since the two olefinic isomers led in the asymmetric hydrogenation to opposite enantiomers in the saturated product, the ee value was strongly dependent on the rate of the isomerization prior to the hydrogenation. In PC a much higher enantioselectivity was noted, indicating that isomerization was almost suppressed. Due to its high non-polarity, the saturated product is highly soluble in *n*-hexane. As a consequence, the product could be extracted with this non-polar solvent. In contrast, the highly polar cationic iridium catalyst remained in the polar PC phase. It could be reused for further six cycles without any loss of activity or stereoselectivity. A precondition for the successful separation was the proper choice of the counter-ion to iridium(I). Only with the weakly coordinating BARF did the ionic properties of the catalyst came sufficiently into play. Other common and more strongly coordinating anions used in hydrogenations with transition metals, such as tetrafluoroborate, hexafluorophosphate, and perchlorate, were less efficient. It is worth noting that the low isomerization tendency described above is in strong contrast to the finding of Behr *et al.*, who noted an accelerating effect of PC in the rhodium-catalyzed isomerizing hydroformylation. [77]

These findings could be considerably broadened by means of the asymmetric hydrogenation of various terminal alkenes using 1,3-oxazolidine phosphite ligands developed by Anderson and Diéguez (Scheme 8.27).[78] In the cases investigated, PC performed similarly to dichloromethane as the solvent, but the former could be

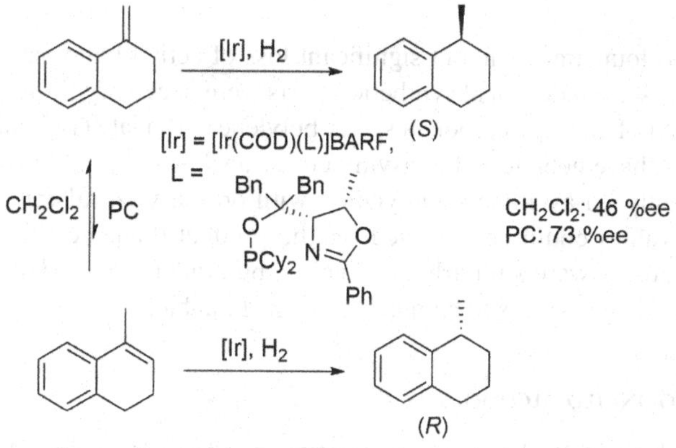

SCHEME 8.26 Suppression of olefin isomerization, due to propylene carbonate as solvent.

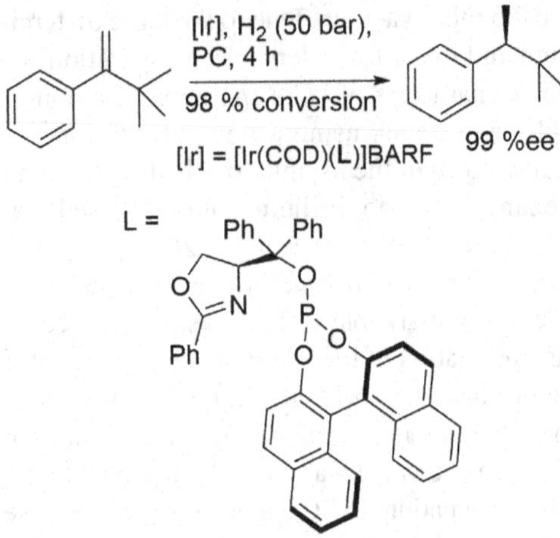

SCHEME 8.27 Iridium-catalyzed asymmetric hydrogenation of non-functionalized olefins in propylene carbonate.

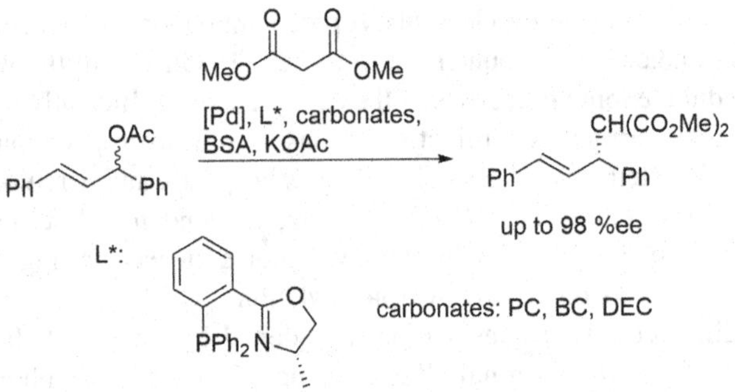

SCHEME 8.28 Asymmetric palladium-catalyzed alkylation in different organic carbonates as solvent.

recycled up to four times without significant loss of activity and enantioselectivity. Also, in this case, no additional synthetic efforts were necessary for ligand synthesis.

In addition to PC, other carbonates, like butylene carbonate (BC) and diethyl carbonate (DEC), have been tested in asymmetric transformations. For example, in the palladium-catalyzed asymmetric allylation with ordinary chiral *P,N*-Phox ligands, excellent ee values could be achieved for the product (Scheme 8.28).[79] Results in methylene chloride were similarly convincing, but, due to the toxicity of the latter, the use of carbonates seems to be much more sustainable.

8.3.4 FLUORINATED ALCOHOLS

Fluorinated alcohols display unique properties as solvents, co-solvents, and additives in catalysis.[80] The first reports on their use were based on biochemical studies. They were first considered due to the beneficial effects of fluorinated alcohols to support the helix-conformation of proteins.[81] The utility of fluorinated alcohols as solvents for non-catalytic oxidations and cyclizations was highlighted for the first time in 2004.[82] In Table 8.3, some physical data are listed together with those of other commonly used solvents.

The most frequently used and cheapest fluorinated alcohols are 2,2,2-trifluoroethanol (TFE) and 1,1,1,3,3,3-hexafluoroisopropanol (HFIP), which are available on a commercial scale. Other fluorinated alcohols, such as perfluoro-*tert*-butanol, 1,3-bis-(1,1,1,3,3,3-hexafluoro-2-hydroxypropyl) benzene, and 2-fluoroethanol, are also available, but their high price often precludes their widespread use. Highly fluorinated alcohols exhibit lower boiling points (b.p.) and higher melting points compared with their non-fluorinated counterparts. It means that TFE (b.p. 74°C) and HFIP (b.p. 59°C) can be easily removed from the reaction mixture by distillation. As with other halogenated compounds, their density is appreciably higher than that of their parent alcohols.

Due to their strong negative inductive effect, fluorine substituents increase the acidity of neighboured hydroxyl groups. The acidity of alcohols increases with the number of fluoro atoms in the molecule ($pKa_{ethanol} = 15.17$, $pKa_{2\text{-fluoroethanol}} = 14.42$, $pKa_{TFE} = 12.4$). Perfluoro-*tert*-butanol exhibits a similarly high acidity ($pKa = 5.2$), like acetic acid. Fluorinated alcohols are highly polar solvents. Despite the presence of the hydroxyl group, fluorinated alcohols are poor hydrogen-bond acceptors.

In 1968, for the first time, a positive effect of fluorinated alcohols on catalysis was noted in the selective hydrogenation of unsaturated carbon–carbon bonds with homogeneous rhodium complexes.[83] TFE as a co-solvent achieved high reaction rates and almost perfect chemoselectivity in the hydrogenation of a mixture of 1-hexyne and 1-octene at 1 bar and 22 °C with the Wilkinson catalysts $Rh(PPh_3)_3Cl$. The relative rate of the hydrogenation of 1-hexyne in benzene/ethanol 1:1 increased by a change to the benzene/TFE ratio from 1 to 12, whereas the rate of the 1-octene hydrogenation remained constant. The same solvent mixture was used in the hydrogenation of but-2-yne-1,4-diol to (Z)-but-2-ene-1,4-diol with $Rh(PPh_3)_3Cl$, which gave a ratio of

TABLE 8.3
Physical Properties of Fluorinated Alcohols and Common Solvents

Solvent	m.p. [°C]	b.p. [°C]	Density [kg·m⁻³]	Self-Association Constant (dm³/mol)
Dichloromethane	−94.9	39.6	1.33	-
Toluene	−95	110.6	0.87	-
Methanol	−97.7	64.5	0.79	-
Ethanol	−117	78	0.80	1.685; 0.89
2-Fluoroethanol	−26.4	103.5	1.10	-
Trifluoroethanol (TFE)	−43.5	74	1.38	1.147; 0.13
Hexafluoropropanol (HFIP)	−4	58.6	1.61	0.65
Trichloroethanol	19	152	1.55	0.476

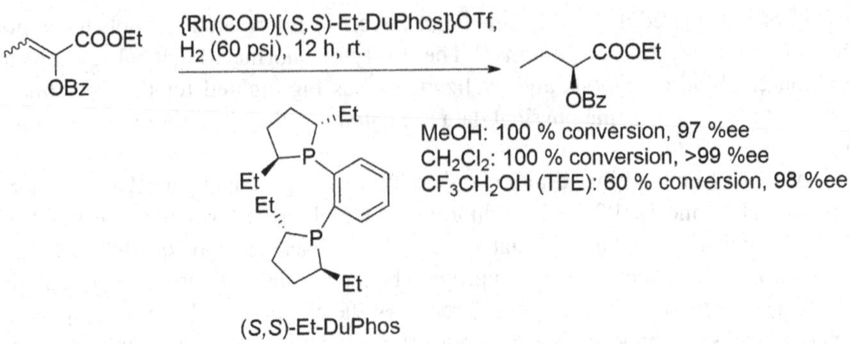

(S,S)-Et-DuPhos

SCHEME 8.29 Asymmetric rhodium-catalyzed hydrogenation in different solvents.

alkene/alkane of 94:4.[84] A similar selectivity was observed by application of Vaska's complex $IrCl(CO)(PPh_3)_2$ as a catalyst and in UV activation. Fluorinated alcohols also played an important role in ruthenium-catalyzed hydrogenation of CO_2 to formic acid.[85] A range of alcohols and other co-catalysts have been tested for their ability to promote this reduction. The yield of formic acid after one hour with HFIP/Et_3N was found to be 5-fold higher than that obtained in methanol/NEt_3. The best results were achieved with a mixture of perfluorophenol/Et_3N in supercritical CO_2 at 120 bar.

The first asymmetric reactions in fluorinated alcohols were reported almost 25 years ago.[86] For example, the asymmetric hydrogenation of methyl α-(benzoyloxy) methylacrylate with a Rh-[(S,S)-Et-DuPhos] catalyst was carried out in TFE with an enantioselectivity of 98 % ee at 60 atm (Scheme 8.29).[87] It is noteworthy that a slower reaction rate was observed, compared with the reaction in common alcohols or methylene chloride.

A palladium catalyst based on (R,R)-Me-DuPhos in TFE was effective in the asymmetric hydrogenation of 2-phthalimido-1-arylethanones, as shown by the Zhou group (Scheme 8.30).[88] Remarkably, various alcohols, such as methanol, ethanol, or isopropanol, failed as solvents in this transformation.

A related Pd-catalyst, based on the chiral ligand (S)-SegPhos, was found to be effective in the asymmetric hydrogenation of a variety of N-phosphinyl imines, in TFE as the solvent.[89] Chiral products were obtained at 87–99 % ee and are valuable precursors of chiral amines.

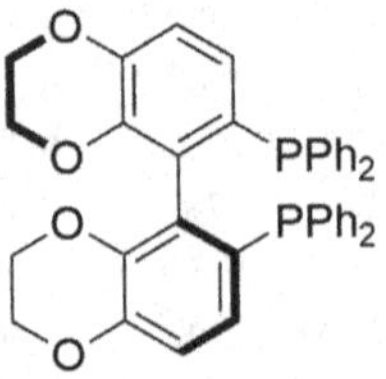

(S)-SegPhos

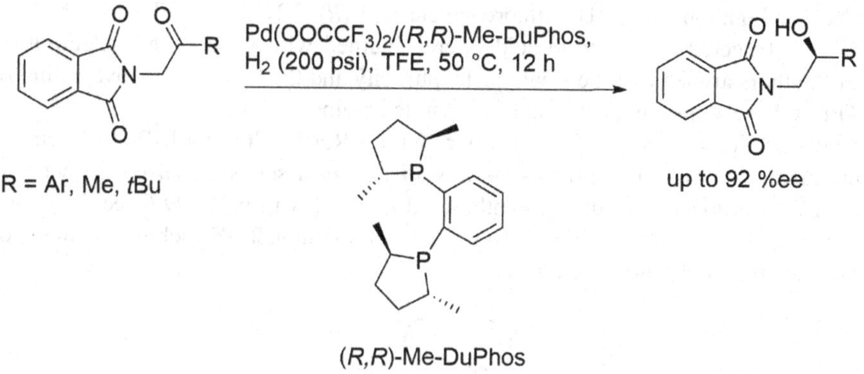

SCHEME 8.30 Asymmetric hydrogenation of 2-phthalimido-1-arylethanones in trifluoroethanol (TFE).

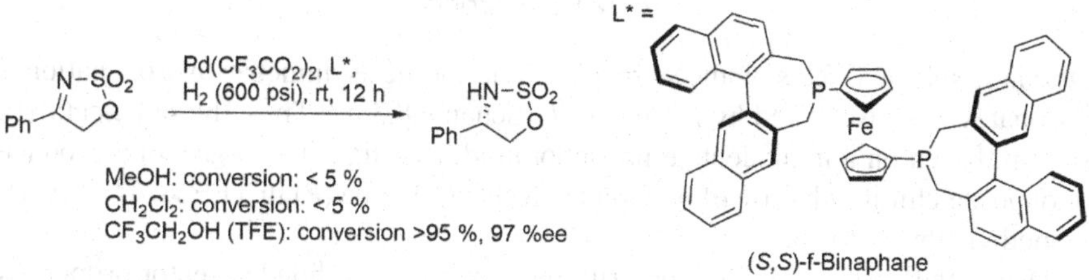

SCHEME 8.31 Asymmetric hydrogenation of cyclic sulfamidates. Comparison of solvents.

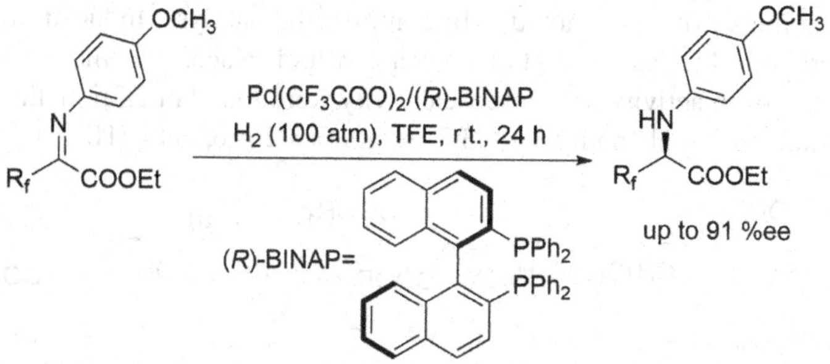

SCHEME 8.32 Asymmetric hydrogenation of imines in trifluoroethanol (TFE).

Superior results were similarly obtained by the same research group in the hydrogenation of cyclic sulfamidates (Scheme 8.31).[90] Under these conditions, high enantioselectivities were noted with f-Binaphane as ligand, whereas other prominent chiral diphosphines, like DuPhos, SynPhos or MeO-Biphep, gave poor results. The same effect happened in methanol or dichloromethane as the solvent.

The pivotal importance of fluorinated alcohol as solvents was shown by Uneyama *et al.* in the asymmetric hydrogenation of α-fluorinated imino esters (Scheme 8.32).[91]

The combination of Pd(II) trifluoroacetate and (R)-BINAP in TFE gave the best results. Targeted (R)-α-amino acids were obtained with good enantioselectivities. The authors argued that the weak nucleophilicity and the low coordinative nature of TFE were responsible for the superior results obtained.

Reductive amination of β-keto esters with a Ru[(R)-Cl-MeO-BIPHEP] catalyst producing β-amino acid precursors was studied in a series of solvents.[92] Various chiral β-amino esters could be synthesized in TFE, with up to 99 % ee. In strong contrast, other commonly used solvents, such as ethanol, THF, dichloromethane, or toluene, did not promote the conversion.

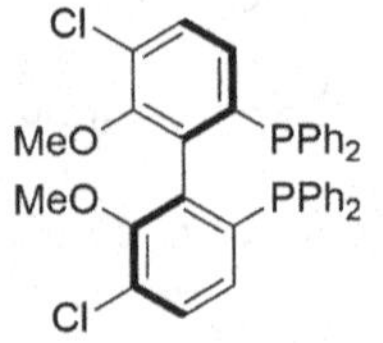

(R)-Cl-MeO-BIPHEP

Dramatic solvent effects on the stereoselectivity of the asymmetric hydrogenation of itaconates was discovered by Zhang et al. (Scheme 8.33).[93] Thus, the same catalytic system, based on a monodentate phosphoramidite as ligand, can generate products of opposite chirality by use of different solvents (99 % ee (R) in TFE vs. 71 % ee (S) in methyl ethyl ketone).

Due to the particularly high polarity but low hydrogen-bond acceptor properties, fluorinated alcohols can mimic the effect of two solvents with different catalytic properties. Thus, in the enantioselective hydrogenation with cationic Rh-complexes based on chiral 2-pyridone-phosphino ligands, high enantioselectivity in dichloromethane as the solvent was noted.[94] In contrast, the catalysts induced poor ee values in methanol. In general, a slow conversion took place. A combination of both superior effects on activity and stereoselectivity were found in 2,2,2-trifluoroethanol (TFE), 2-fluoroethanol, and 1,1,1,3,3,3-hexafluoro-2-propanol (HFIP).[95] ^{31}P-NMR

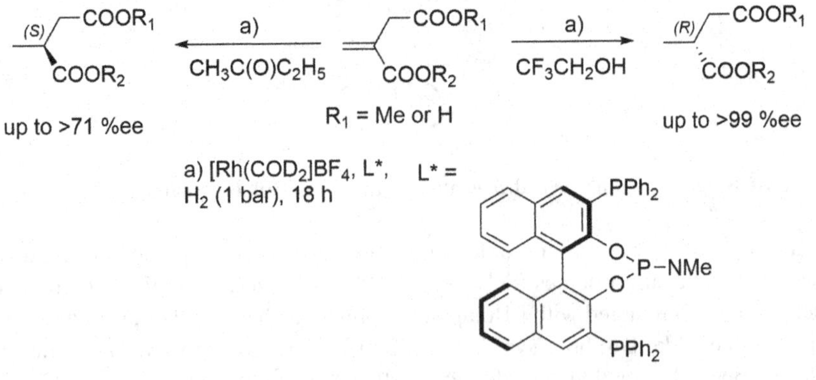

SCHEME 8.33 Rhodium-catalyzed asymmetric hydrogenation of itaconic acid and its dimethyl ester in different solvents.

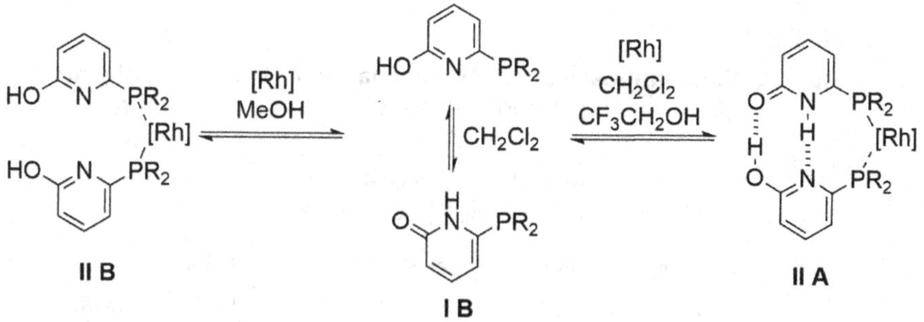

SCHEME 8.34 Pseudo-chelate *versus* monodentate coordinating modus in dependency on the solvent.

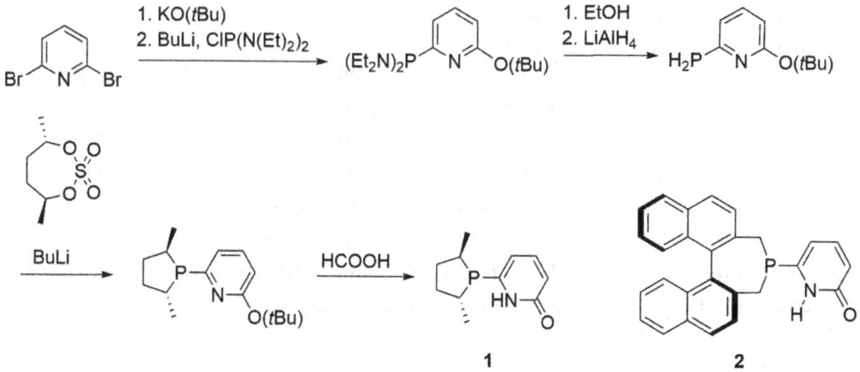

SCHEME 8.35 Synthesis of a chiral phosphorus ligand designed for use in fluorinated alcohols.

spectroscopic investigations showed two resonances in the non-polar solvent as well as in fluorinated alcohols, indicating the presence of two different phosphorus nuclei. This feature gave strong evidence for the hydrogen bonding-based chelating structure of the ligands, where one phosphorus atom is connected to a 2-pyridinol unit and the other is part of a 2-pyridone backbone (Scheme 8.34).

This "pseudo-chelate effect" generated in the fluorinated alcohols was employed in the asymmetric hydrogenation of prochiral olefins with a chiral phospholane **1** and the phosphepine ligand **2**. Both monodentate phosphines have been assembled *via* a multi-step synthesis as shown in the synthesis of **1** (Scheme 8.35).

In the hydrogenation of the standard substrates, methyl Z-α-N-acetamidocinnamate (**a**) and dimethyl succinate (**b**) in methanol with the "small" ligand **1**, low conversion rates and ee values were observed with the amino acid precursor (Table 8.4). However, a change of the solvent to e.g., dichloromethane or TFE, resulted in significant acceleration of the reaction, and the ee values also improved. This effect was still more pronounced for dimethyl itaconate as substrate. By a change to the sterically more demanding phosphine ligand **2**, the enantioselectivity could be

TABLE 8.4

Results of the Enantioselective Hydrogenation in Different Solvents

$$R^3 \diagup\diagdown\,(R^3)(R^1) \xrightarrow[\text{H}_2\ (1\ \text{bar}),\ \text{rt., solvent}]{[\text{Rh(COD)}_2]\text{BF}_4 + 2\text{eq. L*}} R^3 \diagup\diagdown\,(R^3)(R^1)$$

a R^1 = NHAc, R^2 = COOMe, R^3 = Ph
b R^1 = CH$_2$COOMe, R^2 = COOMe, R^3 = H

Ligand	Substrate	Solvent	Conversion [min]	t [min]	ee [%]
1	a	CH$_3$OH	20	1300	13 (R)
1	a	CF$_3$CH$_2$OH	100	1200	68 (R)
1	a	CH$_2$Cl$_2$	100	1200	69 (R)
2	a	CH$_3$OH	100	60	64 (R)
2	a	CF$_3$CH$_2$OH	100	10	95 (R)
2	a	CH$_2$Cl$_2$	100	10	94 (R)
1	b	CH$_3$OH	100	1300	rac
1	b	CF$_3$CH$_2$OH	100	1300	68 (S)
2	b	CH$_3$OH	100	60	64 (S)
2	b	CF$_3$CH$_2$OH	100	10	97 (S)
2	b	CH$_2$Cl$_2$	100	20	99 (S)

furthermore increased to 97 % ee. It should be emphasized that this is one of the rare case of ligand design developing from the initial idea until successful realization.

8.3.5 WATER AS SOLVENT

For the majority of chemical reactions in the laboratory, water must be regarded as an impurity. For good reasons, monographs addressing the purification of chemicals or general synthetic chemistry usually cover the removal of water from organic solvents.[96] Insufficiently dry solvents might have a disastrous effect on reactions. Low-valent or electron-poor organometallic compounds, i.e., metal complexes that might play an important role in catalytic cycles, are no exception. Using water, instead of a typical organic solvent, involves certainly more than just selecting one solvent in place of another. Strong interactions between this solvent, the substrates, and, in particular, the catalyst, are almost certain to happen. Water can act as a σ-donor ligand for cationic and neutral metal centers, and will thus influence the solubility and activity of the catalysts and make the catalytic performances more sensitive to pH changes. On the other hand, there are more advantages for the use of this less common solvent than "simply" convenient catalyst recovery and product separation. Water is inexpensive, readily available at high purity, is non-toxic, and non-flammable. In several cases, water does not only serve as medium, but is actively engaged in the catalytic transformation. One of the main disadvantages of water is its high boiling point and delayed boiling during distillation. In general, activity and enantioselectivity of chiral catalysts in water is strongly dependent on the reaction conditions. In several cases, conversions and ee values in aqueous systems tend to be lower than those found in common, organic solvents.

Numerous papers and reviews have been published about the synthesis of chiral water-soluble ligands, and relevant catalysts, and their use in asymmetric aqueous two-phase catalysis.[97] In general, there are two ways to obtain hydrophilic phosphines: linking a phosphorus compound to a water-soluble moiety, or modifying a phosphorus ligand by introducing hydrophilic substituents.[98] Polar substituents that dissociate in solution, partially or completely, are most common, whereas a second group consists of hydrophilic substituents with hydroxyl or ether functionalities.[99] The most frequently used examples are sulfonic acids and their salts, carboxylic acids and their salts, phosphonic acids and their salts, ammonium groups, hydroxyl, and polyether groups.

Among the vast number of hydrophilic ligands reported in the literature, only sulfonic acids and phosphorus ligands bearing additional hydroxyl groups will be considered in more detail.

8.3.5.1 Phosphorus Ligands Decorated with Sulfonic Acids

A commonly used method to obtain achiral and chiral water-soluble arylphosphines is the direct sulfonation of aryl groups by fuming sulfuric acid. This was first reported in 1958 for the preparation of TPPMS (3-sulfonatophenyl)diphenylphosphine, monosulfonated triphenylphosphine), which is usually obtained as its sodium salt after work-up (Scheme 8.36).[100] Sulfonation under these conditions gives mainly the *meta*-substituted aryl phosphine.

Phosphine oxide by-products or sodium sulphate generally hamper purification of phosphines sulfonated with SO_3/H_2SO_4. Better results can be obtained if such reactions were carried out in the presence of boric acid under an inert gas atmosphere, as suggested by Herrmann *et al.* (Scheme 8.37).[101] Several chiral bidentate phosphine ligands have been sulfonated successfully in this manner.[102]

Coloration experiments by the group of de Vries, with rhodium complexes bearing phosphines with different degree of sulfonation, showed that the monosulfonated Rh-BDPP complex is soluble in organic solvents.[103] A higher degree of sulfonation, commencing with two sulfonate groups, is the precondition for solubility in water.

(S,S)-BDPP$_{MS}$

organic soluble

(S,S)-BDPP$_{DS}$

water soluble

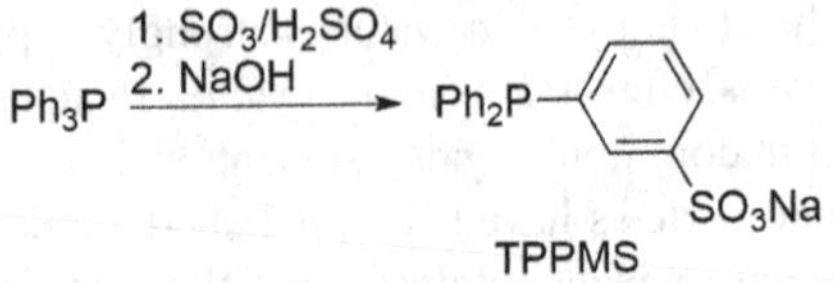

SCHEME 8.36 General sulfonation procedure for aryl phosphines.

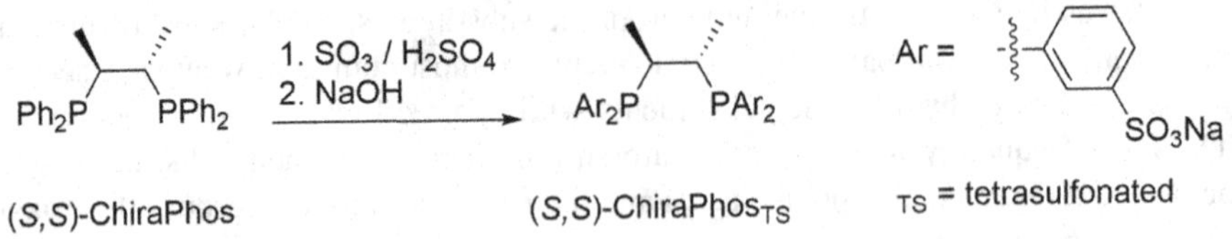

SCHEME 8.37 Direct sulfonation of chiral 1,2-, 1,3-, and 1,4-diphosphino alkanes.

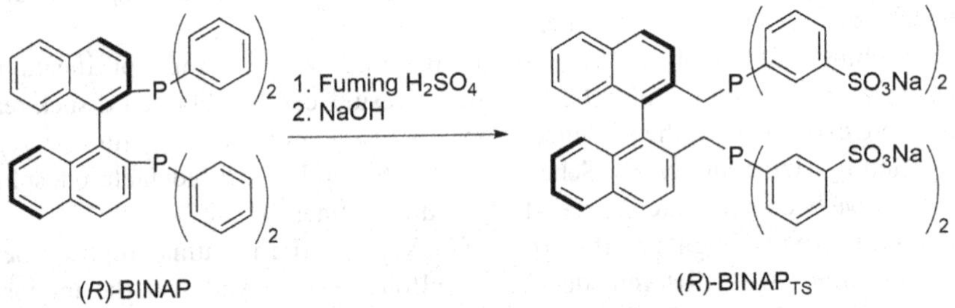

SCHEME 8.38 Sulfonation of BINAP.

Moreover, it was shown, that the ee values in the rhodium-catalyzed hydrogenation of imines decreased with increasing number of sulfonate groups present. In contrast, when BDPP$_{MS}$ was used as ligand, 94 % ee was achieved.

Biaryl-type diphosphine ligands, mainly used for asymmetric hydrogenations, have been subjected to the sulfonation protocol. The most prominent example involved BINAP$_{TS}$, which is a tetrasulphonated sodium salt of BINAP (Scheme 8.38).[104]

BINAS, a homolog of BINAP, called NAPHOS, gave an eight-fold sulphonated diphosphine (Scheme 8.39) under similar conditions.[105]

Asymmetric hydroformylation of styrene in a biphasic toluene/methanol/water mixture with (S)-BINAS as chiral ligand achieved a sound enantiomeric excess of

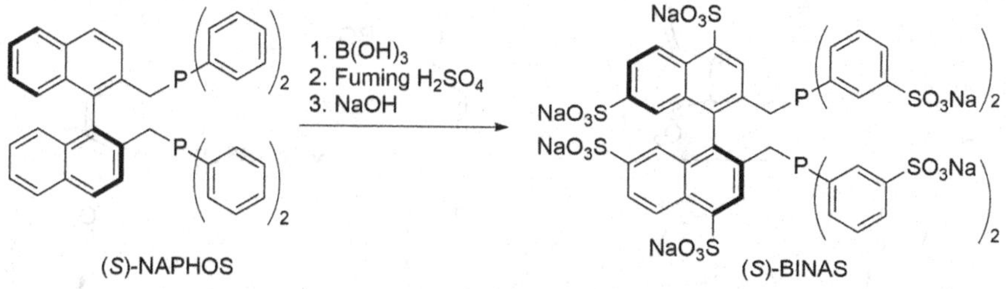

SCHEME 8.39 Sulfonation of NAPHOS in the presence of boric acid.

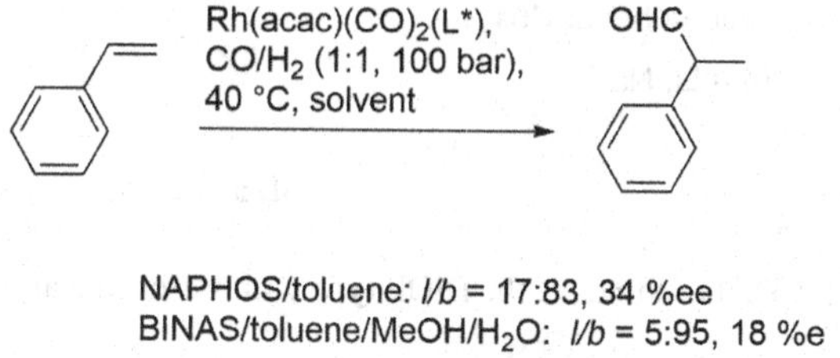

NAPHOS/toluene: l/b = 17:83, 34 %ee
BINAS/toluene/MeOH/H$_2$O: l/b = 5:95, 18 %ee

SCHEME 8.40 Asymmetric hydroformylation of styrene. Comparison of ligands and solvents.

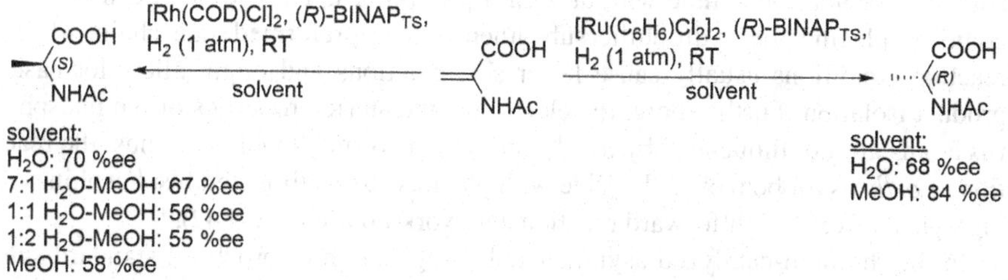

solvent:
H$_2$O: 70 %ee
7:1 H$_2$O-MeOH: 67 %ee
1:1 H$_2$O-MeOH: 56 %ee
1:2 H$_2$O-MeOH: 55 %ee
MeOH: 58 %ee

solvent:
H$_2$O: 68 %ee
MeOH: 84 %ee

SCHEME 8.41 Hydrogenations with (R)-BINAP$_{TS}$ rhodium and ruthenium pre-catalysts in dependency on the solvent used.

the branched aldehyde and a high conversion (92% ee) (Scheme 8.40).[106] The ee value, however, was only 18% as compared to 34% with (S)-NAPHOS under similar conditions, in toluene as the solvent.

In contrast, comparable enantioselectivities were reported for (R,R)-cbDIOP$_{TS}$ and (S,S)-SkewPhos$_{TS}$ rhodium complexes in methanol/water 1:1.[107]

The general tendency, that enantioselectivity decreases significantly when water is used as solvent, could not be confirmed in an example with a rhodium complex of (R)-BINAP$_{TS}$ (Scheme 8.41).[108] Thus, increasing the concentration of methanol affected the ee value in the asymmetric hydrogenation of N-acetylamido acrylic acid in comparison to the reaction in pure water. In strong contrast, the relevant ruthenium catalysts, induced with the same prochiral substrate, not only generated the

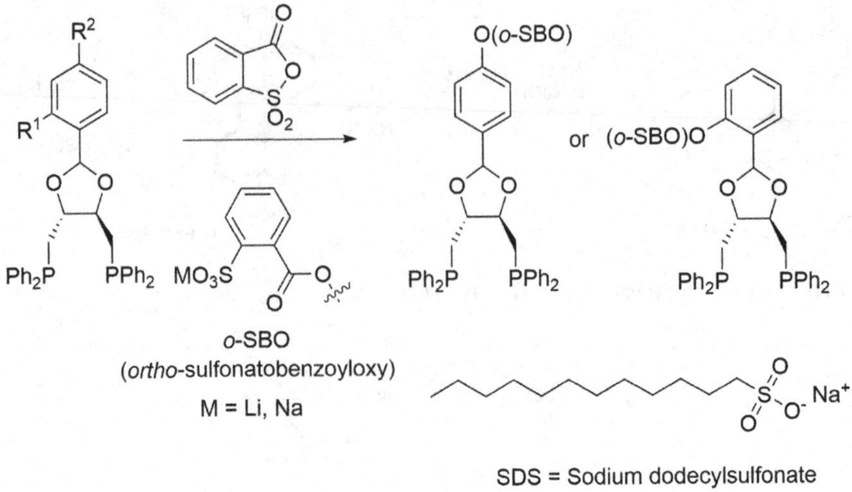

SCHEME 8.42 1,4-Diphosphines of the DIOP-type with sulfonated aryl side chains.

opposite configuration in the product, but the reaction in water was inferior to that in methanol.[109]

The somewhat vigorous reaction conditions used during sulfonation of aryl substituents are not compatible with all chiral phosphines. This problem can be overcome by placing the hydrophilic substituent into a preformed side chain. Milder reaction conditions usually cause fewer side reactions and might allow for easier product isolation. Furthermore, the electronic and steric properties of the phosphorus atoms are not influenced by nearby sulfonic acid functionalities. Thus, the reaction of *ortho*-sulfobenzoic anhydride with phenols gave sulfonated bisphosphines in high yield after straightforward reaction and work-up (Scheme 8.42).[110]

In the rhodium-catalyzed asymmetric hydrogenation of prochiral standard substrates in water, much higher activities were observed in comparison to the transformation with the non-sulfonated ligands. Interestingly, by addition of the amphiphile sodium dodecylsulfonate (SDS), catalysts derived from the latter achieved a faster reaction. Moreover, the combination of sulfonated ligand and equimolar SDS led to a doubling of the reaction rate without affecting the stereoselectivity.

8.3.5.2 Phosphines with Hydroxyl Functionalities

Early investigations with achiral phosphines showed that the water solubility of phosphines can be considerably increased by attachment of hydroxyl groups.[111] Each hydroxyl group almost doubles the water solubility of the relevant catalyst. Protected hydroxyl groups, e.g., methyl ethers, decrease the water solubility.[112]

Hydroxyl groups in the ligand do not only affect the solubility in water but may also have a strong decelerating effect on the catalytic reaction.[113] This was shown in some fundamental studies where, in a butane-diphosphine framework, hydroxyl groups in different positions and numbers were incorporated (Scheme 8.43). For

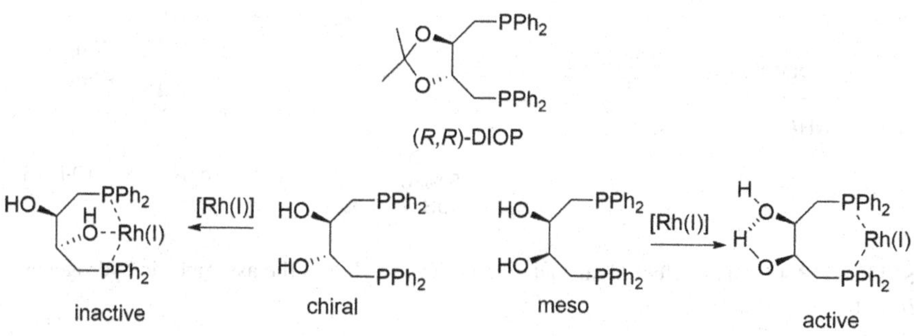

SCHEME 8.43   Hydroxyl groups in the phosphine ligand may interfere with rhodium.

example, when the acetonide of the prominent chiral ligand DIOP is hydrolyzed, a chiral conformationally flexible dihydroxy diphosphine is formed. With this ligand in the rhodium-catalyzed asymmetric hydrogenation, not only was the enantioselectivity of the relevant rhodium catalyst almost destroyed, but the activity was dramatically affected. NMR studies showed that one of the hydroxyl groups coordinates to the metal and therefore becomes electronically saturated. In contrast, by using the *meso*-diphosphine, this effect was suspended by a sterically more favored hydrogenbond between the two hydroxyl groups in the backbone of the catalyst. As a consequence, no decelerating effect was noted, in comparison to the rate with DIOP.

The investigations led finally to the development of more rigid hydroxyl diphosphines, where the hydroxyl group in the ligand were situated far away from the metal, such as HO-NORPhos[114] or its acyclic counterpart 1.[115]

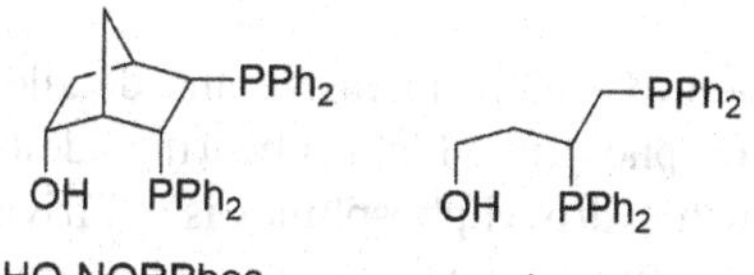

Taking these results into consideration, OH-group-bearing derivatives of DIOP were prepared, which (in combination with Rh(COD)(acac)) was catalytically active in non-aqueous hydrogenation reactions of dehydroamino acids and itaconic acid derivatives.[116]

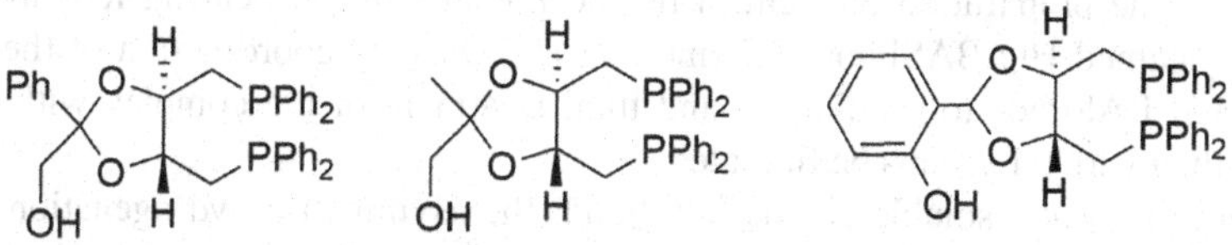

In water as solvent, the relevant rhodium complexes gave only poor ee values of 2–34 % (Scheme 8.44).[117] However, when surfactants, such as SDS or Triton X 100, were added, the ee values increased significantly up to 77%. This effect was noted only with HO-DIOP (**a**), but not with Kagan's DIOP (**b**) itself. Furthermore,

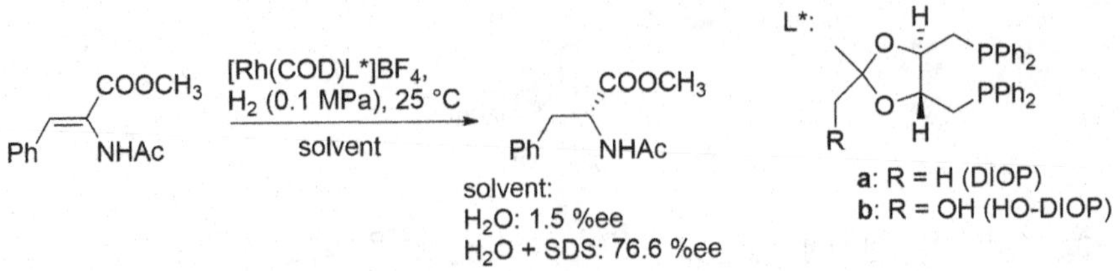

SCHEME 8.44 The effect of surfactants on the ee value in the asymmetric hydrogenation in water.

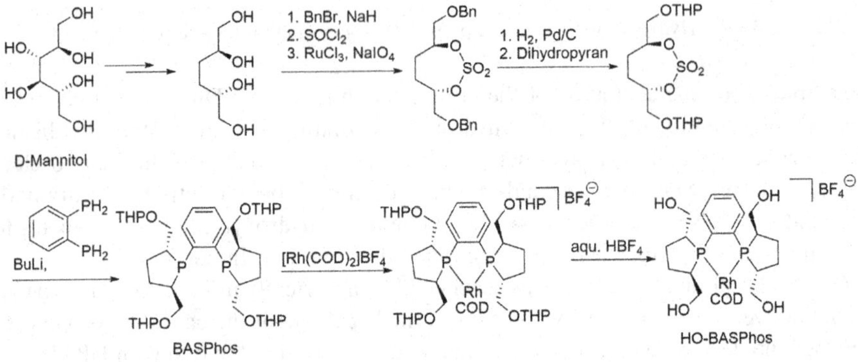

SCHEME 8.45 Total synthesis of the water-soluble HO-BASPhos and its rhodium complex.

the addition of methanol to the latter could not mimic the effect of the internal hydroxyl group.

These observations led to a more intensive and detailed engagement, concerning the synthesis of chiral phosphine ligands bearing additional hydroxyl groups.[118] The total synthesis of such hydroxyl phosphines is not trivial.[119] The challenge consisted especially of the fact that most phosphine groups are introduced into a carbon skeleton by substitution of hydroxyl groups or leaving groups (e.g., tosyl or mesyl) derived from hydroxyl groups. Therefore, those hydroxyl groups, which are not converted, have to be protected over the whole reaction sequence. Frequently, the problem was encountered that commonly applied HO-protective groups could not be removed by established methods at the end of the sequence. A typical example concerns the preparation of a chiral bisphospholane ligand bearing four hydroxyl groups, named HO-BASPhos (Scheme 8.45).[120] Only by coordination of the THP-protected BASPhos to rhodium(I) and final treatment of the complex with HBF$_4$, could the hydroxyl groups be liberated.

With this water-soluble complex in hand, the asymmetric hydrogenation of the water-soluble N-acetylamido acrylic acid could be performed quantitatively with almost perfect enantioselectivity (Scheme 8.46).

RajanBabu assembled other bisphospholanes from D-mannitol in a similar manner and used them in hydrogenations of dehydroamino acids, obtaining ee values up to 99% (Scheme 8.47)[121]

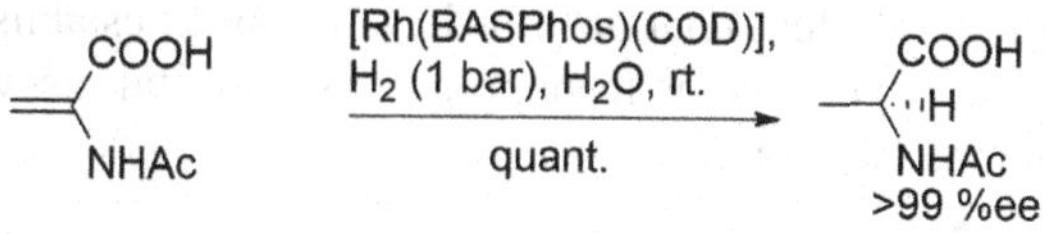

SCHEME 8.46 Asymmetric hydrogenation of a water-soluble substrate with a water-soluble rhodium catalyst.

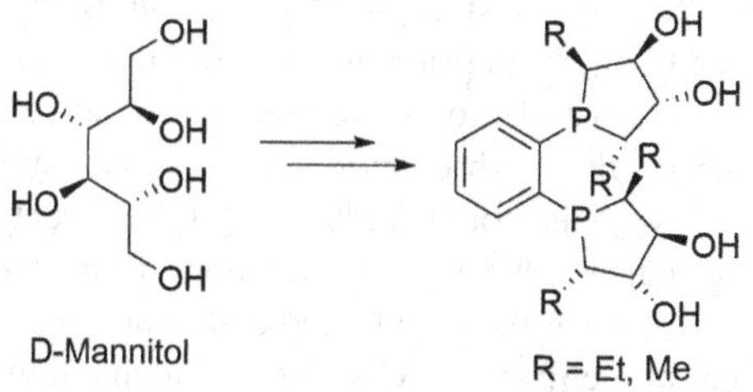

SCHEME 8.47 Synthesis of other bisphospholanes bearing hydroxyl groups.

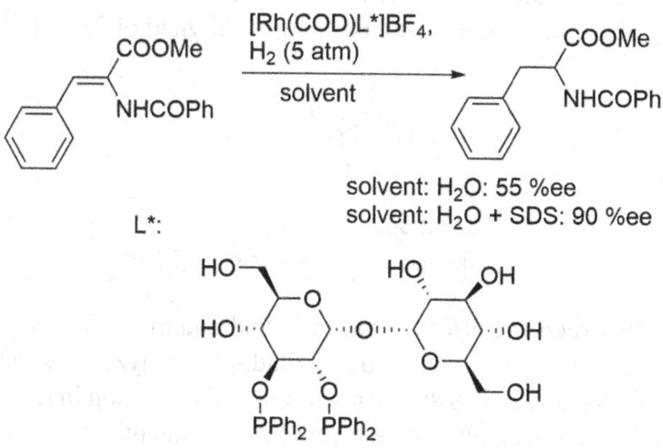

SCHEME 8.48 Rhodium-catalyzed asymmetric hydrogenation with a sugar-based water-soluble diphosphinite ligand.

The authors showed that, after extraction of the product, the aqueous phase could be employed in another catalytic cycle without loss of enantioselectivity.

A trehalose-derived diphospinite ligand was utilized as catalyst in the hydrogenation of dehydroamino acids by Uemura et al. in water.[122] However, only modest ee values were achieved. By addition of SDS, the enantioselectivity could be dramatically increased. The beneficial effect of SDS was also observed for a β,β-trehalose-derived catalyst (Scheme 8.48).[123]

Meanwhile, the research into asymmetric catalysis in water is focused on reactions at the water/oil interface.[124] That means that reactions in emulsions and micellar catalysis play a central role. Unfortunately, the fine structures of relevant chiral amphiphilic catalysts, assembled in emulsions or micelle droplets, are almost unknown and structure–effect relationships have not been established. Intellectual support and new insights may come from progress in biocatalysis, which takes place exclusively in water as medium.

8.3.6   CHIRAL SOLVENTS

The use of chiral solvents in asymmetric catalysis, as support for the effect of a chiral catalyst or as the sole source of stereo-differentiation, is largely unexplored. This fact can be rationalized by the high price of larger quantities of appropriate enantiopure solvents. Moreover, most of those rare studies investigating such effects have reported disappointing results, with enantiomeric excesses typically being ~1%.[125] So far, larger effects have been found only when the solvent becomes covalently bound to a reactant or reagent.[126] This is presumably because noncovalent solvent–solute interactions are usually rather weak, and differences between the interactions of the two enantiomers of a chiral solute with a given enantiomer of the solvent are consequently even smaller.

One of the first examples was given by Seebach and co-workers in 1975.[127] They employed the chiral diamino dialkyl ether (DDB) as solvent for the photolysis of acetophenone, obtaining, in the best case, an optical yield of 23.5 %.

(2S,3S)-2,3-dimethoxy-N^1,N^1,N^4,N^4-
tetramethylbutane-1,4-diamine (DDB)

A comparison between the effect of racemic and enantiomerically pure propylene carbonate was made in the asymmetric palladium-catalyzed alkylation with (S)-PhanePhos as ligand (Scheme 8.49).[80] In relation to the reaction in racemic propylene carbonate, the use of the enantiopure solvent had no benefit.

More promising seems to be the use of enantiopure ionic liquids (ILs), since the ionic interactions with the catalyst–substrate complex are more pronounced. A typical example was given by the Francio group, who employed an achiral disulphonated diaryl diphosphine embedded in a chiral IL (Scheme 8.50).[128] In the rhodium-catalyzed asymmetric hydrogenation of an unsaturated alanine precursor, enantioselectivity up to 69 % ee was achieved.

Another example worthy of mentioning concerns the Sharpless osmium-catalyzed asymmetric dihydroxylation published by Afonso and co-workers (Scheme 8.51).[129] In contrast to the original Sharpless protocol, no cinchona alkaloid as ligand was necessary. Chirality was brought into the system by combination of the tetra-n-hexyl-dimethylguanidinium cation with the naturally occurring and commercially available anion derived from quinic acid. By running the reaction in this IL with 4-methylmorpholine-N-oxide (NMO) as oxidant, similar reactivities and ee values were obtained compared with the parent Sharpless system.

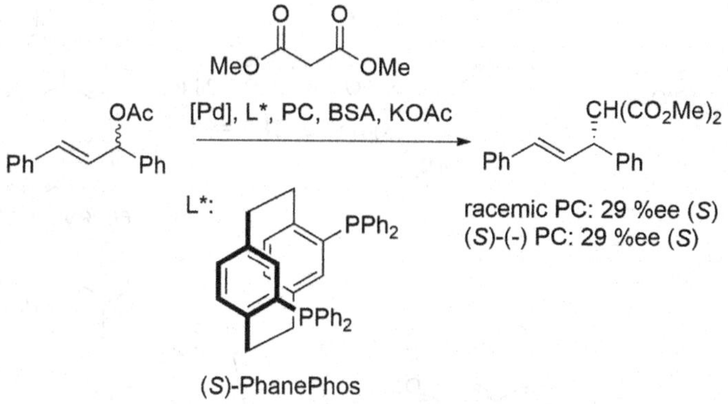

SCHEME 8.49 Asymmetric allylic alkylation in racemic or enantiopure propylene carbonate.

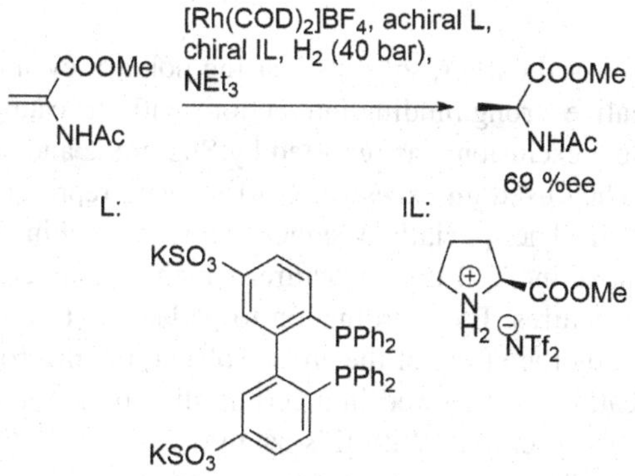

SCHEME 8.50 Asymmetric hydrogenation supported by an enantiopure ionic IL as counter-ion.

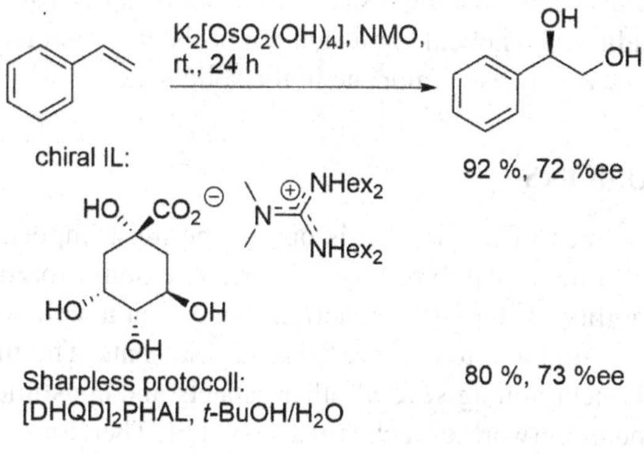

SCHEME 8.51 Asymmetric oxidation supported by an enantiopure ionic liquid.

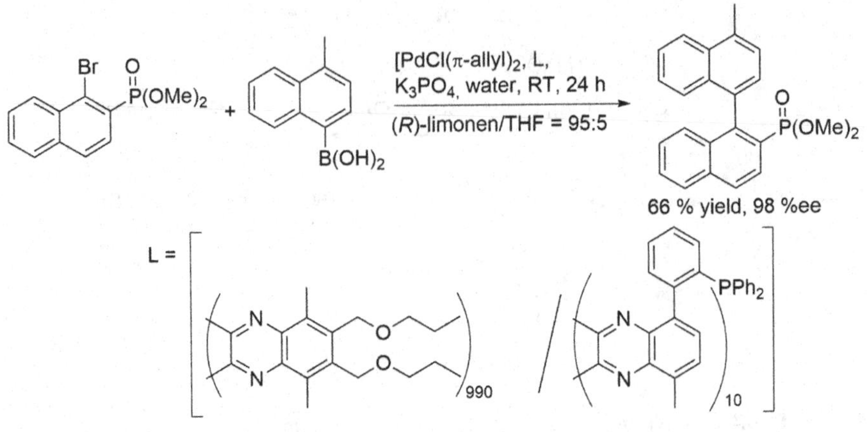

SCHEME 8.52 Asymmetric aryl-aryl coupling supported by a chiral by poly(quinoxaline-2,3-diyl) structure.

In the cases discussed earlier, solvents bearing polar or even acidic groups were used in order to realize strong binding interactions with the catalyst–substrate complexes. A remarkable exclusion was reported by Suginome and co-workers recently (Scheme 8.52).[130] They used an excess of (R)-limonene representing a highly non-polar medium in THF. The specially designed achiral phosphine ligand was embedded in a poly(quinoxaline)-2,3-diyl structure, which is characterized by rigid but dynamic helical structures. By coordination to palladium(I), an achiral catalyst is obtained. However, by the effect of the chiral solvent, the macromolecular scaffold of the palladium catalyst is screwed in a certain direction, and, as a consequence, a helical structure is generated. With this catalyst in hand, in the Suzuki-Miyaura cross coupling, up to 98% ee was induced in the product.

The system could likewise be successfully used in the palladium-catalyzed hydrosilylation of styrene (up to 95 % ee) and in the silaboration (up to 89 % ee). Not only enantiomerically pure (R)-limonene but also limonene with lower enantiomeric excess induced single-handed helical structures with majority-rule-based amplification of homochirality. The helical conformation of the macromolecular catalyst was retained even in the absence of limonene in the solid state.

8.4   CONCLUSIONS

The appropriate choice of the solvent is one of the most important preconditions for the success of a metal-catalyzed asymmetric reaction. Moreover, the need for a higher sustainability of the whole reaction process on a technical scale requires enhanced efforts to find and to employ "greener" solvents. The medium, however, is only one component among several other aspects and constituents in a multidimensional parameter network of a chemical reaction. Therefore, often a change of the solvent may causes dramatic differences in the result of the process. This is a common experience of every chemist working in this area. Efforts to adopt the ligand and/or catalyst by structural modifications to the alternative solvents, are mostly connected with additional synthetic efforts, with no guarantee for a greater efficiency. Due to a change in solvent and ligand, a totally new catalytic system is born, which may only accidentally meet the results obtained in a common solvent with a parent catalytic system. Therefore, a change of the solvent is always connected with extensive procedures for optimization, including tedious synthetic variations of the ligand structure.

From the economic point of view, it seems that only in cases where already-established catalytic processes in common solvents should be improved, a change to alternative solvents could be of some interest. The use of common ligands and catalysts in this respect is recommended first, since the modification of parent structures is time- and cost-consuming. The almost complete absence of such protocols from the relevant literature concerned with asymmetric catalysis on an industrial scale indicates that the way to application is still a long one.[131]

REFERENCES

1. C. Reichardt, T. Welton, *Solvents and Solvent Effects in Organic Chemistry*, Wiley-VCH, Weinheim, 3rd ed., **2003**.
2. G. Cainelli, P. Galetti, D. Giacomini, *Chem. Soc. Rev.* **2009**, *38*, 990–1001.
3. M. Bartók, *Chem. Rev.* **2010**, *110*, 1663–1705.
4. R. Selke, *J. Prakt. Chem.* **1987**, *329*, 717–724.
5. a) L. Panella, B. L. Feringa, J. G. de Vries, A. J. Minnaard, *Org. Lett.* **2005**, *7*, 4177–4180.
6. *Green Reaction Media in Organic Synthesis*, Ed. K. Mikami, Wiley-Blackwell, **2008**.
7. a) *Handbook of Homogeneous Hydrogenation*, Eds. J. G. de Vries, C. J. Elsevier, Wiley-VCH, Weinheim, **2007**; b) S. Nishimura, *Handbook of Heterogeneous Catalytic Hydrogenation for Organic Synthesis*, John Wiley & Sons, New York, **2001**.
8. C. Reichardt and T. Welton, *Solvents and Solvent Effects in Organic Chemistry*, Wiley-VCH, Weinheim, 3rd ed., **2003**.
9. C. Daguenet, P. J. Dyson, I. Krossing, A. Oleinikova, J. Slattery, C. Wakai, H. Weingärtner, *J. Phys. Chem. B*, **2006**, *110*, 12682–12688.
10. a) M. J. Kamlet, J.-L. Abboud, M. H. Abraham, R. W. Taft, *J. Org. Chem.* **1983**, *48*, 2877–2887; b) IUPAC, *Compendium of Chemical Terminology*, 2nd ed. (the "Gold Book"), Compiled by A. D. McNaught and A. Wilkinson, Blackwell Scientific Publications, Oxford (**1997**). Online version (2019–) created by S. J. Chalk. ISBN 0-9678550-9-8. https://doi.org/10.1351/goldbook.
11. P. J. Dyson, P. G. Jessop, *Catal. Sci. Technol.* **2016**, *6*, 3302–3316.
12. D. J. Adam, P. J. Dyson, S. T. Tavener, *Chemistry in Alternative Reaction Media*, John Wiley & Sons, Chichester, **2004**.
13. See e.g. Ref. I. T. Horváth, *Acc. Chem. Res.* **1998**, *31*, 10, 641–650.
14. B. Cornils, *Angew. Chem. Int. Ed. Engl.* **1997**, *36*, 2058–2059.
15. T. Sugiishi, M. Matsugi, H. Hamamoto, H. Amii, *RSC Adv.* **2015**, *5*, 17269–17282.
16. J. Wu, X. Li, F. Wu, B. Wan, *Org. Lett.* **2011**, *13*, 4834–4837.
17. X. Cui, X. Xu, H. Lu, S. Zhu, L. Wojtas, X. P. Zhang, *J. Am. Chem. Soc.* **2011**, *133*, 3304–3307.
18. B. M. Trost, M. Osipov, G. Dong, *Org. Lett.* **2010**, *12*, 1276–1279.

19. For example, a palladium catalyst based on a chiral aminophosphine bearing two fluorous ponytails gave excellent ee values in the asymmetric allylic alkylation in diethyl ether. It was removed by simple solid/liquid separation and could be reused up to five times. Ref. T. Mino, Y. Sato, A. Saito, Y. Tanaka, H. Saotome, M. Sakamoto, T. Fujita, *J. Org. Chem.* **2005**, *70*, 7979–7984.

20. R. P. J. Bronger, J. P. Bermon, J. Herwig, P. C. J. Kamer, P. W. N. M. van Leeuwen, *Adv. Synth. Catal.* **2004**, *346*, 789–799.

21. P. W. N. M. van Leeuwen, P. C. J. Kamer, *Catal. Sci. Technol.* **2018**, *8*, 26–113.

22. a) J. Holz, K. Rumpel, A. Spannenberg, R. Paciello, J. Jiao, A. Börner, *ACS Catal.* **2017**, *7*, 6162–6169; b) J. Holz, G. Wenzel, A. Spannenberg, M. Gandelman, A. Börner, *Tetrahedron* **2020**. https://doi.org/10.1016/j.tet.2020.131142.

23. A. Jourdant, E. González-Zamora, J. Zhu, *J. Org. Chem.* **2002**, *67*, 3163–3164.

24. V. S. Shende, S. K. Shingote, S. H. Deshpande, N. Kuriakose., K. Vanka, A. A. Kelkar, *RSC Adv.* **2014**, *4*, 46351–46356.

25. D. J. Adams, P. J. Dyson, S. J. Taverner, *Chemistry in Alternative Systems*, John Wiley & Sons, Chichester, **2004**.

26. a) D. E. Bergbreiter, Thermomorphic Catalysts, in *Recoverable and Recyclable Catalysts*, Ed. M. Benaglia, Wiley & Sons, Ltd., Chichester, **2009**, pp. 117–147; b) A. Behr, C. Fängewisch, *Chem. Eng. Technol.* **2002**, *25*, 143–147.

27. M. Hejazifar, A. M. Palvoelgyi, J. Bitai, O. Lanaridi, K. Bica-Schroeder, *Org. Proc. Res. Dev.* **2019**, *23*, 1841–1851.

28. G.-J. Deng, Q.-H. Fan, X.-M. Chen, D.-S. Liu, A. S. C. Chan, *Chem. Commun.* **2002**, 1570–1571.

29. a) T. Okano, H. Kumobayashi, S. Akutagawa, J. Kiji, H. Konishi, K. Fukuyama, Y. Shimano, US Pat. **1987**, 4 705 895; b) T. Okano, H. Kumobayashi, S. Akutagawa, J. Kiji, H. Konishi, K. Fukuyama, Y. Shimano, US Pat. **1987**, 4 705 895.

30. M. Berthod, G. Mignani, G. Woodward, M. Lemaire, *Chem. Rev.* **2005**, *105*, 1801–1836.

31. a) D. Heller, H.-J. Drexler, A. Spannenberg, B. Heller, J. You, W. Baumann, *Angew. Chem. Int. Ed.* **2002**, *41*, 777–780; b) E. Alberico, S. Möller, M. Horstmann, H.-J. Drexler, D. Heller, *Catalysts* **2019**, *9*, 582. doi: 10.3390/catal9070582.

32. M. Uhlemann, S. Doerfelt, A. Börner, *Tetrahedron Lett.* **2013**, *54*, 2209–2211.

33. A. De Jong, A. Eftaxias, F. Trabelsi, F. Recasens, J. Sueiras, F. Stuber, *Ind. Eng. Chem. Res.* **2001**, *40*, 3225–3229.

34. P. G. Jessop, R. R. Stanley, R. A. Brown, C. A. Eckert, C. L. Liotta, T. T. Ngo, P. Pollet, *Green Chem.* **2003**, *5*, 123–128.

35. Y. Sun, R. N. Landau, J. Wang, C. LeBlond, D. G. Blackmond, *J. Am. Chem. Soc.* **1996**, *118*, 1348–1353.

36. H. Nishida, N. Takada, M. Yoshimura, T. Sonods, H. Kobayashi, *Bull. Chem. Soc. Jpn.* **1984**, *57*, 2600–2604.

37. M. J. Burk, S. Feng, M. F. Gross, W. Tumas, *J. Am. Chem. Soc.* **1995**, *117*, 8277–8278.

38. a) M. J. Earle, J. M. S. S. Esperanca, M. A. Gilea, J. N. C. Lopes, L. P. N. Rebelo, J. W. Magee, K. R. Seddon, J. A. Widegren, *Nature* **2006**, 831–834; b) P. Wasserscheid, *Nature* **2006**, *439*, 797.

39. P. Wasserscheid, C. M. Gordon, C. Hilgers, M. J. Maldoon, *Chem. Commun.* **2001**, 1186–1187.

40. Y. Chauvin, L. Mussmann, H. Olivier, *Angew. Chem. Int. Ed. Engl.* **1995**, *34*, 2698–2700.

41. A. L. Monteiro, F. K. Zinn, R. F. de Souza, J. Dupont, *Tetrahedron: Asymmetry* **1997**, *8*, 177–179.

42. P. S. Schulz, *Phosphorus Ligands in Asymmetric Catalysis*, Ed. A. Börner, Wiley-VCH, Weinheim, **2008**, pp. 967–984.
43. S. Guernik, A. Wolfson, M. Herskowitz, N. Greenspoon, S. Geresh, *Chem. Commun.* **2001**, 2314–2315.
44. R. Giernoth, M. S. Krumm, *Adv. Synth. Catal.* **2004**, *346*, 989–992.
45. A. Berger, R. F. de Souza, M. R. Delgado, J. Dupont, *Tetrahedron: Asymmetry* **2001**, *12*, 1825–1828.
46. S. G. Lee, Y. J. Zhang, J. Y. Piao, H. Yoon, C. E. Song, J. H. Choi, J. Hong, *Chem. Commun.* **2003**, 2624–2625.
47. B. Pugin, M. Studer, E. Kuesters, G. Sedelmeier, X. Feng, *Adv. Synth. Catal.* **2004**, *346*, 1481–1486.
48. M. Berthod, J.-M. Joerger, G. Mignani, M. Vaultier, M. Lemaire, *Tetrahedron: Asymmetry* **2004**, *15*, 2219–2221.
49. I. Cerna, P. Kluson, M. Bendova, T. Floris, H. Pelantova, T. Pekarek, *Chem. Engl. Process.* **2011**, *50*, 264–272.
50. B. Karimi, M. Tavakolian, M. Akbari, F. Mansouri, *ChemCatChem* **2018**, *13*, 3173–3205.
51. See e.g. Ref. D. Geier, P. Schmitz, J. Walkowiak, W. Leitner, G. Francio, *ACS Catal.* **2018**, *8*, 3297–3303.
52. a) W. Leitner, *Acc. Chem. Res.* 2002, *35*, 746–756; b) D. J. Cole-Hamilton, *Adv. Synth. Catal.* **2006**, *348*, 1341–1351.
53. P. G. Jessop, T. Ikariya, R. Noyori, Ryoji, *Nature (London, United Kingdo*m) **1994**, *368* (6468), 231–233.
54. S. Kainz, A. Brinkmann, W. Leitner, A. Pfaltz, *J. Am. Chem. Soc.* **1999**, *121*, 6421–6429.
55. J. L. Xiao, S. C. A. Nefkens, P. G. Jessop, T. Ikariya, R. Noyori, *Tetrahedron Lett.* **1996**, *37*, 2813–2816.
56. Y. L. Hu, D. J. Birdsall, A. M. Stuart, E. G. Hope, J. L. Xiao, *J. Mol. Catal. A* **2004**, *219*, 57–60.
57. S. Kainz, D. Koch, W. Baumann, W. Leitner, *Angew. Chem., Int. Ed. Engl.* **1997**, *36*, 1628–1630.
58. M. J. Burk, S. Feng, M. F. Gross, W. Tumas, *J. Am. Chem. Soc.* **1995**, *117*, 8277–8278.
59. D. J. Adams, W. Chen, S. Lange, A. M. Stuart, A. West, J. Xiao, *Green Chem.* **2003**, *5*, 118–122.
60. M. van den Berg, A. J. Minnaard, E. P. Schudde, J. van Esch, A. H. M. de Vries, J. G. de Vries, B. L. Feringa, *J. Am. Chem. Soc.* **2000**, *122*, 11539–11540.
61. D. Bonafoux, Z. H. Hua, B. H. Wang, I. Ojima, *J. Fluor. Chem.* **2001**, *112*, 101–108.
62. S. Kainz, W. Leitner, *Catal. Lett.* **1998**, *55*, 223–225.
63. A. Wegner, W. Leitner, *Chem. Commun.* **1999**, 1583–1584.
64. A. Bösmann, G. Franció, E. Janssen, M. Solinas, W. Leitner, P. Wasserscheid, *Angew. Chem. Int. Ed.* **2001**, *40*, 2697–2699.
65. M. V. Escárcega-Bobadilla, L. Rodríguez-Pérez, E. Teuma, P. Serp, A. M. Masdeu-Bultó, M. Gómez, *Catal. Letters* **2011**, *141*, 808–816.
66. a) Y. Chernyak, J. H. Clements, *J. Chem. Eng. Data* **2004**, *49*, 1180–1184; b) S. P. Verevkin A. V. Toktonov, Y. Chernyak, B. Schäffner, A. Börner, *Fluide Phase Equilibria* **2008**, *268*, 1–6; c) V. N. Emel'yanenko, A. V. Toktonov, S. A. Kozlova, S. P. Verevkin, V. Andrushko, N. Andrushko, A. Börner, *J. Phys. Chem. A* **2008**, *112*, 4036–4045; d) S. P. Verevkin, V. N. Emel'yanenko, A. V. Toktonov, Y. Chernyak, B. Schäffner, A. Börner, *J. Chem. Thermodyn.* **2008**, *40*, 1428–1432.
67. B. Schäffner, F. Schäffner, S. Verevkin, A. Börner, *Chem. Rev.* **2010**, *110*, 4551–4581.

68. I. A. Shuklov, N. V. Dubrovina, J. Schulze, W. Tietz, K. Kühlein, A. Börner, *Chem. Eur. J.* **2014**, *20*, 957–960.

69. a) I. A. Shuklov, N. V. Dubrovina, J. Schulze, W. Tietz, K. Kühlein, A. Börner, *Tetrahedron Lett.* **2012**, *53*, 6326–6328; b) I. A. Shuklov, N. V. Dubrovina, J. Schulze, W. Tietz, A. Börner, *Tetrahedron Lett.* **2014**, *55*, 3495–3497; c) I. A. Shuklov, A. D. Shuklov, N. V. Dubrovina, K. Kühlein, A. Börner, *Pure Appl. Chem.* **2018**, *90*, 285–292.

70. a) A. Behr, F. Naendrup, D. Obst, *Eur. J. Lipid Sci. Technol.* **2002**, *104*, 161–166; b) A. Behr, F. Naendrup, D. Obst, *Adv. Synth. Catal.* **2002**, *344*, 1142–1145.

71. a) A. Behr, D. Obst, B. Turkowski, *J. Mol. Catal. A: Chem.* **2005**, *226*, 215–219; b) A. Behr, G. Henze, D. Obst, B. Turkowski, *Green Chem.* **2005**, *7*, 645–649.

72. M. T. Reetz, G. Lohmer, *Chem. Commun.* **1996**, 1921–1922.

73. A. Gamez, J. Köhler, J. Bradley, *Catal. Lett.* **1998**, *55*, 73–77.

74. A. Preetz, H.-J. Drexler, C. Fischer, Z. Dai, A. Börner, W. Baumann, A. Spannenberg, R. Thede, D. Heller, *Chem. Eur. J.* **2008**, *14*, 1445–1451.

75. B. Schäffner, V. Andrushko, J. Holz, S. P. Verevkin, A. Börner, *ChemSusChem* **2008**, *1*, 934–940.

76. J. Bayardon, J. Holz, B. Schäffner, V. Andrushko, S. Verevkin, A. Preetz, A. Börner, *Angew. Chem. Int. Ed.* **2007**, *46*, 5971–5974.

77. A. Behr, D. Obst, B. Turkowski, *J. Mol. Catal. A: Chem.* **2005**, *226*, 215–219.

78. J. Mazuela, J. J. Verendel, M. Coll, B. Schäffner, A. Börner, P. G. Andersson, O. Pàmies, M. Diéguez, *J. Am. Chem. Soc.* **2009**, *131*, 12344–12353.

79. B. Schäffner, J. Holz, S. P. Verevkin, A. Börner, *ChemSusChem* **2008**, *1*, 249–253.

80. I. A. Shuklov, N. V. Dubrovina, A. Börner, *Synthesis* **2007**, 2925–2943.

81. a) J. F. Povey, C. M. Smales, S. J. Hassard, M. J. Howard, *J. Struct. Biol.* **2007**, *157*, 329–338; b) M. Buck, *Q. Rev. Biophys.* **1998**, *31*, 297.

82. J.-P. Bégué, D. Bonnet-Delpon, B. Crousse, *Synlett* **2004**, 18–29.

83. J. P. Candlin, A. R. Oldham, *Discuss. Faraday Soc.* **1968**, *46*, 60–71.

84. a) W. Strohmeier, K. Grünter, *J. Organomet. Chem.* **1975**, *90*, C45–C47; b) W. Strohmeier, K. Grünter, *J. Organomet. Chem.* **1975**, *90*, C48–C50.

85. P. Munshi, A. D. Main, J. C. Linehan, C.-C. Tai, P. G. Jessop, *J. Am. Chem. Soc.* **2002**, *124*, 7963–7971.

86. a) H. Jendralla, *Tetrahedron: Asymmetry* **1994**, *5*, 1183–1186; b) K. Rossen, S. A. Weissman, J. Sager, R. A. Reamer, D. Askin, R. P. Volante, P. J. Reider, *Tetrahedron Lett.* **1995**, 6419–6422.

87. M. J. Burk, C. S. Kalberg, A. Pizzano, *J. Am. Chem. Soc.* **1998**, *120*, 4345–4353.

88. Y.-Q. Wang, S.-M. Lu, Y.-G. Zhou, *Org. Lett.* **2005**, *7*, 3235–3238.

89. Y.-Q. Wang, Y.-G. Zhou, *Synlett* **2006**, 1189–1192.

90. Y.-Q. Wang, C.-B. Yu, D.-W. Wang, X.-B. Wang, Y.-G. Zhou, *Org. Lett.* **2008**, *10*, 2071–2074.

91. a) H. Abe, H. Amii, K. Uneyama, *Org. Lett.* **2001**, *3*, 313–315; b) A. Suzuki, M. Mae, H. Amii, K. Uneyama, *J. Org. Chem.* **2004**, *69*, 5132–5134.

92. T. Bunlaksananusorn, F. Rampf, *Synlett* **2005**, *17*, 2682–2684.

93. W. Zhang, X. Zhang, *J. Org. Chem.* **2007**, *72*, 1020–1027.

94. M.-N. Birkholz, N. V. Dubrovina, H. Jiao, D. Michalik, J. Holz, R. Paciello, B. Breit, A. Börner, *Chem. Eur. J.* **2007**, *13*, 5896–5907.

95. N. V. Dubrovina, I. A. Shuklov, M.-N. Birkholz, D. Michalik, R. Paciello, A. Börner, *Adv. Synth. Catal.* **2007**, *349*, 2183–2187.

96. D. D. Perrin, W. L. F. Armarego, *Purification of Laboratory Chemicals*, 4th edn, Butterworth-Heinemann Ltd., **1996**.

97. B. Cornils, W. A. Herrmann, *Aqueous-Phase Organometallic Catalysis. Concepts and Applications*, Wiley-VCH, Weinheim, **1998**.

98. K. H. Shaughnessy, *Chem. Rev.* **2009**, *109*, 643–710.

99. a) C. Präsang, E. K. Bauer, *Phosphorus Ligands in Asymmetric Catalysis*. Ed. A. Börner, Wiley-VCH, Weinheim, **2008**, pp. 917–954; b) P. B. Webb, D. H. Cole Hamilton, *Phosphorus(III) Ligands in Homogeneous Catalysis*, Eds. P. C. J. Kamer, P. W. N. M. van Leeuwen, John Wiley & Sons, Chichester, **2012**, pp. 499–503.

100. a) S. Ahrland, J. Chatt, N. R. Davies, A. A. Williams, *J. Chem. Soc.* **1958**, 276–288; b) F. Joó, J. Kovács, Á. Kathó, A. C. Bényei, T. Decuir, D. J. Darensbourg, *Inorg. Synth.* **1988**, *32*, 1–45.

101. W. A. Herrmann, G. P. Albanese, R. B. Manetsberger, P. Lappe, H. Bahrmann, *Angew. Chem. Int. Ed. Engl.* **1995**, *34*, 811–813.

102. a) F. Alario, Y. Amrani, Y. Colleuille, T. P. Dang, J. Jenck, D. Morel, D. Sinou, *J. Chem. Soc. Chem. Commun.* **1986**, 202–203; b) Y. Amrani, L. Lecomte, D. Sinou, J. Bakos, I. Toth, B. Heil, *Organometallics* **1989**, 8, 542–547; c) C. Lensink, J. G. de Vries, *Tetrahedron: Asymmetry* **1992**, *3*, 235–238; d) M. D. Fryzuk, B. Bosnich, *J. Am. Chem. Soc.* **1977**, 99, 6262–6267; e) U. Matteoli, V. Beghetto, C. Schiavon, A. Scrivanti, G. Menchi, *Tetrahedron: Asymmetry* **1997**, 8, 1403–1409; f) P. A. MacNeil, N. K. Roberts, B. Bosnich, *J. Am. Chem. Soc.* **1981**, *103*, 2273–2280; g) J. Bakos, I. Tóth, B. Heil, L. Markó, *J. Organomet. Chem.* **1985**, *279*, 23–29; h) M. D. Fryzuk, B. Bosnich, *J. Am. Chem. Soc.* **1978**, *100*, 5491–5494.

103. C. Lensink, E. Rijnberg, J. G. de Vries, *J. Mol. Catal. A* **1997**, *116*, 199–207.

104. K. T. Wan, M. E. Davis, *J. Chem. Soc. Chem. Commun.* **1993**, 1262–1264.

105. H. Bahrmann, K. Bergrath, H.-J. Kleiner, P. Lappe, C. Naumann, D. Peters, D. Regnat, *J. Organomet. Chem.* **1996**, *520*, 97–100.

106. R. W. Eckl, T. Priermeier, W. A. Herrmann, *J. Organomet. Chem.* **1997**, *532*, 243–249.

107. M. D. Miquel-Serrano, A. M. Masdeu-Bultó, C. Claver, D. Sinou, *J. Mol. Catal. A* **1999**, *143*, 49–55.

108. K.-t. Wan, M. E. Davis, *J. Chem. Soc., Chem. Commun.* **1993**, 1262–1264.

109. K.-t. Wan, M. E. Davis, *Tetrahedron: Asymmetry* **1993**, 4, 2461–2468.

110. a) S. Trinkhaus, J. Holz, R. Selke, A. Börner, *Tetrahedron Lett.* **1997**, *38*, 807–808; b) S. Trinkhaus, R. Kadyrov, R. Selke, J. Holz, L. Götze, A. Börner, *J. Mol. Catal. A* **1999**, *144*, 15–26.

111. a) K. Heesche-Wagner, T. N. Mitchell, *J. Organomet. Chem.* **1994**, *468*, 99–106; b) M. Beller, J. G. E. Krauter, A. Zapf, S. Bogdanovic, *Catal. Today* **1999**, *48*, 279–290.

112. T. N. Mitchell, K. Heesche-Wagner, *J. Organomet. Chem.* **1992**, *436*, 43–53.

113. a) S. Borns, R. Kadyrov, D. Heller, W. Baumann, J. Holz, A. Börner, *Tetrahedron: Asymmetry* **1999**, *10*, 1425–1431; b) M. Bühl, W. Baumann, R. Kadyrov, A. Börner, *Helv. Chim. Acta* **1999**, *82*, 811–820.

114. A. Börner, J. Ward, K. Kortus, H. B. Kagan, *Tetrahedron: Asymmetry* **1993**, *4*, 2219–2228.

115. A. Börner, J. Ward, W. Ruth, J. Holz, A. Kless, D. Heller, H. B. Kagan, *Tetrahedron* **1994**, *50*, 10419–10430.

116. a) A. Börner, J. Holz, A. Kless, D. Heller, U. Berens, *Tetrahedron Lett.* **1994**, *35*, 6071–6074; b) J. Holz, A. Börner, A. Kless, S. Borns, S. Trinkhaus, R. Selke, D. Heller, *Tetrahedron: Asymmetry* **1995**, *6*, 1973–1988.

117. R. Selke, J. Holz, A. Riepe, A. Börner, *Chem. Eur. J.* **1998**, *5*, 769–771.

118. J. Holz, M. Quirmbach, A. Börner, *Synthesis* **1997**, 983–1006.

119. See e.g. Ref. 19: a) W. Li, Z. Zhang, D. Xiao, Z. Zhang, *J. Org. Chem.* **2000**, 65, 3489–3496; b) Y.-Y. Yan, T. V. RajanBabu, *J. Org. Chem.* **2000**, *65*, 900–906.

120. J. Holz, D. Heller, R. Stürmer, A. Börner, *Tetrahedron Lett.* **1999**, *40*, 7059–7062.
121. T. V. RajanBabu, Y.-Y. Yan, S. Shin, *J. Am. Chem. Soc.* **2001**, *123*, 10207–10213.
122. K. Yonehara, T. Hashizume, K. Mori, K. Ohe, S. Uemura, *J. Org. Chem.* **1999**, *64*, 5593–5598.
123. K. Yonehara, K. Ohe, S. Uemura, *J. Org. Chem.* **1999**, *64*, 9381–9385.
124. W. Guo, X. Liu, Y. Liu, C. Li, *ACS. Catal.* **2018**, *8*, 328–341.
125. a) A. Faljoni, K. Zinner, R. G. Weiss, *Tetrahedron Lett.* **1974**, 1127–1130; b) D. R. Boyd, D. C. Neill, *J. Chem. Soc., Chem. Commun.* **1977**, 51–52; c) W. H. Laarhoven, T. J. H. M. Cuppen, *J. Chem. Soc., Chem. Commun.* **1977**, 47; d) W. H. Laarhoven, T. J. H. M. Cuppen, *J. Chem. Soc., Perkin Trans.* 2, **1978**, 315–318.
126. a) D. Seebach, W. Langer, *Helv. Chim. Acta* **1979**, *62*, 1701–1709; b) M. Bucciarelli, A. Forni, I. Moretti, G. Torre, *J. Chem. Soc., Perkin Trans.* **1980**, *1*, 2152–2161; c) M. Bucciarelli, A. Forni, I. Moretti, G. Torre, *J. Org. Chem.* **1983**, *48*, 2640–2644.
127. D. Seebach, H. A. Oei, *Angew. Chem. Int. Ed. Engl.* **1975**, *14*, 634–636.
128. M. Schmitkamp, D. Chen, W. Leitner, J. Klankermayer, G. Francio, *Chem. Commun.* **2007**, 4012–4014.
129. L. C. Branco, P. M. P. Gois, N. M. T. Lourenço, V. B. Kurteva, C. A. M. Afonso, *Chem. Commun.* **2006**, 2371–2372.
130. Y. Nagata, R. Takeda, M. Suginome, *ACS Central Science* **2019**, *5*, 1235–1240.
131. a) *Asymmetric Catalysis on Industrial Scale*, Eds. H.-U. Blaser, E. Schmidt, Wiley-VCH, Weinheim, **2004**; b) *Asymmetric Catalysis on Industrial Scale*, Eds. H.-U. Blaser, H.-J. Federsel, Wiley-VCH, Weinheim, **2010**.
132. J. M. Dreimann, H. Warmeling, J. N. Weimann, K. Künnemann, A. Behr, A. J. Vorholt, *AIChE* **2016**, *62*, 4377–4383.

Index

A

B

Printed in the United States
by Baker & Taylor Publisher Services